AF389352

High-Strength Low-Alloy Steels

High-Strength Low-Alloy Steels

Editors

Ricardo Branco
Filippo Berto

MDPI • Basel • Beijing • Wuhan • Barcelona • Belgrade • Manchester • Tokyo • Cluj • Tianjin

Editors

Ricardo Branco
Department of Mechanical
Engineering
University of Coimbra
Coimbra
Portugal

Filippo Berto
Department of Mechanical and
Industrial Engineering
NTNU
Trondheim
Norway

Editorial Office
MDPI
St. Alban-Anlage 66
4052 Basel, Switzerland

This is a reprint of articles from the Special Issue published online in the open access journal *Metals* (ISSN 2075-4701) (available at: www.mdpi.com/journal/metals/special_issues/high_strength_steel).

For citation purposes, cite each article independently as indicated on the article page online and as indicated below:

LastName, A.A.; LastName, B.B.; LastName, C.C. Article Title. *Journal Name* **Year**, *Volume Number*, Page Range.

ISBN 978-3-0365-2040-7 (Hbk)
ISBN 978-3-0365-2039-1 (PDF)

Contents

About the Editors

Ricardo Branco

Ricardo Branco completed his Ph.D. degree in Mechanical Engineering from the University of Coimbra. He is currently Assistant Professor at the Department of Mechanical Engineering of the University of Coimbra. His research interests include the mechanical behavior of materials, fatigue and fracture of engineering materials, and multiaxial fatigue life prediction models.

Filippo Berto

Filippo Berto completed his Ph.D. degree in Mechanical Engineering from the University of Florence. He is currently Chair of Fatigue and Fracture at the Department of Mechanical Engineering of the Norwegian University of Science and Technology. His research interests include local approaches for fatigue assessment, the fatigue and fracture of advanced materials, multiaxial fatigue, the structural integrity of additive, materials, and the numerical modeling of fatigue crack growth.

MDPI

Editorial

High-Strength Low-Alloy Steels

Ricardo Branco [1,*] **and Filippo Berto** [2]

1   CEMMPRE, Department of Mechanical Engineering, University of Coimbra, 3030-78 Coimbra, Portugal
2   Department of Mechanical and Industrial Engineering, Norwegian University of Science and Technology, 7491 Trondheim, Norway; filippo.berto@ntnu.no
*   Correspondence: ricardo.branco@dem.uc.pt

1. Introduction

Modern industry, driven by the recent environmental policies, faces an urgent need for the production of lighter and more environmentally friendly components. High-strength low-alloy steels are key materials in this challenging scenario because they provide a balanced combination of properties, such as strength, toughness, formability, weldability, and corrosion resistance. These features make them ideal for a myriad of engineering applications which experience complex loading conditions and aggressive media, such as aeronautical and automotive components, railway parts, offshore structures, oil and gas pipelines, power transmission towers, construction machinery, among others. The goal of this Special Issue is to foster the dissemination of the latest research devoted to high-strength low-alloy (HSLA) steels from different perspectives.

2. Contributions

The understanding of the microstructure features and their dependence on the mechanical behaviour is of essential importance for the development of safe and durable components as well as to extend the scope of application of the high-strength low-alloy steels. This may justify the intense research conducted on the triangular relationship between the microstructure, the processing techniques, and the final mechanical properties. Solis-Bravo et al. [1] addressed the relationship between the precipitate morphology and dissolution on grain coarsening behaviour in microalloyed linepipe steels with different contents of titanium and niobium. The effect of the hot deformation and the cooling path on the phase transformation kinetics of precipitation-strengthened automotive steels with different contents of titanium and niobium was also examined by Grajcar et al. [2]. Xie et al. [3] studied the effect of nanometre-sized interphase-precipitated carbides on the improvement of monotonic tensile strength in fire resistant hot-rolled steel at room and high temperature. The effect of dissolution and precipitation of different carbides at high temperature on the microstructure of a low-alloy chromium-containing heat-resistant steel was also analysed by Li et al. [4].

Regarding the processing techniques and the evaluation of mechanical properties, different research lines were followed. Guo et al. [5] evaluated the casting process conditions on the mechanical properties of hot-rolled steel and studied the billet quality by developing and optimisation method. Khosravani et al. [6] tackled the microstructural changes that occur during the processing of dual-phase steels by using multiresolution spherical indentation stress–strain tests. Dzioba et al. [7] focused on the effect of temperature on fracture toughness and tensile strength properties of low-carbon high-strength steel. Iob et al. [8] examined the anisotropic mechanical behaviour of high-strength low-alloyed steel based on the micro-void and ductile fracture. The mechanical properties of high-strength low-carbon steels, with different contents of molybdenum and niobium, processed thermomechanically and subjected to direct quench were investigated by Hannula et al. [9]. A numerical study to deal with the cutting problem of ultra-thin steel sheets made of cold rolled steel was developed by Kaczmarczyk et al. [10].

check for **updates**

Citation: Branco, R.; Berto, F. High-Strength Low-Alloy Steels. *Metals* **2021**, *11*, 1000. https://doi.org/10.3390/met11071000

Received: 8 June 2021
Accepted: 16 June 2021
Published: 23 June 2021

Publisher's Note: MDPI stays neutral with regard to jurisdictional claims in published maps and institutional affiliations.

Another active area of research has been welding engineering. It has not been focused on the optimisation of welding techniques but also on the evaluation of microstructure features and mechanical properties. Mičian et al. [11] studied the influence of the cooling rates on mechanical properties of the heat-affected zone, obtained by metal active gas welding, in S960ML high-strength structural steels. The effect of the welding heat input on the heat-affected zone of S960QL high-strength structural steels was also investigated numerically by Gáspár [12]. Moravec et al. [13] examined the effect of grain growth kinetics on the changes of the mechanical properties of the heat-affected zone in a S700MC fine-grained high-strength steel. The fracture toughness response under static and dynamic loading in the same material was examined by Schmidová et al. [14].

The presence of abrupt geometrical changes in conjunction with complex cyclic loads make most mechanical components prone to fatigue failure. This means that engineering design must be able to account for the loading history, the geometrical effects, the environmental effects, and the processing variables, among others. Jiménez-Peña et al. [15] compared the fatigue response of high-strength low-alloy plates with holes manufactured by five processes, namely punching, drilling, waterjet-cut, plasma, and laser-cut. Guo et al. [16] evaluated the variability in the mechanical properties of pipeline steel associated with the centreline segregation in continuously cast slab to meet the requirements of strain-based design. Ślęzak [17] studied the fatigue crack initiation and fatigue crack growth in welded joints made of S960QL high-strength low-alloy steel subjected to strain-controlled conditions. Harun et al. [18] analysed the effect of the localized wall thinning on low-cycle fatigue resistance of elbows, with artificially introduced defects, made of C70600 steel from full-scale tests.

The fatigue design under multiaxial loading is another challenging topic. The development of multiaxial fatigue assessment models as well as the identification of adequate fatigue damage quantifiers remain important objectives for the scientific community. However, it is a very complex task because, in general, multiaxial fatigue response is associated with a huge number of variables. Pawliczek and Rozumek [19] presented an algorithm based on the Palmgren–Miner linear damage rule for calculating the fatigue life in S355J0 steel specimens subjected to multiaxial non-zero mean stress histories. Cruces et al. [20] compared the predictive capabilities of different critical plane-based models for hollow specimens made of S355-J2G3 steel subjected to in-phase and out-of-phase axial–torsional loading in the low-cycle and the high-cycle fatigue regimes.

Within the high-strength low-alloy steels, the third generation plays an important role. The macroscopic mechanical response of this new generation can be further improved by a better understanding of the failure mechanisms on the microstructural level under different service conditions. Shakerifard et al. [21] conducted a comprehensive microstructural characterization of a multiphase low-silicon bainitic steel using a scanning electron microscope (SEM) equipped with an electron backscatter diffraction detector. Concerning metallic structures operating in soils and natural waters, corrosion under the effect of a stray current is among the most hazardous types of damage. Rybkina et al. [22] addressed the effect of sign-alternating cycling polarisation on the localised corrosion of pipelines made of X70 steel subjected to various pH-neutral solutions.

Acknowledgments: This research is sponsored by FEDER funds through the program COMPETE—Programa Operacional Factores de Competitividade—and by national funds through FCT—Fundação para a Ciência e a Tecnologia—under the project UIDB/00285/2020.

References

1. Solis-Bravo, G.; Merwin, M.; Garcia, C.I. Impact of Precipitate Morphology on the Dissolution and Grain-Coarsening Behavior of a Ti-Nb Microalloyed Linepipe Steel. *Metals* **2020**, *10*, 89. [CrossRef]
2. Grajcar, A.; Morawiec, M.; Zalecki, W. Austenite Decomposition and Precipitation Behavior of Plastically Deformed Low-Si Microalloyed Steel. *Metals* **2018**, *8*, 1028. [CrossRef]

3. Xie, Z.; Song, Z.; Chen, K.; Jiang, M.; Tao, Y.; Wang, X.; Shang, C. Study of Nanometer-Sized Precipitation and Properties of Fire Resistant Hot-Rolled Steel. *Metals* **2019**, *9*, 1230. [CrossRef]

4. Li, Z.; Jia, P.; Liu, Y.; Qi, H. Carbide Precipitation, Dissolution, and Coarsening in G18CrMo2–6 Steel. *Metals* **2019**, *9*, 916. [CrossRef]

5. Guo, F.; Wang, X.; Wang, J.; Misra, R.D.K.; Shang, C. The Significance of Central Segregation of Continuously Cast Billet on Banded Microstructure and Mechanical Properties of Section Steel. *Metals* **2020**, *10*, 76. [CrossRef]

6. Khosravani, A.; Caliendo, C.M.; Kalidindi, S.R. New Insights into the Microstructural Changes During the Processing of Dual-Phase Steels from Multiresolution Spherical Indentation Stress–Strain Protocols. *Metals* **2020**, *10*, 18. [CrossRef]

7. Dzioba, I.; Pała, R. Strength and Fracture Toughness of Hardox-400 Steel. *Metals* **2019**, *9*, 508. [CrossRef]

8. Iob, F.; Cortese, L.; Di Schino, A.; Coppola, T. Influence of Mechanical Anisotropy on Micro-Voids and Ductile Fracture Onset and Evolution in High-Strength Low Alloyed Steels. *Metals* **2019**, *9*, 224. [CrossRef]

9. Hannula, J.; Porter, D.; Kaijalainen, A.; Somani, M.; Kömi, J. Mechanical Properties of Direct-Quenched Ultra-High-Strength Steel Alloyed with Molybdenum and Niobium. *Metals* **2019**, *9*, 350. [CrossRef]

10. Kaczmarczyk, J.; Kozłowska, A.; Grajcar, A.; Sławski, S. Modelling and Microstructural Aspects of Ultra-Thin Sheet Metal Bundle Cutting. *Metals* **2019**, *9*, 162. [CrossRef]

11. Mičian, M.; Harmaniak, D.; Nový, F.; Winczek, J.; Moravec, J.; Trško, L. Effect of the $t_{8/5}$ Cooling Time on the Properties of S960MC Steel in the HAZ of Welded Joints Evaluated by Thermal Physical Simulation. *Metals* **2020**, *10*, 229. [CrossRef]

12. Gáspár, M. Effect of Welding Heat Input on Simulated HAZ Areas in S960QL High Strength Steel. *Metals* **2019**, *9*, 1226. [CrossRef]

13. Moravec, J.; Novakova, I.; Sobotka, J.; Neumann, H. Determination of Grain Growth Kinetics and Assessment of Welding Effect on Properties of S700MC Steel in the HAZ of Welded Joints. *Metals* **2019**, *9*, 707. [CrossRef]

14. Schmidová, E.; Bozkurt, F.; Culek, B.; Kumar, S.; Kucharikova, L.; Uhríčik, M. Influence of Welding on Dynamic Fracture Toughness of Strenx 700MC Steel. *Metals* **2019**, *9*, 494. [CrossRef]

15. Jiménez-Peña, C.; Goulas, C.; Preußner, J.; Debruyne, D. Failure Mechanisms of Mechanically and Thermally Produced Holes in High-Strength Low-Alloy Steel Plates Subjected to Fatigue Loading. *Metals* **2020**, *10*, 318. [CrossRef]

16. Guo, F.; Liu, W.; Wang, X.; Misra, R.D.K.; Shang, C. Controlling Variability in Mechanical Properties of Plates by Reducing Centerline Segregation to Meet Strain-Based Design of Pipeline Steel. *Metals* **2019**, *9*, 749. [CrossRef]

17. Ślęzak, T. Fatigue Examination of HSLA Steel with Yield Strength of 960 MPa and Its Welded Joints under Strain Mode. *Metals* **2020**, *10*, 228. [CrossRef]

18. Harun, M.F.; Mohammad, R.; Kotousov, A. Low Cycle Fatigue Behavior of Elbows with Local Wall Thinning. *Metals* **2020**, *10*, 260. [CrossRef]

19. Pawliczek, R.; Rozumek, D. The Effect of Mean Load for S355J0 Steel with Increased Strength. *Metals* **2020**, *10*, 209. [CrossRef]

20. Cruces, A.S.; Lopez-Crespo, P.; Moreno, B.; Antunes, F.V. Multiaxial Fatigue Life Prediction on S355 Structural and Offshore Steel Using the SKS Critical Plane Model. *Metals* **2018**, *8*, 1060. [CrossRef]

21. Shakerifard, B.; Galan Lopez, J.; Kestens, L.A.I. A New Electron Backscatter Diffraction-Based Method to Study the Role of Crystallographic Orientation in Ductile Damage Initiation. *Metals* **2020**, *10*, 113. [CrossRef]

22. Rybkina, A.; Gladkikh, N.; Marshakov, A.; Petrunin, M.; Nazarov, A. Effect of Sign-Alternating Cyclic Polarisation and Hydrogen Uptake on the Localised Corrosion of X70 Pipeline Steel in Near-Neutral Solutions. *Metals* **2020**, *10*, 245. [CrossRef]

 *metals*

Article

Impact of Precipitate Morphology on the Dissolution and Grain-Coarsening Behavior of a Ti-Nb Microalloyed Linepipe Steel

Gregorio Solis-Bravo [1,*] , Matthew Merwin [2] and C. Isaac Garcia [1]

[1] Ferrous Physical Metallurgy Group, University of Pittsburgh, Pittsburgh, PA 15261, USA; cigarcia@pitt.edu
[2] United States Steel Corporation, Pittsburgh, PA 15120, USA; mmerwin@uss.com
*   Correspondence: grs52@pitt.edu; Tel.: +1-412-557-3053

Received: 28 November 2019; Accepted: 1 January 2020; Published: 4 January 2020

Abstract: The relationship between precipitate morphology and dissolution on grain coarsening behavior was studied in two Ti-Nb microalloyed Linepipe (LP) Steels. The developed understanding highlights the importance of the complex relationship between precipitate constitutive make-up, dissolution mechanism and grain boundary (GB) pinning force. Equilibrium-based empirical solubility products were used to calculate precipitate volume fractions and compared to experimental measurements. Scanning Electron Microscopy (SEM), Electron Backscatter Diffraction (EBSD) and Electron Probe Micro-Analysis (EPMA) were conducted on bulk samples. Transmission Electron Microscopy (TEM)-based techniques were used on C-replica extractions and thin-foils. A retardation in the grain-coarsening temperature compared to the predicted coarsening temperature based on equilibrium calculations was noticed. In addition, a consistent NbC epitaxial formation over pre-existing TiN was observed. The resulting reduction in total precipitate/matrix interface area and the low energy of the TiN/NbC interface are pointed to as responsible mechanisms for the retardation in the kinetics of precipitates' dissolution. This dissolution retardation mechanism suggests that a lower Nb content might be effective in controlling the grain coarsening behavior of austenite.

Keywords: grain coarsening; precipitate dissolution; epitaxial precipitation; Ti-Nb microalloyed steel; linepipe

1. Introduction

The understanding of the dissolution and re-precipitation of Nb, Ti and or Ti-Nb carbo-nitrides during the high-temperature processing of austenite is of essential importance for the microstructural refinement of austenite. During the hot deformation of High-Strength Low-Alloy (HSLA) steel, there are a series of well-known steps; (1) reheating, (2) roughing deformation, and (3) finishing deformation. This paper focuses on the impact of the precipitate composition and morphology on two crucial aspects: precipitate dissolution and control of the grain-coarsening behavior of austenite. All are applied to the reheating of a Ti-Nb microalloyed linepipe steel.

The purpose of this work was to study the feasibility of producing similar as-reheated austenitic grain size using two steels with different Nb levels at constant Ti content. The impact of simple precipitation morphology on the grain-coarsening behavior of austenite during reheating follows the kinetics of precipitation dissolution. However, when the morphology of the precipitates is more complex, the dissolution of the precipitates does not follow the standard equilibrium predictions.

It is well-known that the type of precipitation species and precipitation stability are key for Prior-Austenite Grain Size (PAGS) control. Precipitates exert pinning forces on GBs, limiting their mobility [1–6]. Hence, when precipitates dissolve or coarsen, GBs detach from them and grain coarsening occurs. This gives importance to the dissolution of particles and the factors that affect it.

Previous research by Gong et al. have shown the higher stability of Nb-Ti carbo-nitrides as compared to Nb-only carbo-nitrides [7]. Gong's work points out the faster dissolution rate of NbC as compared to TiNb-CN as the reason for this behavior, without considering the structural make-up of the particles as a factor. Gong et al. also point out that PAGS must be calculated considering solute drag effects to get accurate predictions. In a work by Ringer et al, precipitate shape has been proposed as a factor of the GB-pinning force [8]. It is suggested that, in the process of unpinning, the precipitate must first rotate to acquire coherency with the new grain, delaying grain growth in the process. Shape is a clear difference when comparing NbTi carbonitride cuboidal precipitates to Nb-only carbonitrides of more rounded shape. NbC has previously been reported to nucleate at pre-existing TiN [9–11]. If this heterogeneous formation of NbC on TiN particles happens during casting, as suggested by several authors [11,12], the dissolution process during reheating could be retarded. Slower dissolution kinetics, for instance, can be utilized for better PAGS control with lower levels of microalloying elements. Research has already shown epitaxial NbC on TiN as an effective precipitate for grain size control [13,14].

The complexity of TiN-NbC particles' morphology and composition may slow down dissolution kinetics. The stability of NbC interfaces to TiN and TiC has been studied using ab initio calculations; researchers have shown how the NbC/TiC interface is the most stable [15]. They report a smaller mismatch of the NbC/TiC interface when compared to the NbC/TiN interface, as well as a lower interface energy 0.75 J·m^{-2}, 3.6 times smaller than the NbC/TiN interface. This means that as the core-particle loses N and gains C, coherency improves between the layers. Hence, the interface between these two compounds stabilizes as they interdiffuse.

Not-complex particles may dissolve first. Then, microalloy elements in solution can interact with GB by forming solute-rich atmospheres around them. These atmospheres lower GB energy, reducing GB mobility and producing the well-known solute-drag effect.

The major goal of this work was to compare and understand the impact of the precipitate morphology and dissolution behavior on grain coarsening in an HSLA-LP steel.

2. Materials and Methods

The two alloys in Table 1 were studied. The Nb content was the only contrast in chemical composition. The alloys were produced as lab-heats at a partner industrial facility, hot-rolled down to 19 mm thickness and air-cooled. The chemical composition was provided by our partner steelmaking laboratory through Inductively Coupled Plasma Mass Spectroscopy. The initial hot rolled microstructure for both steels was Ferrite (F) and Pearlite (P).

Table 1. Alloys used in this study (wt%).

Alloy	C	Mn	Cu	Ni	Cr	Mo	Nb	V	Ti	N	Al	Si	Fe
High Nb	0.05	1.5	0.1	0.25	0.25	0.15	0.090	0.06	0.01	0.007	0.04	0.25	Base
Low Nb	0.05	1.5	0.1	0.25	0.25	0.15	0.045	0.06	0.01	0.007	0.04	0.25	Base

By assuming equilibrium conditions, the empirical solubility products of the possible precipitates can be used to find their correspondent volume fraction, f_v. Volume fraction was obtained as a function of temperature, based on the chemical compositions. A model that considers the mutual solubility of carbonitride was compared to a simple-precipitates model. See Appendices A and B for details.

Austenitization was done in vacuum quartz capsules containing one piece of each alloy. The capsules were austenitized to 1150, 1200, 1250 and 1300 °C for 1 h, then water quenched. The resulting specimens were cut in half, so half was used for Prior-Austenitic Grain Size measurements, and the other half for precipitate and dissolution assessment. The PAGS was revealed by first tempering the specimens at 600 °C for 24 h, followed by picral solution etching. The picral solution contained 100 mL of distilled de-ionized water, picric acid to saturation, 2 g of sodium dodecylbenzenesulfonate and six drops of hydrochloric acid. Etching was done at 90 °C.

PAGS measurements were done utilizing the free software ImageJ (1.52a, National Institutes of Health, Bethesda, MD, USA), by manually circling grains to obtain the average Feret diameter. Precipitates were quantified based on TEM analysis obtained from random locations of the specimens. Two sets of TEM micrographs were used with magnifications 2000× and 20,000×, to obtain representative samples of the smallest and the biggest particles. The software ImageJ was used to measure particle size and area fraction. The area fraction was converted to volume fraction by assuming the micrographs are representative of a volume as thick as an average precipitate.

Another method for revealing PAGS was reconstruction of austenitic EBSD maps, based on the quenched martensitic specimens. The specimens were ground, polished, vibro-polished for 3 h, then cleaned in an ultrasonic bath of ethanol for 10 min before scanning. EBSD was done using a 20 kV, 13 nA beam, in a hexagonal pattern with 0.3 μm step size. The martensitic EBSD maps were processed to reconstruct prior-austenite maps. MATLAB (R2017b, The Mathworks Inc., Natick, MA, USA) with MTEX toolbox was used, following the procedure proposed by Nyyssönnen, et al. [16,17]. This procedure starts by determining the experimental Orientation Relationship (OR) between martensite and its parent austenite. Kurdjumov-Sachs OR is first assumed, then followed by an iterative comparison of adjacent grains; this determines the experimental OR. This OR is then used for re-constructing the prior-austenite orientation map.

Thin foil specimens for TEM were produced by grinding specimens down to 100 μm thickness with 1200 grit sandpaper. Further grinding to 10–20 μm was done in a Fischione Model 200 dimpling grinder (Fischione Instruments, Export, PA, USA) using 3 nm diamond abrasive paste. Electropolishing was done on the thin foils using a Struers TenuPol-5 (Struers Inc., Cleveland, OH, USA) with 5% perchloric acid solution as electrolyte at −25 to −15 °C. Carbon replica extractions were obtained by evaporating 10–15 nm C-film on 3% Nital etched surfaces of the specimens. The film was then removed by immersion in 10% Nital and supported on copper grids for analysis.

Light microscopy was done in a Keyence Microscope (Keyence Corporation of America, Itasca, IL, USA). SEM analysis was done in a Zeiss Sigma 50vp FEG-SEM (Zeiss International, Oberkochen, Germany) equipped with Oxford Energy Dispersive Spectrometer (EDS) detector (Oxford Instruments, Abingdon, UK). EDS was used as qualitative method, utilizing only the factory-preset standardization of the Aztec software (V3.2 SP1, Oxford Instruments, Abingdon, UK). EBSD was done in a FEI Scios FIB/SEM Dual Beam system (Thermo Fisher, Hillsboro, OR, USA) with EDAX EBSD camera and TEAM software (V4.3, EDAX Inc., Mahwah, NJ, USA). TEM was done in a JEOL JEM2100F (JEOL Ltd., Tokyo, Japan) using 200 kV beam.

3. Results

3.1. Thermodynamic Calculations of Precipitate Volume Fraction

The results of theoretical equilibrium calculations of precipitates' volume fractions (described in Appendix A) are shown in Figure 1. From all eight possible precipitates considered in the calculations: AlN, TiN, NbN, VN, TiC, NbC, VC and MoC, only four presented a significant amount in the Low-Nb steel, and five in the High-Nb steel. The volume fractions of VN, TiC and MoC remained zero at all temperatures of interest. The presence of NbN appeared only in the high-Nb alloy in the range of temperatures where NbC volume fraction declines. This precipitate occurrence is dependent on the amount of N available after all possible TiN and AlN have precipitated. According to this calculation, most simple carbides and nitrides dissolve below 1100 °C; above this temperature, only TiN remains stable.

Considering the mutual solubility of carbides and nitrides (except AlN, see calculations in Appendix B), gives a contrasting result. In Figure 2, the volume fraction of complex (Ti,Nb,V)(C,N) particles is superior to that of TiN in Figure 1, especially in the range of our austenitization experiments, 1150–1300 °C.

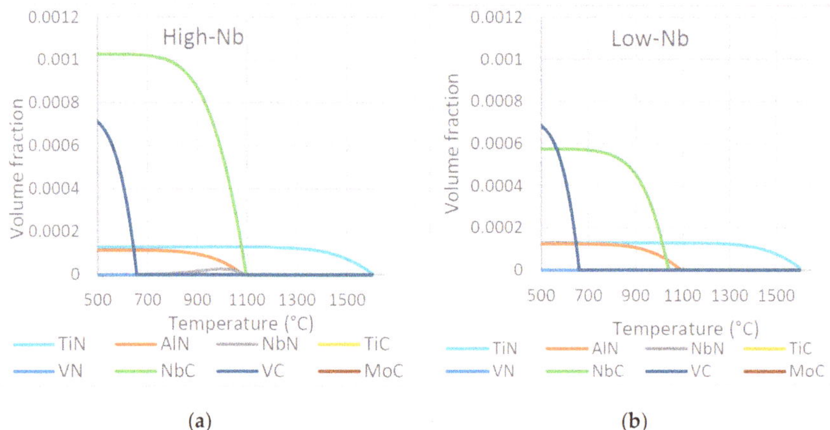

Figure 1. The calculated volume fraction of precipitates considering simple precipitates for: (**a**) high-Nb steel; (**b**) low-Nb steel.

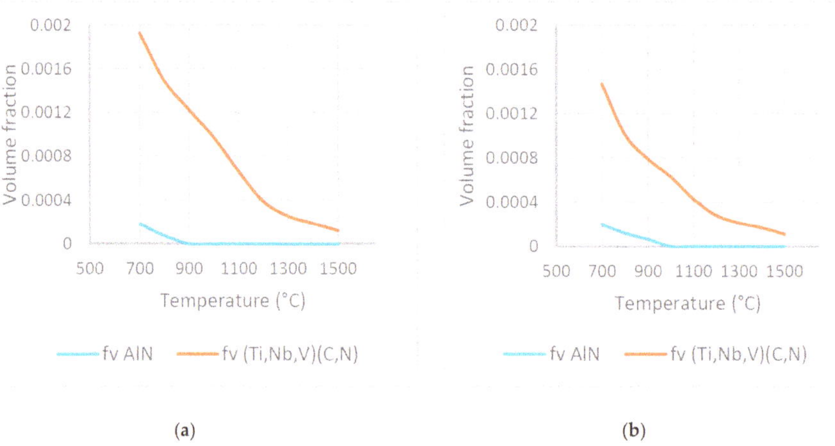

Figure 2. The calculated volume fraction of precipitates considering mutual solubility: (**a**) high-Nb steel; (**b**) low-Nb steel.

Abnormal growth of austenitic grains starts when particles coarsen past a critical size [3] or their volume fraction is drastically reduced, producing a pinning force smaller than the grain growth force. This massive reduction in the volume fraction of precipitates occurs at around 1100 °C according to empirical calculations (Figures 1 and 2). From the calculated volume fraction results, austenitic grain size can be predicted using Zener's model if an average particle size is assumed. For instance, at 1100 °C, an average particle size of 100 nm would allow a grain size of 350 μm. This is a likely scenario; if equilibrium conditions are achieved during reheating, NbC is expected to dissolve at 1100 °C. The presence of a small fraction of relatively coarse TiN particles may allow grain growth above this temperature.

3.2. Experimental Precipitates Volume Fraction

The presence of Nb and Ti-Nb precipitates was verified by SEM, and TEM and their nature was determined by EDS and electron diffraction. Micrographs were taken from random locations to ensure a representative sampling. Based on the TEM micrographs, measurements of the precipitates were obtained, and volume fractions were calculated. Assuming sample thickness equals average precipitate thickness, the area fraction becomes equal to the volume fraction. As expected, the volume fraction decreases as austenitization temperature increases. Average particle size, on the other hand, increased with austenitization temperature. The stability of Nb-rich precipitates was higher than expected at temperatures above 1100 °C.

A good example of Nb-rich particles' stability can be seen in Figure 3a. A micrograph obtained from the Low-Nb steel, austenitized to 1300 °C shows considerable presence of precipitates. These precipitates contain Nb, Ti and exist in a wide range of sizes. The stability of the particles was observed even during austenitization to 1300 °C for 1 h. Figure 3b was obtained from High-Nb steel austenitized to said conditions, and similar examples were seen in Low-Nb steel. EDS provided identification of these particles. Small objective lens aperture was used to selectively analyze every particle in Figure 3b. The chemical compositions of these particles are shown in Table 2. The C content is enhanced due to the C-replica film influence, and the copper due to the grid holding the replica. EDS shows that the particles contain Ti and Nb. V was detected in a fraction below the error of EDS method. The top-left particle in Figure 3b is richer in N and Ti than the other two. Similar particles were observed in both alloys at all tested temperatures, hence their complexity was further studied.

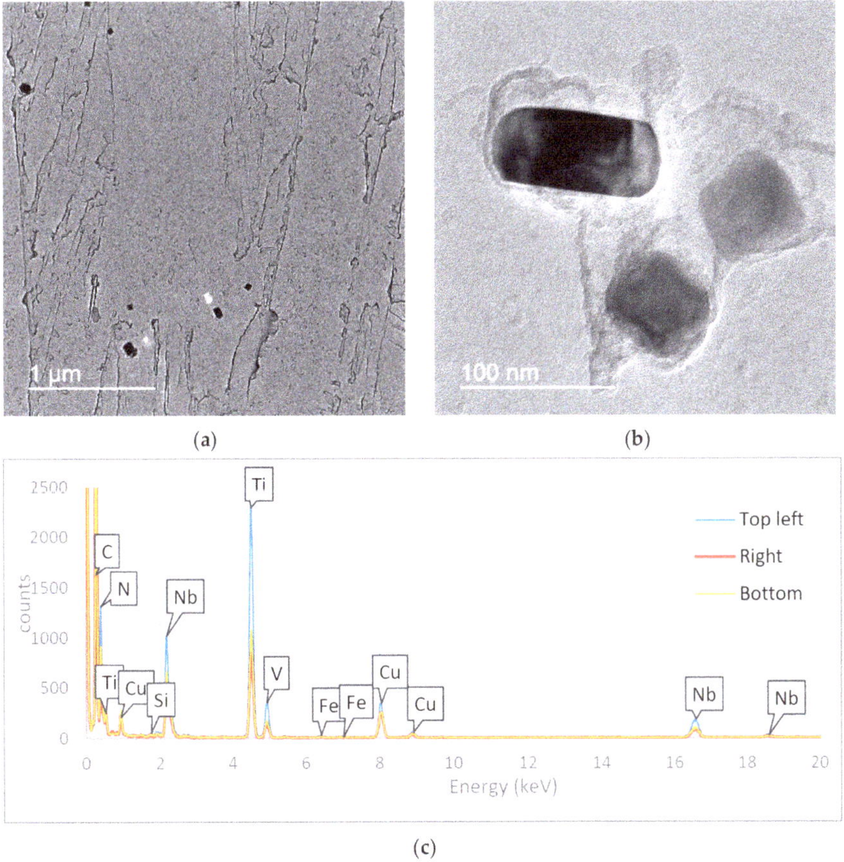

Figure 3. Transmission electron microscopy (TEM) micrograph examples (**a**) low magnification Low-Nb Steel austenitized at 1300 °C for 1 h; (**b**) high magnification from High-Nb Steel austenitized to 1300 °C for 1 h; (**c**) EDS spectra of particles in (**b**).

Table 2. EDS chemical composition measurements from precipitates in Figure 2b (at%/wt%).

Spectra	C	N	Si	Ti	V	Fe	Cu	Nb
Top left	41.82/20.1	32.51/18.2	0.15/0.2	16.03/30.7	0.54/1.1	0.1/0.2	2.93/7.5	5.92/22
Right	78.7/59.4	14.1/12.4		3.97/11.9			1.37/5.5	1.85/10.8
Bottom	73.5/52	17.13/14.1		5.35/15.1			1.61/6	2.34/12.8

The overall results of the dissolution study by TEM are plotted in Figure 4. The average diameter of observed particles is very similar for both alloys up to 1250 °C, then it increases dramatically at 1300 °C amid a volume fraction reduction. The dissolution of small particles and Ostwald ripening of big particles are the phenomena responsible for this average size increase.

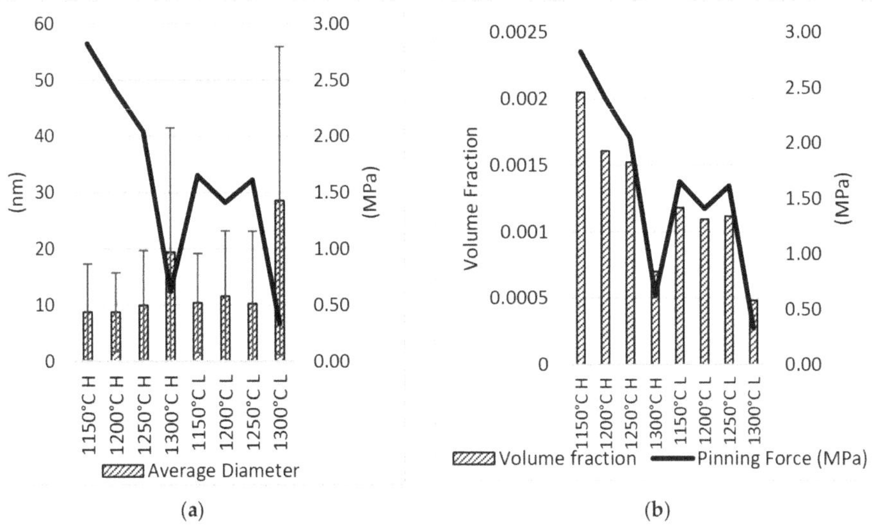

Figure 4. (**a**) Average diameter of precipitates at the different austenitization temperatures with soaking time of 1 h and corresponding pinning force; (**b**) volume fraction and corresponding pinning force. Measurements based on TEM micrography.

The pinning force associated to the particles was calculated using Zener's model as expanded by Gladman [3] for flexible boundaries.

$$F_{PIN} = 4r\sigma N_s \tag{1}$$

where, r, is the mean average particle radius, σ, is the interfacial energy between the particle and austenite per unit area and N_s, is the number of particles per unit area, that for a flexible boundary is given by [18]:

$$N_s = \frac{3f_v^{2/3}}{4\pi r^2} \tag{2}$$

where f_v, is the volume fraction of particles and, r, the mean particle radius. Pinning force calculations showed a trend dominated by volume fraction. The pinning force was high at austenitizing temperatures below 1250 °C for both alloys. At 1250 °C in the low-Nb steel, the pinning force was only 21% below the high-Nb alloy. This allowed both alloys to have homogeneous PAGS below 1250 °C.

The size distribution of precipitates, shown in Figure 5, was analyzed in two categories for better visualization: precipitates smaller than 40 nm, namely small precipitates, and precipitates bigger than 40 nm, namely big precipitates. The size distribution change with temperature illustrates the simultaneous occurrence of Ostwald ripening phenomena and dissolution. Fortunately, the small particles show considerable stability at 1250 °C and below in the low-Nb steel.

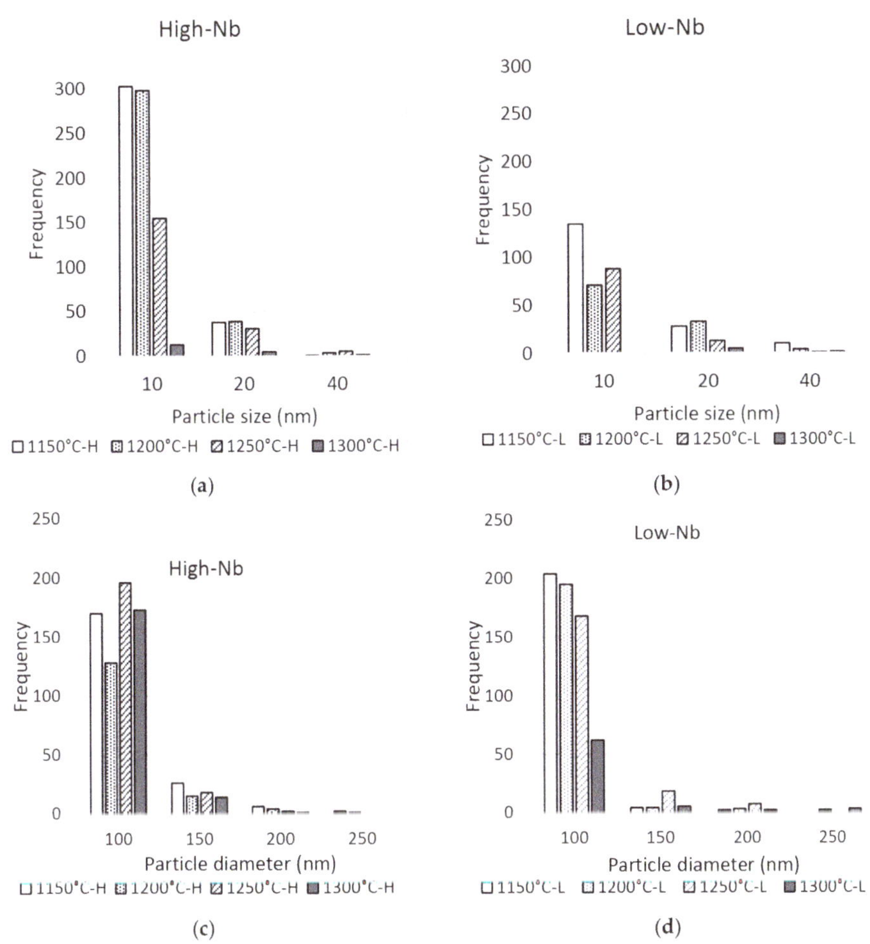

Figure 5. Distribution of particle size: (**a**,**b**) Particles smaller than 40 nm measured from ten 20,000× micrographs; (**c**,**d**) particles bigger than 40 nm, measured from five 5000× micrographs.

Comparing precipitate volume fraction calculations (from Section 3.1.) to experimentally-based volume fraction calculation, a shift in the dissolution behavior is observed, see Figure 6. Experimentally based calculated volume fraction is higher than in equilibrium conditions. After 1 h of soaking at temperature above 1100 °C, the complete dissolution of NbC was theoretically expected, according to the simple-precipitate approach calculations. Furthermore, the volume fraction of complex precipitates was expected to reduce considerably, according to the complex-precipitate approach calculations. Evidently, those equilibrium conditions were not reached, hence, dissolution was delayed. At these temperatures, Nb was found in complex particles containing both Nb and Ti, either in mutual solid solution or in epitaxially grown particles.

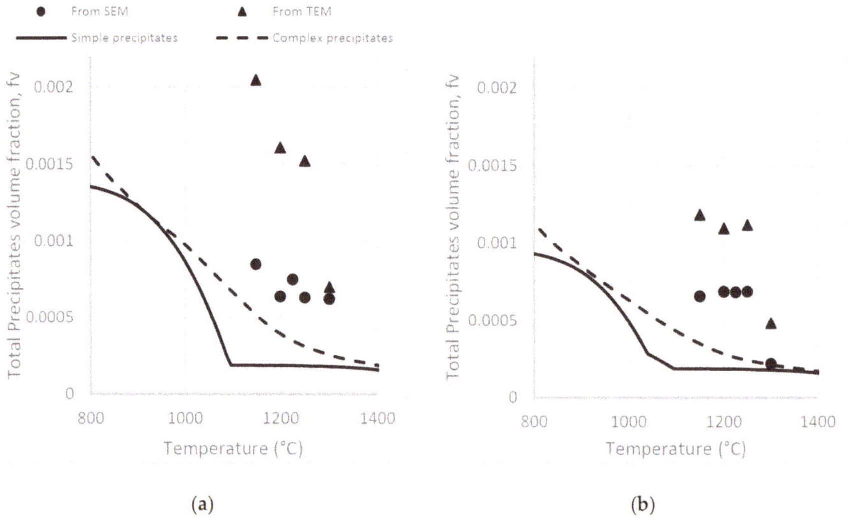

Figure 6. Experimental volume fraction of precipitates, compared to the calculated expectations: (**a**) high-Nb steel; (**b**) low-Nb steel.

3.3. Precipitates Morphology and Complexity Observations

The coarsened particles were complex Nb-Ti(C,N) with various compositions. Particle coarsening by Ostwald ripening is known to occur at temperatures where there is high solute diffusivity and the particles are stable. Many of these particles were single-phase solid solution complex precipitates, whereas others showed diverse epitaxial morphologies.

The higher experimental volume fraction, as compared to the calculated one, may be due to particles' morphology and complexity. Precipitate analysis by TEM confirmed their complexity. The carbides were identified by Dark Field (DF) imaging and EDS. In Figure 7, an NbC particle was identified, adjacent to a TiN particle. The beam selected for DF is consistent with TiN interplanar spacing (004). Analysis of the Selected Area Diffraction Pattern (SADP) shows plane (101) from NbC parallel to (002) from the Fe (α') matrix. This suggests coherence between these two phases at that set of planes. On the other hand, plane (101) Fe (α') holds 6.76° with respect to (002) TiN and 13.6° with respect to ($\bar{2}$02) NbC. EDS analysis, shown at the bottom, did not detect C or N. The main microalloy elements are Ti and Nb. The detected V could be in solid solution in the matrix, or in the particles.

Coherency was found by High Resolution Transmission Electron Microscopy (HRTEM) in big complex particles—an example is presented in Figure 8. Continuity of fringes is observed at the matrix/particle interface. The corresponding Fast Fourier Transform (FFT) evidences the plane (202) from the carbide is parallel to (200) from the iron-based matrix.

NbC growth on pre-existing TiN has been reported in previous research works [9–11] as being formed during the solidification process. In the present study, numerous carbides contained Nb and Ti, and the reheating conditions were not enough to completely dissolve these complex particles. These particles have reportedly been effective at GB pinning and recrystallization inhibition [13].

Precipitates in Figure 9 are examples of the complex precipitates found in both alloys within the 1150–1300 °C austenitization temperature range. Figure 9a is a particle found in low-Nb steel austenitized to 1300 °C for 1 h. Diffraction shows the [010] zone axis from where the beam [402] was used to produce the corresponding DF image. This DF image shows only half of the particle, indicating the bicrystalline nature of the particle. The precipitate in Figure 9b is an example of the presence of these complex precipitates at low austenitization temperatures. It was found in high-Nb steel austenitized to 1150 °C for 1 h. BF-DF of this particle shows its complex morphology.

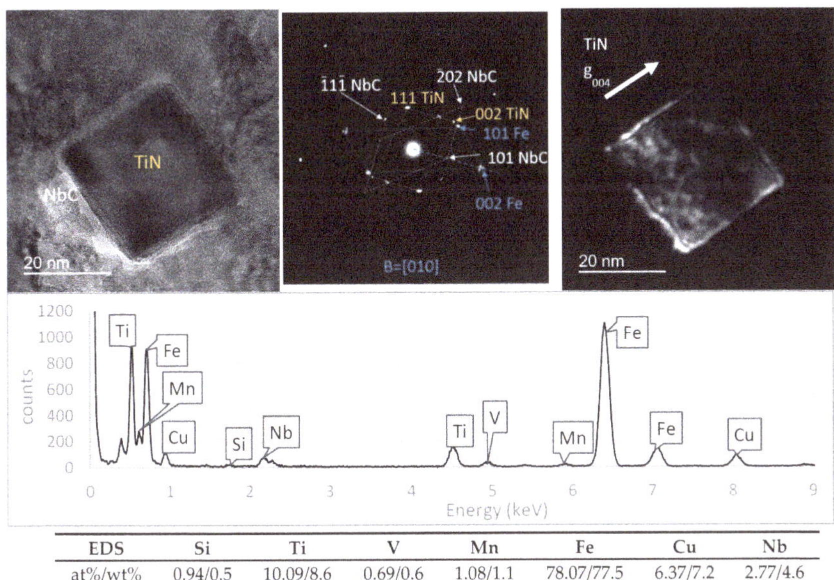

EDS	Si	Ti	V	Mn	Fe	Cu	Nb
at%/wt%	0.94/0.5	10.09/8.6	0.69/0.6	1.08/1.1	78.07/77.5	6.37/7.2	2.77/4.6

Figure 7. Bright field-Dark field (BF-DF) images and SADP from a precipitate in low-Nb steel austenitized at 1200 °C for 1 h.

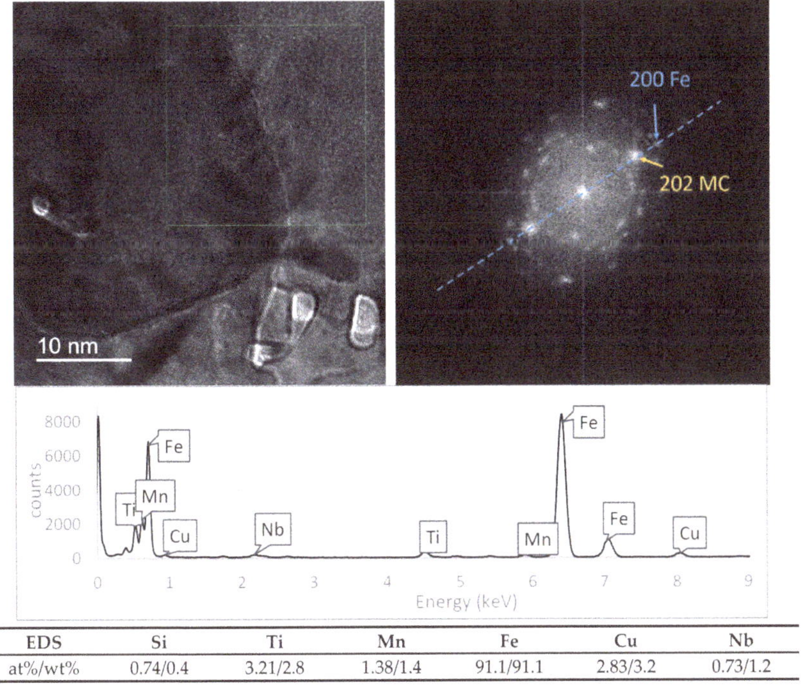

EDS	Si	Ti	Mn	Fe	Cu	Nb
at%/wt%	0.74/0.4	3.21/2.8	1.38/1.4	91.1/91.1	2.83/3.2	0.73/1.2

Figure 8. HRTEM, Fast Fourier Transform, and EDS results of a (Ti,Nb)(C,N) precipitate (big) in low-Nb steel austenitized at 1200 °C for 1 h.

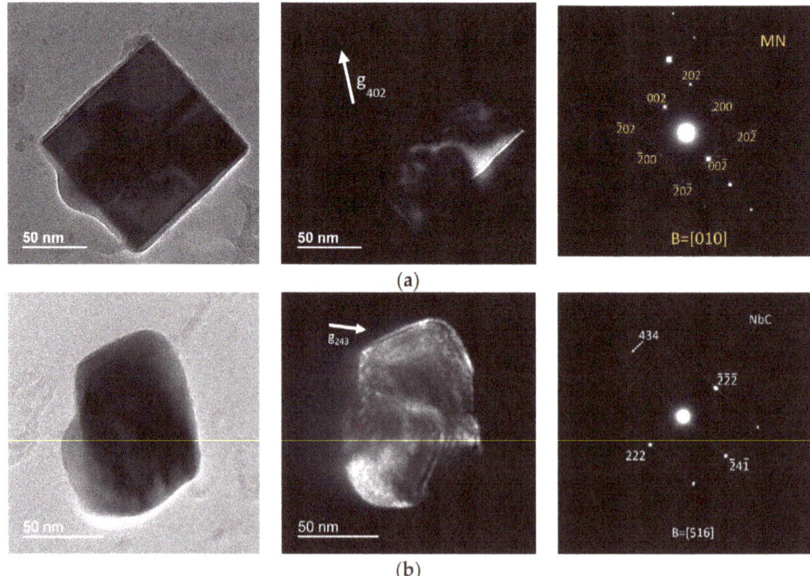

Figure 9. BF–DF and SADP of complex bicrystalline particles found in: (**a**) low-Nb steel austenitized at 1300 °C for 1 h.; (**b**) high-Nb steel austenitized to 1150 °C for 1 h.

Given the TEM evidence, it is highly likely that the epitaxial formation of precipitates is responsible for retarding dissolution. Two phenomena can be directly associated to the difficult dissolution of epitaxially grown precipitates: the reduction in the matrix/particle surface area and the low energy of particle/particle interfaces. Both phenomena hurdle the dissolution process. This may be responsible for the rich volume fraction of precipitates at high temperatures in both steels.

3.4. PAGS Measurements and Predictions

Measurements of PAGS are shown in Figure 10, where the average diameter of 50 grains is shown. A gradual increase in the average PAGS is observed as the temperature increases. The standard deviation at 1250 °C is broad for the low Nb steel, indicating that this steel has entered the abnormal growth stage. At 1300 °C the standard deviation is narrower, denoting homogeneous grain growth.

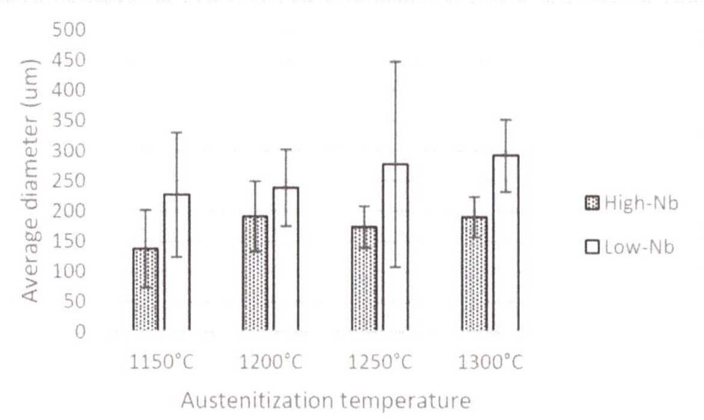

Figure 10. Average Feret diameter of Prior-Austenitic Grains. Measured from light microscopy after picral etching.

The grain size distribution is shown in Figure 11. The low-Nb steel maintains a stable size at 1150 and 1200 °C, peaking close to 250 μm on both cases. At 1250 °C, however, grain size distribution is spread, denoting abnormal growth. At 1300 °C, the small grains have yield way to the grown ones, hence the population is more normal.

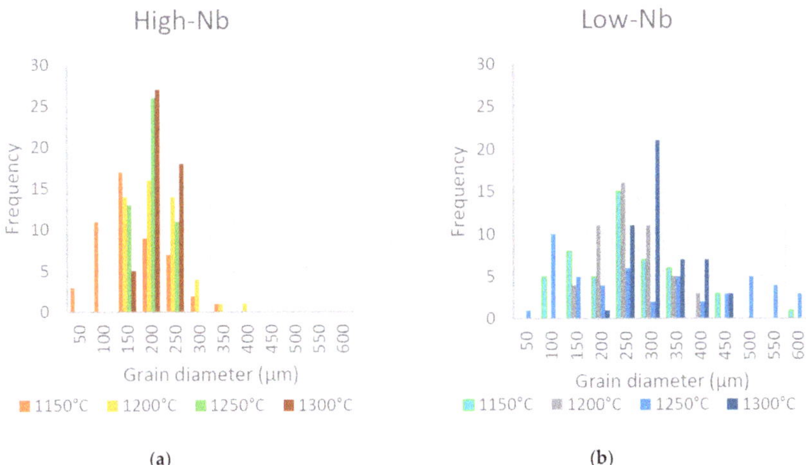

Figure 11. Grain size distribution of (**a**) high-Nb steel; and (**b**) Low-Nb steel.

Zener's model for precipitate-limited grain size was utilized to predict PAGS. Results in Figure 12 show coarsening starting at 1250 °C for the Low-Nb alloy and 1300 °C for the High-Nb alloy. Three other models by Gladman-Hillert [3], Rios [4,5], and Nishizawa, et al. [6] were used, predicting PAGS one order of magnitude smaller. However, a bigger mismatch is present if the measured precipitate volume fraction is used for estimating the PAGS. The estimation is too little compared to experimental results; these models need further reviewing for the present alloys.

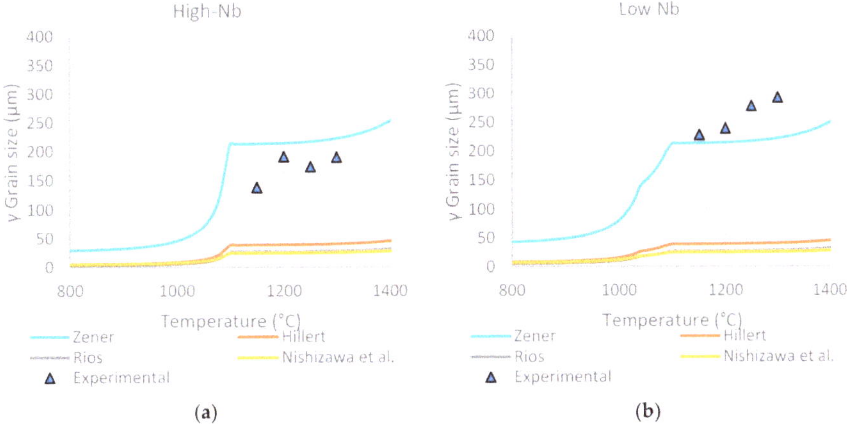

Figure 12. Experimentally determined PAGS compared to calculated predictions: (**a**) High-Nb steel; and (**b**) Low-Nb steel.

3.5. EBSD Reconstruction of PAGB

EBSD mapping was done on the samples that were quenched after austenitization. From the martensitic structure EBSD scans, maps of prior-austenite were reconstructed. Figure 13 shows the correspondence between PAGS revealed by etching, the martensitic EBSD Orientation Imaging Microscopy (OIM) map and reconstructed parent-phase austenite OIM map. Comparing Figure 13a to Figure 13c, a grain boundary correspondence is observed.

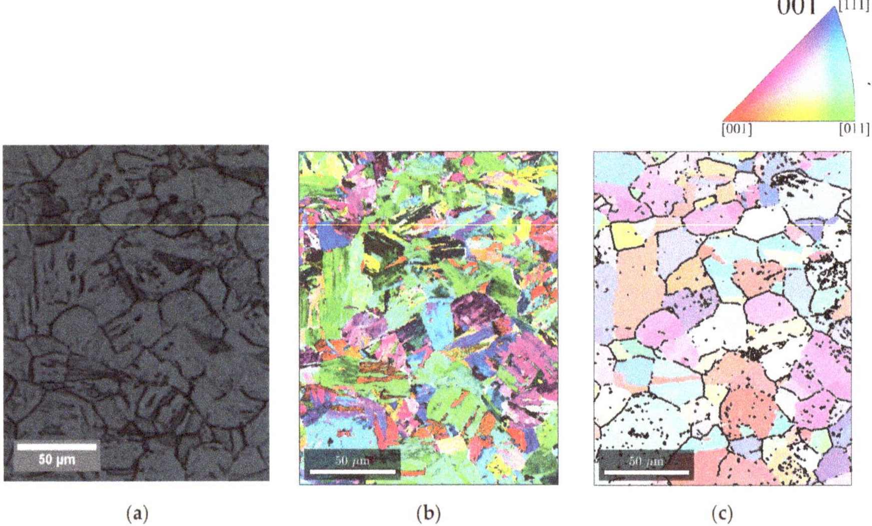

Figure 13. Comparison of the reconstructed PAGS maps to a picral etched micrograph. (**a**) Micrograph after picral etching; (**b**) EBSD OIM map from quenched specimen; (**c**) reconstructed OIM map of austenite, based on (**b**). High-Nb sample austenitized to 1150 °C for 1 h.

The resulting austenite maps showed similar grain sizes to the measurements presented in Figure 10. In Figure 14, the top row shows the grain coarsening behavior of the high-Nb alloy. Grain coarsening was avoided below 1300 °C. Meanwhile, the low-Nb alloy, shown in the bottom row of Figure 14, behaves according to the measurements by light microscopy on Figures 10 and 11.

Grain Boundary Misorientation Distribution changes due to reheating temperature were observed thanks to the reconstructed austenitic maps. GB misorientation data was collected from the maps and its distribution is plotted in Figure 15. A clear trend of an increasing population of GBs with misorientation above 50° occurs in the low-Nb alloy, at the expense of GBs with misorientations 20–50°, whereas on the high-Nb alloy there is not a clear trend. This decrease in boundaries between 20° and 50° may be due to grain coarsening; it has been proposed by other authors [19,20] that GBs in this range have high energy and high mobility. A consequence of this high mobility is that these boundaries are more susceptible to disappear, as they close in on each other, devouring grains. This behavior is not observed in the high-Nb steel, where coarsening did not take place.

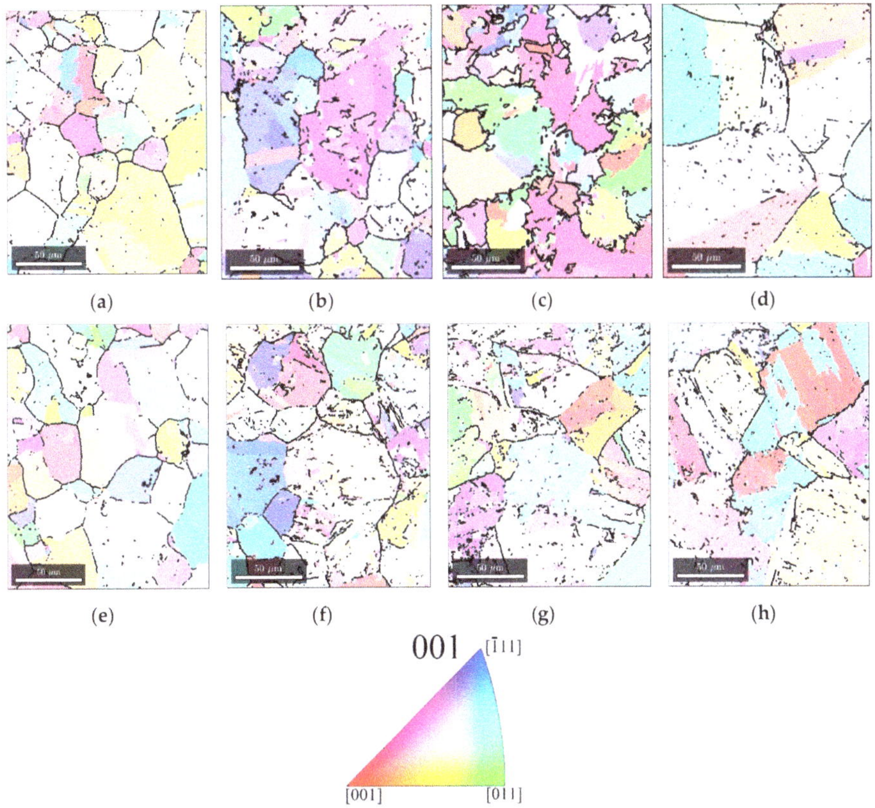

Figure 14. Reconstructed austenite EBSD OIM maps from austenitizing high-Nb steel to: (**a**) 1150 °C, (**b**) 1200 °C, (**c**) 1250 °C, (**d**) 1300 °C for 1 h; and from austenitizing Low-Nb Steel to: (**e**) 1150 °C, (**f**) 1200 °C, (**g**) 1250 °C and (**h**) 1300 °C for 1 h. All images are $150 \times 190\ \mu m^2$.

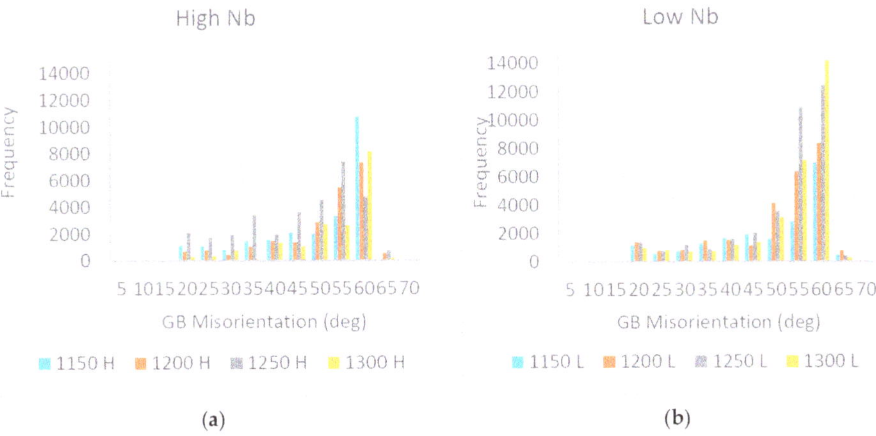

Figure 15. Distribution of GB misorientation for various temperatures on: (**a**) High-Nb steel and (**b**) low-Nb steel.

4. Discussion

The empirical volume fractions of precipitates, compared to the experimentally-based volume fractions, provided a better understanding of the observed retardation of the grain coarsening behavior. The experimentally based volume fraction exceeds calculated predictions, including calculations that consider the mutual solubility of precipitates, which happened experimentally. Mutual carbo-nitrides solubility produced complex and more stable particles.

The presence of complex precipitates explains the present experiments—the grain-coarsening temperature is higher than calculated. The jump in grain size of Low-Nb steel can be observed in Figure 14, where it happens between 1200 and 1250 °C. Moreover, the drop in experimental precipitate volume fraction is observed to happen after 1250 °C in Figure 6. This evidence strongly suggests that lower Nb contents are efficient to control PAGS, under the experimental conditions used in this study.

Grain-coarsening studies in similar alloys agree on the effectiveness of Ti-Nb microalloyed steels [12,21–23] to keep fine PAGS at high temperature. Alogab, et al. [21,24] have shown how Nb adds resistance to austenitic grain coarsening during carburization, through different heating rates and temperatures below 1100 °C. Their work treats TiN and NbC separately. A study by Zou and Kirkaldy [25] on very similar alloys agrees that Nb-Ti complex particles nucleate on Ti-rich particles, but, as they coarsen, the Nb content increases. Another study [23] used a very similar alloy to High-Nb steel and found abnormal growth after 1 h of soaking at 1050 °C. However, they did not report the occurrence of epitaxial NbC on TiN. Hence, their results agree with our hypothesis that Nb-Ti complex particles are responsible for the retardation of dissolution as compared to plain TiN or plain NbC.

Other studies involving very similar steels agree that (Nb,Ti)C precipitates are highly likely to form [22]. Zhang, et al. [15] have done first-principles calculations of interfacial energies between NbC, TiC and TiN, and have justified the heterogeneous nucleation of NbC at these particles' surfaces. Graux, et al. [14] have modeled the PAGS in a Ti-Nb microalloyed steel, and have considered Ti-Nb particles. They have found (Ti,Nb)C stable after 30 min at 1200 °C. Furthermore, a theoretical study by Hutchinson, et al. [26] suggests that as NbC dissolves, Nb remains an important GB deterrent by solute drag.

The results of this work support the feasibility of using lower Nb contents while still maintaining good control of PAGS. The present work has shown that grain-coarsening behavior of a low-Nb alloy is controlled by a delay in the dissolution of the precipitates. However, further work is required to fully understand the relationship between the epitaxial growth of NbC on TiN and the retardation of the dissolution behavior. Future work should utilize an Nb-free alloy compared to two Nb-Ti bearing alloys. The mechanisms of dissolution of epitaxial particles need to be verified for these alloys. The role of factors like interfacial energy, morphology and phenomena like epitaxial growth, Ostwald ripening and GB-particle interactions need to be studied in alloys containing these precipitate types.

5. Conclusions

A low-Nb LP steel demonstrated good PAGS control up to 1200 °C with the holding times used in this study. This behavior exceeded empirical equilibrium-based expectations. The thermodynamic equilibrium equations used for predicting precipitates dissolution temperature need to be revised for the case of binary precipitates containing additional epitaxial precipitation, i.e., TiN + NbC.

Experimental evidence showed a higher dissolution temperature than calculated models predict, especially for complex precipitates. The epitaxial presence of NbC on TiN led to the conclusion of slower dissolution due to interface energy and area reduction. The reconstruction of prior austenite maps through the Nyyssönnen, et al. method proved useful and in close agreement with the traditional PAGS-revealing method.

Author Contributions: Conceptualization, G.S.-B., M.M. and C.I.G.; methodology, G.S.-B.; validation, G.S.-B., M.M. and C.I.G.; formal analysis, G.S.-B.; investigation, G.S.-B. and M.M.; resources, M.M. and C.I.G.; data curation, G.S.-B.; writing—original draft preparation, G.S.-B.; writing—review and editing, C.I.G.; visualization, G.S.-B.; supervision, C.I.G.; project administration, C.I.G.; funding acquisition, C.I.G. All authors have read and agreed to the published version of the manuscript.

Funding: This research was funded by the Pennsylvania Department of Community and Economic Development CMU Subcontract 1060145-406196—Pitt I# 0062204; Programa para el Desarrollo Profesional Docente, PRODEP DSA/103.5/15/9555; United States Steel Corporation, and the APC was funded by The Ferrous Physical Metallurgy research group at University of Pittsburgh.

Acknowledgments: We are grateful for the support of: PRODEP México, Universidad Veracruzana, Roberto Rocca Education Program and Pennsylvania Department of Community and Economic Development. The authors want to thank the Ferrous Physical Metallurgy Research Group fellows and collaborators at University of Pittsburgh, as much as to collaborators at United States Steel Corporation: Justin Bryan, Debra Giansante, Andrew Warble and Christopher Snyder.

Appendix A

The calculations for volume fractions were done in the following manner. Let us first consider the solubility products of the precipitates involved. The solubility product denotes the equilibrium concentrations of the alloying elements at a given temperature. For instance, the concentrations of M and N in equilibrium with the precipitate $M_m N_n$ are given by

$$[M]^m [N]^n = K \tag{A1}$$

[M] and [N] are mole fractions of M and N in solution and K is the solubility product [27]. The solubility product changes with temperature, allowing for dissolution at higher temperature.

$$K = K_0 \exp \frac{-\Delta H}{RT} \tag{A2}$$

where ΔH represents the precipitate's enthalpy of formation in the matrix and K_0 is a constant. Empirical adaptations use weight percentages instead of molar fractions of the solutes, and express the solubility product, Ks, as

$$\log Ks = A - \frac{B}{T} \tag{A3}$$

where T, is the temperature of the alloy and the constants A, and B, are experimentally determined. These values were collected for the involved precipitates in an austenitic matrix and, combined with mass balance equations, a series of quadratic equations was produced. Then, these equations were solved in order of precipitate stability as expected from Gibbs' free energy of formation.

As the alloy cools down, for instance, the preferable nitride is TiN. After TiN forms, the remaining N is available to form AlN, NbN and VN, in that preferential order. Similarly, for carbides, the preferential order for carbon would be TiC, NbC, VC, MoC. Following that order, the system of equations was solved for the mass fraction of every element that is in precipitate form at temperature intervals of 10 °C, from 200 to 1500 °C.

Cooling from high temperatures, the first precipitate to be formed would be TiN. Hence, after all the possible TiN has been formed and equilibrium is reached, the following equations must hold true

$$N_T = [N] + N_{TiN} \tag{A4}$$

$$Ti_T = [Ti] + Ti_{TiN} \tag{A5}$$

where X_T is the total mass fraction content in the bulk composition of species X, [X] is the mass fraction of X that remains in solid solution and X_{TiN} is the mass fraction of the species X in the form of TiN. Now, considering the stoichiometric mass ratio of TiN and following the same nomenclature, Equation (A4) can be re-written as

$$N_T = [N] + \frac{14}{48} Ti_{TiN} \tag{A6}$$

The solid solution contents of N and Ti in equilibrium with TiN are related through the solubility product, K_{TiN}. So, substituting Ti_{TiN} from Equation (A1) into Equation (A6), we can write

$$N_T = [N] + \frac{14}{48}\left(Ti_T + \frac{K_{TiN}}{[N]}\right) \tag{A7}$$

Now, notice that N_T and Ti_T are known fractions from our bulk composition, and K_{TiN} is known as a function of temperature from experimental reports in the literature (See Table A1).

Table A1. Solubility products of precipitates of interest in Austenite.

K_s	$\log K_s = A - B/T$		Reference
	A	**B**	
$[Nb][C]$	3.42	7900	Gladman [28]
$[Nb][N]$	2.8	8500	Gladman [28]
$[Ti][C]$	5.33	10,475	Gladman [28]
$[Ti][N]$	3.82	15,020	Gladman [28]
$[Al][N]$	1.8	7750	Gao-Baker [2]
$[V][C]^{0.75}$	4.45	6560	Turkdogan [29,30]
$[V][N]$	2.86	7700	Turkdogan [29,30]
$[Mo][C]$	1.29	523	Pavlina, et al. [31]

Hence, Equation (A7) can be rearranged and solved for $[N]$ as a quadratic equation for every temperature of interest.

$$[N]^2 + \left(\frac{7}{24}Ti_T - N_T\right)[N] - \frac{7}{24}\left(\log(2.8) - \frac{8500}{T}\right) = 0 \tag{A8}$$

Once $[N]$ is found as the solid solution concentration of nitrogen in equilibrium with TiN, the simple mass balance from Equation (A6) helps us find N_{TiN}, the amount of N in TiN. This way, the amount of TiN present at every temperature during cooling is found using the stoichiometric mass ratio.

Calculations for the mass fraction of every precipitate followed this method. Every new calculation considered new mass balance equations, accounting for the mass already precipitated in the higher temperature stable precipitates. For instance, when calculating MoC mass fraction, the mass balance equation for carbon must consider that precipitations of TiC, NbC and VC have already happened. Hence it must be written

$$C_T = C_{TiC} + C_{NbC} + C_{VC} + C_{MoC} + [C] \tag{A9}$$

When MoC precipitates, however, all other considered carbides are already present, so the amount of carbon present in each is known. Hence, the only unknowns in this equation are C_{MoC} and $[C]$.

Appendix B

Solubility product models that consider the complexity of precipitates mutual solubility have been proposed elsewhere [32–34]. The present work used a simplified system of equations, inspired by Xu, et al.'s work. A system of 21 equations was reduced to eight equations and eight unknowns. The eight main equations came from two equations of activity of the precipitates, and six equations of the mass balance of C, N, Ti, Nb, V and Al. These eight equations were fed with the other 13 equations, relating the solubility products of seven precipitates (AlN, TiC, TiN, NbC, NbN, VC and VN) and the activities of the six involved components.

The first two main equations state assumptions of activities equal to one, for a stable precipitate in the matrix. The mutual solubility of TiC, TiN, NbC, NbN, VC and VN is considered, so the combined activity of this complex precipitate is 1, whereas AlN is insoluble, so its activity remains 1:

$$a_{TiC} + a_{TiN} + a_{NbC} + a_{NbN} + a_{VC} + a_{VN} = 1 \tag{A10}$$

$$a_{AlN=1} \tag{A11}$$

Mass balance provided the other six main equations. For every element, the amount present in every phase must add up to the total bulk content of that component, M_T. Hence, the molar fractions, X, of three phases were considered: the matrix, the volume fraction of aluminum nitride, X_{AlN}, and the molar fraction of the mutually soluble precipitates, X_{cmplx}.

$$C_T = \left(1 - X_{cmplx} - X_{AlN}\right)[C] + (a_{TiC} + a_{NbC} + a_{VC})X_{cmplx} \tag{A12}$$

$$N_T = \left(1 - X_{cmplx} - X_{AlN}\right)[N] + (a_{TiN} + a_{NbN} + a_{VN})X_{cmplx} + a_{AlN}X_{AlN} \tag{A13}$$

$$Ti_T = \left(1 - X_{cmplx} - X_{AlN}\right)[Ti] + (a_{TiC} + a_{TiN})X_{cmplx} \tag{A14}$$

$$Nb_T = \left(1 - X_{cmplx} - X_{AlN}\right)[Nb] + (a_{NbC} + a_{NbN})X_{cmplx} \tag{A15}$$

$$V_T = \left(1 - X_{cmplx} - X_{AlN}\right)[V] + (a_{VC} + a_{VN})X_{cmplx} \tag{A16}$$

$$Al_T = \left(1 - X_{cmplx} - X_{AlN}\right)[Al] + a_{AlN}X_{AlN} \tag{A17}$$

At this point, there are 14 unknowns (six activities, two molar fractions and six concentrations) that outnumber the eight equations, but the following equations can be considered to simplify the system. The activities of all considered nitrides and carbides can be replaced by functions of temperature and the activities of the individual elements, utilizing the solubility products, K.

$$a_{AlN} = \frac{(a_{Al}a_N)}{K_{AlN}} \tag{A18}$$

$$a_{TiC} = \frac{(a_{Ti}a_C)}{K_{TiC}} \tag{A19}$$

$$a_{TiN} = \frac{(a_{Ti}a_N)}{K_{TiN}} \tag{A20}$$

$$a_{NbC} = \frac{(a_{Nb}a_C)}{K_{NbC}} \tag{A21}$$

$$a_{NbN} = \frac{(a_{Nb}a_N)}{K_{NbN}} \tag{A22}$$

$$a_{VC} = \frac{(a_V a_C)}{K_{VC}} \tag{A23}$$

$$a_{VN} = \frac{(a_V a_N)}{K_{VN}} \tag{A24}$$

Therefore, the activity of every element may be replaced by the product of the activity coefficient, γ, of the element, and that element's concentration in solution $[M]$. γ can be written as a function of concentrations and Wagner interaction coefficients, e_{M1}^{M2}, of every component on a given element. These coefficients can be found in the literature

$$a_C = [C] * \left(e_C^C * [C] + e_C^N * [N] + e_C^{Ti} * [Ti] + e_C^{Nb} * [Nb] + e_C^V * [V] + e_C^{Al} * [Al]\right) \tag{A25}$$

$$a_N = [N] * \left(e_N^C * [C] + e_N^N * [N] + e_N^{Ti} * [Ti] + e_N^{Nb} * [Nb] + e_N^V * [V] + e_N^{AL} * [Al] \right) \tag{A26}$$

$$a_{Ti} = [Ti] * \left(e_{Ti}^C * [C] + e_{Ti}^N * [N] + e_{Ti}^{Ti} * [Ti] + e_{Ti}^{Nb} * [Nb] + e_{Ti}^V * [V] + e_{Ti}^{AL} * [Al] \right) \tag{A27}$$

$$a_{Nb} = [Nb] * \left(e_{Nb}^C * [C] + e_{Nb}^N * [N] + e_{Nb}^{Ti} * [Ti] + e_{Nb}^{Nb} * [Nb] + e_{Nb}^V * [V] + e_{Nb}^{AL} * [Al] \right) \tag{A28}$$

$$a_V = [V] * \left(e_V^C * [C] + e_V^N * [N] + e_V^{Ti} * [Ti] + e_V^{Nb} * [Nb] + e_V^V * [V] + e_V^{AL} * [Al] \right) \tag{A29}$$

$$a_{Al} = [Al] * \left(e_{Ti}^C * [C] + e_{Ti}^N * [N] + e_{Ti}^{Ti} * [Ti] + e_{Ti}^{Nb} * [Nb] + e_{Ti}^V * [V] + e_{Ti}^{AL} * [Al] \right) \tag{A30}$$

Substituting activities from Equations (A25)–(A30) into Equations (A18)–(A24), and then substitute (A18)–(A24) into (A10)–(A17) transforms Equations (A10)–(A17) into a system of eight equations with eight unknowns. Furthermore, if the solid solution is regarded as dilute, due to the very small amounts of solutes, the activities of the individual components can be approximated to their concentration in solid solution. This transforms Equations (A10)–(A17) into a much simpler system of eight equations and eight unknowns (six concentrations and two molar fractions):

$$\frac{[Ti][C]}{K_{TiC}} + \frac{[Ti][N]}{K_{TiN}} + \frac{[Nb][C]}{K_{NbC}} + \frac{[Nb][N]}{K_{NbN}} + \frac{[V][C]}{K_{VC}} + \frac{[V][N]}{K_{VN}} = 1 \tag{A31}$$

$$\frac{[Al][N]}{K_{AlN}} = 1 \tag{A32}$$

$$C_T = \left(1 - X_{cmplx} - X_{AlN} \right)[C] + \left(\frac{[Ti][C]}{K_{TiC}} + \frac{[Nb][C]}{K_{NbC}} + \frac{[V][C]}{K_{VC}} \right) X_{cmplx} \tag{A33}$$

$$N_T = \left(1 - X_{cmplx} - X_{AlN} \right)[N] + \left(\frac{[Ti][N]}{K_{TiN}} + \frac{[Nb][N]}{K_{NbN}} + \frac{[V][N]}{K_{VN}} \right) X_{cmplx} + X_{AlN} \tag{A34}$$

$$Ti_T = \left(1 - X_{cmplx} - X_{AlN} \right)[Ti] + \left(\frac{[Ti][C]}{K_{TiC}} + \frac{[Ti][N]}{K_{TiN}} \right) X_{cmplx} \tag{A35}$$

$$Nb_T = \left(1 - X_{cmplx} - X_{AlN} \right)[Nb] + \left(\frac{[Nb][C]}{K_{NbC}} + \frac{[Nb][N]}{K_{NbN}} \right) X_{cmplx} \tag{A36}$$

$$V_T = \left(1 - X_{cmplx} - X_{AlN} \right)[V] + \left(\frac{[V][C]}{K_{VC}} + \frac{[V][N]}{K_{VN}} \right) X_{cmplx} \tag{A37}$$

$$Al_T = \left(1 - X_{cmplx} - X_{AlN} \right)[Al] + X_{AlN} \tag{A38}$$

This system has numerous sets of solutions for a given temperature. Many of these solutions include complex and negative numbers, others fail to provide concentration values between zero and the bulk content of a component. Discretion was used to choose the right solution and obtain a reliable trend. The system of equations was solved with Wolfram Mathematica computing software.

References

1. Cuddy, L.J.; Raley, J.C. Austenite grain coarsening in microalloyed steels. *Metall. Trans. A* **1983**, *14*, 1989–1995. [CrossRef]
2. Gao, N.; Baker, T.N. Austenite Grain Growth Behaviour of Microalloyed Al–V–N and Al–V–Ti–N Steels. *ISIJ Int.* **1998**, *38*, 744–751. [CrossRef]
3. Gladman, T. On the theory of the effect of precipitate particles on grain growth in metals. *Proc. R. Soc. Lond. Ser. A Math. Phys. Sci.* **1966**, *294*, 298–309.

4. Rios, P.R. On the relationship between pinning force and limiting grain radius. *Scr. Mater.* **1996**, *34*, 1185–1188. [CrossRef]

5. Rios, P.R. A theory for grain boundary pinning by particles. *Acta Metall.* **1987**, *35*, 2805–2814. [CrossRef]

6. Nishizawa, T.; Ohnuma, I.; Ishida, K. Examination of the Zener relationship between grain size and particle dispersion. *Mater. Trans. JIM* **1997**, *38*, 950–956. [CrossRef]

7. Gong, P.; Palmiere, E.J.; Rainforth, W.M. Dissolution and precipitation behaviour in steels microalloyed with niobium during thermomechanical processing. *Acta Mater.* **2015**, *97*, 392–403. [CrossRef]

8. Ringer, S.P.; Li, W.B.; Easterling, K.E. On the interaction and pinning of grain boundaries by cubic shaped precipitate particles. *Acta Metall.* **1989**, *37*, 831–841. [CrossRef]

9. Hong, S.G.; Jun, H.J.; Kang, K.B.; Park, C.G. Evolution of precipitates in the Nb-Ti-V microalloyed HSLA steels during reheating. *Scr. Mater.* **2003**, *48*, 1201–1206. [CrossRef]

10. Ma, X.; Miao, C.; Langelier, B.; Subramanian, S. Suppression of strain-induced precipitation of NbC by epitaxial growth of NbC on pre-existing TiN in Nb-Ti microalloyed steel. *Mater. Des.* **2017**, *132*, 244–249. [CrossRef]

11. Craven, A.J.; He, K.; Garvie, L.A.J.; Baker, T.N. Complex heterogeneous precipitation in titanium–niobium microalloyed Al-killed HSLA steels—I. (Ti,Nb)(C,N) particles. *Acta Mater.* **2000**, *48*, 3857–3868. [CrossRef]

12. Wang, R.; Garcia, C.I.; Hua, M.; Cho, K.; Zhang, H.; DeArdo, A.J. Microstructure and Precipitation Behavior of Nb, Ti Complex Microalloyed Steel Produced by Compact Strip Processing. *ISIJ Int.* **2006**, *46*, 1345–1353. [CrossRef]

13. Subramanian, S.V.; Xiaoping, M.; Rehman, K. Austenite Grain Size Control in Upstream Processing of Niobium Microalloyed Steels by Nano-Scale Precipitate Engineering of TiN-NbC Composite. In *Energy Materials 2014*; Springer: Cham, Switzerland, 2014; pp. 639–650.

14. Graux, A.; Cazottes, S.; De Castro, D.; San Martín, D.; Capdevila, C.; Cabrera, J.M.; Molas, S.; Schreiber, S.; Mirković, D.; Danoix, F.; et al. Precipitation and grain growth modelling in Ti-Nb microalloyed steels. *Materialia* **2019**, *5*, 100233. [CrossRef]

15. Zhang, H.; Huihui, X. First-principles study of NbC heterogeneous nucleation on TiC vs. TiN in microalloy steel. *Ironmak. Steelmak.* **2018**. [CrossRef]

16. Nyyssönen, T.; Isakov, M.; Peura, P.; Kuokkala, V.-T. Iterative Determination of the Orientation Relationship Between Austenite and Martensite from a Large Amount of Grain Pair Misorientations. *Metall. Mater. Trans. A* **2016**, *47*, 2587–2590. [CrossRef]

17. Nyyssönen, T.; Peura, P.; Kuokkala, V.-T. Crystallography, Morphology, and Martensite Transformation of Prior Austenite in Intercritically Annealed High-Aluminum Steel. *Metall. Mater. Trans. A* **2018**, *49*, 6426–6441. [CrossRef]

18. Palmiere, E.J.; Garcia, C.I.; DeArdo, A.J. The influence of niobium supersaturation in austenite on the static recrystallization behavior of low carbon microalloyed steels. *Metall. Mater. Trans. A* **1996**, *27*, 951–960. [CrossRef]

19. Blancas, V. *A New View of the Grain-Coarsening Behavior of Austenite in Ti-Microalloyed Low-Carbon Steels*; University of Pittsburgh: Pittsburgh, PA, USA, 2016.

20. Hayakawa, Y.; Szpunar, J.A. The role of grain boundary character distribution in secondary recrystallization of electrical steels. *Acta Mater.* **1997**, *45*, 1285–1295. [CrossRef]

21. Alogab, K.A.; Matlock, D.K.; Speer, J.G.; Kleebe, H.J. The Influence of Niobium Microalloying on Austenite Grain Coarsening Behavior of Ti-modified SAE 8620 Steel. *ISIJ Int.* **2007**, *47*, 307–316. [CrossRef]

22. Charleux, M.; Poole, W.J.; Militzer, M.; Deschamps, A. Precipitation Behavior and its Effect on Strengthening of an HSLA-NbTi Steel. *Metall. Mater. Trans. A* **2001**, *32*, 1635–1648. [CrossRef]

23. Karmakar, A.; Kundu, S.; Roy, S.; Neogy, S.; Srivastava, D.; Chakrabarti, D. Effect of microalloying elements on austenite grain growth in Nb–Ti and Nb–V steels. *Mater. Sci. Technol.* **2014**, *30*, 653–664. [CrossRef]

24. AlOgab, K.; Matlock, D.; Speer, J.; Kleebe, H. The Effects of Heating Rate on Austenite Grain Growth in a Ti-modified SAE 8620 Steel with Controlled Niobium Additions. *ISIJ Int.* **2007**, *47*, 1034–1041. [CrossRef]

25. Zou, H.; Kirkaldy, J.S. Carbonitride precipitate growth in titanium/niobium microalloyed steels. *Metall. Trans. A* **1991**, *22*, 1511–1524. [CrossRef]

26. Hutchinson, C.R.; Zurob, H.S.; Sinclair, C.W.; Brechet, Y.J.M. The comparative effectiveness of Nb solute and NbC precipitates at impeding grain-boundary motion in Nb steels. *Scr. Mater.* **2008**, *59*, 635–637. [CrossRef]

27. Porter, D.A.; Easterling, K.E.; Sherif, M. *Phase Transformations in Metals and Alloys, (Revised Reprint)*; CRC press: New York, NY, USA, 2009.

28. Gladman, T. Precipitation hardening in metals. *Mater. Sci. Technol.* **1999**, *15*, 30–36. [CrossRef]

29. Turkdogan, E. Causes and effects of nitride and carbonitride precipitation during continuous casting. *Iron Steelmak.* **1989**, *16*, 61.

30. Homsher, C.N. *Determination of the Non-Recrystallization Temperature (Tnr) in Multiple Microalloyed Steels*; Colorado School of Mines: Denver, CL, USA, 2013.

31. Pavlina, E.J.; Speer, J.G.; Van Tyne, C.J. Equilibrium solubility products of molybdenum carbide and tungsten carbide in iron. *Scr. Mater.* **2012**, *66*, 243–246. [CrossRef]

32. Xu, K.; Thomas, B.G.; O'malley, R. Equilibrium Model of Precipitation in Microalloyed Steels. *Metall. Mater. Trans. A* **2011**, *42*, 524–539. [CrossRef]

33. Wang, W.; Wang, H.R. A simple method to determine the complex carbonitride composition in multicomponent microalloyed austenite. *Mater. Lett.* **2007**, *61*, 2227–2230. [CrossRef]

34. Xiaodong, L.; Solberg, J.K.; Gjengedal, R.; Kluken, A.O. An expression for solubility product of complex carbonitrides in multicomponent microalloyed austenite. *Scr. Metall. Mater.* **1994**, *31*, 1607–1612. [CrossRef]

Article

Austenite Decomposition and Precipitation Behavior of Plastically Deformed Low-Si Microalloyed Steel

Adam Grajcar [1],*, **Mateusz Morawiec [1]** and **Wladyslaw Zalecki [2]**

[1] Institute of Engineering Materials and Biomaterials, Silesian University of Technology, 18A Konarskiego Street, 44-100 Gliwice, Poland; mateusz.morawiec@polsl.pl

[2] Department of Process Simulation, Institute for Ferrous Metallurgy, 12-14 K. Miarki Street, 44-100 Gliwice, Poland; wzalecki@imz.pl

* Correspondence: adam.grajcar@polsl.pl; Tel.: +48-32-237-2933

Received: 21 November 2018; Accepted: 3 December 2018; Published: 6 December 2018

Abstract: The aim of the present study is to assess the effects of hot deformation and cooling paths on the phase transformation kinetics in a precipitation-strengthened automotive 0.2C–1.5Mn–0.5Si steel with Nb and Ti microadditions. The analysis of the precipitation processes was performed while taking into account equilibrium calculations and phase transitions resulting from calculated time–temperature–transformation (TTT) and continuous cooling transformation (CCT) diagrams. The austenite decomposition was monitored based on thermodynamic calculations of the volume fraction evolution of individual phases as a function of temperature. The calculations were compared to real CCT and DCCT (deformation continuous cooling transformation) diagrams produced using dilatometric tests. The research included the identification of the microstructure of the nondeformed and thermomechanically processed supercooled austenite products formed at various cooling rates. The complex interactions between the precipitation process, hot deformation, and cooling schedules are linked.

Keywords: austenite decomposition; high-strength steel; microstructure evolution; phase transformation

---

1. Introduction

The strengthening of advanced ferrous alloys can be achieved in many ways, including solid solution strengthening, grain refinement, cold deformation, precipitation phenomena, phase transformations, high-pressure methods, etc. [1–3]. One of the most interesting and evolving groups of advanced steels for the automotive industry is the multiphase transformation-induced plasticity (TRIP)-aided steels consisting of ferrite, bainite, and retained austenite [4–6]. These steels utilize different strengthening mechanisms to improve their strength level, but the most important is the strain-induced transformation of retained austenite into martensite [7–9]. This transformation takes place during the cold forming of steel sheets and results in improving both strength and plasticity. This phenomenon is called the transformation-induced plasticity (TRIP) effect [10–13].

A special feature of TRIP steels is the presence of carbide-free bainite, which can be formed during a multistep heat treatment using increased Si and/or Al additions [14]. These elements are necessary to prevent carbide precipitation under conditions of isothermal bainitic transformation. Hence, carbon enriches an austenitic phase instead of forming carbide precipitation and subsequently allows the stabilization of a retained austenite amount at room temperature in equilibrium with carbide-free bainite. A typical addition of silicon to the steel is approximately 1.5%. Unfortunately, such a high amount of this element deteriorates the sheet wettability by liquid zinc during hot-dip galvanizing [15]. Therefore, industrial requirements prefer low-Si grades in TRIP-aided steels [16,17]. The decrease of the solid solution potential of silicon and a resulting decrease of the strength level (Al has a smaller solid solution strengthening effect) can be compensated by Nb, Ti, and/or V microalloyed additions [18–22]. These elements form stable nitrides and carbonitrides during hot working in the austenite region. The increase of strength is due to grain refinement and precipitation strengthening by dispersive particles of MX-type phases [23–27]. These microalloyed multiphase steels are of special interest to the automotive industry because of the weight reduction potential [11,12,15]. Since the most investigated TRIP-aided steels are manufactured by cold deformation and subsequent intercritical annealing [15,23,25], this work is focused on hot-deformed TRIP-aided steels with low Si content.

The stabilization of retained austenite is affected by three major factors: an enrichment of the γ phase in C and Mn, its grain size, and its dislocation density [9,28]. In low-Si grades, the driving force for carbon enrichment of the austenite is smaller. Thus, the two other factors, i.e., the grain size reduction and increased dislocation density, become of great importance for the potential stabilization of retained austenite. It is assumed that both factors can be obtained for hot-rolled microalloying-aided sheet products. Therefore, the aim of this work is to monitor precipitation processes in the austenite and to assess undeformed and deformed austenite decomposition under conditions of continuous cooling of a low-Si microalloyed steel.

2. Materials and Methods

The chemical composition of the investigated steel is presented in Table 1. The silicon addition was minimized to 0.5%, and this low level was compensated for by microadditions of Nb and Ti. The liquid metal was vacuum-melted in an induction furnace VSG-50 (Balzers, Asslar, Germany) and subsequently cast as a 25 kg ingot, which was forged between 1200 °C and 900 °C to a thickness of 22 mm.

Table 1. Chemical composition of the investigated steel.

C	Mn	Si	Al	S	P	Nb	Ti	N
0.2 ± 0.01	1.41 ± 0.03	0.50 ± 0.03	0.020 ± 0.002	0.008 ± 0.001	0.014 ± 0.002	0.027 ± 0.001	0.010 ± 0.001	0.0047 ± 0.0002

The precipitation behavior of MX-type phases with Nb and Ti microadditions formed under thermodynamic equilibrium conditions was analyzed as the first step. The temperature of TiN precipitation was determined using Equation (1). In the same way, the solubility product of niobium carbonitride was calculated based on Equation (2). This equation takes into account the real concentrations of Mn and Si in the steel. Thermodynamic calculations of the phase transformations of undercooled austenite were simulated using the JMatPro software (version 8.0, Sente Software Ltd, Guildford, UK) package. The CCT (continuous cooling transformation) diagram and TTT (time–temperature–transformation) diagram were designated for continuous cooling and for isothermal cooling, respectively.

The analysis of the real phase transformation kinetics was performed using dilatometric tests. For these tests, $\Phi 4$ mm × $\Phi 3$ mm × 7 mm tubular samples and $\Phi 4$ mm × 7 mm cylindrical specimens were prepared. The dilatometric experiments were conducted with the use of a DIL805 dilatometer produced by Bähr Thermoanalyse GmbH (Hüllhorst, Germany), equipped with an LVDT-type measuring head with a theoretical resolution of ± 0.057 µm. The study involved classical heat treatment tests and thermomechanical tests, i.e., for plastically deformed samples after austenitization. For nondeformed tests, the tubular samples after the austenitization step at 1100 °C were cooled at a rate of 5 °C/s to the temperature of 875 °C, isothermally held for 20 s, and finally cooled to room temperature at cooling rates ranging from 260 to 0.17 °C/s. The plastically deformed samples were subjected to a similar heat treatment procedure, but after cooling the samples to 875 °C and before final cooling they were plastically compressed with a reduction of 50% at a strain rate of 1 s^{-1}. Microscopic tests after dilatometric analyses included classical metallographic procedures using a light microscopy (Leica MEF 4A, Wetzlar, Germany) technique with 3% nital etching.

3. Results and Discussion

3.1. Thermodynamic Calculations of Precipitation Processes

A first step of the study included a thermodynamic analysis of precipitation processes occurring in the austenite containing microalloyed additions. To determine the equilibrium precipitation temperatures of MX-type phases with Nb and Ti, Equations (1) and (2) were used. The calculations were based on the following constant values: A = 8000, B = 0.32 for TiN and A = 6440, B = 2.26 for Nb(CN) [24].

$$log[Ti][N] = 0.32 - 8000/T \tag{1}$$

$$log[Nb]\left[C + \frac{12}{14}N\right] = 2.26 + \frac{838[Mn]^{0.246} - 1730[Si]^{0.594} - 6440}{T} \tag{2}$$

Tables 2 and 3 present the solubility products for the analyzed MX-type phases as a function of temperature. The parameter k_s determines the total concentrations of microalloying elements and metalloids possibly dissolved at temperature T under thermodynamic equilibrium conditions. It is clearly seen that the values of solubility products for compounds TiN and Nb(CN) increase with increasing temperature. Comparing the solubility products of the TiN and Nb(CN) phases, it can be seen that higher stability is retained by TiN because it has much lower values of parameter k_s.

Table 2. Solubility products for TiN at various temperatures in the austenite.

Temperature [K]	Log k_s	k_s
1423	−5.30193	0.00000499
1373	−5.50666	0.00000311
1323	−5.72686	0.00000188
1273	−5.96437	0.00000109
1223	−6.22129	0.00000060
1173	−6.50012	0.00000032
1123	−6.80378	0.00000016

Table 3. Solubility products for Nb(CN) at various temperatures in the austenite calculated by taking into account Mn and Si contents in the analyzed steel.

Temperature [K]	Log k_s	k_s
1423	−2.43025	0.003713
1373	−2.60105	0.002506
1323	−2.78476	0.001641
1273	−2.98291	0.00104
1223	−3.19725	0.000635
1173	−3.42987	0.000372
1123	−3.68321	0.000207

On the basis of the data in Tables 2 and 3, the corresponding mutual solubility limit diagrams for individual MX-type interstitial phases were prepared (Figures 1 and 2). On their basis, the chemical composition of the solid solution can be calculated, i.e., concentrations of microadditions and metalloids in solid solution and concentrations of microadditions and metalloids fixed in the stable MX phase, as well as a mass fraction of the particles precipitated from the matrix [24]. Figure 1 presents the solubility diagram for titanium and nitrogen in the austenite for two selected temperatures. The first temperature of 1100 °C corresponds to the austenitizing temperature, whereas the second one (875 °C) is the deformation temperature before cooling. Thus, the initial and final chemical compositions of the solid solution can be determined under equilibrium conditions. The chemical composition of the analyzed steel is also marked on the diagram. The line that represents a stoichiometric ratio of titanium to nitrogen of 48:14 is marked on the diagram. By moving the stoichiometric line parallel to the point indicating the chemical composition of the analyzed steel, an intersection point with the curve indicating the solubility product at a given temperature can be obtained.

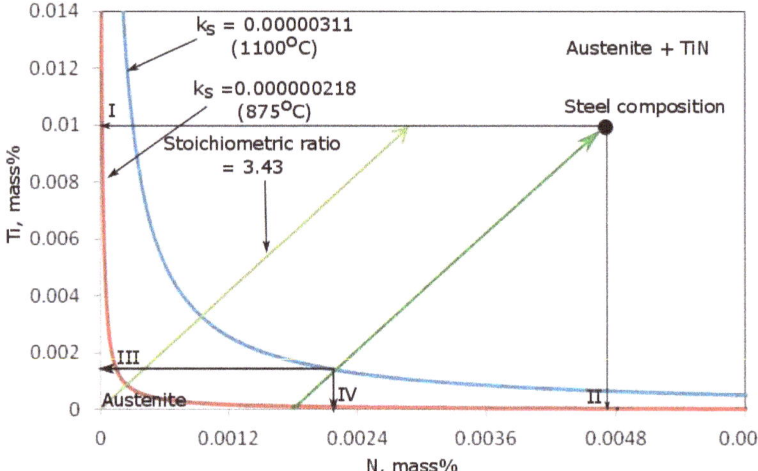

Figure 1. Limits of mutual solubility of titanium and nitrogen at 1100 °C and 875 °C in the austenite with the stoichiometric ratio between Ti and N.

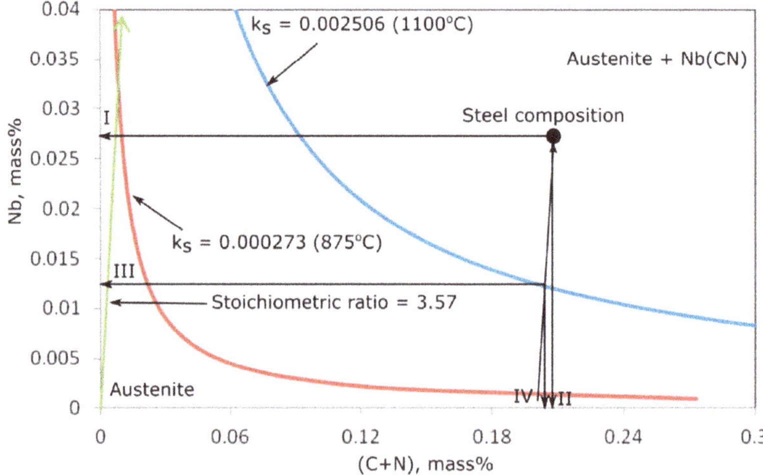

Figure 2. Limits of mutual solubility of niobium as well as carbon and nitrogen at 1100 °C and 875 °C in the austenite range with the stoichiometric ratio between Nb and (C,N).

It can be seen that at the temperature of 1100 °C the amount of dissolved Ti in the austenite is equal to 0.0014 mass% (point III on the diagram), and that circa 0.0021 mass% of nitrogen was dissolved in the austenite (point IV). This means that only a small fraction of titanium and nitrogen is dissolved in the solid solution whereas the rest is fixed into TiN, limiting grain growth during the austenitization step. When the temperature of the steel decreases to 875 °C, the amount of Ti dissolved in the solid solution is close to 0. All Ti is bound in TiN. At the same time, the difference in Ti and N contents between the analyzed steel composition and the corresponding intersection lines signify the mass fractions of titanium Ti_{TiN} and nitrogen N_{TiN} bound in TiN, respectively.

A similar analysis for niobium, carbon, and nitrogen was performed, and the results are presented in Figure 2. At 1100 °C, the amount of niobium dissolved in the solid solution is 0.012 mass% together with 0.21 mass% of combined (C+N). This means that ca. 50% of the total niobium is dissolved. When the temperature of the steel decreases to 875 °C, there is only a small concentration of Nb dissolved in the austenite. Practically all niobium is bound in Nb(CN). These results illustrate that between the austenitization step and the plastic deformation temperature, the Nb(C,N) precipitation process should occur. This may lead to grain refinement because the nitrides and carbonitrides slow down grain growth considerably [23,29].

Tables 4 and 5 present the calculated chemical compositions of the austenite as well as TiN and Nb(C,N) mass% fractions, which stay in equilibrium at particular temperatures. Figures 3 and 4 show the corresponding results of the performed calculations of the solubility of TiN and Nb(C,N). In the case of TiN we can see a gradual decrease in the contents of nitrogen and titanium dissolved in solid solution and a subsequent increase in the TiN fraction occuring with decreasing temperature (Figure 3). At the temperature of 1100 °C, there is ca. 14% of the titanium dissolved in solid solution together with 46% of the nitrogen. When the temperature decreases to 850 °C, the titanium is entirely bound in TiN. The amount of dissolved nitrogen at 850 °C is 0.0018 mass%, which corresponds to the results in Figure 1. The amount of Ti required to bind all nitrogen into TiN is 3.4 times the mass percent value of N in the steel. In the case of the analyzed steel, the amount of Ti needed for the complete binding of N is 0.016 mass%. Taking into account the limited titanium concentration in the steel (0.01%), this means that some of the nitrogen content remains dissolved. In practice, this part should be fixed in Nb(C,N). The available content of N at 850 °C is about 0.018 mass% (Figure 3b).

Table 4. Contents of [Ti] and [N] dissolved in the austenite and mass% (TiN) separated from the matrix at various temperatures.

Temperature		Log ks	ks	[Ti], mass%	[N], mass%	(TiN), mass%
[°C]	[K]					
1150	1423	−5.30193	0.00000499	0.00200	0.002386	0.010314
1100	1373	−5.50666	0.00000311	0.00140	0.002191	0.011109
1050	1323	−5.72686	0.00000188	0.00092	0.002043	0.011739
1000	1273	−5.96437	0.00000109	0.00056	0.001939	0.012201
950	1223	−6.22129	0.00000060	0.000321	0.001869	0.012510
900	1173	−6.50012	0.00000032	0.000173	0.001825	0.012702
850	1123	−6.80378	0.00000016	0.0000086	0.001800	0.012891

Table 5. Contents of [Nb] and [C,N] dissolved in the austenite and mass% (NbCN) separated from the matrix at various temperatures.

Temperature		Log ks	ks	[Nb], mass%	[C,N], mass%	(NbCN), mass%
[°C]	[K]					
1150	1423	−2.43025	0.003713	0.0390	0.211679	0
1100	1373	−2.60105	0.002506	0.0250	0.208265	0.003735
1050	1323	−2.78476	0.001641	0.0164	0.205846	0.015390
1000	1273	−2.98291	0.001040	0.0100	0.204198	0.022802
950	1223	−3.19725	0.000635	0.0059	0.203120	0.027980
900	1173	−3.42987	0.000372	0.0033	0.202447	0.031253
850	1123	−3.68321	0.000207	0.00175	0.202046	0.033204

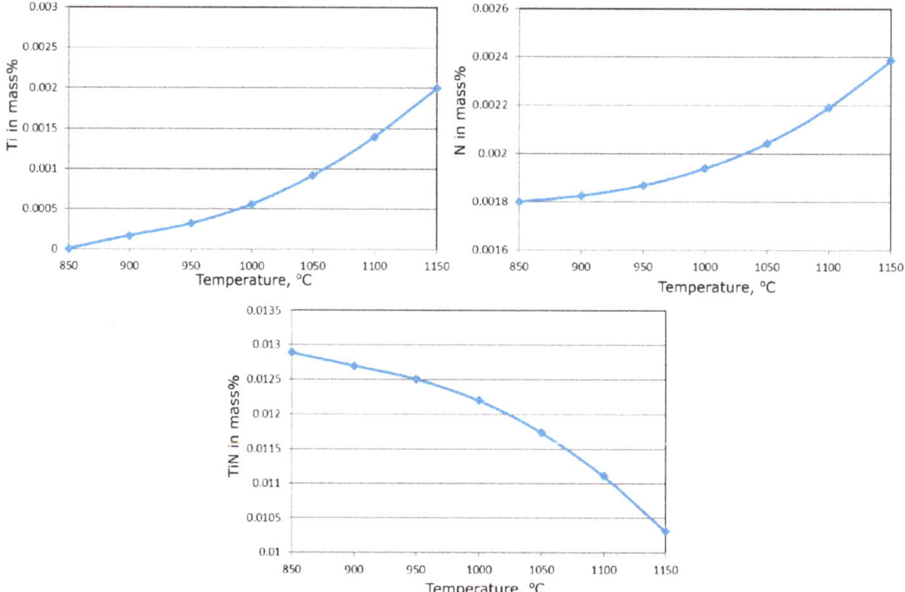

Figure 3. Temperature range of TiN precipitation in the austenite: (**a**) mass% Ti dissolved in the solid solution; (**b**) mass% N dissolved in the solid solution; (**c**) mass% fraction of (TiN) compound precipitated in the austenite.

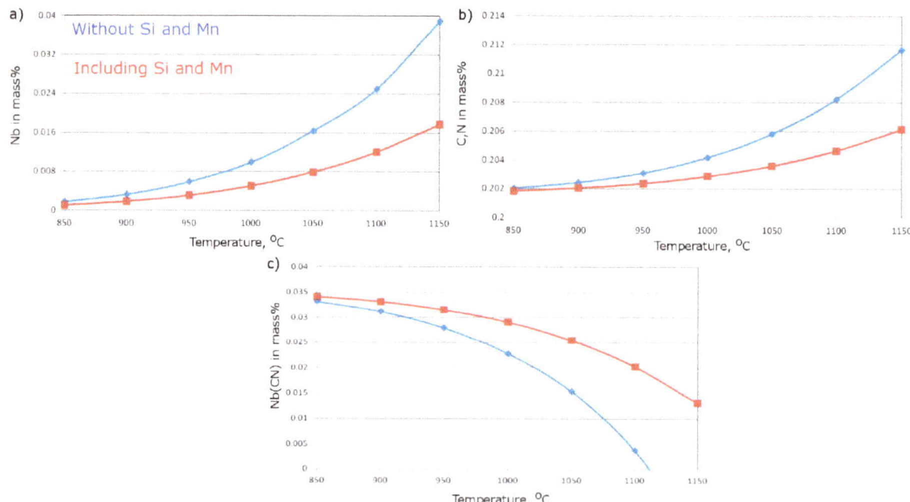

Figure 4. Temperature range of Nb(CN) precipitation in the austenite: (**a**) mass% Nb dissolved in the solid solution; (**b**) combined mass% (C,N) dissolved in the solid solution; (**c**) mass% fraction of Nb(CN) compound precipitated in the austenite.

The second analysis concerned Nb(CN). The corresponding results are presented in Figure 4. First, the calculations for the precipitation of Nb(CN) without Mn and Si were carried out (blue lines in Figure 4). The precipitation of Nb(CN) starts at around 1110 °C, and its amount increases with decreasing temperature. It is accompanied by a decrease in the amount of Nb dissolved in the solid solution. At the temperature of 850 °C, almost all niobium is bound in Nb(CN). After that, the calculations included the Si and Mn concentrations in the steel. It can be seen that after taking into account the Si and Mn concentrations, the amount of precipitated Nb(CN) increases at a particular temperature. The difference in the amount of precipitated Nb(CN) is higher at high temperatures and decreases along with the temperature drop. Moreover, the precipitation process starts at a higher temperature. Siciliano and Jonas [29] observed that in the case of steel including microadditions of Nb, an increase in the Si content leads to an acceleration of the precipitation of Nb(CN). On the other hand, manganese decreases Nb diffusivity in the austenite and inhibits the precipitation of Nb(CN) [30]. These results were obtained for the equilibrium conditions. Under real cooling conditions, some higher part of the microadditions stays supersaturated in the solid solution due to undercooling [27,31]. Comparing the relative effects of Si and Mn in the analyzed steel, one can note that the accelerating effect of the addition of 0.5% Si is greater than the decelerating effect of 1.4% Mn addition.

3.2. Thermodynamic Calculations of Phase Transformations

To analyze the phase transformations of the steel under equilibrium conditions, JMatPro software was used [32]. Figure 5 presents the evolution of the phases as a function of temperature. It represents a change in the microstructure state during a very slow cooling corresponding to thermodynamic equilibrium conditions. The graph enables us to identify temperatures of particular phase transitions and to assess a volume fraction of the particular phases at a given temperature. The liquidus temperature is equal to ca. 1510 °C, and the solidus temperature is equal to 1460 °C. The equilibrium temperature (A_{e1}) for the alpha phase is 688 °C, and the area of austenite phase is stable from 688 °C to 1460 °C. The calculated amount of austenite during the austenitzation at 1100 °C is almost 100%. It can be seen that the remaining part is M(C,N)-type carbides, which is in accordance with the previously described precipitation behaviour. The analysis shows that at room temperature under equilibrium conditions, some fraction of cementite can exist in the microstructure.

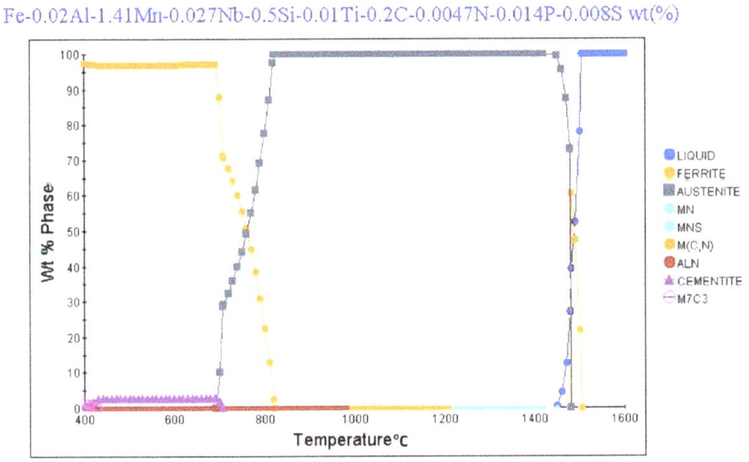

Figure 5. Thermodynamic evolution of the phases determined using JMatPro software.

Figure 6 shows the calculated continuous cooling transformation (CCT) diagram of the analyzed steel using the JMatPro software. The ferritic transformation starting temperature (F_s) is equal to 822 °C for equilibrium conditions. It decreases with increasing cooling rate, which was kept constant for the whole temperature range. Based on the obtained graph, cooling rates faster than 10 °C/s are required to prevent a pearlitic transformation. The pearlite would consume carbon, which is needed for the stabilization of retained austenite in TRIP steels. The bainite is formed below 500 °C, whereas the martensite start temperature (M_s) is about 390 °C.

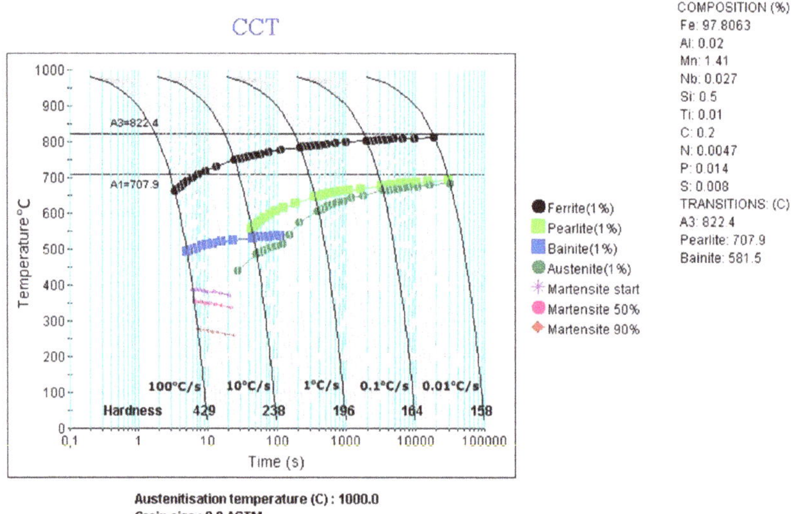

Figure 6. A continuous cooling transformation (CCT) diagram of the analyzed steel.

The results of the calculation of the time–temperature–transformation (TTT) diagram are presented in Figure 7. It can be seen that the pearlite trasformation is delayed to longer times at temperatures above 650 °C. There is a risk of pearlite precipitation after about 10 s at 600 °C. The ferritic and bainitic transformations are left-shifted compared with the CCT diagram. The time needed for a complete bainitic transformaton under isothermal conditions is relatively short, i.e., ca. 20 s. The generated diagrams indicate that there are conditions to design a cooling path consisting of ferrite and bainite, but it requires a multistep design instead of continuous cooling [11].

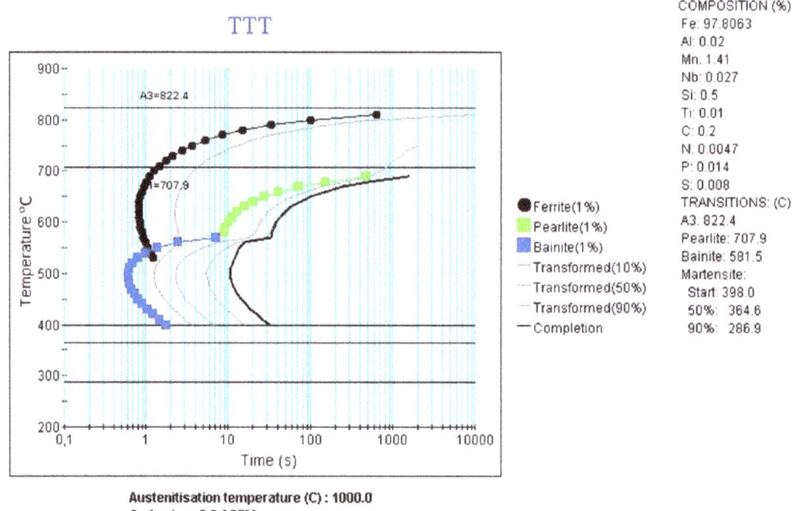

Figure 7. A time–temperature–transformation (TTT) diagram of the analyzed steel.

3.3. CCT and DCCT Diagrams

The calculated diagrams were experimentally verified in dilatometric tests. Two types of the diagrams were determined: the CCT diagram for undeformed specimens (Figure 8) and the DCCT (deformation continuous cooling transformation) diagram for deformed specimens (Figure 9). In general, argon cooling was applied. Helium was used for cooling rates faster than 60 °C/s. Hence, the cooling rates covered the entire austenite decomposition range. They were kept stable during the experiment. These diagrams are very important tools for designing a multistep cooling schedule [4,7].

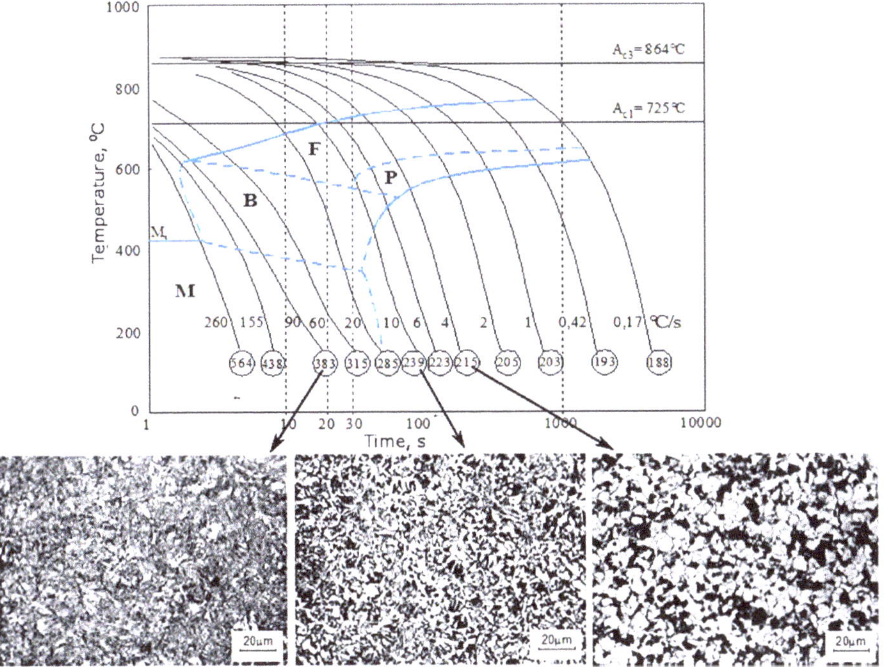

Figure 8. CCT diagram of supercooled austenite transformations of the analyzed steel, and microstructures obtained for different cooling rates from 875 °C.

The first important information obtained from these diagrams includes critical temperatures: A_{c1} is equal to 725 °C, A_{c3} = 864 °C, and M_s = 418 °C. These temperatures slighty differ fom the calculated ones. It can be seen that the calculated cooling rate (using JMatPro) preventing a pearlite transformation coresponds well to that obtained on the real CCT diagram (about 10 °C/s). The microstructure of undeformed steels changes from ferritic–pearlitic for cooling rates lower than 6 °C/s, through ferritic–bainitic in the range 10–60 °C/s and bainitic–martensitic (90–155 °C/s), to pure martensitic microstructures for cooling rates higher than 260 °C/s. (Figure 8). The obtained microstructures of the steel are relatively fine-grained, presumably due to the precipitation of TiN and Nb(CN).

The DCCT diagram obtained for the specimens deformed at 875 °C put on the CCT diagram is shown in Figure 9. A distinct change of phase transformation kinetics is visible after plastic deformation. One can see that a major effect is the displacement of all phase transformations to higher temperatures and shorter times. The reason for these shifts is the increased diffusity of elements in the deformed austenite and a higher density of preferrable nucleation places for diffusional transformation products. Such places are especially shear bands and grain boundaries containing increased dislocation density [11–13]. The shift of ferritic and bainitic transformations creates better conditions for the realization of multistep cooling because these products can be produced for shorter times. It is especially important in industrial practice due to the short cooling sections available [15,18].

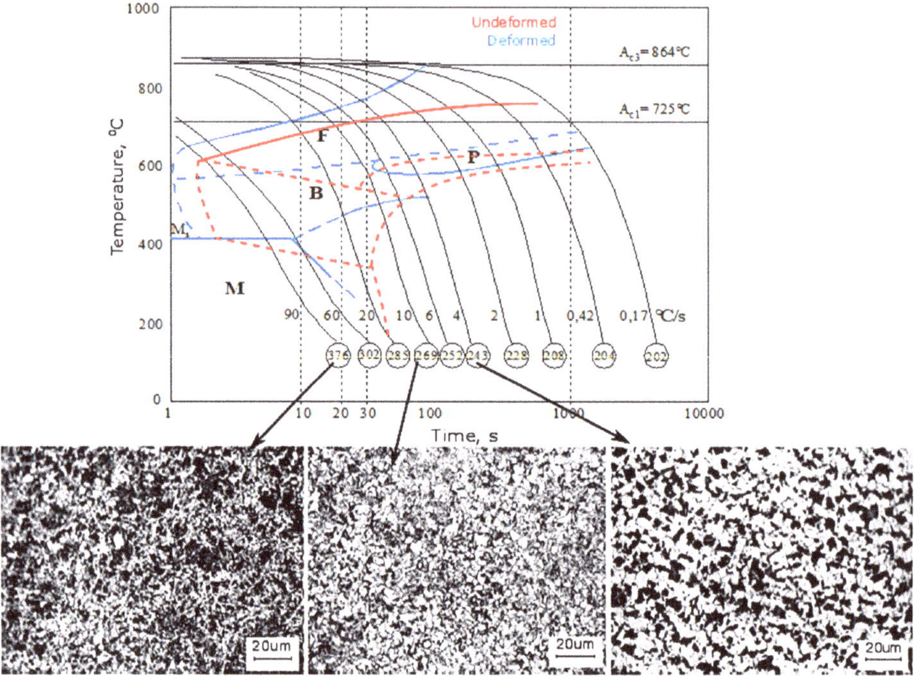

Figure 9. DCCT diagram of supercooled austenite transformations for deformed specimens superimposed on the CCT diagram, and microstructures obtained for different cooling rates from the deformation temperature of 875 °C.

The plastically deformed microstructures are more fine-grained when compared to the undeformed specimens. This is due to the higher number of nucleation places for ferrite and bainite. The size of grains is an important factor for the stability of retained austenite [6,10]. Its decrease favors the stability of the γ phase. Jimenez-Melero et al. [28] reported that there is a minimum volume size of austenite grains, equal to 5 μm^3, below which the martensitic transformation of austenite does not occur.

The results indicate that the plastic deformation significantly affects the technological windows required for producing multiphase microstructures consisting of ferrite, bainite, and retained austenite. The major positive effects are shifting the ferrite and bainite regions to shorter times and higher temperatures. In practice, this means the faster production of a sufficient amount of ferrite (ca. 60%–70%) in typical TRIP steels [4,10]. This is especially important for short run-out tables in technological hot rolling lines [22,25]. The detailed microstructures and mechanical properties for such a designed thermomechanical approach can be found elsewhere [11].

4. Conclusions

The work addressed the thermodynamic calculations of precipitation processes and austenite decomposition in a low-Si Nb–Ti-microalloyed automotive steel. The thermodynamic calculations of precipitation processes indicated that TiN particles should limit the grain growth during the austenitizing step at 1100 °C, whereas Nb(C,N) particles should be effectively precipitated during hot deformation in the austenite range. The effects of alloying (Mn and Si) and plastic deformation on the austenite decomposition were identified. The DCCT diagram possessed a large ferrite region beneficial for the production of ferrite within a reasonable time after finishing rolling. The technological windows for the realization of ferrite and bainite transformations from the deformed austenite were identified. The plastically deformed samples showed more fine-grained microstructures compared to nondeformed samples. This should favor the stabilization of some fraction of retained austenite through grain refinement stabilization during incomplete bainitic transformation.

Author Contributions: A.G. conceived, designed, and performed the experiments and reviewed the paper; M.M. performed the thermodynamic calculations and wrote the paper; W.Z. performed dilatometric experiments, calculated CCT and TTT diagrams, analysed the data, and reviewed the paper.

Funding: The work was financially supported by statutory funds from the Faculty of Mechanical Engineering of Silesian University of Technology in 2018.

Conflicts of Interest: The authors declare no conflict of interest.

References

1. Baker, T.N. Microalloyed steels. *Ironmak. Steelmak.* **2016**, *43*, 264–307. [CrossRef]
2. Muszka, K.; Dymek, S.; Majta, J.; Hodgson, P. Microstructure and properties of a C-Mn steel subjected to heavy plastic deformation. *Arch. Metall. Mater.* **2010**, *55*, 641–645.
3. Smith, D.; Joris, O.P.J.; Sankaran, A.; Weekes, H.E.; Bull, D.J.; Prior, T.J.; Dye, D.; Errandonea, D.; Proctor, J.E. On the high-pressure phase stability and elastic properties of β-titanium alloys. *J. Phys. Condens. Matter* **2017**, *29*, 155401. [CrossRef] [PubMed]
4. Zhang, M.; Li, L.; Fu, R.Y.; Krizan, D.; De Cooman, B.C. Continuous cooling transformation diagrams and properties of micro-alloyed TRIP steels. *Mater. Sci. Eng. A* **2006**, *438–440*, 296–299. [CrossRef]
5. Grajcar, A.; Różański, M.; Kamińska, M.; Grzegorczyk, B. Effect of gas atmosphere on the non-metallic inclusions in laser-welded TRIP steel with Al and Si additions. *Mater. Technol.* **2016**, *50*, 945–950. [CrossRef]
6. Pereloma, E.V.; Al-Harbi, F.; Gazder, A.A. The crystallography of carbide-free bainites in thermo-mechanically processed low Si transformation-induced plasticity steels. *J. Alloys Compd.* **2014**, *615*, 96–110. [CrossRef]
7. Kaijalainen, A.; Vahakuopus, N.; Somani, M.; Mehtonen, S.; Porter, D.; Komi, J. The effects of finish rolling temperature and niobium microalloying on the microstructure and properties of a direct quenched high-strength steel. *Arch. Metall. Mater.* **2017**, *62*, 619–626. [CrossRef]
8. Grajcar, A.; Radwański, K. Microstructural comparison of the thermomechanically treated and cold deformed Nb-microalloyed TRIP steel. *Mater. Technol.* **2014**, *48*, 679–683.
9. Caballero, F.B.; Roelofs, H.; Hasler, S.; Capadevila, C.; Chao, J.; Cornide, J.; Garcia-Mateo, C. Influence of bainite morphology on impact toughness of continuously cooled cementite free bainitic steels. *Mater. Sci. Technol.* **2012**, *28*, 95–102. [CrossRef]
10. Ranjan, R.; Beladi, H.; Singh, S.B.; Hodgson, P.D. Thermo-mechanical processing of TRIP-aided steels. *Metall. Mater. Trans. A* **2015**, *46*, 3232–3247. [CrossRef]
11. Adamczyk, J.; Grajcar, A. Structure and mechanical properties of DP-type and TRIP-type sheets obtained after the thermomechanical processing. *J. Mater. Process. Technol.* **2005**, *162–163*, 267–274. [CrossRef]
12. Tsukatani, I.; Hashimoto, S.; Inoue, T. Effects of silicon and manganese addition on mechanical properties of high-strength hot-rolled sheet steel containing retained austenite. *ISIJ Int.* **1991**, *31*, 992–1000. [CrossRef]
13. Kawulok, R.; Schindler, I.; Kawulok, P.; Rusz, S.; Opela, P.; Kliber, J.; Solowski, Z.; Cmiel, K.M.; Podolinsky, P.; Malis, M.; et al. Transformation kinetics of selected steel grades after plastic deformation. *Metalurgija* **2016**, *55*, 357–360.
14. Grajcar, A. Thermodynamic analysis of precipitation processes in Nb-Ti-microalloyed Si-Al TRIP steel. *J. Therm. Anal. Calorim.* **2014**, *118*, 1011–1020. [CrossRef]
15. De Ardo, A.J.; Garcia, J.E.; Hua, M.; Garcia, C.I. A new frontier in microalloying: Advanced high strength, coated sheet steels. *Mater. Sci. Forum* **2005**, *500–501*, 27–38. [CrossRef]
16. Nasr El-Din, H. Effect of cold deformation and multiphase treatment conditions on the characterization of low-carbon, low-silicon multiphase steel. *Steel Grips* **2005**, *3*, 196–205.
17. Radwanski, A. Structural characterization of low-carbon multiphase steels merging advanced research methods with light optical microscopy. *Arch. Civ. Mech. Eng.* **2016**, *16*, 282–293. [CrossRef]

18. Opiela, M. Thermodynamic analysis of the precipitation of carbonitrides in microalloyed steels. *Mater. Technol.* **2015**, *49*, 395–401. [CrossRef]

19. Kurc-Lisiecka, A.; Lisiecki, A. Laser welding of the new grade of advanced high-strength steel Domex 960. *Mater. Technol.* **2017**, *51*, 199–204.

20. Opiela, M.; Grajcar, A. Hot deformation behavior and softening kinetics of Ti-V-B microalloyed steels. *Arch. Civ. Mech. Eng.* **2012**, *12*, 327–333. [CrossRef]

21. Kucerova, L.; Opatova, K.; Kana, J.; Jirkova, H. High versatility of niobium alloyed AHSS. *Arch. Metall. Mater.* **2017**, *62*, 1485–1491. [CrossRef]

22. Uranga, P.; Lopez, B.; Rodriguez-Ibabe, J.M. Microstructural modelling of Nb microalloyed steels during thin slab direct rolling processing. *Steel Res. Int.* **2007**, *78*, 199–209. [CrossRef]

23. Krizan, D.; Spiradek-Hahn, K.; Pichler, A. Relationship between microstructure and mechanical properties in Nb-V microalloyed TRIP steel. *Mater. Sci. Technol.* **2015**, *31*, 661–668. [CrossRef]

24. Gladman, T. *The Physical Metallurgy of Microalloyed Steels*; The Institute of Materials, University Press: Cambridge, UK, 1997.

25. Jung, J.; Lee, S.J.; Kim, S.; De Cooman, B.C. Effect of Ti additions on micro-alloyed Nb TRIP steel. *Steel Res. Int.* **2011**, *82*, 857–865. [CrossRef]

26. Irvine, K.J.; Pickering, F.B.; Gladman, T. Grain refined C-Mn steels. *J. Iron Steel Inst.* **1967**, *205*, 161–182.

27. Gorka, J. Microstructure and properties of the high-temperature HAZ of thermo-mechanically treated S700MC high-yield-strength steel. *Mater. Technol.* **2016**, *50*, 617–621. [CrossRef]

28. Jimenez-Melero, E.; van Dijk, N.H.; Zhao, L.; Sietsma, J.; Offerman, S.E.; Wright, J.P.; van der Zwaag, S. Martensitic transformation of individual grains in low-alloyed TRIP steels. *Scr. Mater.* **2007**, *56*, 421–424. [CrossRef]

29. Siciliano, F.; Jonas, J.J. Mathematical modeling of the hot strip rolling of microalloyed Nb, multiply-alloyed Cr-Mo, and plain C-Mn steels. *Metall. Mater. Trans. A* **2000**, *31A*, 511–530. [CrossRef]

30. Siciliano, F.; Poliak, E.I. Modeling of the resistance to hot deformation and the effects of microalloying in high-Al steels under industrial conditions. *Mater. Sci. Forum* **2005**, *500–501*, 195–202. [CrossRef]

31. Nowotnik, A.; Siwecki, T. The effect of TMCP parameters on the microstructure and mechanical properties of Ti-Nb microalloyed steel. *J. Microsc.* **2010**, *237*, 258–262. [CrossRef]

32. Saunders, N.; Guo, Z.; Li, X.; Miodownik, A.P.; Schille, J.P. Using JMatPro to model materials properties and behavior. *JOM* **2003**, *55*, 60–65. [CrossRef]

Article

Study of Nanometer-Sized Precipitation and Properties of Fire Resistant Hot-Rolled Steel

Zhenjia Xie [1], Zhendong Song [2], Kun Chen [1], Minghong Jiang [1], Ya Tao [2], Xuemin Wang [1] and Chengjia Shang [1,*

[1] Collaborative Innovation Center of Steel Technology, University of Science and Technology Beijing, Beijing 100083, China; zjxie@ustb.edu.cn (Z.X.); chenkunqing@126.com (K.C.); 17801051733@163.com (M.J.); wxm@mater.ustb.edu.cn (X.W.)

[2] Technical Center of Inner Mongolia Baotou Steel Union Co., Ltd., Baotou 014010, China; szhd1982@126.com (Z.S.); Taoya2002@sina.com (Y.T.)

* Correspondence: cjshang@ustb.edu.cn; Tel.: +86-10-6233-4910

Received: 27 October 2019; Accepted: 13 November 2019; Published: 18 November 2019

Abstract: Nanometer-sized precipitated carbides in a low carbon Ti-V-Mo bearing steel were obtained through hot rolling and air cooling and were investigated by transmission electron microscopy (TEM). The nanometer-sized interphase-precipitated carbides have been found to exhibit an average diameter of ~6.1 ± 2.7 nm, with an average spacing of ~24–34 nm. Yield strength of 578 ± 20 MPa and tensile strength of 813 ± 25 MPa were achieved with high elongation of 25.0 ± 0.5% at room temperature. The nanometer-sized precipitation exhibited high stability after annealing at high temperatures of 600 °C and 650 °C for 3 h. Average diameters of carbides were statistically measured to be ~6.9 ± 2.3 nm and 8.4 ± 2.6 nm after tempering at high temperatures of 600 °C and 650 °C, respectively. The micro-hardness was ~263–268 HV0.1 after high temperature tempering, which was similar to the hot-rolled sample (273 HV0.1), and yield strength of 325 ± 13 MPa and 278 ± 4 MPa was achieved at elevated temperatures of 600 °C and 650 °C, respectively. The significant decrease of yield strength at 650 °C was attributed to the large decrease in shear modulus.

Keywords: interphase precipitation; high temperature strength; stability; low carbon; low alloy steel

1. Introduction

As reference a material, high-strength low-alloy (HSLA) steel is of great importance for construction materials. With buildings tending to be high-rise and large-scale, there is an increased demand in higher strength steel for consideration of self-weight reduction, green, and safety [1]. Moreover, fire-resistant performance of structural steel has gained more attention since the sudden collapse of the Word Trade Center towers due to the 9/11 terrorist attack in 2001. Therefore, for the fire-resistant HSLA steel, the strictest requirement is that the yield strength at 600 °C should be guaranteed to be higher than two-thirds of the yield-strength value specified at room temperature [2]. However, strength of steel inevitably decreases at high temperature in fire. The rapid decrease in strength has been suggested to be attributed to thermal activation processes, including atomic diffusion, coarsening of precipitates, and dislocation recovery and annihilation [3].

In order to develop fire-resistant steel, plenty of studies have been carried out by researchers. Studies [4,5] suggested that Mo addition was an effective approach for achieving high strength at high temperature for low alloy steels. It was suggested that there was a strength increment of ~13.7 MPa per 0.1% Mo addition at 600 °C when the total Mo content was lower than 0.5% by solid-solution strengthening and bainite strengthening [6–8]. However, as an expensive alloying element, high-addition of Mo will greatly increase the cost. Therefore, an alternative method by microalloying of Nb, V, and Ti and controlled accelerated cooling was introduced to low carbon low Mo steels for fire-resistant applications [1,9,10]. In this method, bainite strengthening was obtained by controlled accelerated cooling, and remained microalloying elements of Nb, V, and Ti in solid-solution precipitates as nanometer-sized MC-type carbides at elevated temperature in fire to provide high temperature strength. However, accelerated cooling process generates higher energy-consumption and water-pollution in comparison with a hot rolling process. In addition, the remaining microalloying elements in solid-solution waste resources.

Interphase precipitation strengthening is generally recognized as an effect and economic approach in developing high-strength low-alloy hot-rolled steels [11–14]. Previous published works have shown that the nanometer-sized interphase precipitated carbides could contribute 300–400 MPa to yield strength of hot-rolled steels [15,16]. The objective of the present work is to develop a high-strength low-alloy fire-resistant steel by introducing nanometer-sized interphase precipitation of microalloying element carbides. Transmission electron microscopy (TEM) characterization investigated the stability of interphase precipitated carbides including size, morphology, and distribution. Moreover, the effect of nanometer-sized interphase precipitated carbides on strength at elevated temperatures was discussed. The findings from the present study may provide an alternative approach for developing high-strength fire-resistant hot-rolled steels.

2. Experimental Material and Procedure

The chemical composition of the experimental steel based on low carbon, high titanium, and vanadium micro-alloying design is in weight percent (wt.%): 0.08C, 1.43Mn, 0.21Si, <1.0(Ni+Cr+Cu), 0.27Mo, <0.25(Nb+Ti+V). Low carbon and low manganese alloy design was good for weldability [17]. The combined addition of Ni, Cr, and Cu aimed to obtain good weather-resistant performance [18]. The steel was melted in a high frequency induction vacuum furnace (ZGIL0.1-200-2.5, Jinzhou, China) and cast into 25 kg ingot in a cylindrical shape with a diameter of ~120 mm. The ingot was re-heated to 1200 °C for homogenization, then hot rolled to a 20-mm thick plate through seven passes. The starting temperature of hot rolling was ~1020 °C and the finishing temperature was ~860 °C. During the entire hot rolling, a minimum reduction rate of 20% per pass was given. Lastly, the plate with a thickness of 20 mm was air cooled to an ambient temperature.

Tensile specimens in a dog-bone-shape with a gauge length of 50 mm and a diameter of 10 mm were prepared from the hot-rolled plate along the longitudinal direction. The tensile tests were conducted at room temperature and at elevated temperatures (600 °C and 650 °C), according to the Chinese standards GB/T228-2002 and GB/T4338-2015, respectively. Before high temperature tensile tests, tensile specimens were held at elevated temperatures for 3 h. Two samples were tested for each testing temperature and the average values were taken for the results of tensile tests. Specimens for microstructure examinations were cut from the edge of the tensile tested samples. The specimens for scanning electron microscopy (SEM) and microhardness tests were etched in 3% nital after mechanical grounding and polishing. SEM was performed using the TESCAN MIRA 3 LMH (Brno, Czech Republic) field-emission scanning electron microscope (FE-SEM). Vickers microhardness was measured at a load of 100 g. More than 30 measurements in ferrite were conducted for each sample. Samples obtained after annealing at 600 °C for 3 h and a uniform deformed section at 600 °C, respectively, were prepared for electron backscatter diffraction (EBSD) by metallographic mechanical polishing and electrolytic polishing. EBSD analysis was conducted using TESCAN MIRA 3 LMH FE-SEM equipped with an Oxford Symmetry EBSD detector at an acceleration voltage of 20 kV and a step size of 0.2 μm. EBSD data was post-processed by HKL CHANNEL 5 (Oxford, UK) flamenco software to acquire the necessary information. Thin foils with ~400 μm thickness were cut from the edge of samples after tensile tests at room temperature and elevated temperatures. Then, the thin foils were mechanically ground to ~60 μm thickness. Next, TEM disks (3 mm in diameter) in diameter were punched from the foils and twin-jet polished using an electrolyte consisting of 5% perchloric acid and 95% methanol at ~−20 °C. TEM observations were performed using FEI Tecnai F20 (Hillsboro, OR, USA) field-emission transmission electron microscope (FE-TEM) and FEI Tecnai G20 (Hillsboro, OR, USA) transmission electron microscope both combined with an energy dispersive X-ray spectrometer (EDS) detector at 200 kV.

3. Results and Discussion

3.1. Tensile Properties

Tensile properties of the hot-rolled steel were measured at room temperature and elevated temperatures. The obtained results are summarized in Table 1. After hot-rolling, the experimental steel exhibited high yield strength of 578 ± 20 MPa, high tensile strength of 812 ± 25 MPa, and high elongation of 25.0 ± 0.5% at room temperature. After holding at 600 °C for 3 h, the yield strength and tensile strength of the studied steel were 325 ± 13 MPa. It was found that the studied hot-rolled steel met two-thirds of 460 MPa grade yield strength for fire-resistant structural application. When further elevating the test temperature to 650 °C, the yield strength and tensile strength were further decreased to 278 ± 4 MPa.

Table 1. Tensile properties of the hot rolled steel at room temperature and elevated temperatures.

Testing Temperature	Yield Strength, MPa	Tensile Strength, MPa	Elongation, %
Room temperature	578 ± 20	812 ± 25	25.0 ± 0.5
600 °C	325 ± 13	-	-
650 °C	278 ± 4	-	-

3.2. Microstructure and Microhardness

Figure 1 presents the optical microstructure of the hot-rolled steel. It can be seen that multi-phase microstructure consisting of polygonal ferrite and bainite was obtained in the hot-rolled steel. By statistical analysis of more than 10 optical images, the volume fraction of ferrite and bainite was determined to be ~84% and ~16%, respectively. The grain size of ferrite was not uniform and has been found to range from ~2 to ~30 μm. To study the grain size of ferrite, EBSD analysis was performed, and the obtained band contrast (BC) image is given in Figure 2a. In Figure 2a, grain boundaries with misorientation higher than 5° were highlighted by black lines. It was easy to find out polygonal ferrite and bainite due to their different morphologies. After getting rid of bainite manually, the ferrite grain size was determined using HKL CHANNEL 5 flamenco software. The obtained results are plotted in Figure 2b. It can be seen that wide range distribution of grain size was obtained. The majority of ferrite grains ranged from 2–10 μm with the peak at ~4 μm. A few ferrite grains had a size of ~10–30 μm with a small peak at ~13 μm. During continuous air cooling, ferrite transformation took place continuously at different temperatures. The grain size of ferrite is related to the transformation temperature, which determines the carbon atom diffusivity and the austenite/ferrite grain boundary migration rate [19]. Based on this, the grain size of ferrite formed at high temperature should be larger than that formed at a low temperature [20]. However, ferrite grain size also depends on the density of nucleation sides. In this work, austenite grains were flattened after final rolling at 860 °C. During subsequent air cooling, high dense ferrite grains nucleated at austenite grain boundaries, grain corners, and deformation bands. These tiny ferrite grains were not coalesced during the following cooling, as reported in literature [21]. In the late period of air cooling, a few ferrite grains nucleated surrounding the tiny ferrite grains and grew into residual austenite, and the large size of ferrite grains was obtained. From the EBSD image in Figure 2a, it can be seen that large ferrite grains were mainly adjacent to bainite, which transformed from carbon-enriched residual austenite in the final stage of air cooling. This observation gives evidence to support that the later mechanism was the reason for the wide range distribution of grain sizes observed in the present work. Moreover, the hot deformation was helpful for refining ferrite grains by promoting the nucleation of ferrite by introducing crystal defects within parent grains [20], which resulted in a fine average grain size of ~5.06 ± 3.93 μm for ferrite.

Microstructure and microhardness for the studied steel after hot rolling and high temperature holding were investigated. The obtained results are presented in Figure 3. From Figure 3a–c, it can be seen that both SEM microstructure and microhardness were not changed clearly for the hot-rolled steel before and after holding at 600 °C and 650 °C for 3 h. The average Vickers microhardness of the hot-rolled steel was HV0.1 276 ± 10. After holding at 600 °C and 650 °C for 3 h, the average Vickers microhardness of the studied steel decreased slightly to HV0.1 268 ± 12 and HV0.1 263 ± 9, respectively. It can be found that the ferrite obtained through hot-rolling presented high resistance to soften after high temperature holding.

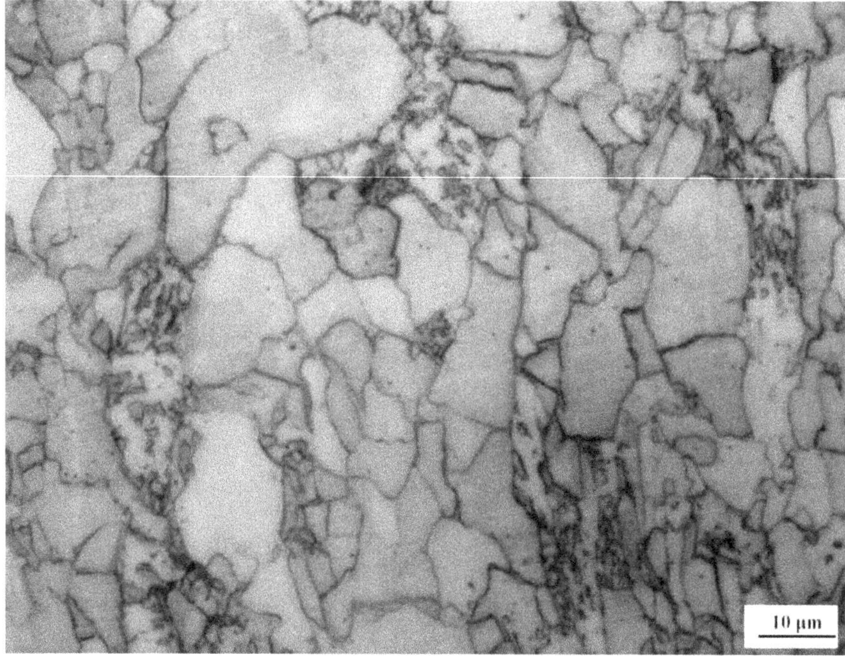

Figure 1. Optical microstructure of the hot-rolled experimental steel.

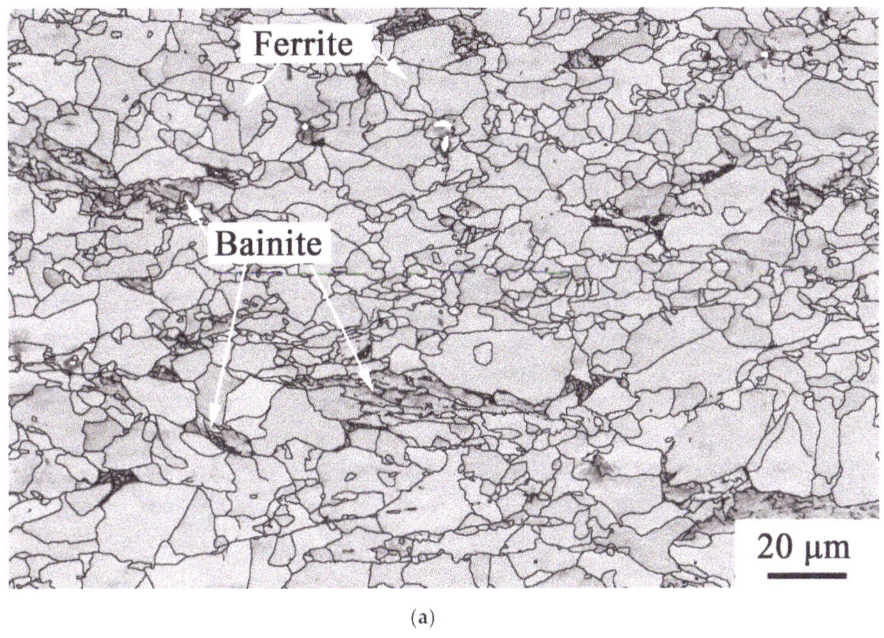

(a)

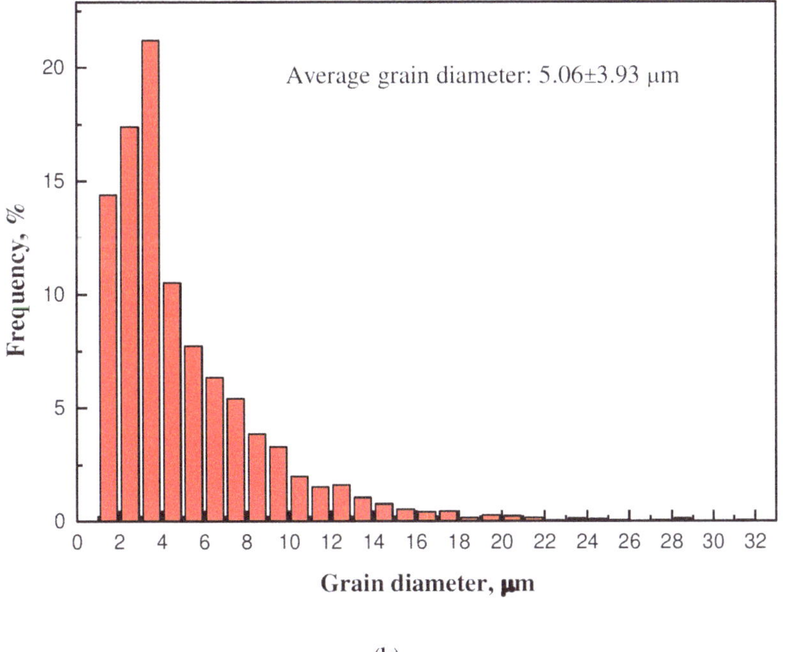

(b)

Figure 2. (**a**) EBSD image showing ferrite grains and bainitic microstructure. (**b**) The ferrite grain size distribution profile obtained from the EBSD image using HKL CHANNEL 5 flamenco software after getting rid of bainite manually.

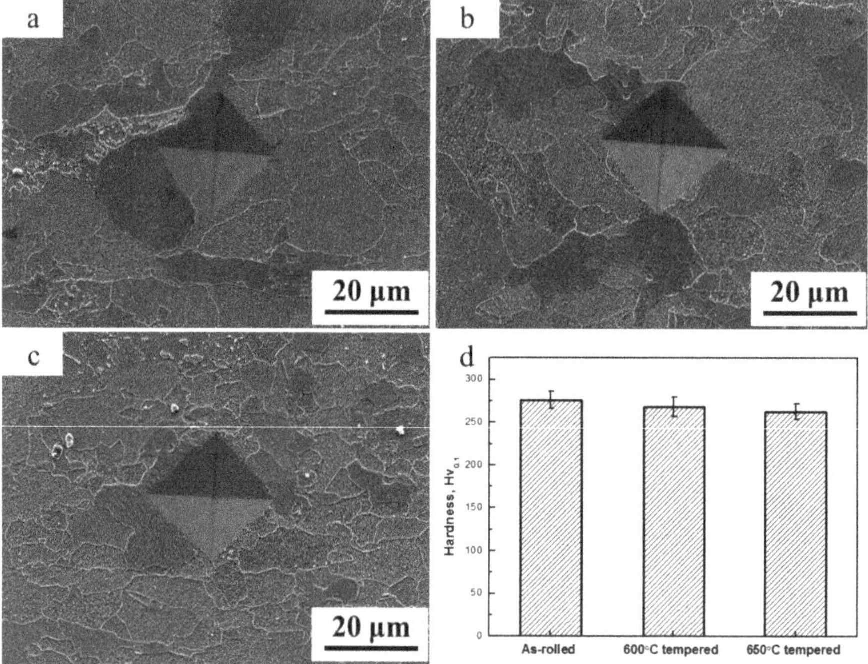

Figure 3. SEM images by secondary electron image mode showing a ferrite microstructure for samples after hot-rolling (**a**), holding at 600 °C (**b**) and 650 °C (**c**) for 3 h. (**d**) Microhardness plots for the studied steel.

3.3. Nanometer-Sized Precipitation

TEM was employed to study the nanometer-sized precipitation in ferrite before and after holding at high temperatures of 600 °C and 650 °C. The observed results for the hot-rolled sample are presented in Figure 4. Both random precipitates and interphase precipitates were obtained in ferrite after hot rolling and air cooling. It can be found that the row space of interphase precipitates was not uniform within the observed ferrite grain. As pointed out in Figure 4, the row spacing increased from ~24 nm to ~34 nm along the white dash arrow direction. Then, the precipitates existed in random distribution. EDS analysis indicated that these precipitates were (Ti, V)C complex carbides. A ledge mechanism proposed by Honeycombe and Mehl [22] is well accepted for interphase precipitation. During transformation from austenite (γ) to ferrite (α), a terrace γ/α plane with relatively low energy and a step γ/α plane with relatively high energy co-exist. The high-energy step γ/α plane moves too quickly to precipitate carbides during transformation. Yet, the low-energy terrace γ/α plane is immovable, such that carbides precipitate in the plane. When the electron beam is parallel to the precipitated terrace plane, row-arranged interphase precipitates are observed. The row spacing is related to the height of the step. Bhadeshia [23] suggested that the step height was proportional to the interface energy of the terrace plane and was inversely proportional to the driving force for ferrite transformation. This means reducing the driving force will increase the step height. In this study, carbon was enriched in austenite during ferrite growth. Liu et al. [24] revealed that carbon enrichment in austenite decreased the transformation temperature due to the reduction of driving force for ferrite transformation. This may be the reason for the increase in row spacing in the present work. Moreover, Chen et al. [25] suggested that γ→α transformation becomes sluggish in the late ferrite transformation, the growth of carbides nucleated in the step plane was compatible with the movement of the transformation front. Therefore, randomly dispersed carbide precipitation can be obtained in this instance.

Figure 4. Bright-field TEM image obtained by FEI Tecnai F20 FE-TEM showing nanometer-sized precipitates in the hot-rolled sample.

Figure 5a,b present bright-field TEM images of nanometer-sized precipitates for samples after holding at 600 °C and 650 °C for 3 h, respectively. From Figure 5, interphase carbides were observed in both samples holding at 600 °C and 650 °C for 3 h, which is similar to those in the hot-rolled sample. In addition to interphase precipitated carbides, nanometer-sized carbides can be observed between interphase carbide rows, as pointed out by white arrows in Figure 5. To investigate the thermal stability of precipitation, the size of carbides for samples before and after holding at high temperatures was estimated statistically from hundreds of particles. As a matter of convenience, bright-field TEM images with random and dispersed carbides obtained by tilting the samples were used for the statistical estimation. The obtained results are plotted in Figure 6. It is clear that the complex (Ti, V)C precipitates have great resistance to growth when holding at high temperatures of 600 °C and 650 °C. The diameter of carbides in the hot-rolled sample was mainly in the range of 2–8 nm, and the average diameter was ~6.1 ± 2.7 nm. After holding at 600 °C for 3 h, more than 80% of carbides had a diameter of 4–10 nm, and the average diameter increased slightly to ~6.9 ± 2.3 nm. The diameter of carbides in the sample after holding at 650 °C for 3 h ranged from 4 to 15 nm with a peak at 8–10 nm. The average diameter was further increased to ~8.4 ± 2.6 nm. Dunlop and Honeycombe [26] studied the aging characteristics of (V, Ti)C dispersions in ferrite. Their results also indicated high resistance of (V, Ti)C precipitates to coarsening. The high stability of (V, Ti)C precipitates was suggested to be attributed to their high chemical bonding energy and low solubility in a ferritic matrix.

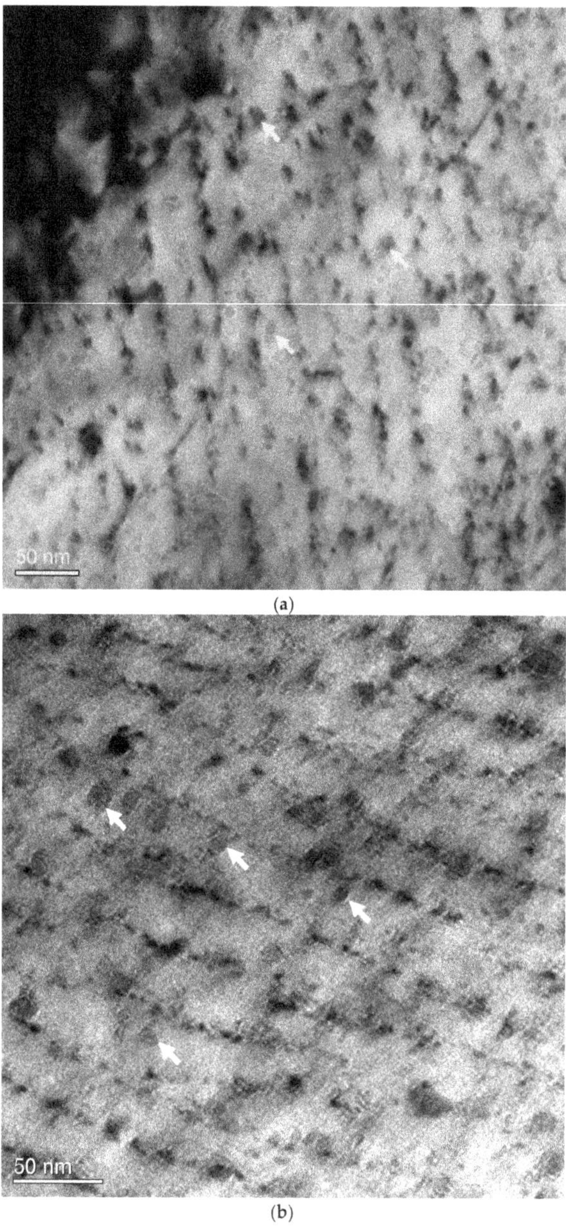

(a)

(b)

Figure 5. Bright-field TEM images showing nanometer-sized precipitates in samples after holding at 600 °C (**a**) and 650 °C (**b**) for 3 h, obtained by FEI Tecnai G20 TEM and FEI Tecnai F20 FE-TEM, respectively.

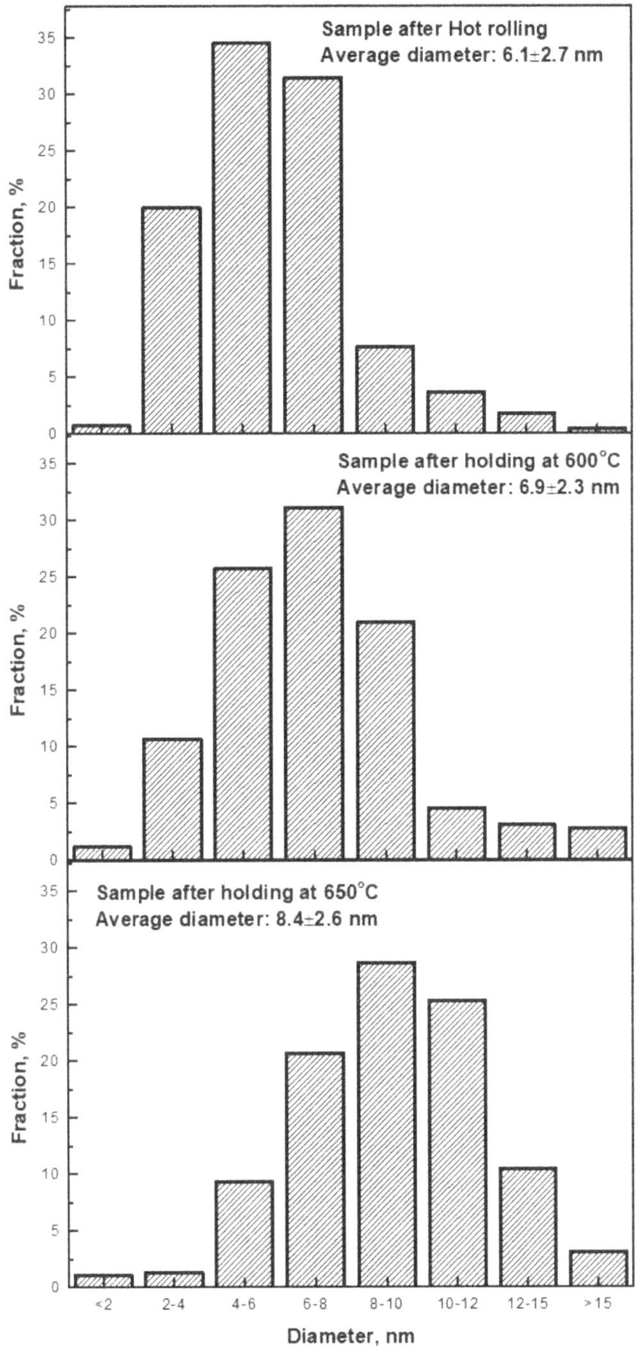

Figure 6. Size distributions of carbides for samples before and after holding at high temperatures.

3.4. Strengthening Mechanism

Generally, the yield strength of metals at room temperature is attributed to intrinsic friction stress, solid solute strengthening, grain boundary strengthening, dislocation strengthening, and precipitation hardening. For low carbon, low alloy ferritic steel, the strength contribution from precipitation hardening may be the greatest among all these strengthening mechanisms. The Ashby-Orowan model based on the mechanism of dislocations bypassing particles is normally accepted for evaluating precipitation strengthening provided by nanometer-sized hard (Ti, V)C carbides. Precipitation strengthening from MC-type carbides, according to the Ashby-Orowan model, is expressed by Equation (1) below [27,28].

$$\sigma_{ppt,\ Ashby\text{-}Orowan} = \frac{0.8MGb}{2\pi\sqrt{1-\upsilon}L_{MC}}\ \ln\left(\frac{x}{2b}\right)\ (\text{MPa}) \tag{1}$$

$$L_{MC} = \sqrt{\frac{2}{3}}\left(\sqrt{\frac{\pi}{f}}-2\right)\cdot r_{MC}.\ (\text{m}) \tag{2}$$

$$x = 2\sqrt{\frac{2}{3}}\cdot r_{MC}\ (\text{m}) \tag{3}$$

where M and υ is the Taylor factor and Poisson's ratio, taken as 2.75 and 0.29 for body-centered cubic metal with a random texture, respectively [29,30]. G is the shear modulus, b is the Burgers vector taken as 0.248 nm [31], r_{MC} is the average radius, and f is the volume fraction of MC-type carbides. It can be seen that the average diameter and volume fraction of carbides play determining roles in the strengthening effect. In the present work, the equilibrium volume fractions of MC-type carbides at different temperatures were calculated using commercial Thermo-Calc software with the TC-FE7 database [32,33]. The calculated results are plotted in Figure 7. The volume fraction of MC-type carbides increases when the temperature decreases, and reaches a maximum value of ~0.5% at 500 °C.

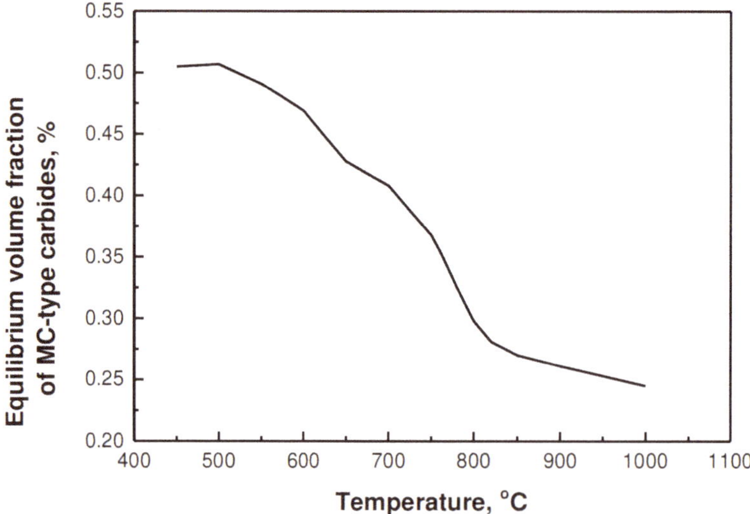

Figure 7. Equilibrium volume fraction of MC-type carbides vs. temperature plots calculated from commercial Thermo-Calc software with its TC-FE7 database.

At elevated temperatures, strength from precipitation strengthening is reduced due to the decrease in the shear modulus (G). The shear modulus (G) is related to Young's modulus (E) and Poisson's ratio (v) by the following equation [30].

$$G = \frac{E}{2(1 + v)} \tag{4}$$

In the present work, the Young's modulus E of the studied steel at 600 °C and 650 °C was estimated by fitting the data of the elastic deformation stage from the stress-strain curves, as shown in Figure 8. The obtained Young's modulus E at room temperature and elevated temperatures of 600 °C and 650 °C were ~197 GPa, ~92 GPa, and ~69 GPa, respectively. Hence, the shear modulus (G) was ~76 GPa, ~36 GPa, and ~27 GPa for the studied steel at room temperature, at 600 °C, and at 650 °C, respectively. Assuming that the volume fraction (f) and the average diameter of MC-type carbides for the hot-rolled sample was respectively ~0.5% and 6.1 nm, the contribution from precipitation strengthening was estimated to be ~318 MPa to yield strength at room temperature, according to the Ashby-Orowan model. Given that the volume fraction of MC-type carbides was ~0.47% and ~0.43% from Figure 7 and the average diameter was 6.9 nm and 8.4 nm from Figure 6, the strength contribution from precipitation strengthening can be calculated to be ~133 MPa and ~84 MPa at 600 °C and 650 °C, respectively. It can be seen that the large volume fraction of nanometer-sized carbide precipitation provided a significant contribution to yield strength at room temperature and at an elevated temperature. The dramatic decrease in yield strength at 650 °C was attributed to the great loss in shear modulus at a higher temperature.

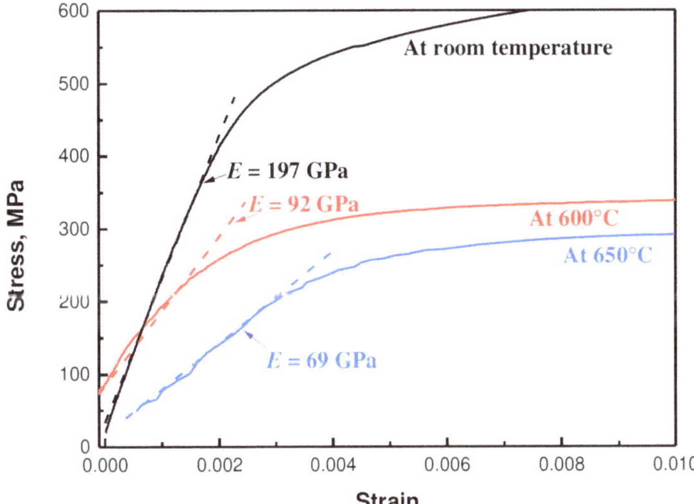

Figure 8. Fitted Young's modulus E at room temperature and elevated temperatures of 600 °C and 650 °C from stress-strain curves.

4. Conclusions

In this study, a multi-phase microstructure consisting of ferrite and a small fraction of bainite was obtained in a low-carbon low-alloy hot-rolled steel. Nanometer-sized precipitation and mechanical properties were investigated. The conclusions are summarized as follows.

(1) Nanometer-sized interphase precipitates were obtained in ferritic matrix. The interphase precipitated carbides have been found to exhibit an average diameter of 6.1 ± 2.7 nm, with an average distance of ~24–34 nm by TEM observation. EDX results indicated that the precipitates were (Ti, V)C complex carbides.

(2) The nanometer-sized precipitation exhibited high stability against tempering at high temperatures of 600 °C and 650 °C for 3 h. Average diameters of carbides were measured to be equal to ~6.9 ± 2.3 nm and 8.4 ± 2.6 nm after annealing at high temperatures of 600 °C and 650 °C for 3 h, respectively.

(3) Yield strength of 578 ± 20 MPa and tensile strength of 813 ± 25 MPa were achieved with high elongation of 25.0 ± 0.5% at room temperature. In addition, yield strength of 325 ± 13 MPa and 278 ± 4 MPa was achieved at elevated temperatures of 600 °C and 650 °C, respectively. Nanometer-sized precipitation contributed ~318 MPa to yield strength at room temperature, and the yield strength contributions decreased to ~133 MPa and ~84 MPa at 600 °C and 650 °C, respectively. The significant decrease of yield strength at 650 °C was attributed to the large decrease in the shear modulus.

Author Contributions: Writing—original draft preparation, Z.X.; investigation, Z.S., K.C., M.J. and Y.T.; methodology, X.W.; supervision, C.S.

Funding: The National Key R&D Program of China (No. 2017YFB0304700, 2017YFB0304701) and the National Natural Science Foundation of China (No. 51701012) funded this research.

Conflicts of Interest: The authors declare no conflict of interest.

References

1. Zhang, Z.Y.; Yong, Q.L.; Sun, X.J.; Li, Z.D.; Kang, J.Y.; Wang, G.D. Microstructure and mechanical properties of precipitation strengthened fire resistant steel containing high Nb and low Mo. *J. Iron Steel Res. Int.* **2015**, *22*, 337–343. [CrossRef]
2. Chijiiwa, R.; Yoshida, Y.; Uemori, R.; Tamehiro, H.; Funato, K.; Horii, Y. Development and practical application of fire-resistant steel for buildings. *Nippon Steel Tech. Rep.* **1993**, *58*, 47–55.
3. Gross, C.T.; Isheim, D.; Vaynman, S.; Fine, M.E.; Chung, Y. Design and development of lightly alloyed ferritic fire-resistant structural steels. *Metall. Mater. Trans. A* **2019**, *50*, 209–219. [CrossRef]
4. Miyata, K.; Sawaragi, Y. Effect of Mo and W on the phase stability of precipitates in low Cr heat resistant steels. *ISIJ Int.* **2001**, *41*, 281–289. [CrossRef]
5. Mizutani, Y.; Ishibashi, K.; Yoshii, K.; Watanabe, Y.; Chijhwa, R.; Yoshida, Y. 590MPa class of fire-resistant steel for building structural use. *Nippon Steel Tech. Rep.* **2004**, *90*, 38–44.
6. Wan, R.C.; Sun, F.; Zhang, L.T.; Shan, A.D. Effects of Mo on high-temperature strength of fire-resistant steel. *Mater. Des.* **2012**, *35*, 335–341. [CrossRef]
7. Wan, R.C.; Sun, F.; Zhang, L.T.; Shan, A.D. Effect of Mo addition on strength of fire-resistant steel at elevated temperature. *J. Mater. Eng. Perform.* **2014**, *23*, 2780–2786. [CrossRef]
8. Lee, W.B.; Hong, S.G.; Park, C.G.; Park, S.H. Carbide precipitation and high-temperature strength of hot-rolled high-strength, low-alloy steels containing Nb and Mo. *Metall. Mater. Trans. A* **2002**, *33*, 1689–1698. [CrossRef]
9. Wan, R.C.; Sun, F.; Zhang, L.T.; Shan, A.D. Development and study of high-strength low-Mo fire-resistant steel. *Mater. Des.* **2012**, *36*, 227–232. [CrossRef]
10. Assefpour-Dezfuly, M.; Hugaas, B.A.; Brownrigg, A. Fire resistant high strength low alloy steels. *Mater. Sci. Technol.* **1990**, *6*, 1210–1214. [CrossRef]
11. Funakawa, Y.; Shiozaki, T.; Tomita, K.; Yamamoto, T.; Maeda, E. Development of high strength hot-rolled sheet steel consisting of ferrite and nanometer-sized carbides. *ISIJ Int.* **2004**, *44*, 1945–1951. [CrossRef]
12. Khalid, F.A.; Edmonds, D.V. Interphase precipitation in microalloyed engineering steels and model alloy. *Mater. Sci. Technol.* **1993**, *9*, 384–396. [CrossRef]
13. Jang, J.H.; Heo, Y.U.; Lee, C.H.; Bhadeshia, H.K.D.H.; Suh, D.W. Interphase precipitation in Ti–Nb and Ti–Nb–Mo bearing steel. *Mater. Sci. Technol.* **2013**, *29*, 309–313. [CrossRef]
14. Miyamoto, G.; Hori, R.; Poorganji, B.; Furuhara, T. Interphase precipitation of VC and resultant hardening in V-added medium carbon steels. *ISIJ Int.* **2011**, *51*, 1733–1739. [CrossRef]
15. Yen, H.W.; Chen, P.Y.; Huang, C.Y.; Yang, J.R. Interphase precipitation of nanometer-sized carbides in a titanium–molybdenum-bearing low-carbon steel. *Acta Mater.* **2011**, *59*, 6264–6274. [CrossRef]

16. Bu, F.Z.; Wang, X.M.; Yang, S.W.; Shang, C.J.; Misra, R.D.K. Contribution of interphase precipitation on yield strength in thermomechanically simulated Ti–Nb and Ti–Nb–Mo microalloyed steels. *Mater. Sci. Eng. A* **2015**, *620*, 22–29. [CrossRef]

17. Ferreira, D.; Alves, A.; Cruz Neto, R.; Martins, T.; Brandi, S. A new approach to simulate HSLA steel multipass welding through distributed point heat sources model. *Metals* **2018**, *8*, 951. [CrossRef]

18. Wang, S.T.; Yang, S.W.; Gao, K.W.; He, X.L. Corrosion resistance of low alloying weathering steels in environment containing chloride ion. *Trans. Mater. Heat Treat.* **2008**, *4*, 170–175.

19. Bhadeshia, H.; Honeycombe, R. *Steels: Microstructure and Properties*, 3rd ed.; Butterworth-Heinemann: Oxford, UK, 2006.

20. Chen, C.Y.; Chen, S.F.; Chen, C.C.; Yang, J.R. Control of precipitation morphology in the novel HSLA steel. *Mater. Sci. Eng. A* **2015**, *634*, 123–133.

21. Tsai, S.P.; Tsai, Y.T.; Chen, Y.W.; Yang, J.R.; Chen, C.Y.; Wang, Y.T.; Huang, C.Y. Precipitation behavior in bimodal ferrite grains in a low carbon Ti-V-bearing steel. *Scr. Mater.* **2018**, *143*, 103–107. [CrossRef]

22. Honeycombe, R.W.K.; Mehl, R.F. Transformation from austenite in alloy steels. *Metall. Trans. A* **1976**, *7*, 915–936. [CrossRef]

23. Bhadeshia, H.K.D.H. Diffusional transformations: A theory for the formation of superledges. *Phys. Stat. Sol. (a)* **1982**, *69*, 745–750. [CrossRef]

24. Liu, Z.Q.; Miyamoto, G.; Yang, Z.G.; Furuhara, T. Direct measurement of carbon enrichment during austenite to ferrite transformation in hypoeutectoid Fe–2Mn–C alloys. *Acta Mater.* **2013**, *61*, 3120–3129. [CrossRef]

25. Chen, M.Y.; Yen, H.W.; Yang, J.R. The transition from interphase-precipitated carbides to fibrous carbides in a vanadium-containing medium-carbon steel. *Scr. Mater.* **2013**, *68*, 829–832. [CrossRef]

26. Dunlop, G.; Honeycombe, R. Ageing characteristics of VC, TiC, and (V, Ti) C dispersions in ferrite. *Metal Sci.* **1978**, *12*, 367–371. [CrossRef]

27. Ashby, M.F. *Physics of Strength and Plasticity*; The MIT Press: Cambridge, MA, USA, 1969.

28. Kelly, A.; Nicholson, R.B. Precipitation hardening. *Prog. Mater. Sci.* **1963**, *10*, 151–391.

29. Chin, G.Y.; Mammel, W.L. Computer solutions of the Taylor analysis for axisymmetric flow. *Trans. Metall. Soc. AIME* **1967**, *239*, 1400–1405.

30. Kaye, G.W.C.; Laby, T.H. *Tables of Physical and Chemical Constants and Some Mathematical Functions*, 14th ed.; Longman: London, UK, 1973.

31. Kamikawa, N.; Sato, K.; Miyamoto, G.; Murayama, M.; Sekido, N.; Tsuzaki, K.; Furuhara, T. Stress–strain behavior of ferrite and bainite with nano-precipitation in low carbon steels. *Acta Mater.* **2015**, *83*, 383–396. [CrossRef]

32. Andersson, J.O.; Helander, T.; Höglund, L.; Shi, P.F.; Sundman, B. Thermo-Calc & DICTRA, computational tools for materials science. *Calphad* **2002**, *26*, 273–312.

33. Thermo-Calc Software. Available online: http://www.thermocalc.com/ (accessed on 1 October 2019).

 *metals*

Article

Carbide Precipitation, Dissolution, and Coarsening in G18CrMo2–6 Steel

Zhenjiang Li [1,2,*] **, Pengju Jia** [1,2]**, Yujing Liu** [1,2] **and Huiping Qi** [1,2]

1   School of Materials Science and Engineering, Taiyuan University of Science and Technology,
    Taiyuan 030024, China
2   Shanxi Key Laboratory of Metal Forming Theory and Technology, Taiyuan 030024, China
*   Correspondence: lizj@tyust.edu.cn; Tel.: +86-0351-2161126

Received: 25 July 2019; Accepted: 19 August 2019; Published: 22 August 2019

Abstract: The precipitation, dissolution, and coarsening of different carbides at 680 °C in G18CrMo2–6 steel was investigated experimentally combined with Jmatpro simulation. The G18CrMo2–6 steel was normalized at 940 °C, followed by tempering at different times at a constant temperature of 680 °C. During the tempering process, there are mainly two kinds of carbide, namely M_3C and $M_{23}C_6$. Through characterization of microstructural evolution, thermodynamic calculation, and kinetic simulation, it was observed that during the tempering process, the stable $M_{23}C_6$ carbide was growing, whereas the metastable M_3C carbide was disappearing. At the end, the M_3C carbide was dissolved and the $M_{23}C_6$ carbide was in equilibrium with the matrix.

Keywords: G18CrMo2–6 steel; phase diagrams; carbide; kinetic simulation

1. Introduction

Low-alloy chromium-containing heat-resistant steel is widely used in the power generation industry of pressure vessels and superheater tubes due to its excellent mechanical properties, good weldability, and good creep properties at high temperatures. To increase the efficiency of steam turbine equipment, these steels have been improved to provide higher creep strength. The creep strength of these steels usually decreases rapidly during long-term use because the microstructure becomes unstable at high temperatures. The coarsening of carbide is detrimental for creep strength. Therefore, it is very important to carefully investigate the evolution behavior of carbide. Actually, the alloying element molybdenum is usually added to these steels to reduce the segregation of the impurity elements at the grain boundaries and to cause solid solution strengthening. In addition, the alloying element vanadium is added to form fine grain strengthening and precipitation hardening.

Many authors have studied the crystal structure, chemical composition, and distribution of carbides in these steels. The evolution and distribution of carbides during heat treatment and service are the main factors affecting the creep resistance of such low-alloy steels. Tsai et al. [1] and Tsai and Yang [2] found that in the heat affected zone of 2.25Cr1Mo steels, M_2C, M_3C, M_7C_3, and $M_{23}C_6$ carbides precipitated at different times during tempering at 700 °C, and the final equilibrium precipitate is $M_{23}C_6$ carbide. Yang and Kim [3] studied the mechanical properties and thermal embrittlement of 2.25Cr1Mo steel after a long period of high-temperature aging treatment and observed that some carbides were transformed during that treatment, leading to thermal embrittlement. Janovec et al. [4,5] found that the composition of steel, tempering temperature, and tempering holding time had an influence on the composition, type, and distribution position of the precipitate carbides. Tao et al. [6] investigated the carbide evolution of 2.25Cr1Mo weld metal during tempering at 700 °C and found that the precipitation strengthening effect was weakened due to carbide coarsening, and then the hardness of the tempered sample decreased. Jiang et al. [7,8] studied the evolution of 2.25Cr1Mo0.25V during tempering at 700 °C, and found that the decomposition of the martensite–austenite island and the coarsening of the MC carbide led to a decrease in material strength and an increase in ductile–brittle transition temperature (DBTT).

A method for simulating diffusion reactions in a multi-component system was developed by Ågren [9,10], which can be used to calculate nucleation, dissolution, and coarsening, etc. It is assumed that there is no difference in chemical composition and chemical potential at the interface between the matrix and the carbide. In this case, the problem is a simple diffusion problem. Therefore, the concentration of the components at the interface can be estimated by thermodynamic calculation. Based on the methods, many researchers [11–18] have studied the evolution of precipitated phases in steel.

Related research work has been done to study the influence of alloying elements on precipitation behavior. The growth and coarsening of $M_{23}C_6$ carbide was the subject of many related investigations. However, only a few investigations focus on the dissolution of carbides. The dissolution, precipitation, and growth of carbides often occur at the same time. Simultaneously simulating the evolution of several precipitates can reduce the simulation error. The aim of this paper was to investigate the precipitation, coarsening, and possible dissolution of carbides in matrix at 680 °C, which is a normal tempering temperature for low-alloy chromium-containing heat-resistant steels.

2. Experimental Section

2.1. Material Preparation and Heat Treatments

G18CrMo2–6 steel was melted into an ingot of 20 kg using a vacuum induction furnace (Wanfeng Inc., Luoyang, China). The chemical composition (wt%) of G18CrMo2–6 steel is listed in Table 1. Jmatpro calculation provides useful guidelines for the evolution and volume fractions of phases in G18CrMo2–6 steel. The thermodynamic calculations were carried out under equilibrium conditions and the kinetic calculations were carried out at a temperature of 680 °C. The steel was diffusion annealed at 1200 °C for 10 h and then air cooled to room temperature. Afterward, $10 \times 10 \times 10 \text{ mm}^3$ blocks were cut down from the diffusion-annealed material using a DK7740 wire cutting machine (Star Peak Inc., Taizhou, China), and then the specimen was placed in a glass tube filled with argon for subsequent heat treatments. All specimens were cooled in a furnace after being heated at 940 °C for 2 h to simulate the cooling rate of the actual production process. Then, all samples were tempered at 680 °C; this tempering temperature is commonly used to make large castings and forgings. In order to study the evolution of the microstructure and precipitates during tempering, the tempering time varied from 0 to 1000 h.

Table 1. Steel compositions (wt%).

Fe	Cr	Mn	Mo	Ni	Si	C
Bal.	0.6	0.75	0.6	0.46	0.45	0.16

2.2. Sample Preparation and Characterization

After the heat treatments, metallographic specimens were ground, polished, and then etched with 4 vol % nital liquid for 5–10 s for microstructure observation. An AXIOVERT 200MAT Optical microscope (OM) (Zeiss, Oberkochen, Germany), an S-3400N scanning electron microscope (SEM) (JEOL Ltd, Tokyo, Japan), and a transmission electron microscope (TEM) (Oxford Instruments, Abingdon UK) were used to observe the microstructure. Thin foil samples for TEM were prepared by mechanical polishing, and twin jet electro-polishing was performed with a 10% perchloric acid-glacial acetic acid solution with 20 V voltage at −20 °C. The composition of carbides in the samples was detected by an energy-dispersive X-ray (EDX) analyzer (JEOL Ltd, Tokyo, Japan) which was attached to the SEM and TEM. The carbide powder obtained by electrolytic extraction method was identified by X-ray diffraction (XRD) to investigate the carbides' evolution during tempering.

3. Results and Discussion

3.1. Phase Diagram Calculation

The equilibrium phase diagram of G18CrMo2–6 steel is calculated as shown in Figure 1. The abscissa is the temperature and the ordinate is the weight fraction of the equilibrium phase. According to the equilibrium phase diagram, there is only one precipitate phase of $M_{23}C_6$ (M = Fe, Cr, Mn, Mo) at 680 °C. The results of the phase diagram calculation are inconsistent with the experimental results reported in the literature [6,19]. This is mainly because the equilibrium phase diagram is only the result of thermodynamic calculations, and the calculation process ignores the effects of kinetics. In addition, the state of the experimental materials is metastable under normal conditions.

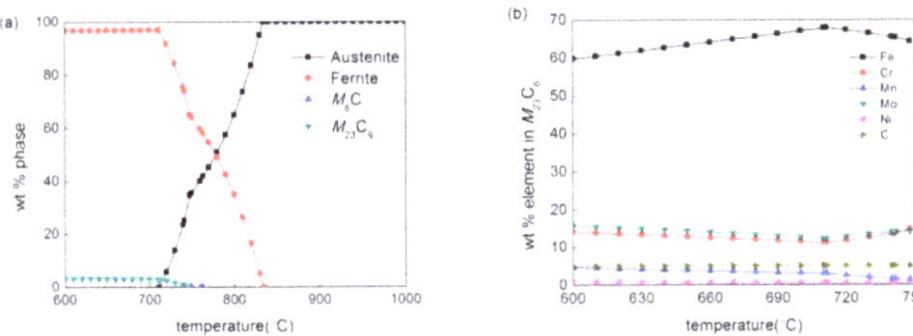

Figure 1. (**a**) Equilibrium phase diagram of G18CrMo2–6 steel; (**b**) composition of $M_{23}C_6$ carbide.

3.2. Characterization of Microstructural Evolution

3.2.1. Characterization of Microstructural Evolution by OM

Figure 2 presents the metallographic photographs of the samples after tempering at 680 °C for different tempering times. After the above heat treatment, it can be seen that the microstructure of the samples is composed of two phases of ferrite and bainite; the white area is ferrite and the gray and dark areas are bainite. With the extension of tempering time, the participation of white parts on the matrix increased gradually, and the color difference between ferrite and bainite decreased. It seems that the participation of ferrite increased and the participation of bainite decreased. However, tempering heat treatment cannot change the crystal structure of ferrite and bainite in the matrix. This may be the illusion caused by the evolution of the precipitates on the bainite matrix during tempering. However, the morphology, distribution, and type of carbides may change during tempering.

3.2.2. Characterization of Microstructural Evolution by SEM and XRD

In the initial stages of tempering (Figure 3a,b), many long rod-shaped precipitates in the bainite grains can be observed, and almost no precipitates appear at the grain boundaries. When the tempering time is extended, the length of the rod-shaped precipitate in the bainite decreases, and some precipitates appear and gradually grow at the grain boundaries (Figure 3c,d). After 1000 h of tempering, the precipitates in the bainitic matrix become finely dispersed and the aspect ratio becomes smaller. The carbides in the bainite become spherical and their number decreases; this indicates that the carbides have spheroidized and dissolved. Additionally, there are much more precipitates at the grain boundaries.

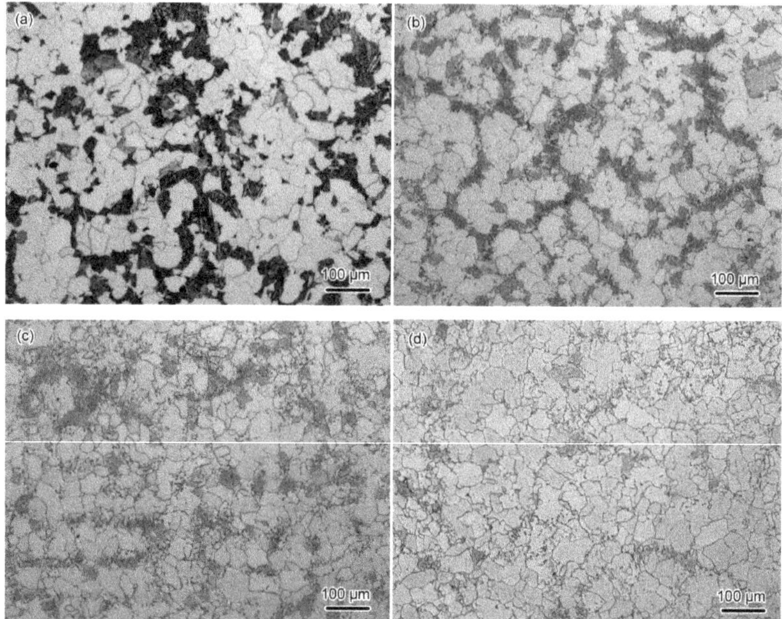

Figure 2. Metallographic photographs of the microstructure of G18CrMo2–6 steel tempered at 680 °C for different times: (**a**) 0 h, (**b**) 100 h, (**c**) 500 h, (**d**) 1000 h.

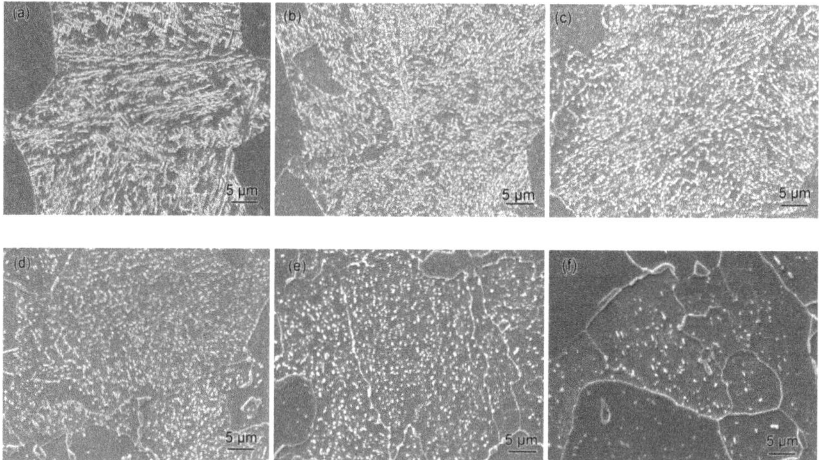

Figure 3. Images of bainite region after tempering at 680 °C for (**a**) 0 h, (**b**) 10 h, (**c**) 50 h, (**d**) 100 h, (**e**) 500 h, and (**f**) 1000 h.

The precipitate of the samples after tempering at 680 °C was obtained by extraction, and the main precipitated phase was found to be M_3C and $M_{23}C_6$ by XRD (Figure 4). During tempering for a short time (2 h), the dominant precipitate in the sample is M_3C. In the case of long-time tempering (250 h), the precipitate in the sample is composed of M_3C and $M_{23}C_6$. This indicates that $M_{23}C_6$ gradually precipitates during the tempering process.

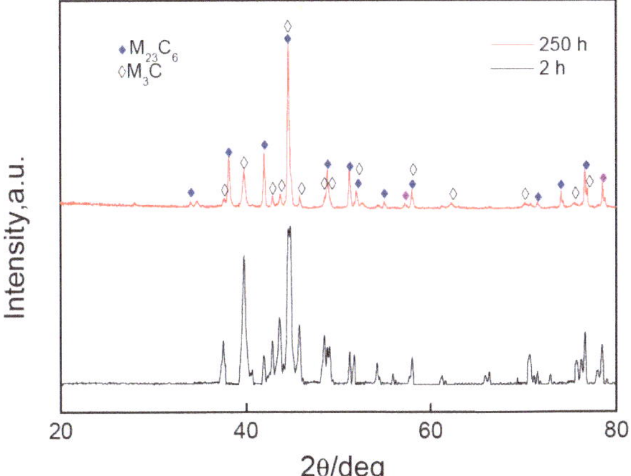

Figure 4. Patterns of precipitate extracted from samples tempered at 680 °C for 2 h and 250 h, reproduced from [19], with copyright permission from Elsevier, 2014.

3.2.3. Characterization of Microstructural Evolution by TEM

Figure 5a,b shows the TEM images of the precipitates in the bainite grains. When the tempering time is extended, the length of the rod-shaped precipitate in the bainite is shortened (Figure 5b) and the precipitates in the bainitic matrix become finely dispersed. Both the rod and spherical particles in the bainitic matrix are M_3C-type carbides and the corresponding selected area electron diffraction (SAED) is shown in Figure 5c. The result of EDX analysis indicates that M_3C is rich in Fe, Mn, Mo, and Cr (Figure 5d). As the tempering time is prolonged, the aspect ratio of M_3C becomes lower, indicating that the M_3C has spheroidized. The spheroidization mechanism for M_3C has been proposed [20]. The flat surface is inconsistent with the chemical potential of the curved surface. This difference sets up a chemical potential gradient and thus provides a driving force for the diffusion of carbon atoms and alloy elements. The diffusion causes the spalling and spheroidization of the flat surface cementite. Therefore, as the tempering time is extended, the aspect ratio of M_3C is gradually reduced, eventually forming a granular cementite.

Figure 6a shows the TEM images of the precipitates at grain boundary. The SAED reveals that the precipitates at the grain boundary are $M_{23}C_6$ carbides, which is shown in Figure 6b. $M_{23}C_6$ carbide precipitates and coarsens at the grain boundary with the extension of tempering time.

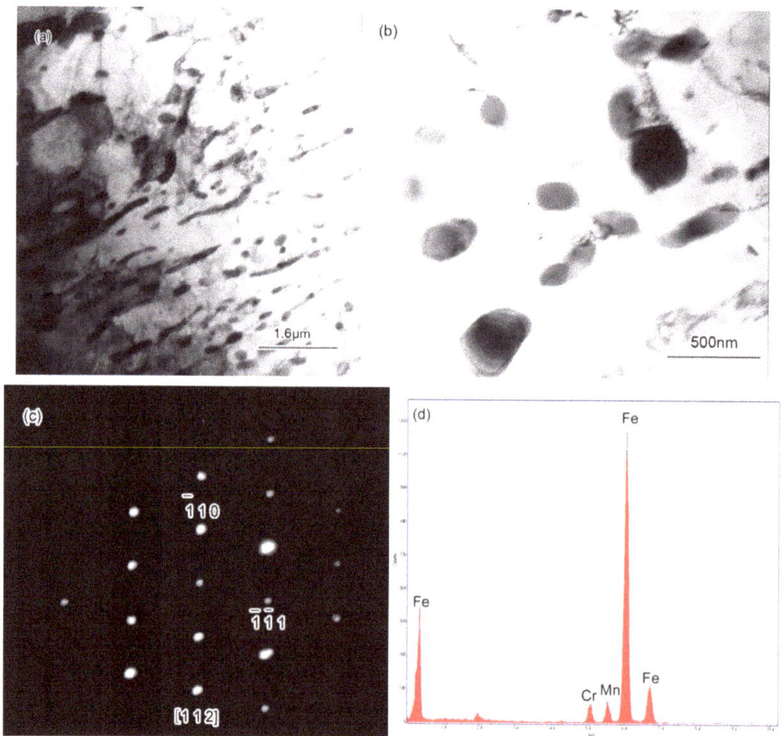

Figure 5. Micrographs of particles within bainite: (**a**) 680-2, (**b**) 680-100, (**c**) SAED pattern, and (**d**) EDX.

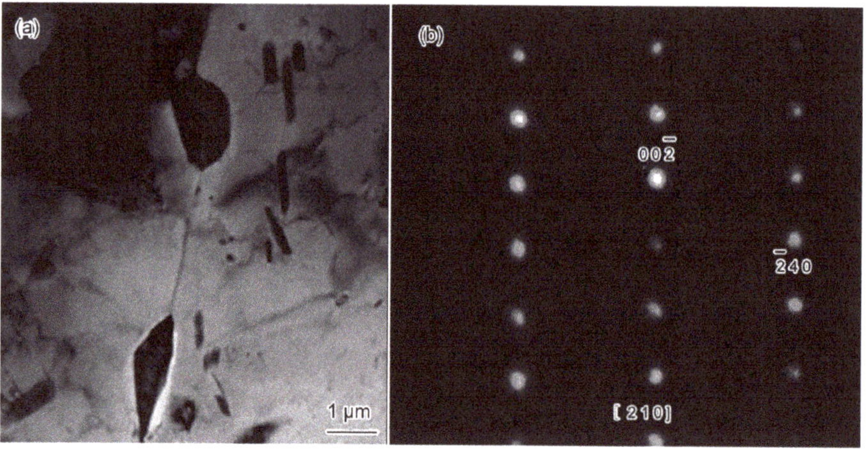

Figure 6. (**a**) Micrographs of particles at the grain boundary and (**b**) SAED pattern.

3.3. Kinetic Simulation

According to the equilibrium phase diagram in Figure 1, only $M_{23}C6$ carbide at 680 °C is stable. The results of the equilibrium phase diagram are inconsistent with the experimental results. The experiments show that there are two kinds of carbides, namely, M_3C in the bainite matrix and $M_{23}C_6$ at the grain boundary. There seem to be two different evolutionary processes, one with dissolution of the metastable M_3C within the bainite grains and one with precipitation of the stable $M_{23}C_6$ at grain boundaries.

The results of the kinetic simulation are shown in Figure 7. Within about 10 s, the percentage of M_3C carbides decreases at a low rate, and the percentage of $M_{23}C_6$ carbides remains unchanged. After about 10 s, the percentage of M_3C carbides decreases faster than previously, and the percentage of $M_{23}C_6$ carbides begins to increase. This increase of $M_{23}C_6$ carbides between 10 and 1000 s is obviously coupled with the decrease of the M_3C carbides. After the M_3C carbides are dissolved, the percentage of $M_{23}C_6$ carbides keep increasing and then the system eventually approaches equilibrium between the $M_{23}C_6$ carbides and the matrix. As equilibrium is approached, the percentage of $M_{23}C_6$ carbides remains unchanged. Although the dissolution of M_3C is beneficial to the creep strength of the material, the coarsening of $M_{23}C_6$ carbide is detrimental for creep strength.

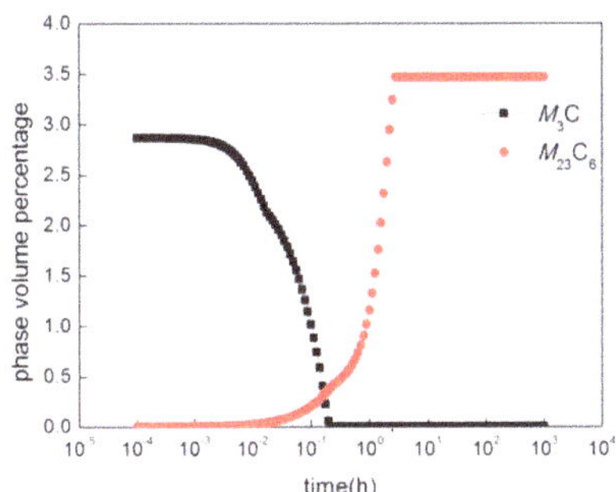

Figure 7. Simulation of the growth of $M_{23}C_6$ simultaneously with the dissolution of M_3C.

4. Conclusions

The evolution of carbides in G18CrMo2–6 steel during 680 °C tempering was investigated experimentally combined with Jmatpro simulation.

(1) During the tempering process, there are mainly two kinds of carbides, namely M_3C in the bainite matrix and $M_{23}C_6$ at the grain boundary.

(2) The experimental, thermodynamic and kinetic results show that when the tempering time is short, the stable $M_{23}C_6$ carbide precipitates and coarsens, whereas the metastable M_3C carbide disappears. At the end, the M_3C carbide was dissolved and the $M_{23}C_6$ carbide was in equilibrium with the matrix.

Author Contributions: Data curation, Z.L.; writing—original draft preparation, Z.L., P.J. and Y.L.; writing—review and editing, Z.L. and H.Q.; supervision, Z.L.

Funding: This research was funded by the National Natural Science Foundation of China (No. 51875383) and the China Scholarship Council (No. 201708140137), as well as the Doctoral Research Foundation of Taiyuan University of Science and Technology, China (20162011).

Conflicts of Interest: The authors declare no conflict of interest.

References

1. Tsai, M.; Chiou, C.; Yang, J. Microstructural evolution of simulated heat-affected zone in modified 2.25Cr–1Mo steel during high temperature exposure. *J. Mater. Sci.* **2003**, *38*, 2373–2391. [CrossRef]
2. Tsai, M.; Yang, J. Microstructural degeneration of simulated heat-affected zone in 2.25Cr–1Mo steel during high-temperature exposure. *Mater. Sci. Eng. A* **2003**, *340*, 15–32. [CrossRef]
3. Yang, H.; Kim, S. A study on the mechanical strength change of 2.25Cr–1Mo steel by thermal aging. *Mater. Sci. Eng. A* **2001**, *319*, 316–320. [CrossRef]
4. Janovec, J.; Svoboda, M.; Kroupa, A.; Výrostková, A. Thermal-induced evolution of secondary phases in Cr–Mo–V low alloy steels. *J. Mater. Sci.* **2006**, *41*, 3425–3433. [CrossRef]

5. Janovec, J.; Svoboda, M.; Výrostková, A.; Kroupa, A. Time–temperature–precipitation diagrams of carbide evolution in low alloy steels. *Mater. Sci. Eng. A* **2005**, *402*, 288–293. [CrossRef]

6. Tao, P.; Zhang, C.; Yang, Z.-G.; Hiroyuki, T. Evolution and Coarsening of Carbides in 2.25Cr–1Mo Steel Weld Metal During High Temperature Tempering. *J. Iron Steel Res. Int.* **2010**, *17*, 74–78. [CrossRef]

7. Jiang, Z.; Wang, P.; Li, D.; Li, Y. The evolutions of microstructure and mechanical properties of 2.25Cr-1Mo-0.25V steel with different initial microstructures during tempering. *Mater. Sci. Eng. A* **2017**, *699*, 165–175. [CrossRef]

8. Jiang, Z.; Wang, P.; Li, D.; Li, Y. Influence of the decomposition behavior of retained austenite during tempering on the mechanical properties of 2.25Cr-1Mo-0.25 V steel. *Mater. Sci. Eng. A* **2019**, *742*, 540–552. [CrossRef]

9. Agren, J. Local equilibrium and prediction of diffusional transformations. *Scand. J. Metall.* **1991**, *20*, 86–92.

10. Ågren, J. Computer simulations of diffusional reactions in complex steels. *ISIJ Int.* **1992**, *32*, 291–296. [CrossRef]

11. Bjärbo, A.; Hättestrand, M. Complex carbide growth, dissolution, and coarsening in a modified 12 pct chromium steel—An experimental and theoretical study. *Metall. Mater. Trans. A* **2001**, *32*, 19–27. [CrossRef]

12. Prat, O.; Garcia, J.; Rojas, D.; Carrasco, C.; Kaysser-Pyzalla, A. Investigations on coarsening of MX and M23C6 precipitates in 12% Cr creep resistant steels assisted by computational thermodynamics. *Mater. Sci. Eng. A* **2010**, *527*, 5976–5983. [CrossRef]

13. Xia, Z.; Zhang, C.; Yang, Z. Control of precipitation behavior in reduced activation steels by intermediate heat treatment. *Mater. Sci. Eng. A* **2011**, *528*, 6764–6768. [CrossRef]

14. Xiao, X.; Liu, G.; Hu, B.; Wang, J.; Ma, W. Coarsening behavior for M23C6 carbide in 12% Cr-reduced activation ferrite/martensite steel: experimental study combined with DICTRA simulation. *J. Mater. Sci.* **2013**, *48*, 5410–5419. [CrossRef]

15. Zhu, N.; He, Y.; Liu, W.; Li, L.; Huang, S.; Vleugels, J.; van der Biest, O. Modeling of nucleation and growth of M23C6 carbide in multi-component Fe-based alloy. *J. Mater. Sci. Technol.* **2011**, *27*, 725–728. [CrossRef]

16. Dahlgren, R. Prediction of Near Surface M3C after Hard Turning of SAE 52100 Steel. Master's Thesis, Chalmers University of Technology, Gothenburg, Sweden, 2012.

17. Gustafson, Å.; Hättestrand, M. Coarsening of precipitates in an advanced creep resistant 9% chromium steel—quantitative microscopy and simulations. *Mater. Sci. Eng. A* **2002**, *333*, 279–286. [CrossRef]

18. Yin, Y.F.; Faulkner, R.G. Simulations of precipitation in ferritic steels. *Met. Sci. J.* **2003**, *19*, 91–98. [CrossRef]

19. Li, Z.; Xiao, N.; Li, D.; Zhang, J.; Luo, Y.; Zhang, R. Effect of microstructure evolution on strength and impact toughness of G18CrMo2–6 heat-resistant steel during tempering. *Mater. Sci. Eng. A* **2014**, *604*, 103–110. [CrossRef]

20. Tian, Y.L.; Kraft, R.W. Mechanisms of pearlite spheroidization. *Metall. Trans. A* **1987**, *18*, 1403–1414. [CrossRef]

 *metals*

Article

The Significance of Central Segregation of Continuously Cast Billet on Banded Microstructure and Mechanical Properties of Section Steel

Fujian Guo [1], Xuelin Wang [1], Jingliang Wang [1], R. D. K. Misra [2] and Chengjia Shang [1,3,*]

[1] Collaborative Innovation Center of Steel Technology, University of Science and Technology Beijing, Beijing 100083, China; carterguo2013@163.com (F.G.); xuelin2076@163.com (X.W.); jlwang@ustb.edu.cn (J.W.)

[2] Department of Metallurgical and Materials Engineering, University of Texas EI Paso, EI Paso, TX 799968, USA; dmisra2@utep.edu

[3] State Key Laboratory of Metal Materials for Marine Equipment and Applications, Anshan 114021, China

* Correspondence: cjshang@ustb.edu.cn; Tel.: +86-10-6233-2428

Received: 11 December 2019; Accepted: 26 December 2019; Published: 2 January 2020

Abstract: The solidification structure and segregation of continuously cast billets produced by different continuous casting processes are investigated to elucidate their effect on segregated bands in hot-rolled section steel. It suggested that segregated spots are mainly observed in the equiaxed crystal zone of a billet. The solidification structure is directly related to superheating and the intensities of secondary cooling. To a certain extent, the ratio of the columnar crystal increases with the increase of superheating and secondary cooling. Moreover, the number of spot segregations decreases with the decrease of the equiaxed crystal ratio. After hot rolling, the segregation spots are deformed to form segregated bands in steels. The severe segregation of Mn in segregated bands corresponds with that in the segregation spots. The elongation ratio and low temperature toughness deteriorate significantly by a high fraction of degenerate pearlite caused by central segregation. With a decrease of central segregation, the total elongation is increased by 10% and the ductile–brittle transition temperature (DBTT) is also reduced from −10 to −40 °C. According to the experimental results, columnar crystal in billets is preferred to effectively reduce the degree of central segregation and further improve low temperature toughness and the elongation ratio.

Keywords: continuous casting process; solidification structure; central segregation; banded microstructure; mechanical properties

---

1. Introduction

With the increasing application of section steel, the service environment is becoming harsher and the requirements for low temperature toughness are getting higher, e.g., in arctic regions. Compared to plates, the rolling reduction and, consequently, deformation accumulation of section steels are relatively low, which lead to insufficient recrystallization and coarse prior austenite grains [1]. Moreover, the microstructures of ferrite and perlite are difficult to refine due to the lack of controlled cooling process [2,3]. Therefore, low temperature toughness is always a challenge for section steel [4]. For high strength low alloy section steel, besides grain refinement [5], a banded microstructure is considered to be an important factor that affects mechanical properties [6]. A banded microstructure is always related to the cooling rate of ferrite/perlite phase transformation [7]. However, the central segregation of C, Mn, and other elements during solidification also result in the formation of kinds of abnormal band microstructures, like pearlite [8], bainite [9] and martensite [10,11].

The quality of a continuously cast billet is closely related to the continuous casting processes, such as superheating, secondary cooling [12], electromagnetic stirring [13] and soft reduction [14].

For large size billets, the continuous casting process directly affects their solidification structures (equiaxed/columnar crystal ratio). Due to the influence of various factors, such as selective crystallization and the solidification structure, the segregation of C, Mn and other alloys in the billet are inevitable [15, 16]. Previous studies have shown [10,11] that central segregation in a billet drastically deteriorate the billet's mechanical properties—low temperature toughness, in particular. Meanwhile, a banded microstructure is also an internal defect of steel that deteriorates mechanical properties. Though the relationship between central segregation and continuous casting parameters has been reported in previous studies [12–14], the heredity effect of central segregation on segregated bands and mechanical properties is still a problem that has not yet been solved. A few works have been done to examine the relative influences of central segregation of billets on the banded microstructures and mechanical properties of final products. Thus it is more important to study the influence of the central segregation of billets on banded microstructures than the rolling and cooling process [17].

In this paper, the macrostructure and microstructure of billets and steels under two continuous casting process conditions are studied; the relationship between central segregation, banded structure and the mechanical properties are analyzed; and the optimization method of billet quality is discussed.

2. Materials and Methods

The materials used in this study were U-section steels, which were obtained from the continuously cast billet with a cross-section of 280×380 mm^2. The composition of the steel in weight percent (wt%) was 0.15C–0.28Si–1.5Mn–0.07V–0.01N. The design goal of this experiment was to study the effect of solidification structures on the degree of segregation, homogeneity of microstructure, and low temperature toughness. As such, just two set of parameters were provided, and the relevant parameters of the continuous casting process are shown in Table 1 (high superheating and high secondary cooling intensity led to a low proportion of equiaxed structure; stop using EMS (electro magnetic stirring) promoted columnar crystal growth and enlarged the columnar crystal region). Two corresponding billets are referred to as Billet 1 and Billet 2. After rolling, the corresponding samples were renamed Steel 1 and Steel 2. The billet samples were cut after continuous casting under a stable state, and the continuously cast billet was rolled by using the same rolling process that excluded influence by rolling factors.

Table 1. The parameters of the continuous casting process of the experimental steel.

Billet	Superheating/°C	Speed/m/min	Secondary Cooling Kg/L	EMS
1	24	0.7	0.25	with
2	35	0.7	0.4	without

The billet and steel samples were taken from the head of the experimental billet and the corresponding U-section steel. The continuously cast billet was evaluated in terms of macrostructure, and its subsequent effect on the mechanical properties and microstructure of the U-section steel was studied. The sampling region for the billets and the U-section steels from the neighboring region was selected to study the severity and distribution of segregation. The central segregation of the billet and the U-section steel were revealed by 50 pct HCl at 80 °C for 0.5 h.

Samples for microstructure observation were cut from the central area of the black rectangle on the steel and then mounted and mechanically polished by using a standard metallographic procedure. The specimens were etched with 3% nital for optical microscopy (OM) and scanning electron microscopy (SEM) observations. Segregation behavior was analyzed by electron probe micro-analysis (EPMA, JXA-8530F, Tokyo, Japan). Prior to EPMA analysis, the segregation bands were revealed with a solution that contained picric acid, detergent, carbon tetrachloride, and sodium chloride. The etching temperature was ~60 °C, and the soaking time was 2 min. The CCT diagram was calculated with the JMatPro 7.0 according to the alloy content that was measured by EPMA in segregation area.

The temperature in the center and quarter positions of the billet was calculated by the Procast software (version 7.5, ESI Group, Paris, France) according to the continuous casting parameters shown in Table 1.

CVN impact tests were conducted by using full-size Charpy specimens with dimensions of 55 × 10 × 10 mm that were machined according to the ISO 148-1 standard from the geometrical center of the black rectangle on the steel along the longitudinal direction. CVN impact tests were conducted at temperatures of 0, −20 and −40 °C. Dog bone-shaped round tensile specimens (gage length: 110 mm; diameter: 8 mm; and orientation: longitudinal) were prepared, and the test was conducted at room temperature. The Vickers micro-hardness of the segregation band was measured by using a load of 1 Kg and a dwell time of 15 s. The load was selected to ensure that the indent could simultaneously overlap both ferrite and pearlite.

3. Results

3.1. Macro-Etching Results of Billets and U-Section Steels

The macro-etching images of Billets 1 and 2 are shown in Figure 1a,b, respectively. The obvious difference was present in the solidification structure. The equiaxed crystal ratio of Billet 1 was higher (19.1%), and the small segregation spots (black spots) were dispersed in the equiaxed crystal region; the equiaxed crystal ratio of Billet 2 was only 2.1%, and several shrinkage holes were observed in the equiaxed crystal region. Figure 1c,d shows a magnified view of the center region of Figure 1a,b, where the characteristics can be seen more clearly. Many black spots were distributed in the equiaxed crystal area of Billet 1, while the segregated spot in the equiaxed crystal area of Billet 2 become smaller, with the exception of the presence of several shrinkage holes.

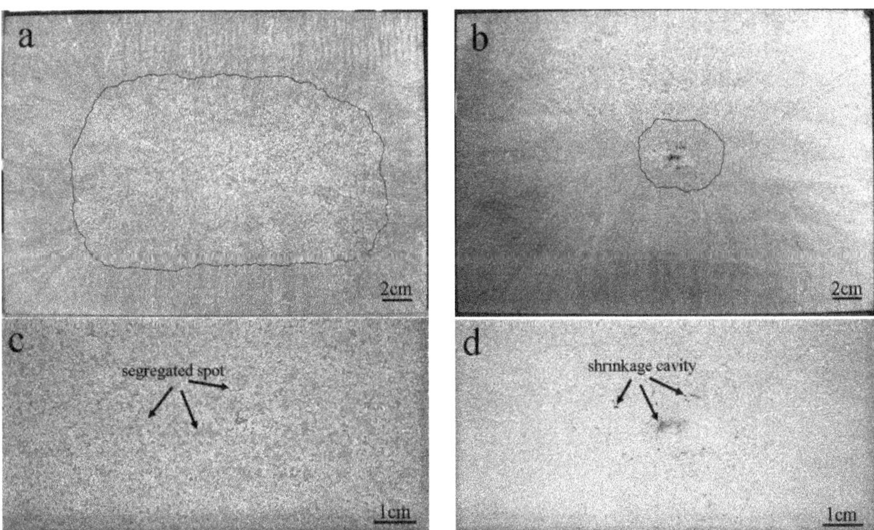

Figure 1. *Cont.*

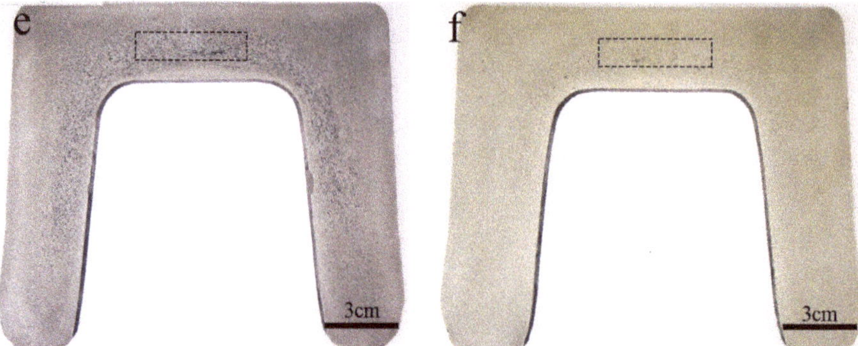

Figure 1. Macro-etching results of continuously cast billet and U-section steel. (**a**,**b**) Billets 1 and 2; (**c**,**d**) the enlarged image of the center region of Billets 1 and 2; (**e**,**f**) Steels 1 and 2. (Samples for microstructure observation and mechanical properties were cut from the central area of the black rectangle on the steel, shown in Figure 1e,f).

After hot rolling, the segregation spots in the equiaxed crystal region were compressed to the inner arc of U-section steel, as shown in Figure 1e,f. Dense segregation spots were severely present in the cross-section of Steel 1, while just a small number of segregated spots were observed in Steel 2, and the shrinkage holes disappeared. Thus, the segregation region could not be eliminated in the subsequent process, but the shrinkage holes could be welded during the rolling process.

3.2. Enrichment of Elements in Continuous Casting Billets and U-Section Steel

The segregating spots in the equiaxed crystal region of the billet were analyzed by EPMA. The test samples were cut from an identical position in the central area of the billets. The mapping images of the segregating elements C and Mn are depicted in Figure 2, which suggests the black spots in the equiaxed crystal region were enriched by C and Mn and the degree of segregation of Billet 1 (Figure 2b,c) was higher than that of Billet 2 (Figure 2e,f).

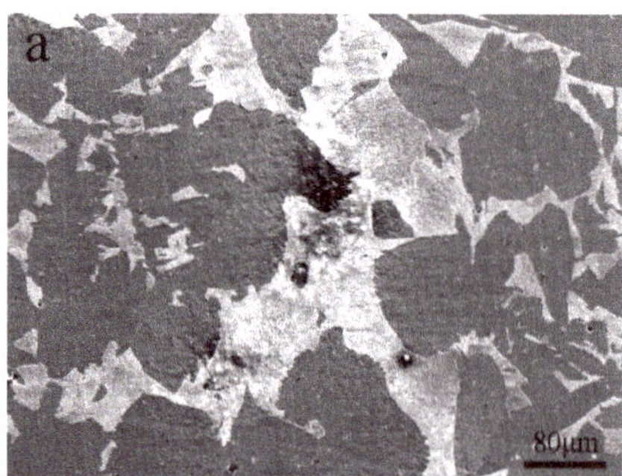

Figure 2. *Cont.*

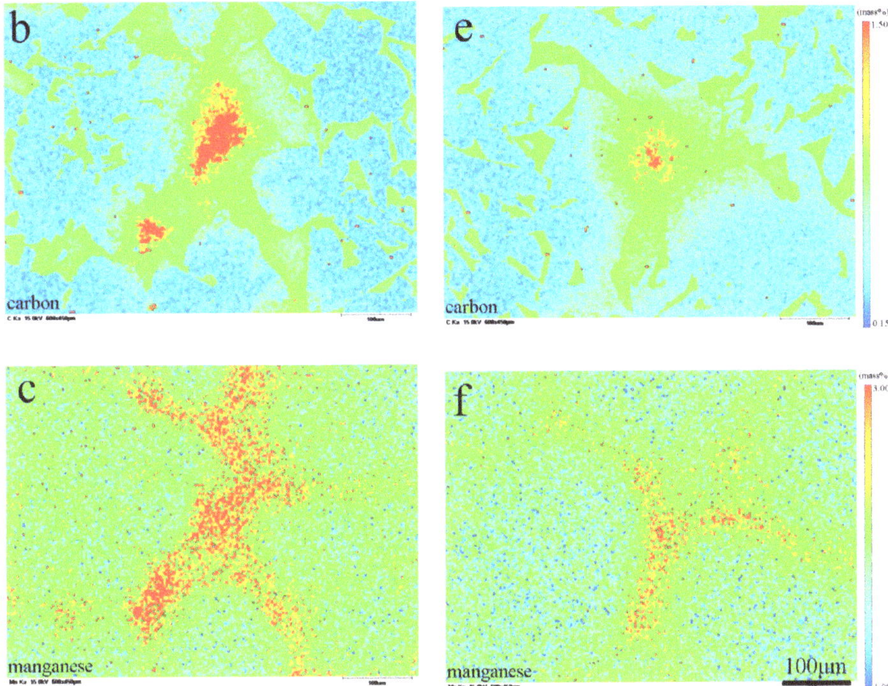

Figure 2. Two-dimensional mapping of solute elements in the billet sample: (**a**–**c**) Billet 1; (**d**–**f**) Billet 2.

After hot rolling, the segregating spots in the billet were compressed to the segregating band. In order to investigate the distribution of the enriched elements in the microstructure, the composition of the microstructure was measured by the map scanning of EPMA. The segregated band was etched by picric acid, as shown in Figure 3a,d. The black bands were identified as segregation bands, which were inherited from the central segregation in the equiaxed crystal region of the billets. The number of segregation bands in Steel 1 was larger than that of Steel 2, with the width of segregation band in Steel 1 being larger than that of Steel 2. The carbon segregation (Figure 3b,e) in the microstructure showed that carbon was enriched in pearlite, while Figure 3c,f shows that the black bands were caused by Mn enrichment. The above results suggest that the degree of segregation in Steel 1 was more severe than that of Steel 2 and was consistent with the degree of central segregation in the billets.

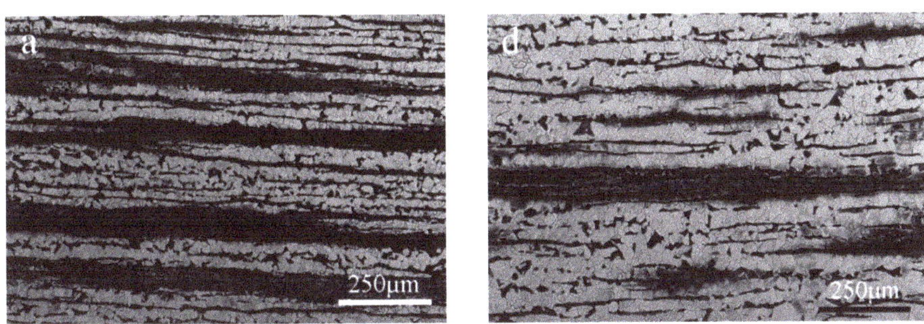

Figure 3. *Cont.*

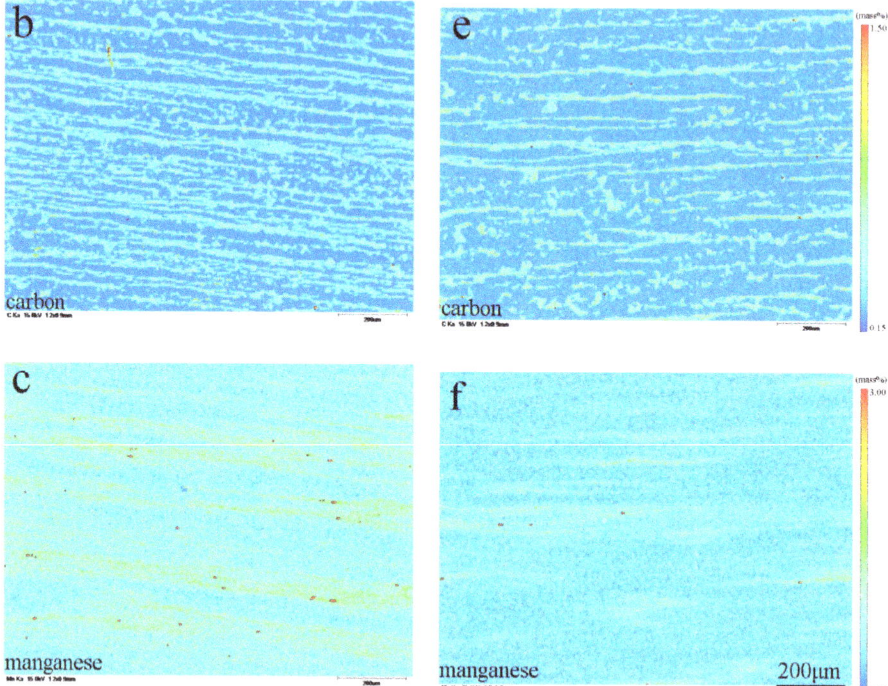

Figure 3. Segregation studies by electron probe micro-analysis (EPMA) mapping in the segregation region of steel: (**a**–**c**) Steel 1; (**d**–**f**) Steel 2.

EPMA line scanning was used to further study the segregation behavior of carbon and manganese, and the results are shown in Figure 4. The microstructures obtained in Figure 4a,b are pearlite and ferrite and are displayed as gray and dark morphologies, respectively. Line scanning results (Figure 4c,d) indicated that the peak of carbon enrichment emerged at the region of pearlite. Though there was no difference in the carbon enrichment of both samples, the peak and fluctuation of Mn enrichment suggested that the degree of Mn segregation in Sample 1 was significantly larger than that of Sample 2. The mean contents of Mn in Samples 1 and 2 were 1.98% and 1.81%, respectively.

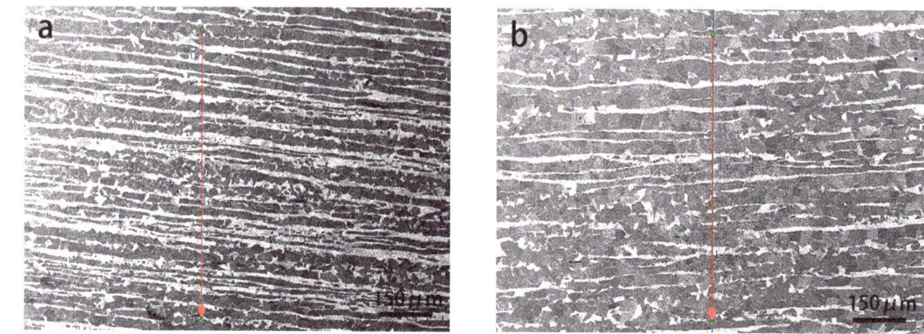

Figure 4. *Cont.*

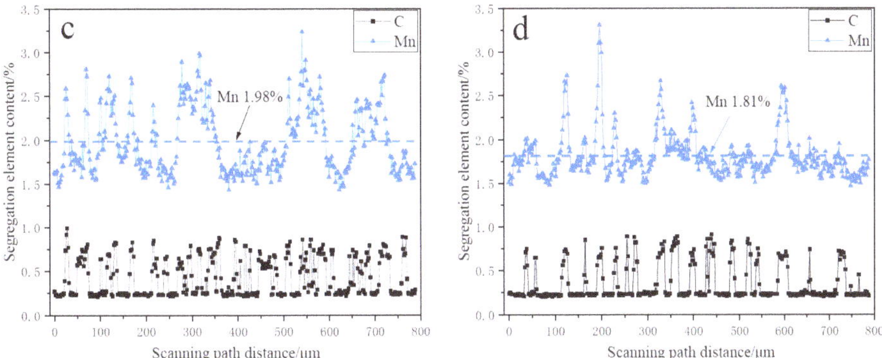

Figure 4. Segregation element content measured by EPMA line scanning in the segregation region of steel: (**a**,**c**) Steel 1; (**b**,**d**) Steel 2.

3.3. Microstructure in U-Section Steel

Samples for microstructure observation were cut from the center of the black rectangle (Figure 1e,f) on a cross-section. The microstructure of this steel is shown in Figure 5. It consisted of ferrite and pearlite in the black rectangle region (Figure 1e,f). However, the difference in microstructure was obvious. The fraction of pearlite in Steel 1 (Figure 5a,c) was high (41%), while the fraction of pearlite in Steel 2 (Figure 5b,d) was just 14%. The morphogenesis of pearlite is shown clearer in a magnification image (Figure 5e–h). The microstructures in the rectangular areas of Figure 5c,d are zoomed in Figure 5e,f. The pearlite in the two steels was different, as the spherical and discontinuous pearlite (Figure 5e) was degenerate pearlite in Steel 1 while the pearlite in Steel 2 was uniformly aligned as lamellar (Figure 5f). In addition, Figure 5e,f are further zoomed in to observe the morphology of different pearlite structures, as shown in Figure 5g,h.

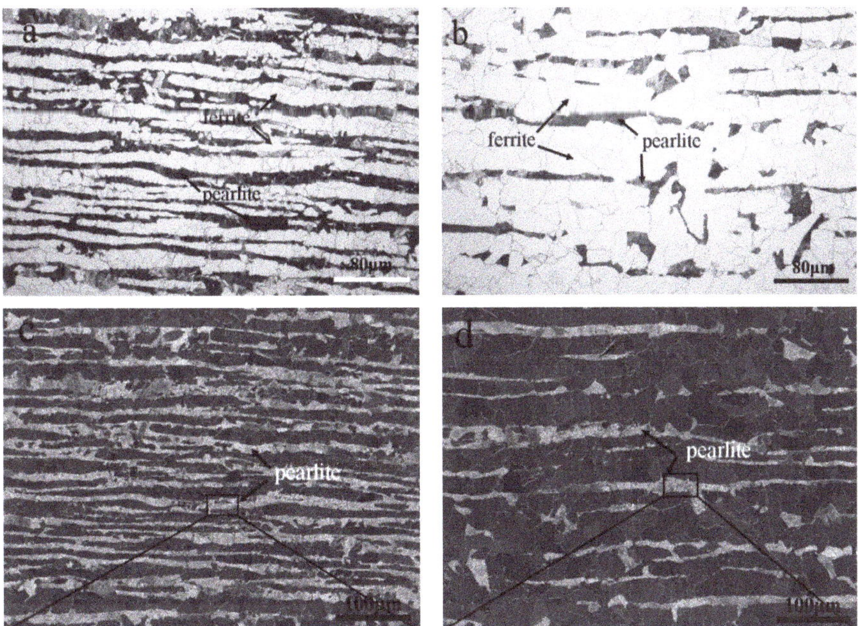

Figure 5. *Cont.*

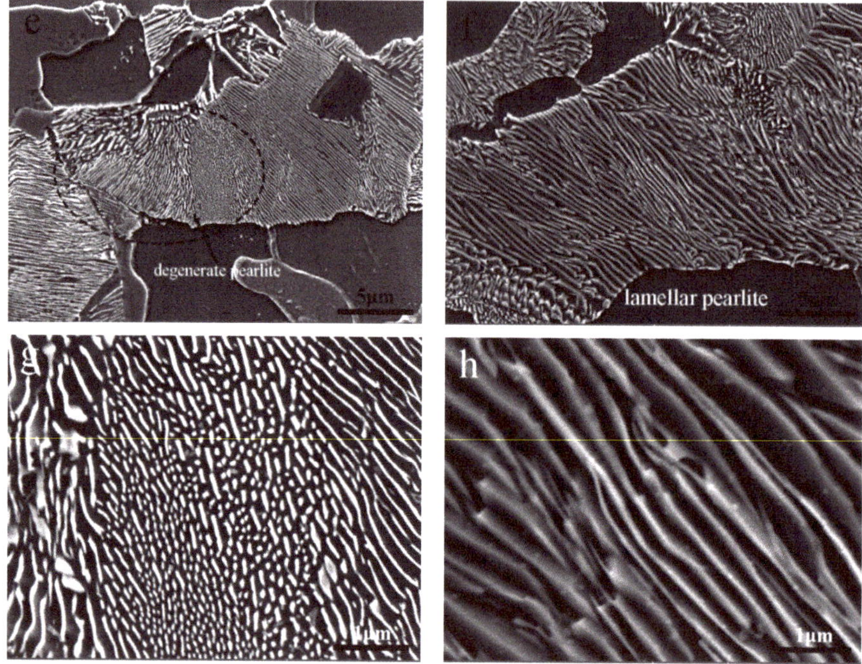

Figure 5. The microstructures of the U-section steel seen by an optical microscope (**a**,**b**) and scanning electron microscope (**c**–**h**). (**a**,**c**,**e**,**g**) Steel 1; (**b**,**d**,**f**,**h**) Steel 2.

3.4. Mechanical Properties

The tensile strength (TS), yield strength (YS), uniform elongation (Agt), and total elongation (A) of the test specimens were measured, and the results are shown in Table 2. By comparing the mechanical properties of Steel 1 and Steel 2, it may be noted that yield strength of Steel 1 was higher than that of Steel 2 by ~110 MPa, and the tensile strength of Steel 1 was higher than that of Steel 2 by ~20 MPa. The uniform elongation of Steel 1 was lower than that of Steel 2 by ~6.5%, and the total elongation of Steel 1 was lower than that of Steel 2 by ~10%. The above results suggest that the yield strength of Steel 1 was significantly higher than that of Steel 2, and the elongation of Steel 1 was obviously lower than that of Steel 2. The microstructure of the U-section steel was composed of ferrite and pearlite bands. The microhardness of the segregation area was measured at the geometrical center of the black rectangle (Figure 1e,f) in the U-section steel. A total of 10 measurements were taken to obtain an average value. In order to ensure the microhardness accuracy, 1 Kg was selected for the overlapping of both ferrite and pearlite. The microhardness values of Steels 1 and 2 were ~190 and 162 HV, respectively, as shown in Figure 6a. It can be concluded that the microhardness of Steel 1 was higher than Steel 2, as could be seen from the tensile strength and the corresponding microstructures.

Table 2. Tensile properties of experimental steel.

Steel	YS/MPa	TS/MPa	Agt/%	A/%
1	504.5 (±9.2)	572.0 (±4.2)	8.53 (±1.05)	25.36 (±0.79)
2	386.0 (±2.8)	552.5 (±0.7)	15.04 (±0.30)	34.40 (±0.10)

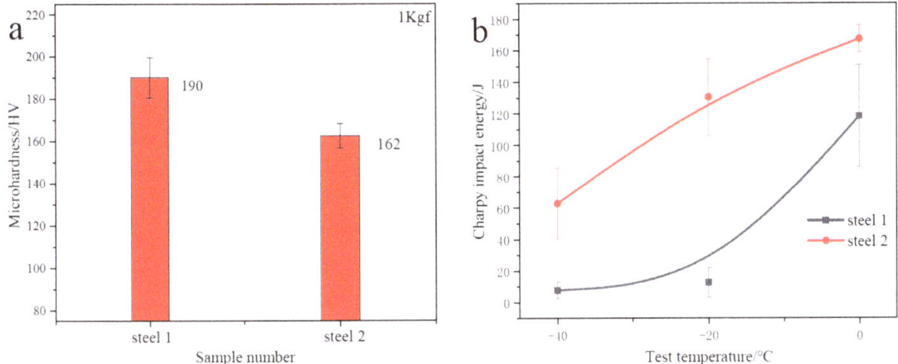

Figure 6. Microhardness (**a**) and Charpy impact toughness (**b**) of the U-section steel.

As U-section structural steels were used in a low temperature environment, the superior low temperature toughness had to satisfy the target requirements. The ductile–brittle transition temperature (DBTT) of the two steels was measured, and the experimental results are shown in Figure 6b. It was clear that Steel 2 showed an excellent low temperature toughness with a DBTT of −40 °C; the toughness of Steel 1 did not meet the requirements of the low temperature, as its DBTT was −10 °C.

4. Discussion

4.1. Occurrence of Degenerate Pearlite and Its Influence on Mechanical Properties

The central segregation in the continuously cast billet could not be eliminated by subsequent processing and was directly inherited in the subsequently rolled steel, and this issue was confirmed in the macro-etched results (Figure 1e,f) of the rolled steel. When the central region of the continuously cast billet had severe segregation (Figure 1a), Mn segregation was also inherited in the microstructure of the rolled steel (Figures 2 and 3). Previous studies [9,18–20] have suggested that the segregation of solute elements in a slab or billet affects the transition temperature (Ar3) of austenite to ferrite during the cooling process and becomes the source of banded segregation. The CCT diagram was calculated by the JMatPro software according to the alloy content that was measured by EPMA in the segregation area (Steel 1, 0.15C–1.98Mn–0.28Si–0.07V–0.01N; Steel 2, 0.15C–1.81Mn–0.28Si–0.07V–0.01N). As shown in Figure 7, the phase transformation temperature of Steel 1 was obviously lower than that of Steel 2. Mn is an austenite stabilizing element and can reduce Ar3, so the Ar3 in a rich-Mn region is lower than that in a poor-Mn region [21]. As temperature decreases, proeutectoid ferrite preferentially nucleates in the poor-Mn dendrites. As ferrite grains grow during the longitudinal extension of the deformation process, adjacent ferrites form bands during the rolling process. In addition, since the solubility of C in ferrite is much lower than that of austenite, with the formation and growth of ferrite, C is continuously rejected to the area around austenite that has not been transformed. At the same time, it is known that the concentration of the Mn element in this region is high, and Mn will reduce the activity of C and hinder its diffusion [9]. Therefore, in the subsequent transformation process, the austenite in the segregation region easily forms pearlite, and, here, the proportion of pearlite in Steel 1 was higher than that of Steel 2 (Figure 5). In addition, the formation of degenerate pearlite was caused by the enrichment of C and Mn in the segregated region. During continuous cooling phase transformation, the degenerated pearlite is transformed from carbon-rich austenite. In general, during the phase transformation from austenite to ferrite, carbon is enriched to untransformed austenite. Carbon-rich austenite has a higher stability and a lower transformation temperature. At lower transformation temperatures, the transformation driving force is larger, resulting in the separate formation of cementite and ferrite. Finally, the morphology of pearlite is spherical or discontinuously lamellar. The morphology of pearlite has also been reported in the literature [22], where it has been suggested that at high reaction temperatures, group nodules of pearlite form that consist of many colonies of pearlite; at lower reaction temperatures, pearlite nodules form as hemispheres or as sectors of spheres.

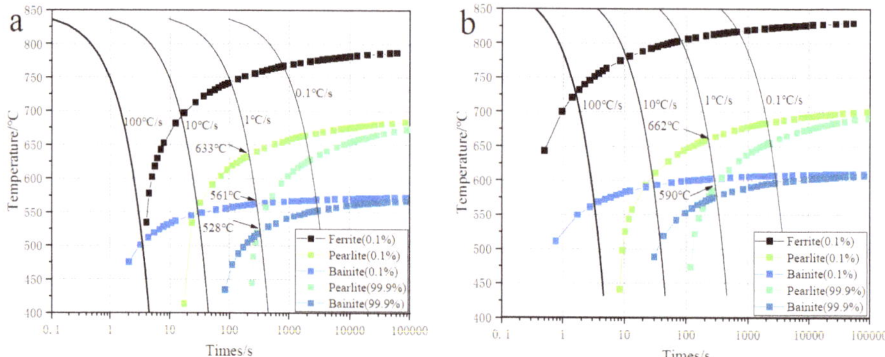

Figure 7. CCT diagram of the central segregation region of steel. (**a**) Steel 1; (**b**) Steel 2.

Mechanical properties are caused by the microstructure of material. In our experiments, the elongation and toughness of pearlite was lower than that of ferrite, and the strength and hardness of pearlite was higher than that of ferrite. In addition, the hardness of the degenerate pearlite was higher than the lamellar pearlite. The degenerate pearlite did not have good plastic deformation. Therefore, the strength of Steel 1 was significantly higher than that of Steel 2. On the other hand, when an external load was applied, stress concentration tended to occur at the interface between the degenerate pearlite and ferrite. The more severe the stress concentration, the easier it was for the crack to nucleate and propagate. Thus, the low temperature toughness of Steel 2 was obviously higher than that of Steel 1.

4.2. Effect of Continuous Casting Process on Solidification Structure and Central Segregation of Billet

In our present study of low alloy steels, as the superheating and secondary cooling intensity increased, the area affected by the central segregation was reduced (Figure 1a,b), and the segregation spots in the equiaxed crystal zone significantly decreased (Figure 1c,d). The result was consistent with Ji's study [21], where segregating spots decreased with the decreased round bloom of the equiaxed crystal zone. Previous studies [23] have suggested that factors affecting solidification structures include high superheating temperatures that lead to a high proportion of columnar structures, stop using EMS (electro magnetic stirring) that promotes columnar crystal growth and enlarges the columnar crystal region, and a high secondary cooling intensity that increases the columnar crystal ratio. In addition, according to the model given by Equation (1) [24] of the columnar-equiaxed transformation, the higher the secondary cooling intensity and superheating temperature, the larger temperature gradient; meanwhile, the other values remain unchanged, and the ratio of G_n/V more easily exceeds the critical value of the columnar crystal region, such the columnar crystal region proportion is enlarged. Thus, two kinds of solidification structures were obtained in Billets 1 and 2, and their equiaxed crystal ratios were ~19.1% and 2.1%, respectively.

$$G_n/V > C_{st} = a(8.6\Delta T_0 \frac{N_0^{\frac{1}{3}}}{n+1})^n \tag{1}$$

where G_n is the temperature gradient, V is the grain growth rate, a is the alloy material constant, ΔT_0 is the superheating value, N_0 is the nucleation density, and n is the alloy material related constant.

Central segregation is the result of the selective crystallization of solute elements in the solidification front during the continuous casting process. The solubility of solute elements in the solid phase is lower than that of the liquid phase. The diffusion of solute elements during solidification and selective crystallization is directly related to the cooling rate during the continuous casting process. The cooling rate in solidification provides kinetic and thermodynamic conditions for segregation behavior. The temperature distribution in the center and quarter positions of the billet was calculated by the Procast software according to the continuous casting parameters in Table 1. The temperature profile in the center and quarter positions of the billet is shown in Figure 8. The cooling rate of Billet 1 is lower than that of Billet 2. That is, when the cooling rate was low, the solute element precipitated in the solidification front and dissolved into the liquid phase, eventually forming the central segregation region in the center of the continuously cast billet; on the contrary, a small amount or no solute element diffused at the solidification front from the solid to liquid phases at a high cooling rate, and the central segregation of the continuously cast billet was suppressed. In addition, the segregation spots were formed due to the segregation of elements that were transported to the small cells or the free spaces in the mushy zone [8]. The center of the equiaxed crystal zone may have provided an opportunity to increase the volume of the small cells, which presented the black spots in the equiaxed crystal zone. Therefore, in the continuously cast billet of the low alloy steel, the formation of central segregation was suppressed by the decrease of the equiaxed crystal zone.

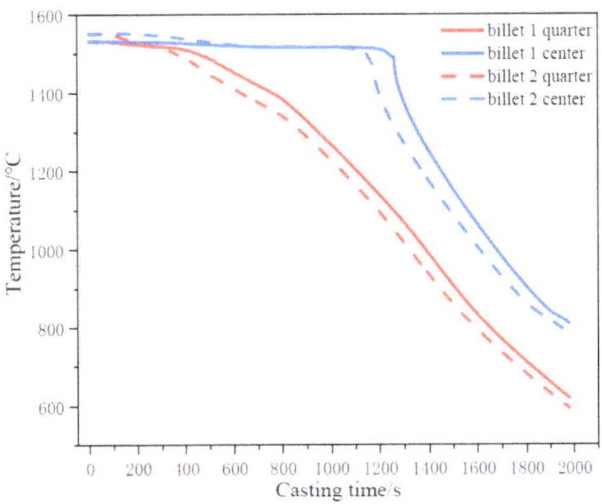

Figure 8. The temperature of the quarter and center positions of the billet in the solidification process.

5. Conclusions

A comparative study of U-section steels rolled from continuously cast billets with different internal qualities showed that solidification structure had a significant effect on the degree of central segregation in billets, and the segregation affected these billets' microstructures, distributions of solute elements and mechanical properties. The conclusions are as follows:

(1) The central crystal zone of a continuously cast billet presented segregated spots that were caused by the concentration of Mn.

(2) With the application of high superheating and secondary cooling temperatures without EMS (electro magnetic stirring), the fraction of the equiaxed crystal zone effectively decreased. The columnar crystal structure was more effective to control central segregation in the continuously casting billet.

(3) The microstructure of the steel that was rolled from the billet with severe segregation consisted of ferrite and degenerate pearlite, while the microstructure of the steel that was rolled from the optimized billet was dominated by ferrite and lamellar pearlite. The fraction of pearlite in Steel 1 (41%) was obviously higher than that of Steel 2 (14%).

(4) Controlling the degree of central segregation could achieve an excellent low temperature toughness, as the DBTT decreased from −10 to −40 °C, and the elongation increased by ~10%.

Author Contributions: C.S., R.D.K.M., J.W. and X.W. designed and supervised the research; F.G. performed the experiments, analyzed the data, and wrote the paper. All authors have read and agreed to the published version of the manuscript.

Funding: This work is financially supported by National Key Research and Development Project of China (2017YFB0304900 and 2017YFB0304700), Anshan Iron and Steel Group Company, State Key Laboratory of Metal Materials for Marine Equipment and Applications.

Conflicts of Interest: The authors declare no conflict of interest.

References

1. Miao, C.L.; Shang, C.J.; Zhang, G.D. Recrystallization and strain accumulation behaviors of high Nb-bearing line pipe steel in plate and strip rolling. *Mater. Sci. Eng. A* **2010**, *527*, 4985–4992. [CrossRef]

2. Nie, Y.; Shang, C.; You, Y. 960 MPa grade high performance weldable structural steel plate processed by using TMCP. *J. Iron Steel Res. Int.* **2010**, *17*, 63–66. [CrossRef]

3. Nie, Y.; Shang, C.; Song, X. Properties and homogeneity of 550-MPa grade TMCP steel for ship hull. *Int. J. Min. Metall. Mater.* **2010**, *17*, 179–184. [CrossRef]

4. Chen, S.; An, Y.G.; Lahaije, C. Toughness improvement in hot rolled HSLA steel plates through asymmetric rolling. *J. Mater. Sci. Eng. A* **2015**, *625*, 374–379. [CrossRef]

5. Chaudhuri, A.; Raghupathy, Y.; Srinivasan, D.; Suwas, S.; Srivastava, C. Microstructural evolution of cold-sprayed Inconel 625 superalloy coatings on low alloy steel substrate. *J. Acta Mater.* **2017**, *129*, 11–25. [CrossRef]

6. Ronevich, J.A.; Somerday, B.P.; San Marchi, C.W. Effects of microstructure banding on hydrogen assisted fatigue crack growth in X65 pipeline steels. *Int. J. Fatigue* **2016**, *82*, 497–504. [CrossRef]

7. Krauss, G. Solidification, segregation, and banding in carbon and alloy steels. *Metall. Mater. Trans. B* **2003**, *34*, 781–792. [CrossRef]

8. Shibanuma, K.; Nemoto, Y.; Hiraide, T.; Suzuki, K.; Sadamatsu, S.; Adachi, Y.; Aihara, S. A strategy to predict the fracture toughness of steels with a banded ferrite–pearlite structure based on the micromechanics of brittle fracture initiation. *Acta Mater.* **2018**, *144*, 386–399. [CrossRef]

9. Bo, L.; Zhonghua, Z.; Huasong, L. Characteristics and Evolution of the Spot Segregations and Banded Defects in High Strength Corrosion Resistant Tube Steel. *Acta Metall. Sin.* **2019**, *55*, 762–772.

10. Guo, F.; Wang, X.; Liu, W. The Influence of Centerline Segregation on the Mechanical Performance and Microstructure of X70 Pipeline Steel. *Steel Res. Int.* **2018**, *89*, 1800407. [CrossRef]

11. Guo, F.; Liu, W.; Wang, X. Controlling Variability in Mechanical Properties of Plates by Reducing Centerline Segregation to Meet Strain-Based Design of Pipeline Steel. *Metals* **2019**, *9*, 749. [CrossRef]

12. Sahoo, P.P.; Kumar, A.; Halder, J. Optimisation of electromagnetic stirring in steel billet caster by using image processing technique for improvement in billet quality. *ISIJ Int.* **2009**, *49*, 521–528. [CrossRef]

13. Nakata, H.; Inoue, T.; Mori, H.; Lshiguro, S.; Nakata, H.; Murakami, T.; Kominami, T. Improvement of billet surface quality by ultra-high-frequency electromagnetic casting. *J. ISIJ Int.* **2002**, *42*, 264–272. [CrossRef]

14. An, H.; Bao, Y.P.; Wang, M.; Zhao, L.H. Reducing macro segregation of high carbon steel in continuous casting bloom by final electromagnetic stirring and mechanical soft reduction integrated process. *Metall. Res. Technol.* **2017**, *114*, 405. [CrossRef]

15. Choudhary, S.K.; Ganguly, S. Morphology and segregation in continuously cast high carbon steel billets. *ISIJ Int.* **2007**, *47*, 1759–1766. [CrossRef]

16. Choudhary, S.K.; Ganguly, S.; Sengupta, A. Solidification morphology and segregation in continuously cast steel slab. *J. Mater. Process. Tech.* **2017**, *243*, 312–321. [CrossRef]

17. Xiu, Z.G.; Wang, X.H. Effect of superheat on the center segregation in continuous casting slab of X65 steel. *Steel Mak.* **2014**, *30*, 36–44.

18. Mukherjee, M.; Chintha, A.R.; Kundu, S. Development of stretch flangeable ferrite–bainite grades through thin slab casting and rolling. *Mater. Sci. Technol.* **2016**, *32*, 348–355. [CrossRef]

19. Zhang, H.; Cheng, X.; Bai, B. The tempering behavior of granular structure in a Mn-series low carbon steel. *Mater. Sci. Eng. A* **2011**, *528*, 920–924. [CrossRef]

20. Rui, F. Research on Internal Quality and Its Influencing Factors of Ferrite Pearlite Steel Plate. Ph.D. Thesis, Shandong University, Jinan, Shandong, China, 30 October 2013.

21. Ji, Y.; Lan, P.; Geng, H. Behavior of spot segregation in continuously cast blooms and the resulting segregated band in oil pipe steels. *J. Steel Res. Int.* **2018**, *89*, 1700331. [CrossRef]
22. Aaronson, H.I.; Enomoto, M.; Lee, J.K. *Mechanisms of Diffusional Phase Transformations in Metals and Alloys*; CRC Press: Boca Raton, FL, USA, 2016; pp. 576–577. ISBN 978-1-4398-8253-5.
23. Zhang, J.; Chen, D.; Wang, S.; Long, M. Compensation control model of superheat and cooling water temperature for secondary cooling of continuous casting. *Steel Res. Int.* **2011**, *82*, 213–221. [CrossRef]
24. Feng, J.; Cheng, W.Q. Effect of high secondary cooling intensity on solidification structure and center carbon segregation of high carbon steel billet. *Spec. Steel* **2006**, *27*, 42–44.

 *metals*

Article

New Insights into the Microstructural Changes During the Processing of Dual-Phase Steels from Multiresolution Spherical Indentation Stress–Strain Protocols

Ali Khosravani [1], Charles M. Caliendo [2] and Surya R. Kalidindi [1,2,*] 

[1] The George W. Woodruff School of Mechanical Engineering, Georgia Institute of Technology, Atlanta, GA 30332, USA; alikhosravani@gatech.edu

[2] School of Materials Science and Engineering, Georgia Institute of Technology, Atlanta, GA 30332, USA; ccaliendo@gatech.edu

* Correspondence: surya.kalidindi@me.gatech.edu; Tel.: +1-404-385-2886

Received: 1 December 2019; Accepted: 19 December 2019; Published: 21 December 2019

Abstract: In this study, recently established multiresolution spherical indentation stress–strain protocols have been employed to derive new insights into the microstructural changes that occur during the processing of dual-phase (DP) steels. This is accomplished by utilizing indenter tips of different radii such that the mechanical responses can be evaluated both at the macroscale (reflecting the bulk properties of the sample) and at the microscale (reflecting the properties of the constituent phases). More specifically, nine different thermo-mechanical processing conditions involving different combinations of intercritical annealing temperatures and bake hardening after different amounts of cold work were studied. In addition to demonstrating the tremendous benefits of the indentation protocols for evaluating the variations within each sample and between the samples at different material length scales in a high throughput manner, the measurements provided several new insights into the microstructural changes occurring in the alloys during their processing. In particular, the indentation measurements indicated that the strength of the martensite phase reduces by about 37% when quenched from 810 °C compared to being quenched from 750 °C, while the strength of the ferrite phase remains about the same. In addition, during the 10% thickness reduction and bake hardening steps, the strength of the martensite phase shows a small decrease due to tempering, while the strength of the ferrite increases by about 50% by static aging.

Keywords: dual-phase steels; spherical indentation; multi length-scale mechanical testing

1. Introduction

Dual-phase (DP) steels with a combination of high tensile strength, high work-hardening rate, and good ductility are being evaluated for the lightweighting of critical structural components in automobiles [1–5]. The desired combinations of properties are achieved in DP steels through the use of multiple thermo-mechanical processing steps. These processing steps typically include intercritical annealing at 730–830 °C for a few minutes up to an hour, quenching at different cooling rates, cold working to different deformation levels, and aging at 100–250 °C up to few hours. The last two steps are typically referred to as bake hardening [6–23]. During the intercritical annealing treatment, the material is heated up to a temperature where the austenite and the ferrite phases are stable. During the subsequent quenching to the room temperature, the austenite transforms to the much harder martensite phase [17,23–29], which essentially controls the properties of the DP steel. It is evident from the Fe–C phase diagram (see Figure 1) that the different intercritical annealing temperatures will result in different volume fractions of the martensite phase [24]. Furthermore, the amount of carbon content in the martensite (as a solid solution) varies significantly with the chosen intercritical annealing temperature (for example, it can change from 0.17 to 0.77 wt.% when the intercritical annealing temperature is reduced from 830 and 730 °C), while the corresponding change in the ferrite phase is insignificant (only about 0.01–0.02 wt.%). Furthermore, the relevant section of the phase diagram (see Figure 1) suggests that intercritical annealing at lower temperatures results in a smaller volume fraction of martensite but with a higher carbon content. As already mentioned, another important component in the processing of DP steels is the bake hardening (BH) [6–23] step, which includes cold working followed by aging heat treatment. This step is known to impart DP steels with a characteristic property known as continuous yielding, which is generally attributed to the production and pinning of dislocations in the ferrite component, especially in the vicinity of ferrite/martensite interfaces [1,7–14,16,18,19,22,28,30–34].

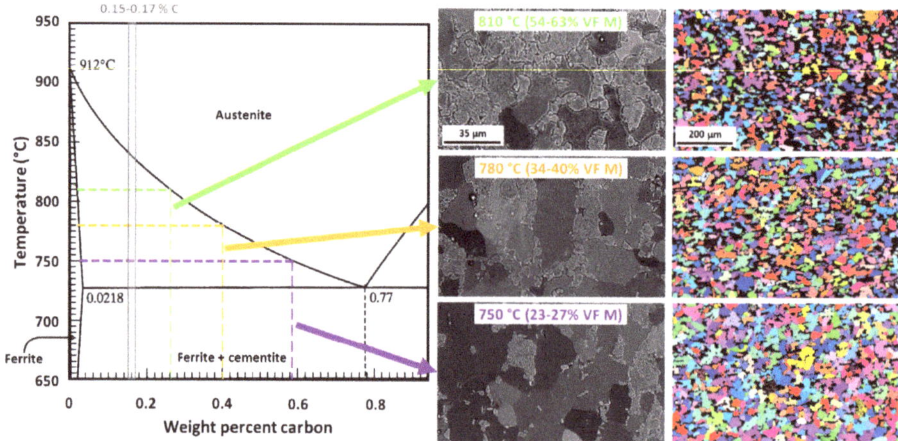

Figure 1. Small section of the binary Fe–C phase diagram showing the different compositions and microstructures produced from three different intercritical annealing treatments on the initial low carbon steel with 0.15–0.17 wt.% carbon content. The intercritical annealing temperatures selected for this study were 750 °C, 780 °C, and 810 °C. The microstructures are shown as back-scattered electrons (BSE) and electron-backscattered diffraction (EBSD) maps.

The discussion above points out the difficulties encountered in the optimization of the processing of DP steels to meet the desired combination of properties [1,35,36]. This is mainly because of the need to explore a very large and complex process space (each element of this space should specify the entire process history, including the complete sequence of substeps described earlier). One of the central bottlenecks comes from the lack of reliable information on the changes in the properties of the constituent phases (i.e., martensite and ferrite) as a function of the process parameters. A number of prior studies [37–52] have employed indentation techniques for this task. However, most of these studies have employed sharp indenters and reported large variances in the measured values. A summary of such measurements in DP steels is presented in Table 1, which shows a range of 2–7 GPa for the hardness of the ferrite phase and a range of 3–13 GPa for the hardness of the martensite phase. The large variations in the reported hardness data have hindered attempts aimed at extracting quantitative physical insights that could guide the rational design of DP process histories to achieve desirable combinations of bulk properties. In recent work [53–61], it has been demonstrated that indentation yield strength is a much more reproducible and reliable measure of the intrinsic plastic strength of the microscale constituents in a heterogeneous material, and could be estimated from the recently established spherical indentation stress–strain protocols [62,63]. Using this analysis method, it has been shown [57] that the indentation yield strength is very sensitive to the carbon content in the lath martensite; increasing C content from 0.13 to 0.30 wt.% improves the indentation yield strength by 42–48% and the indentation work hardening by 27–47%. In this work, we extend and employ these techniques to provide quantitative insights into the changes in the yield strengths of the martensite and ferrite phases in DP steels, especially during the intercritical annealing and the bake hardening steps.

Table 1. Summary of hardness measurements in martensite and ferrite from prior literature.

Alloy Grade/Composition	Indenter Type	Max Depth (nm)	Max Load (mN)	Martensite Hardness (GPa)	Ferrite Hardness (GPa)	Reference
DP780	Berkovich	-	5	6.2 ± 0.11	3.7 ± 0.03	37
DP1300	Berkovich	-	5	4.5-10	2-5	38
0.1C5Mn3Al	Berkovich	-	5	4.7 ± 0.4	4.1 ± 0.3	39
0.1C5Mn3Al (60% TR)	Berkovich	-	5	5.9 ± 0.7	5.0 ± 0.5	
DP980	Berkovich	-	2.5	3-10	1.5–5.5	40
DP980 (7% strain)	Berkovich	-	2.5	3-13	2–7	
(0.04, 0.07, 0.1) C1.2Mn0.15Si	spherical (R = 2.8, 5.7 μm)	-	15	-	1.8–2, 1.3–1.6	41
DP980	Berkovich	40	-	4.5–9	3–4.75	42
API-X100	Spherical (R = 0.5, 3 μm)	-	15, 30	-	3.4–4.1, 1.9–2.4	43
0.16C1.5Mn1Si	cube-corner	-	1	6.3–7.9	2.8	44
0.19C1.6Mn0.2Si	Berkovich	-	-	3–10.8	2.8-6.8	45
0.18C0.75Mn0.4Si	Berkovich	-	10	7.6	2.2 ± 0.2	46
0.38C0.67Mn0.2Si	Berkovich	-	10	4.9–7.3	-	47
DP980	Berkovich	50	-	6–11	4–5.5	48
DP980	Berkovich	-	0.8	8.4 ± 0.9	4.1 ± 0.3	49
0.08C1.74Mn0.75Si	Berkovich	-	0.05	4.5–5.5	3–3.5	50
DP980	Berkovich	-	3	6.3–8.1	2.5–3.5	51
DP590	Berkovich	50	-	3.5–4.1	1.5–1.8	52

2. Materials and Method

2.1. Sample Preparation

A four mm thick strip of low carbon steel with a chemical composition (in wt.%) of 0.16C, 1.4Mn, 0.04P, and 0.04S was used to produce the DP steel samples needed for this study. Small coupons of the low carbon steel with dimensions of 10 mm × 20 mm × 4 mm were cut, and heat-treated at 450 °C for 2 h to obtain a starting annealed microstructure. Three intercritical annealing temperatures of 750 °C, 780 °C, and 810 °C were selected to produce different volume fractions of the martensite phase after quenching. The heat treatment was carried out in a molten salt bath, LIQUID HEAT 168 from Houghton International (Houghton International Inc., Norristown, PA, USA), to ensure quick heating and uniform temperature inside the small coupons. After holding three samples for 4 min at each selected intercritical annealing temperature, they were quenched in an oil bath to room temperature. Of each set of three samples thus produced, one sample was retained without bake hardening. The other two samples from each set were subjected to thickness reductions of 5% and 10% by cold rolling, respectively, and heat-treated at 170 °C for 20 min and quenched in water. As a result of the protocols described above, a total of nine samples with nine different processing conditions were produced. The sample labeling was designed to reflect the processing history in the form "intercritical annealing temperature-thickness reduction percentage-bake hardening temperature". As an example, sample 810-10-170 indicates quenching after an intercritical annealing temperature of 810 °C followed by 10% thickness reduction and bake hardening at 170 °C for 20 min. Likewise, sample 810-00-000 indicates that the sample was subjected to only intercritical annealing at 810 °C followed by quenching.

Samples were prepared for microscopy and indentation using standard metallography procedures. This included grinding with silicon carbide papers down to a grade of 4000 and polishing sequentially with suspensions of 3 μm and 1 μm diamond particles. The final step of polishing included vibro-polishing (Struers Inc., Cleveland, OH, USA) for 24 h using colloidal silica suspension. After polishing, SEM (scanning electron microscopy) images and EBSD (electron backscatter diffraction) (EDAX Inc., Mahwah, NJ, USA) maps were obtained from all samples using a TESCAN MIRA3 (TESCAN USA Inc., Warrendale, PA, USA) scanning electron microscope with a field emission gun set at 20 kV. High contrast in electron channeling contrast image (ECCI) at sub-micron resolution was achieved with a working distance of 5–6 mm and a voltage of 30 kV.

2.2. Spherical Nano-Indentation Stress–Strain Protocols

After imaging, spherical nanoindentation tests were carried out in an Agilent G200 Nanoindenter (KLA Inc., Milpitas, CA, USA). As these measurements were aimed at obtaining responses from the ferrite and martensite regions in the sample, smaller indenter tips of radii 1 μm and 16 μm were employed in these tests. At least 20–30 indentation measurements were conducted for each phase (i.e., ferrite and martensite) on each sample. A constant strain rate of 0.05 s^{-1} was used with a maximum depth of 200 nm and 350 nm for the indenter tips with radii of 1 μm and 16 μm, respectively. The Agilent G200 Nanoindenter used in this study had an XP head and CSM (continuous stiffness measurement) module. The CSM superimposes small sinusoidal load/unload cycles on the monotonic loading history with a frequency of 45 Hz and an amplitude of 2 nm. The CSM capability allows an accurate estimation of contact radius used in calculating indentation stress and indentation strain [62,63]. The analysis protocols used in the nanoindentation tests are briefly presented next.

Let P, h_e, E_{eff}, and R_{eff} denote the indentation load, the elastic indentation depth, the effective modulus of the indenter-sample system, and the effective radius of the indenter-sample system, respectively (see Figure 2). These variables can be related to each other using Hertz's theory [64] for elastic contact between two isotropic bodies as

$$P = \frac{4}{3} E_{eff} \sqrt{R_{eff} h_e^3} \, , \tag{1}$$

$$\frac{1}{E_{eff}} = \frac{1 - v_i^2}{E_i} + \frac{1 - v_s^2}{E_s}, \tag{2}$$

$$\frac{1}{R_{eff}} = \frac{1}{R_i} + \frac{1}{R_s}, \tag{3}$$

where E_i , v_i and E_s , v_s are Young's modulus and Poisson ratio for the indenter and the sample, respectively. For the initial elastic loading when the surface of the sample is still flat, the effective radius, R_{eff}, is equal to the indenter radius, R_i. Therefore, it would be possible to extract E_{eff} from the measured load–displacement using standard regression techniques. One of the central challenges in this analysis comes from the need for a highly accurate estimation of the effective point of the initial contact (i.e., zero-point correction) [62,63,65–71]. This zero-point correction helps in dealing with many of the unavoidable issues encountered at initial contact, including imperfections in indenter shape and non-ideal surface conditions (e.g., oxide layer, surface roughness). In the protocols used in this work, the initial contact point was determined by finding load and displacement corrections (P^* and h^*) from the following relation derived from Equation (1) [62]:

In Equation (4), S is the measured elastic unloading stiffness obtained using the CSM capability mentioned earlier, and $\widetilde{P}$ and $\widetilde{h}$ are the raw measurements of load and displacement, respectively. To estimate P^* and h^*, Equation (4) is re-cast as

$$S = \frac{3P}{2h_e} = \frac{3}{2}\frac{(\widetilde{P} - P^*)}{(\widetilde{h} - h^*)}. \tag{4}$$

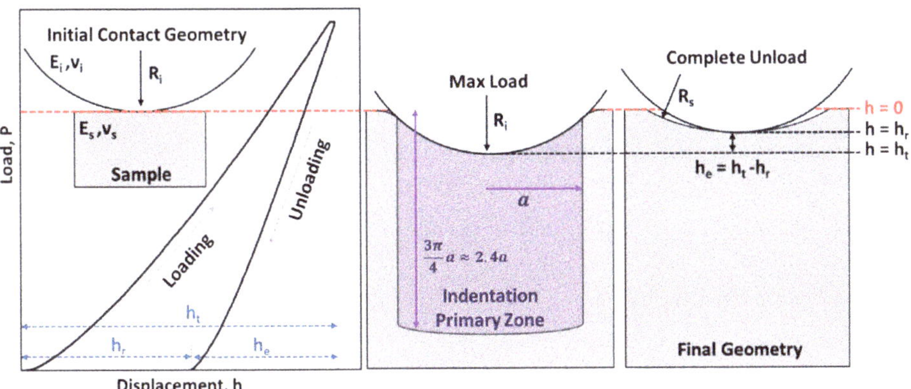

Figure 2. Schematic of a typical measured spherical indentation load–displacement curve and their corresponding initial and final contact geometries. The indentation primary zone within which the majority of deformation occurs is highlighted as a purple cylinder with a radius and height of a and 2.4 a, respectively.

The form of Equation (5) allows an accurate estimation of the zero-point corrections (i.e., values of P^* and h^*) by performing linear regression between the measured values of $\widetilde{P} - \frac{2}{3}S\widetilde{h}_e$ and S;

$$\widetilde{P} - \frac{2}{3}S\widetilde{h}_e = -\frac{2}{3}h^*S + P^*. \tag{5}$$

Figure 3a–c illustrates the main steps involved in the analyses of the nanoindentation measurement reports in this study. An example measured raw load–displacement data as shown in Figure 3a. The application of the zero-point correction described in Equation (5) is illustrated in Figure 3b, where the expected linear portion based on Hertz's theory is shown in yellow (the corresponding segment in the load–displacement curve is shown in the inset in Figure 3a). After the zero-point correction, E_{eff} can be estimated by performing a linear regression between P and $h^{3/2}$ in the initial elastic portion of the measured load–displacement curve (highlighted in yellow in Figure 3a) [62]. The estimated value of E_{eff} is then used to estimate the continuously evolving contact radius, a, using the following relationship derived from Hertz's theory [64]:

$$a = \frac{S}{2E_{eff}}. \tag{6}$$

An important aspect of Equation (6) is that it is applicable at any point in the complex elastic-plastic loading-unloading cycles applied to the sample. By Equation (6), the effective indentation modulus measured from the initial elastic loading segment is assumed to remain constant throughout the elastic–plastic loading applied to the sample. Estimation of the continuously evolving contact radius allows the estimation of indentation stress and indentation strain [62] defined as follows:

$$\sigma_{ind} = \frac{P}{\pi a^2}, \tag{7}$$

$$\varepsilon_{ind} = \frac{4}{3\pi}\frac{h_t}{a}. \tag{8}$$

An example indentation stress–strain curve extracted using the above protocols presented in Figure 3c. The spherical nanoindentation stress–strain protocols described above have been validated extensively in both experiments [53–61,72–95] and numerical simulations (performed using finite element models) [96–99]. As a result of these prior validations, we are now fairly confident in obtaining highly reproducible indentation stress–strain curves on a broad variety of material samples.

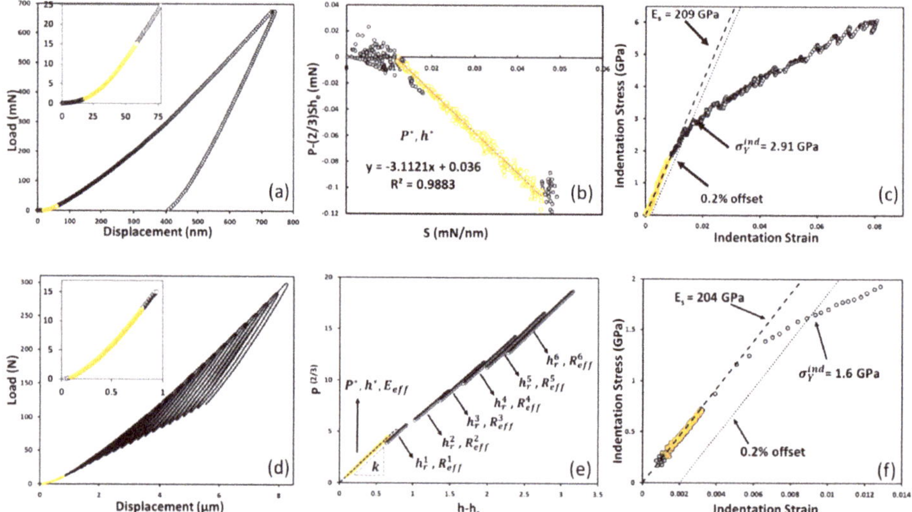

Figure 3. (**a**) Typical load–displacement curve from the nanoindentation test and the initial elastic segment (highlighted in yellow) identified in (b). (**b**) The identification of the effective zero-point by linear regression analyses of the linear portion of the curve between $P - \frac{2}{3}Sh_e$ and S. This linear regression allows the estimation of the zero-point corrections, P^* and h^*. (**c**) Indentation stress–strain curve extracted from the corrected nanoindentation data. (**d**) Typical load–displacement curve from the microindentation test in absence of CSM (continuous stiffness measurement) measurements, using a multitude of unloading segments. The initial elastic segment is highlighted in yellow. (**e**) The identification of the effective zero-point by linear regression analyses of the straight line between $P^{2/3}$ and $h - h_r$. This analysis allows the estimation of P^* and h^*. (**f**) Indentation stress–strain curve extracted from the corrected data for the microindentation test.

2.3. Spherical Micro-Indentation Stress–Strain Protocols

As already noted earlier, our interest here also includes bulk plastic properties of the various samples produced for the study. The bulk mechanical response of each sample was measured using a customized Zwick-Roell Z2.5 hardness (microindentation) tester (ZwickRoell Group, Ulm, Germany). For these tests, a spherical indenter with a 6350 μm radius indenter tip was used. The indenter is made of tungsten carbide to ensure high rigidity of the indenter. A total of 10–15 tests were conducted at randomly selected locations on each sample. A constant crosshead speed of 0.1 mm/min was used in all microindentation tests reported in this study.

As the Zwick-Roell Z2.5 tester does not have CSM capability, one needs a different strategy to extract indentation stress–strain curves from the spherical microindentation measurements. In recent work, it was shown that one can estimate the contact radius reliably from superimposed unloading segments (corresponding to about 30–50% of the peak force) when the CSM option is not available [63]. Figure 3d shows an example measurement of the load–displacement data in multiple loading-unloading cycles in the spherical microindentation protocols employed in this work. In these protocols, each unloading segment is assumed to be purely elastic and is analyzed using Hertz's theory [64]:

$$\left(\widetilde{h}_e - h^*\right) = k\left(\widetilde{P} - P^*\right)^{\frac{2}{3}}, \tag{9}$$

$$k = \left[\frac{3}{4}\frac{1}{E_{eff}}\frac{1}{\sqrt{R_{eff}}}\right], \tag{10}$$

where P^* and h^* represent once again the zero-point corrections. To estimate the zero-point load and displacement corrections, P^* and h^*, a least-squares regression was applied to the initial measured load–displacement data in the elastic regime in Equation (9). As the indentation starts on a flat surface, R_{eff} in the initial elastic regime (highlighted in yellow in Figure 3d) is equal to the indenter radius, R_i. Thus, one can estimate E_{eff} from Equation (10), which is assumed to remain constant even after the specimen has undergone plastic deformation.

Plastic deformation under the indenter leads to a continuous evolution of R_{eff}. Each subsequent unloading (after the estimation of E_{eff} from the initial elastic loading segment) is analyzed using Hertz's theory, but this time with a focus on estimating the evolving values of R_{eff} and the values of the indentation contact radius, a. This is accomplished by regressing each unloading curve in Figure 3d to Hertz's theory expressed as

$$h - h_r = k(P)^{\frac{2}{3}}, \tag{11}$$

where h_r denotes residual displacement (after complete unloading; see Figure 2). This regression analysis allows the determination of the values of h_r and R_{eff} (see Equation (10)) corresponding to each unloading segment obtained in the test. Figure 3e illustrates the above protocol for six selected unloading segments (out of a much larger number of unloading segments depicted in Figure 3d). The values of the contact radius a are then estimated using Hertz's theory as

$$a = \sqrt{R_{eff}(h_{s,max} - h_r)}, \tag{12}$$

where $h_{s,max}$ is the indentation displacement in the sample at the peak of each unload. Once the contact radius is estimated, the values of indentation stress and indentation strain can be computed using Equations (7) and (8). It should be noted that each unloading segment in these protocols results in the estimation of one point on the microindentation stress–strain curve. Consequently, multiple load–unloading cycles are needed to produce a reasonable indentation stress–strain curve. Twenty unloading segments were incorporated in each microindentation test reported in this study. An example microindentation stress–strain curve extracted in our study is shown in Figure 3f.

3. Results and Discussion

Example SEM and EBSD images obtained from the DP steel samples produced in this work are presented in Figure 1. SEM images reveal the presence of martensite islands in a matrix of ferrite grains. For the selected low carbon alloy with carbon content in the range of 0.15–0.17 wt.%, samples quenched from 750 °C, 780 °C, and 810 °C were expected to produce 23–27 vol.%, 35–40 vol.%, and 56–64 vol.% of martensite, respectively. These predicted values were confirmed from several large EBSD scans obtained at different locations on each sample (presented in Table 2). In the EBSD maps, multicolored regions are ferrite grains whose lattice orientations have been mapped out with a 1 μm spatial resolution and martensite regions are colored black based on low IQ (image quality) compared to the ferrite regions. As seen from the phase diagram in Figure 1, quenching from a higher intercritical annealing temperature will result in a higher volume fraction of lower carbon content martensite in the microstructure. Some prior studies have shown the significant effect of carbon content on the hardness of the martensite by systematically changing C content in the tested materials [100,101]. As already mentioned, the protocols used in prior literature using sharp indenters result in high variances between different studies (see Table 1). In the present work, we employed the spherical indentation stress–strain protocols that have been shown to produce reliable and highly reproducible values for indentation yield strengths at multiple material length scales.

Table 2. Summary of the microindentation measurements obtained in this study. The martensite volume fractions were estimated from large EBSD scans collected on each sample.

Sample Code	Martensite Volume Fraction (%)	Average Elastic Modulus (GPa)	Average Indentation Yield Strength (MPa)	Contact Area Diameter at Yield Point (µm)
750-00-000	25.4	174.5 ± 25.3	899.5 ± 62.9	284.6 ± 14.0
750-05-170	27.5	193.4 ± 23.3	950.4 ± 29.7	333.4 ± 26.4
750-10-170	23.5	200.9 ± 22.3	1100.9 ± 130.3	352.4 ± 23.0
780-00-000	34.8	216.8 ± 23.3	1097.7 ± 41.6	331.4 ± 29.6
780-05-170	35.3	209.6 ± 15.4	1299.4 ± 68.7	356.2 ± 35.4
780-10-170	38.9	201.9 ± 16.7	1336.3 ± 69.5	362.6 ± 28.4
810-00-000	56.3	188.5 ± 14.7	1168.6 ± 186.8	306.6 ± 21.8
810-05-170	59.6	219.2 ± 14.5	1340.4 ± 130.9	338.6 ± 20.4
810-10-170	60.4	207.6 ± 9.7	1506.5 ± 132.3	361.0 ± 30.2

3.1. Microindentation and Results

About 10–15 microindentation tests were conducted at randomly selected locations on each of the nine differently processed samples. From each microindentation test, values of Young's modulus and indentation yield strength (defined using a 0.2% plastic strain offset as shown in Figure 3f) were extracted; these are summarized in Figure 4 and Table 2. It is seen from Table 2 that increasing the intercritical annealing temperature increased the bulk indentation yield strength of the sample (e.g., the indentation yield strength increased by ~30%, from 899.5 ± 62.9 MPa for sample 750-00-000 to 1168.6 ± 186.8 MPa for sample 810-00-000). In addition, the same trend can be detected when cold work and bake hardening processes are applied. The indentation yield strength increased by 17–29%, when 10% thickness reduction and bake hardening were applied, when compared to the quenched sample without bake hardening. As a specific example, the indentation yield strength increased from 1168.6 ± 186.8 MPa for sample 810-00-000 to 1506.5 ± 109.5 MPa for sample 810-10-170.

In the indentation tests, the contact diameter $(2a)$ provides an estimate of the length scale of the material under the indenter that has been subjected to significant plastic deformation (see Figure 2). The estimated contact diameters at the indentation yield estimated by Hertz's theory (Equation (12)) are summarized in Table 2. As an example, the estimated contact diameter at the yield point for one of the microindentations conducted in this study is shown as a blue dashed circle on a sample micrograph in Figure 4d. Furthermore, the primary deformation zone under the indenter is estimated to extend to about $2.4a$ underneath the indenter [62]. It is therefore estimated that the indentation zone at yield is approximately 300 µm in diameter and 400 µm in depth for the samples studied here. The primary indentation zone at yield in the microindentation tests reported in this study contained ~300–700 grains (based on an approximate grain size of 30 µm). Therefore, the microindentation measurements presented here can be assumed to reflect the bulk (macroscale) properties of the samples.

The microindentation measurements presented here compare well with the values reported in the literature for samples with similar compositions and processing histories. The reported values of Young's moduli for similar DP steels are in the range of 170–210 GPa [102–104]. Furthermore, tensile yield strengths have been reported in the range of 345–482 MPa [24,105–107], 413–622 MPa [24,106,108,109], and 520–670 MPa [110–113] for the intercritical annealing temperatures of 750 °C, 780 °C, and 810 °C, respectively. Considering a scaling factor of 2 (see Patel et al. [97]) for converting indentation yield strength to the tensile yield strength, the values reported in Table 2 are in good agreement with the values reported in the literature.

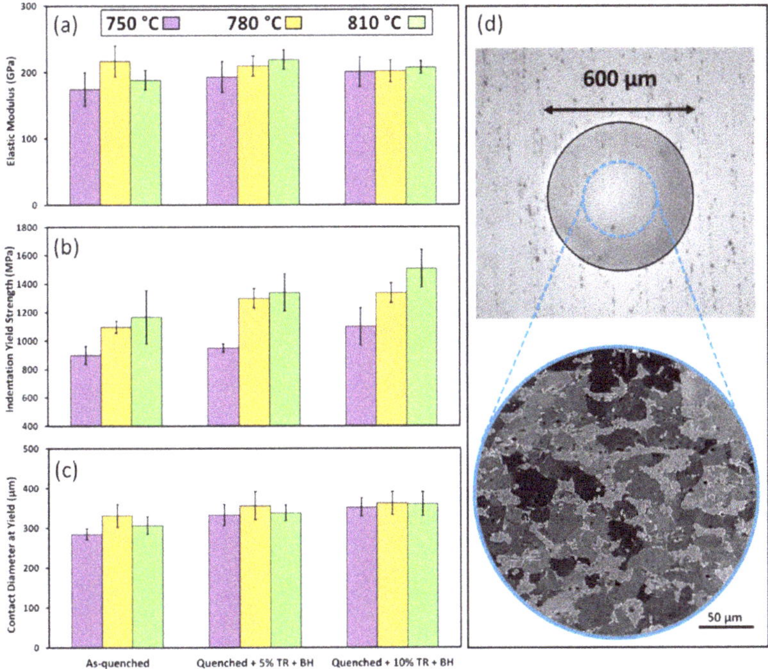

Figure 4. (**a–c**) Histograms summarizing the extracted values of Young's moduli, indentation yield strengths, and the contact diameters at yield from microindentation tests on samples produced by different processing histories. The different processing histories included intercritical annealing at 750 °C, 780 °C, and 810 °C followed by quenching. The samples were then cold-worked (5% and 10%) and bake hardened (BH). (**d**) An optical micrograph and BSE image showing the indentation size at yield as a dashed blue circle.

The results from the microindentation measurements raise two important questions. First, when the martensite volume fraction doubled (from ~25% to ~58%), the indentation yield strength increased only by ~30%. Given the expected high strength of the martensite compared to the ferrite, one should expect a bigger increase in the macroscale yield strength. Second, the bake hardening produced an increase in the strength by 1729%. Given that the martensite volume fraction remains the same during the bake hardening process, and a possible reduction of the martensite strength during the bake hardening step, the physical processes responsible for this significant increase in the strength are not clear. These questions can be addressed by measuring the mechanical responses at the length scale of the constituent phases using the nanoindentation protocols described earlier.

3.2. Nanoindentation and Results

As discussed earlier, DP steel microstructures exhibit rich heterogeneity over a hierarchy of material length scales (see Figure 5). At the mesoscale, these include different thermodynamic phases with different crystal structures (Figure 5a,b) and local differences in chemical compositions (especially in carbon and alloying elements). At lower length scales, there exist other heterogeneities within each phase. In martensite, one encounters grains of different crystal orientations, low and high angle grain/phase boundaries, and possibly small carbides (Figure 5c,d). In ferrite, one observes heavily deformed ferrite regions, especially at the vicinity of the ferrite/martensite interface (Figure 5e). Studies of the mechanical responses of DP steels at different spatial resolutions are critically needed to obtain new physical insights into the overall response of the alloy. Given the heterogeneity involved, it is also important to perform multiple measurements at randomly selected locations to obtain statistically meaningful data.

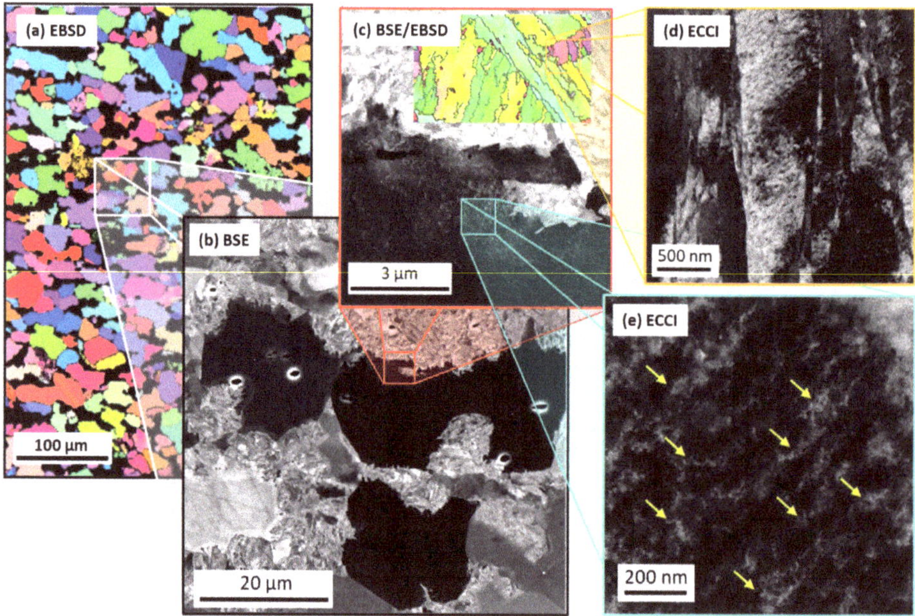

Figure 5. A sampling of microstructures of the sample 810-00-000 using different techniques showing the heterogeneous and hierarchical microstructure of DP steel at different length scale. (**a**) An EBSD map showing several ferrite grains (in different colors) and martensite regions (in black color). (**b**) A BSE-SEM image of the microstructure at slightly higher magnification shows details of the martensite regions that contain blocks of martensite reflected with high contrast inside each island. (**c**) The magnified image shows more detail at the martensite/ferrite interface. The colored image is a high-resolution EBSD map on the martensite island showing the orientations of martensite blocks. (**d**,**e**) High-resolution maps of the martensite and ferrite regions obtained using the electron channeling contrast imaging (ECCI) technique. The ECCI micrograph on the martensite shows a highly dislocated structure. The ECCI micrograph on the ferrite grain were collected at the vicinity of the martensite/ferrite interface, where a higher dislocation network (highlighted by yellow arrows) was observed compared to the regions in the center of the ferrite grains.

In this study, we investigated the local mechanical responses in the DP steel samples using two different spherical indenter tips of radii 1 μm and 16 μm, respectively. In prior work on fully martensitic Fe-Ni alloys [57], the contact diameters in indentations conducted with a 1 μm radius tip were estimated to be ~100–150 nm at yield. The primary deformation zone in these indentations is likely to have included only a handful of martensite laths, as their thickness was reported to be in the range of 50–200 nm [114,115]. Furthermore, it was reported [57] that the measured average indentation yield strengths with both the 1 μm and the 16 μm tip radii were in good agreement with each other, while the standard deviations were smaller for the measurements with the 16 μm tip. This is reasonable because one would expect a larger number of martensite laths in the primary indentation zone with the larger indenter tip. For the DP steel samples studied here, it was observed that the martensite indentation yield strengths measured with the 16 μm tip were systematically lower compared to the measurements with the 1 μm tip. For example, for the case of sample 810-10-170, the measured values of the Young's modulus and indentation yield strength on martensite regions using the 16 μm tip radius were 179.5 ± 10.6 GPa and 2.15 ± 0.26 GPa, respectively. However, the values of Young's modulus and indentation yield strength measured using the 1 μm tip radius were 241.4 ± 20.3 GPa and 3.0 ± 0.46 GPa, respectively. This observation is particularly surprising as neither the fully martensitic alloy [57] nor other metals we have previously tested using similar protocols [57,59–61] showed any strong effects of indenter tip size on the measured indentation yield strengths. We believe that the lower values measured with the 16 μm tip are a consequence of the fact that the stress fields under the indenter extend far beyond the primary indentation deformation zone (defined to be of the order of the contact diameter). Indeed, finite element simulations conducted in prior studies from our research group [62] have revealed that the stress field underneath the indenter can extend as far as 10*a* to 15*a* before reducing to negligible levels. This is also the primary reason why the existing Vickers, Knoop, and Rockwell hardness standards [116,117] recommend that the thickness of the tested sample should be at least 10 times the indentation depth to minimize any influence of the substrate on the measured indentation properties. In the DP steel samples studied here, the martensite particles are surrounded by soft ferrite grains. Given the typical martensite particle size of 10 μm (see Figure 1) the average thickness of the martensite region in the indentation direction is likely to be only ~5 μm (i.e., if one assumes that half the particle has been polished to reveal the martensite region on the sample). Based on the above discussion, we estimate that the stress field under the 16 μm tip in the indentations on martensite regions is likely to extend to about 10 μm, which is significantly larger than the expected remaining thickness of the martensite plate at the indentation site in our measurements. The above discussion is schematically illustrated in Figure 6 where the indentation primary zone (volume of material underneath indentation where the majority of plasticity occurs) is highlighted in red and the extent of the indentation stress field is highlighted in yellow for the 1 μm and 16 μm tip radii on a BSE-SEM image of a typical sample studied in this work. Although the primary indentation zone with the 16 μm tip is expected to lie within the martensite region, the corresponding stress field should be expected to extend into the soft ferrite region underneath the indented martensite particle. For the indentations performed with the 1 μm tip radius, both the primary indentation zone and the indentation stress field are much more likely to lie within the martensite phase. It is also important to note that the local stresses in the primary indentation zone in the martensite are expected to be high because of the higher hardness of the martensite phase. Indeed, the softer ferrite regions at the martensite-ferrite boundary under the 16 μm tip indenter might even experience some plastic deformation. Because of the factors discussed above, we believe that the lower values of Young's moduli and the indentation yield strengths measured on the martensite regions with the 16 μm tip are largely a consequence of the softer ferrite underneath the indented martensite particles. Consequently, only the indentation measurements on martensite regions obtained using the 1 μm tip are reported in this work.

For the nanoindentation on ferrite grains, it was observed that the measurements obtained with the 1 μm tip exhibited high levels of variance within a single sample. We believe this is because of the heterogeneously deformed regions present in the ferrite grains (see Figure 5b,c,e). In particular, higher levels of dislocation density have been noted in the vicinity of the martensite/ferrite interface [1,11,118,119]. To obtain reliable measurements of indentation properties from the ferrite regions, we identified relatively large ferrite grains and used the 16 μm tip. The softer ferrite limits the stress values in the primary indentation zone, further mitigating any influence of the harder martensite regions below the indented ferrite grains.

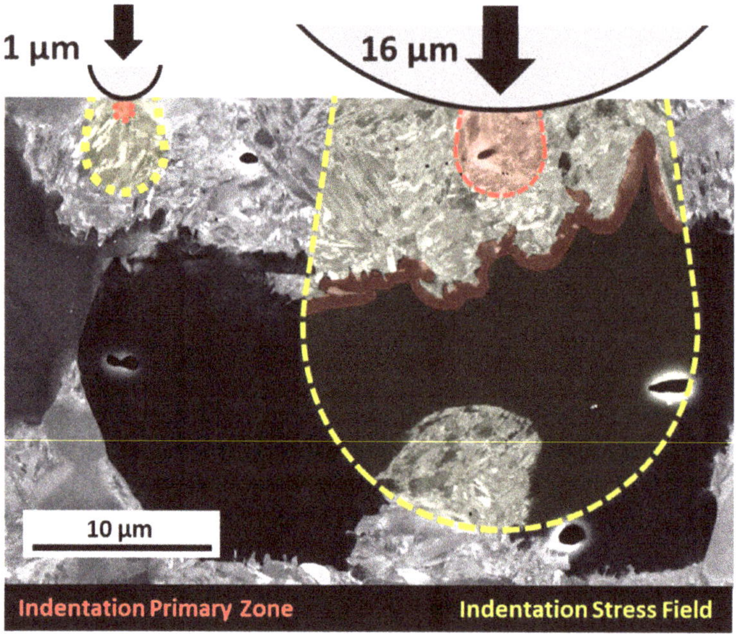

Figure 6. Schematic illustration of the primary indentation deformation zone (volume of the material where the majority of deformation occurs) and the extent of the indentation stress field for both the 1 μm and 16 μm tip radii, superimposed on a representative BSE-SEM image of DP steel studied in this work.

Examples of the indentation load–displacement and indentation stress–strain curves obtained in this work are presented in Figure 7 for both the martensite and ferrite regions in sample 750-00-000. As discussed before, the transition from the elastic regime to the elastic-plastic regime is clearly discernable in the indentation plots. The values of Young's modulus for the martensite regions were found to be higher compared to those measured in the ferrite regions (in Figure 7, these values were 226 GPa and 176 GPa, respectively). The indentation yield strength is usually defined using a 0.002 plastic strain offset. This protocol has been used extensively in prior studies [54–59], especially in the absence of pop-in events. A pop-in is identified as a sudden jump in the indentation displacement at a roughly constant load (the tests are performed in load control) and appears as a strain burst in the indentation stress–strain curves (see the measurement for the ferrite region in Figure 7). Pop-in events have been observed extensively in previous work [54,120–125] and were attributed to the difficulty of activating dislocation sources in the very small primary indentation zone. These pop-ins make it difficult to accurately estimate the indentation yield strength. In prior work [54,55,58,61], it was shown that a back-extrapolation method (see Figure 7) provides a reasonable estimate of the indentation yield strength in such cases. This same strategy was employed in the present work.

As discussed earlier, the intercritical annealing temperature and bake hardening steps are the critical processing steps in manufacturing DP steels. The effect of the quenching temperature on the strength of the individual ferrite and martensite regions was studied using samples 750-00-000, 780-00-000, and 810-00-000. As already mentioned, these samples produce martensite regions with significant differences in C content, offering an opportunity to quantitatively study the effect of C content on the yield strength of martensite. Comparing the nanoindentation measurements in the martensite and ferrite regions in samples 810-00-000 and 810-10-170 will provide quantitative insights into how bake hardening influences the yield strengths of these constituents. About 25–30 nanoindentation tests were performed in centers of randomly selected, relatively large, ferrite, and martensite regions on each sample identified above.

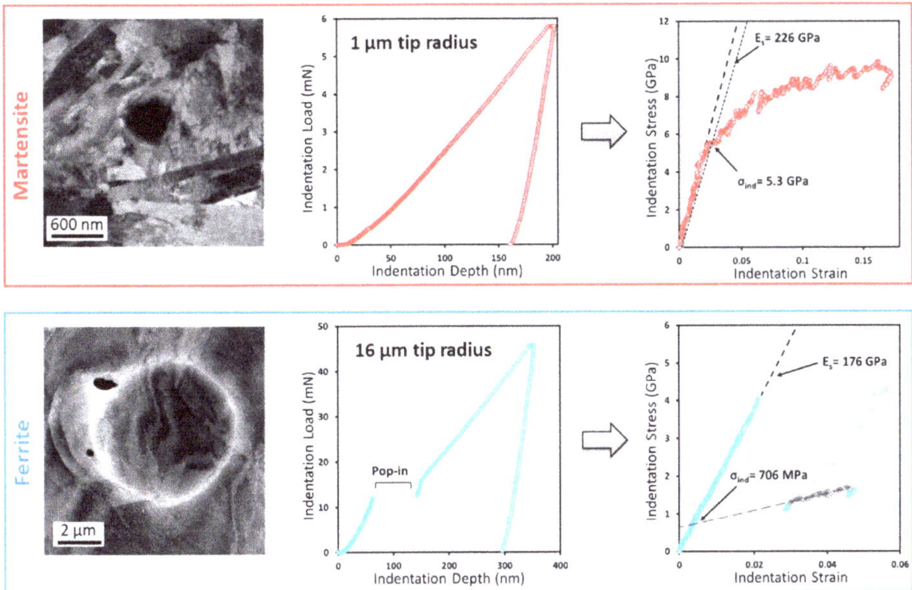

Figure 7. Example load–displacement, indentation stress–strain curves, and SEM micrographs of residual indentations from the martensite (top in red color) and ferrite (bottom in blue color) regions.

Figure 8 summarizes all of the indentation stress–strain curves extracted from the tests conducted on the four samples identified above. For reasons already discussed, the 1 μm tip was used on martensite regions, the 16 μm tip was used on ferrite regions, and the 6.35 mm tip (microindentation) was used for evaluating the bulk properties of the samples. The Young's moduli for martensite and ferrite regions were found to be significantly different from each other in these samples. The Young's modulus in the martensite regions varied in the 220–252 GPa range, while it varied in a much narrower range of 171–179 GPa for the ferrite regions. The larger variance in the extracted values of Young's moduli for the martensite regions can be attributed to the fact that the smallest tip (1 μm radius) was used in these studies. The limited number of laths in the indentation zone combined with the expected significant effect of the martensite lattice orientation on the indentation measurements can explain the observed larger variance. Notwithstanding the larger variance, the measurements clearly indicate that the Young's modulus of the martensite is higher compared to the Young's modulus of the ferrite.

The average indentation yield strength (± one standard deviation) is shown as a colored band on each plot in Figure 8. The indentation yield strengths measured from all four samples mentioned earlier are summarized in Figure 9. As expected, the measured variances decrease with an increasing indenter tip radius. This is because the larger indenter tips provide averaged responses over larger material volumes. Figure 9 also presents percentage changes in the indentation yield strengths measured in the martensite and ferrite regions in samples 780-00-000 and 810-00-000 using sample 750-00-000 as the baseline. It is seen that the bulk indentation yield strength increased by 27% when the intercritical annealing temperature was increased from 750 °C to 810 °C. As already noted, there is a significant increase in the martensite volume fraction (25% for 750-00-000 samples and 56% for 810-00-000 samples; see Table 2). Figure 9 also shows that the martensite yield strength has decreased by ~37% in the 810-00-000 sample, compared to the 750-00-000 sample. This reduction is attributed to the decrease in the C content in the martensite. Based on the Fe–C phase diagram, the C content in martensite (C_M)

quenched from 750 °C is expected to be ~0.58 wt.%, while it is ~0.26 wt.% for the sample quenched at 810 °C. The strong influence of C content on the martensite strength has been also discussed in prior studies [57,100,101,126]. Notably, the carbon content in the ferrite phase is expected to exhibit only a negligible increase, which does not appear to influence significantly the indentation yield strength of the ferrite.

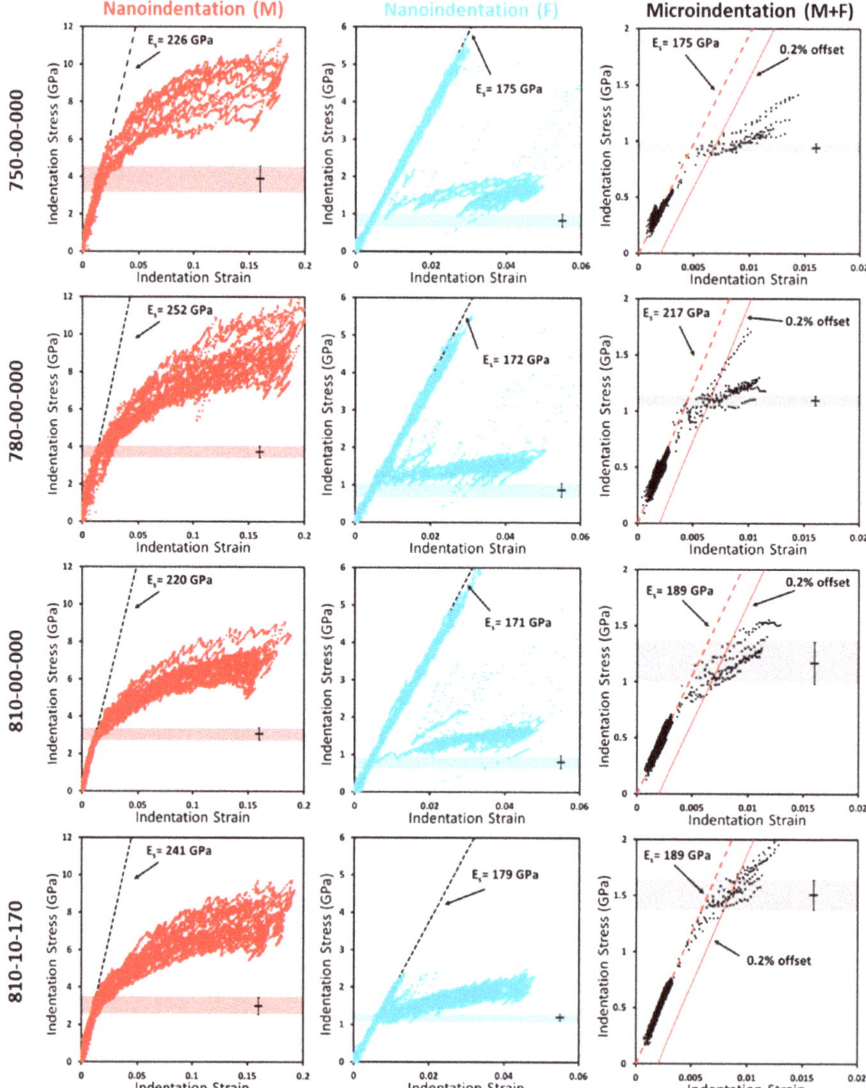

Figure 8. Multiple indentation stress–strain measurements at different lengths scales on samples 750-00-000, 780-00-000, 810-00-000, and 810-10-170. The left column shows the indentation stress–strain curves from multiple tests on martensite using the 1 µm tip radius. The center column shows the indentation stress–strain curves from multiple tests on ferrite grains using the 16 µm tip radius. The right column shows the micro-indentation stress–strain curves from multiple tests using the 6.35 mm tip radius. The highlighted horizontal bands inside each plot show the average values of the indentation yield strength and their variations.

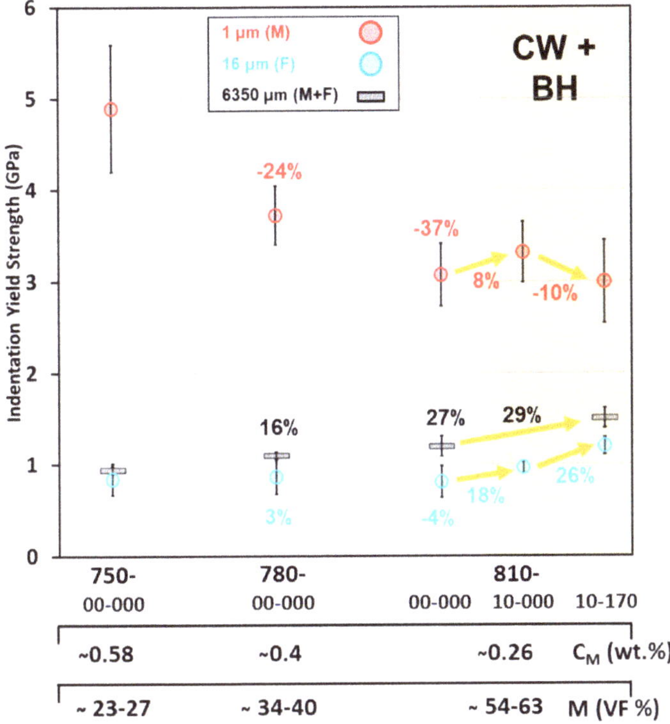

Figure 9. Extracted indentation yield strengths for samples 750-00-000, 780-00-000, 810-00-000, 810-10-000, and 810-10-170 at different length scales in martensite (M) and ferrite (F) regions. Bulk measurements (M + F) are obtained using microindentation tests. The numbers in the white region show the percentage changes with respect to the values from sample 750-00-000. The numbers in the yellow box show the percentage changes during the cold work and tempering steps in the bake hardening process for the 810-10-170 sample. The percentage of the martensite volume fractions and the carbon concentration in martensite are estimated from the Fe–C phase diagram in Figure 1.

Comparing the measurements on the martensite regions in sample 810-00-000 with those on sample 810-10-170 in Figure 9, it is seen that the average indentation yield strength decreased by a negligible amount (~2.5%) after the bake hardening step. This observation is consistent with other studies where it was reported that martensite hardness remains unchanged when the bake hardening temperature is kept below ~250 °C [126–130]. To separate the effects of hardening during cold work and softening by tempering in the bake hardening process, another sample was prepared right after cold work and without applying bake hardening. This sample is labeled as 810-10-000 in Figure 9. The indentation results show an increase of 8% in the indentation yield strength of the martensite regions during the cold work, followed by 10% softening during the tempering process in the bake hardening step. It was observed that during the bake hardening process, carbon atoms diffuse out of martensite islands into low carbon content ferrite grains and tend to build a carbon Cottrell atmosphere around dislocation cores, which immobilizes the dislocations [11,34,131–133]. Figure 10 shows an example of a ferrite grain surrounded by martensite islands in sample 810-10-170. As seen in Figure 10a, the ferrite grain has a highly contrasted matrix, which makes it very difficult to identify its interface with neighboring martensite islands. The ferrite grain boundaries are highlighted by dashed lines colored in yellow in this figure. Higher resolution images from the ECCI technique (see Figure 10b,c) reveal the presence of highly dense dislocation networks inside the contrasting regions. Few dislocations are highlighted by blue arrows in Figure 10c, which appear as white contrast on the black-background matrix. As shown in the highlighted yellow area in Figure 9, the indentation yield strength of ferrite increased by 50% when cold work and bake hardening were applied (from 0.81 ± 0.17 GPa for sample 810-00-000 to 1.2 ± 0.09 for sample 810-10-170). Indeed, one should expect an increase in the yield strength of the ferrite due to the imposed cold work. However, ferrite exhibits a low work hardening rate and the applied cold work only increased the indentation yield strength by 18% in the sample 810-10-000. A further 26% increase was observed in the indentation yield strengths in the ferrite between the samples 810-10-000 and 810-10-170, which is attributed to static aging resulting from the pinning of the dislocations by the diffused carbon atoms from the martensite.

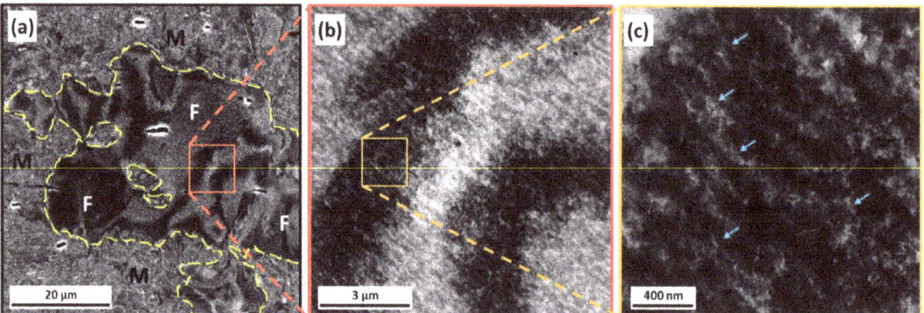

Figure 10. (**a**) BSE-SEM micrograph obtained on sample 810-10-170. The high contrast in both ferrite and martensite makes it difficult to identify their interface. For this purpose, each phase is labeled as martensite (M) or Ferrite (F) and their interfaces are highlighted as yellow dashed lines. (**b**) The ECCI technique at higher magnifications reveals a dense network of dislocations inside a ferrite grain. (**c**) The dislocation cores can be distinguished at sub-micron magnifications (see blue arrows).

Indirect evidence of a significant role of dislocation pinning on the strengthening of the ferrite can be found in indentation results. All the collected indentation load–displacement curves from ferrite regions on sample 810-10-170 revealed significant pop-in events while no pop-ins were observed from ferrite regions on sample 810-10-000. Figure 11a shows an example of a 3.6 nm pop-in (highlighted by a red arrow) in the indentation depth range of 30–40 nm. In prior studies using the spherical indentation protocols described earlier, pop-ins were observed exclusively on fully-annealed metal samples when small indenter tips (e.g., a spherical tip with 1 μm radius) were utilized [54,122,134–136]. In those studies, pop-ins disappeared after the introduction of small amounts of plastic deformation [136]. Therefore, in the prior studies, it was concluded that the pop-ins were a consequence of the difficulty of activating dislocation sources within the primary indentation deformation zone [54,120–122,124,125]. In the present work, the lack of pop-ins in the 810-10-000 sample is consistent with the previous observations described above. However, the 810-10-170 sample showed significant pop-ins, even though the ferrite grains have multiple dislocations within the indentation zone size of the 16 μm radius tip (see Figure 10). In the indentations performed on ferrite regions in this study, the ratio of the pop-in stress to the back-extrapolated yield strength was found to be 1.63 ± 0.35. This means that the initiation of plastic deformation was found to be significantly more difficult than continued plastic deformation after the initial yield. Therefore, the pop-ins observed in the tests reported here on the ferrite regions are attributed to the pinning of dislocations by static aging described earlier. Furthermore, as indentation proceeds, the indentation primary zone expands and encompasses new pinned dislocations. If an insufficient number of mobile dislocations exist inside the indentation primary zone, one should expect the occurrence of additional pop-ins. In our experiments, we often found such additional pop-ins (see the one marked by the orange arrow at the depth of 40 nm in Figure 11a). Of course, the subsequent pop-ins do not produce as large drops in stresses as the response is averaged over a much larger volume. It is important to note that such additional pop-ins were rarely observed in prior studies on fully annealed metal samples [54,58,61] because the dislocation sources are easily established in these samples at the larger indentation zone sizes created after the first pop-in. The presence of these pop-ins in the 810-10-170 and their absence in 810-10-000 provides the strongest direct evidence supporting the hypothesis that the ferrite regions in the DP steels experience a significant amount of dislocation pinning during the bake hardening step.

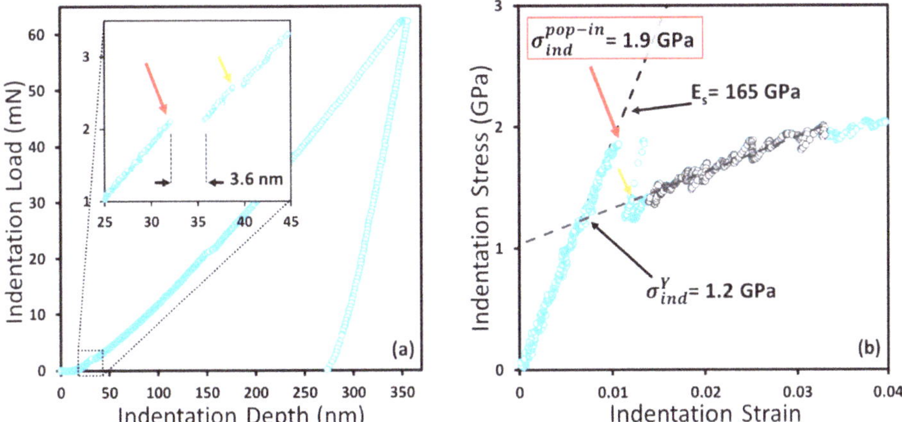

Figure 11. Example (**a**) load–displacement and (**b**) indentation stress–strain curves on the ferrite region from sample 810-10-170. The observed pop-in is attributed to the higher stress needed to unpin the dislocations from the carbon atmosphere.

4. Conclusion

The mechanical properties of DP steels were investigated at different length scales using spherical indentation stress–strain protocols. A total of nine samples processed with different combinations of intercritical annealing temperatures and different amounts of cold-work in the bake hardening step were studied. Bulk property measurements were carried out using a 6.35 mm tip radius showed results consistent with the relevant existing tensile data from the literature. However, the spherical indentation protocols used in this study require very small material volumes. An increase of the martensite volume fraction from 25% to 56% (for samples quenched from 750 °C and 810 °C, respectively) resulted in a 27% increase of the bulk indentation yield strength. Furthermore, the addition of the bake hardening step to the sample subjected to intercritical annealing at 810 °C produced a further 29% increase in the indentation yield strength.

Nanoindentation measurements were performed separately on both martensite and ferrite regions to obtain quantitative insights into how the different processing parameters affected their individual properties (at the scale of individual phases). It was first established that the measurements on martensite are best conducted using the 1 μm radius indenter tip, while the measurements on ferrite needed the 16 μm radius indenter tip for producing the most reliable data. The measurements conducted in this study revealed that the martensite regions exhibited on average a 36% higher Young's modulus compared to the ferrite regions. Intercritical annealing at 810 °C instead of at 750 °C decreased the indentation yield strength of martensite by ~37%, while the indentation yield strength of the ferrite regions remains more or less unchanged. Cold work and bake hardening of the sample intercritically annealed at 810 °C, increased the indentation yield strength in the ferrite regions by 50% without a significant change in the indentation yield strength of the martensite. The effects of cold work and tempering treatments employed in bake hardening were investigated. The results showed that the cold work increased the indentation yield strength of both phases, while the tempering treatment softened the martensite and hardened the ferrite. Most importantly, the nature of the pop-ins observed in the nanoindentations performed on the ferrite regions suggested that the increase in the indentation yield strength during bake hardening can be attributed to the pinning of dislocations by the diffused carbon.

Author Contributions: A.K. and S.R.K. conceived and designed the experiments; A.K. and C.M.C. performed the experiments; A.K. analyzed the data; A.K. and S.R.K. wrote the paper. All authors have read and agreed to the published version of the manuscript.

Funding: The authors gratefully acknowledge support from the National Science Foundation through the funding grant 1761406.

Conflicts of Interest: The authors declare no conflict of interest–exists.

References

1. Tasan, C.C.; Diehl, M.; Yan, D.; Bechtold, M.; Roters, F.; Schemmann, L.; Zheng, C.; Peranio, N.; Ponge, D.; Koyama, M.; et al. An Overview of Dual-Phase Steels: Advances in Microstructure-Oriented Processing and Micromechanically Guided Design. *Annu. Rev. Mater. Res.* **2015**, *45*, 391–431. [CrossRef]

2. Bian, J.; Mohrbacher, H.; Zhang, J.-S.; Zhao, Y.-T.; Lu, H.-Z.; Dong, H. Application potential of high performance steels for weight reduction and efficiency increase in commercial vehicles. *Adv. Manuf.* **2015**, *3*, 27–36. [CrossRef]

3. Fonstein, N. Dual-phase steels. In *Automotive Steels: Design, Metallurgy, Processing and Applications*; Rana, R., Brat Singh, S., Eds.; Woodhead Publishing: Cambridge, UK, 2017; pp. 169–216.

4. Horvath, C.D. Advanced steels for lightweight automotive structures. In *Materials, Design and Manufacturing for Lightweight Vehicles*; Mallick, P.K., Ed.; Woodhead Publishing: Cambridge, UK, 2010; pp. 35–78.

5. Rashid, M.S. Dual phase steels. *Annu. Rev. Mater. Sci.* **1981**, *11*, 245–266. [CrossRef]

6. Robertson, L.T.; Hilditch, T.B.; Hodgson, P.D. The effect of prestrain and bake hardening on the low-cycle fatigue properties of TRIP steel. *Int. J. Fatigue* **2008**, *30*, 587–594. [CrossRef]

7. Ramazani, A.; Bruehl, S.; Gerber, T.; Bleck, W.; Prahl, U. Quantification of bake hardening effect in DP600 and TRIP700 steels. *Mater. Des.* **2014**, *57*, 479–486. [CrossRef]

8. Kuang, C.-F.; Zhang, S.-G.; Li, J.; Wang, J.; Li, P. Effect of temper rolling on the bake-hardening behavior of low carbon steel. *Int. J. Miner. Metall. Mater.* **2015**, *22*, 32–36. [CrossRef]

9. Waterschoot, T.; Verbeken, K.; de Cooman, B.C. Tempering Kinetics of the Martensitic Phase in DP Steel. *ISIJ Int.* **2006**, *46*, 138–146. [CrossRef]

10. Berbenni, S.; Favier, V.; Lemoine, X.; Berveiller, M. A micromechanical approach to model the bake hardening effect for low carbon steels. *Scr. Mater.* **2004**, *51*, 303–308. [CrossRef]

11. Timokhina, I.B.; Hodgson, P.D.; Pereloma, E.V. Transmission Electron Microscopy Characterization of the Bake-Hardening Behavior of Transformation-Induced Plasticity and Dual-Phase Steels. *Metall. Mater. Trans.* **2007**, *38*, 2442–2454. [CrossRef]

12. Lindqvist, K. Bake Hardening Effect in Advanced High-Strength Steels. Master's Thesis, Chalmers University of Technology, Gothenburg, Sweden, 2013.

13. Kuang, C.-F.; Wang, J.; Li, J.; Zhang, S.-G.; Liu, H.-F.; Yang, H.-L. Effect of Continuous Annealing on Microstructure and Bake Hardening Behavior of Low Carbon Steel. *J. Iron Steel Res. Int.* **2015**, *22*, 163–170. [CrossRef]

14. Wang, H.; Shi, W.; He, Y.-L.; Lu, X.-G.; Li, L. Effect of Overaging on Solute Distributions and Bake Hardening Phenomenon in Bake Hardening Steels. *J. Iron Steel Res. Int.* **2012**, *19*, 53–59. [CrossRef]

15. Durrenberger, L.; Lemoine, X.; Molinari, A. Effects of pre-strain and bake-hardening on the crash properties of a top-hat section. *J. Mater. Process. Technol.* **2011**, *211*, 1937–1947. [CrossRef]

16. Kilic, S.; Ozturk, F.; Sigirtmac, T.; Tekin, G. Effects of Pre-strain and Temperature on Bake Hardening of TWIP900CR Steel. *J. Iron Steel Res. Int.* **2015**, *22*, 361–365. [CrossRef]

17. Šebek, M.; Horňak, P.; Zimovčák, P. Effect of Annealing on the Microstructure Evolution and Mechanical Properties of Dual Phase Steel. *Mater. Sci. Forum* **2014**, *782*, 111–116. [CrossRef]

18. Asadi, M.; de Cooman, B.C.; Palkowski, H. Influence of martensite volume fraction and cooling rate on the properties of thermomechanically processed dual phase steel. *Mater. Sci. Eng.* **2012**, *538*, 42–52. [CrossRef]

19. Vasilyev, A.A.; Lee, H.-C.; Kuzmin, N.L. Nature of strain aging stages in bake hardening steel for automotive application. *Mater. Sci. Eng.* **2008**, *485*, 282–289. [CrossRef]

20. Jeong, W.C. Relationship between mechanical properties and microstructure in a 1.5% Mn–0.3% Mo ultra-low carbon steel with bake hardening. *Mater. Lett.* **2007**, *61*, 2579–2583. [CrossRef]

21. Cao, Y.; Ahlström, J.; Karlsson, B. The influence of temperatures and strain rates on the mechanical behavior of dual phase steel in different conditions. *J. Mater. Res. Technol.* **2015**, *4*, 68–74. [CrossRef]

22. Kuang, C.F.; Li, J.; Zhang, S.G.; Wang, J.; Liu, H.F.; Volinsky, A.A. Effects of quenching and tempering on the microstructure and bake hardening behavior of ferrite and dual phase steels. *Mater. Sci. Eng.* **2014**, *613*, 178–183. [CrossRef]

23. Li, C.-S.; Li, Z.-X.; Cen, Y.-M.; Ma, B.; Huo, G. Microstructure and mechanical properties of dual phase strip steel in the overaging process of continuous annealing. *Mater. Sci. Eng.* **2015**, *627*, 281–289. [CrossRef]

24. Hüseyin, A.; Havva, K.Z.; Ceylan, K. Effect of Intercritical Annealing Parameters on Dual Phase Behavior of Commercial Low-Alloyed Steels. *J. Iron Steel Res. Int.* **2010**, *17*, 73–78. [CrossRef]

25. Zeytin, H.K.; Kubilay, C.; Aydin, H. Investigation of dual phase transformation of commercial low alloy steels: Effect of holding time at low inter-critical annealing temperatures. *Mater. Lett.* **2008**, *62*, 2651–2653. [CrossRef]

26. Demir, B.; Erdoğan, M. The hardenability of austenite with different alloy content and dispersion in dual-phase steels. *J. Mater. Process. Technol.* **2008**, *208*, 75–84. [CrossRef]

27. Sudersanan, N.K.P.D.; Aprameyan, S.; Kempaiah, D.U.N. The Effect of Carbon Content in Martensite on the Strength of Dual Phase Steel. *Bonfring Int. J. Ind. Eng. Manag. Sci.* **2012**, *2*, 1–4.

28. Seyedrezai, H. Thermo-Mechanical Processing of Dual-Phase Steels and its Effects on the Work Hardening Behaviour. Ph.D. Thesis, Queen's University, Kingston, ON, Canada, 2014.

29. Schemmann, L.; Zaefferer, S.; Raabe, D.; Friedel, F.; Mattissen, D. Alloying effects on microstructure formation of dual phase steels. *Acta Mater.* **2015**, *95*, 386–398. [CrossRef]

30. Timokhina, I.B.; Pereloma, E.V.; Ringer, S.P.; Zheng, R.K.; Hodgson, P.D. Characterization of the Bake-hardening Behavior of Transformation Induced Plasticity and Dual-phase Steels Using Advanced Analytical Techniques. *ISIJ Int.* **2010**, *50*, 574–582. [CrossRef]

31. Allain, S.Y.P.; Bouaziz, O.; Pushkareva, I.; Scott, C.P. Towards the microstructure design of DP steels: A generic size-sensitive mean-field mechanical model. *Mater. Sci. Eng.* **2015**, *637*, 222–234. [CrossRef]

32. Korzekwa, D.A.; Matlock, D.K.; Krauss, G. Dislocation substructure as a function of strain in a dual-phase steel. *Metall. Trans.* **1984**, *15*, 1221–1228. [CrossRef]

33. Calcagnotto, M.; Ponge, D.; Demir, E.; Raabe, D. Orientation gradients and geometrically necessary dislocations in ultrafine grained dual-phase steels studied by 2D and 3D EBSD. *Mater. Sci. Eng.* **2010**, *527*, 2738–2746. [CrossRef]

34. Cottrell, A.H.; Bilby, B.A. Dislocation theory of yielding and strain ageing of iron. *Proc. Phys. Soc.* **1949**, *62*, 49. [CrossRef]

35. Pan, Z.; Gao, B.; Lai, Q.; Chen, X.; Cao, Y.; Liu, M.; Zhou, H. Microstructure and Mechanical Properties of a Cold-Rolled Ultrafine-Grained Dual-Phase Steel. *Materials* **2018**, *11*, 1399. [CrossRef]

36. Gerbig, D.; Srivastava, A.; Osovski, S.; Hector, L.G.; Bower, A. Analysis and design of dual-phase steel microstructure for enhanced ductile fracture resistance. *Int. J. Fract.* **2018**, *209*, 3–26. [CrossRef]

37. Kundu, J.; Ray, T.; Kundu, A.; Shome, M. Effect of the laser power on the mechanical performance of the laser spot welds in dual phase steels. *J. Mater. Process. Technol.* **2019**, *267*, 114–123. [CrossRef]

38. Samei, J.; Zhou, L.; Kang, J.; Wilkinson, D.S. Microstructural analysis of ductility and fracture in fine-grained and ultrafine-grained vanadium-added DP1300 steels. *Int. J. Plast.* **2019**, *117*, 58–70. [CrossRef]

39. Zhang, L.; Liang, M.; Feng, Z.; Zhang, L.; Cao, W.; Wu, G. Nanoindentation characterization of strengthening mechanism in a high strength ferrite/martensite steel. *Mater. Sci. Eng.* **2019**, *745*, 144–148. [CrossRef]

40. Ghassemi-Armaki, H.; Maaß, R.; Bhat, S.P.; Sriram, S.; Greer, J.R.; Kumar, K.S. Deformation response of ferrite and martensite in a dual-phase steel. *Acta Mater.* **2014**, *62*, 197–211. [CrossRef]

41. Seok, M.-Y.; Kim, Y.-J.; Choi, I.-C.; Zhao, Y.; Jang, J.-I. Predicting flow curves of two-phase steels from spherical nanoindentation data of constituent phases: Isostrain method vs. non-isostrain method. *Int. J. Plast.* **2014**, *59*, 108–118. [CrossRef]

42. Taylor, M.D.; Choi, K.S.; Sun, X.; Matlock, D.K.; Packard, C.E.; Xu, L.; Barlat, F. Correlations between nanoindentation hardness and macroscopic mechanical properties in DP980 steels. *Mater. Sci. Eng.* **2014**, *597*, 431–439. [CrossRef]

43. Choi, B.-W.; Seo, D.-H.; Yoo, J.-Y.; Jang, J.-I. Predicting macroscopic plastic flow of high-performance, dual-phase steel through spherical nanoindentation on each microphase. *J. Mater. Res.* **2009**, *24*, 816–822. [CrossRef]

44. Hayashi, K.; Miyata, K.; Katsuki, F.; Ishimoto, T.; Nakano, T. Individual mechanical properties of ferrite and martensite in Fe-0.16mass% C-1.0mass% Si-1.5mass% Mn steel. *J. Alloy. Compd.* **2013**, *577*, S593–S596. [CrossRef]

45. Scott, C.P.; Amirkhiz, B.S.; Pushkareva, I.; Fazeli, F.; Allain, S.Y.P.; Azizi, H. New insights into martensite strength and the damage behaviour of dual phase steels. *Acta Mater.* **2018**, *159*, 112–122. [CrossRef]

46. Farivar, H.; Richter, S.; Hans, M.; Schwedt, A.; Prahl, U.; Bleck, W. Experimental quantification of carbon gradients in martensite and its multi-scale effects in a DP steel. *Mater. Sci. Eng.* **2018**, *718*, 250–259. [CrossRef]

47. Banadkouki, S.S.G.; Fereiduni, E. Effect of prior austenite carbon partitioning on martensite hardening variation in a low alloy ferrite—Martensite dual phase steel. *Mater. Sci. Eng.* **2014**, *619*, 129–136. [CrossRef]

48. Cheng, G.; Zhang, F.; Ruimi, A.; Field, D.P.; Sun, X. Quantifying the effects of tempering on individual phase properties of DP980 steel with nanoindentation. *Mater. Sci. Eng.* **2016**, *667*, 240–249. [CrossRef]

49. Zhang, F.; Ruimi, A.; Field, D.P. Phase Identification of Dual-Phase (DP980) Steels by Electron Backscatter Diffraction and Nanoindentation Techniques. *Microsc. Microanal.* **2016**, *22*, 99–107. [CrossRef] [PubMed]

50. Li, H.; Gao, S.; Tian, Y.; Terada, D.; Shibata, A.; Tsuji, N. Influence of Tempering on Mechanical Properties of Ferrite and Martensite Dual Phase Steel. *Mater. Today Proc.* **2015**, *2*, S667–S671. [CrossRef]

51. Hernandez, V.H.B.; Panda, S.K.; Kuntz, M.L.; Zhou, Y. Nanoindentation and microstructure analysis of resistance spot welded dual phase steel. *Mater. Lett.* **2010**, *64*, 207–210. [CrossRef]

52. Azuma, M.; Goutianos, S.; Hansen, N.; Winther, G.; Huang, X. Effect of hardness of martensite and ferrite on void formation in dual phase steel. *Mater. Sci. Technol.* **2012**, *28*, 1092–1100. [CrossRef]

53. Weaver, J.S.; Pathak, S.; Reichardt, A.; Vo, H.T.; Maloy, S.A.; Hosemann, P.; Mara, N.A. Spherical nanoindentation of proton irradiated 304 stainless steel: A comparison of small scale mechanical test techniques for measuring irradiation hardening. *J. Nucl. Mater.* **2017**, *493*, 368–379. [CrossRef]

54. Pathak, S.; Michler, J.; Wasmer, K.; Kalidindi, S.R. Studying grain boundary regions in polycrystalline materials using spherical nano-indentation and orientation imaging microscopy. *J. Mater. Sci.* **2012**, *47*, 815–823. [CrossRef]

55. Pathak, S.; Stojakovic, D.; Kalidindi, S.R. Measurement of the local mechanical properties in polycrystalline samples using spherical nanoindentation and orientation imaging microscopy. *Acta Mater.* **2009**, *57*, 3020–3028. [CrossRef]

56. Vachhani, S.J.; Doherty, R.D.; Kalidindi, S.R. Studies of grain boundary regions in deformed polycrystalline aluminum using spherical nanoindentation. *Int. J. Plast.* **2016**, *81*, 87–101. [CrossRef]

57. Khosravani, A.; Morsdorf, L.; Tasan, C.C.; Kalidindi, S.R. Multiresolution mechanical characterization of hierarchical materials: Spherical nanoindentation on martensitic Fe-Ni-C steels. *Acta Mater.* **2018**, *153*, 257–269. [CrossRef]

58. Weaver, J.S.; Priddy, M.W.; McDowell, D.L.; Kalidindi, S.R. On capturing the grain-scale elastic and plastic anisotropy of alpha-Ti with spherical nanoindentation and electron back-scattered diffraction. *Acta Mater.* **2016**, *117*, 23–34. [CrossRef]

59. Weaver, J.S.; Kalidindi, S.R. Mechanical characterization of Ti-6Al-4V titanium alloy at multiple length scales using spherical indentation stress-strain measurements. *Mater. Des.* **2016**, *111*, 463–472. [CrossRef]

60. Pathak, S.; Kalidindi, S.R.; Mara, N.A. Investigations of orientation and length scale effects on micromechanical responses in polycrystalline zirconium using spherical nanoindentation. *Scr. Mater.* **2016**, *113*, 241–245. [CrossRef]

61. Vachhani, S.J.; Kalidindi, S.R. Grain-scale measurement of slip resistances in aluminum polycrystals using spherical nanoindentation. *Acta Mater.* **2015**, *90*, 27–36. [CrossRef]

62. Kalidindi, S.R.; Pathak, S. Determination of the effective zero-point and the extraction of spherical nanoindentation stress-strain curves. *Acta Mater.* **2008**, *56*, 3523–3532. [CrossRef]

63. Pathak, S.; Shaffer, J.; Kalidindi, S.R. Determination of an effective zero-point and extraction of indentation stress-strain curves without the continuous stiffness measurement signal. *Scr. Mater.* **2009**, *60*, 439–442. [CrossRef]

64. Hertz, H. *Miscellaneous Papers*; MacMillan and Co. Ltd.: New York, NY, USA, 1896.

65. Fischer-Cripps, A.C. Critical review of analysis and interpretation of nanoindentation test data. *Surf. Coat. Technol.* **2006**, *200*, 4153–4165. [CrossRef]

66. Chudoba, T.; Griepentrog, M.; Dück, A.; Schneider, D.; Richter, F. Young's modulus measurements on ultra-thin coatings. *J. Mater. Res.* **2004**, *19*, 301–314. [CrossRef]

67. Chudoba, T.; Schwarzer, N.; Richter, F. Determination of elastic properties of thin films by indentation measurements with a spherical indenter. *Surf. Coat. Technol.* **2000**, *127*, 9–17. [CrossRef]

68. Richter, F.; Herrmann, M.; Molnar, F.; Chudoba, T.; Schwarzer, N.; Keunecke, M.; Bewilogua, K.; Zhang, X.W.; Boyen, H.G.; Ziemann, P. Substrate influence in Young's modulus determination of thin films by indentation methods: Cubic boron nitride as an example. *Surf. Coat. Technol.* **2006**, *201*, 3577–3587. [CrossRef]

69. Grau, P.; Berg, G.; Fränzel, W.; Meinhard, H. Recording hardness testing. Problems of measurement at small indentation depths. *Phys. Status Solidi* **1994**, *146*, 537–548. [CrossRef]

70. Ullner, C. Requirement of a robust method for the precise determination of the contact point in the depth sensing hardness test. *Measurement* **2000**, *27*, 43–51. [CrossRef]

71. Lim, Y.Y.; Chaudhri, M.M. Indentation of elastic solids with a rigid Vickers pyramidal indenter. *Mech. Mater.* **2006**, *38*, 1213–1228. [CrossRef]

72. Fernandez-Zelaia, P.; Nguyen, V.; Zhang, H.; Kumar, A.; Melkote, S.N. The effects of material anisotropy on secondary processing of additively manufactured CoCrMo. *Addit. Manuf.* **2019**, *29*, 100764. [CrossRef]

73. Weaver, J.S.; Jones, D.R.; Li, N.; Mara, N.; Fensin, S.; Gray, G.T. Quantifying heterogeneous deformation in grain boundary regions on shock loaded tantalum using spherical and sharp tip nanoindentation. *Mater. Sci. Eng.* **2018**, *737*, 373–382. [CrossRef]

74. Leung, J.F.W.; Bedekar, V.; Voothaluru, R.; Neu, R.W. Mechanical Properties of White Etching Areas in Carburized Bearing Steel Using Spherical Nanoindentation. *Metall. Mater. Trans.* **2019**, *50*, 4949–4954. [CrossRef]

75. Pathak, S.; Weaver, J.S.; Sun, C.; Wang, Y.; Kalidindi, S.R.; Mara, N.A. *Spherical Nanoindentation Stress-Strain Analysis of Ion-Irradiated Tungsten*; Minerals. Metals and Materials Series; Springer International Publishing: New York, NY, USA, 2019; pp. 617–635.

76. Fernandez-Zelaia, P.; Joseph, V.R.; Kalidindi, S.R.; Melkote, S.N. Estimating mechanical properties from spherical indentation using Bayesian approaches. *Mater. Des.* **2018**, *147*, 92–105. [CrossRef]

77. Weaver, J.S.; Sun, C.; Wang, Y.; Kalidindi, S.R.; Doerner, R.P.; Mara, N.A.; Pathak, S. Quantifying the mechanical effects of He, W and He + W ion irradiation on tungsten with spherical nanoindentation. *J. Mater. Sci.* **2018**, *53*, 5296–5316. [CrossRef]

78. Iskakov, A.; Yabansu, Y.C.; Rajagopalan, S.; Kapustina, A.; Kalidindi, S.R. Application of spherical indentation and the materials knowledge system framework to establishing microstructure-yield strength linkages from carbon steel scoops excised from high-temperature exposed components. *Acta Mater.* **2018**, *144*, 758–767. [CrossRef]

79. Pathak, S.; Kalidindi, S.R.; Weaver, J.S.; Wang, Y.; Doerner, R.P.; Mara, N.A. Probing nanoscale damage gradients in ion-irradiated metals using spherical nanoindentation. *Sci. Rep.* **2017**, *7*, 11918. [CrossRef] [PubMed]

80. Khosravani, A.; Cecen, A.; Kalidindi, S.R. Development of high throughput assays for establishing process-structure-property linkages in multiphase polycrystalline metals: Application to dual-phase steels. *Acta Mater.* **2017**, *123*, 55–69. [CrossRef]

81. Zhang, Q.; Mochalin, V.N.; Neitzel, I.; Knoke, I.Y.; Han, J.; Klug, C.A.; Zhou, J.G.; Lelkes, P.I.; Gogotsi, Y. Fluorescent PLLA-nanodiamond composites for bone tissue engineering. *Biomaterials* **2011**, *32*, 87–94. [CrossRef] [PubMed]

82. Weaver, J.S.; Livescu, V.; Mara, N.A. A comparison of adiabatic shear bands in wrought and additively manufactured 316 L stainless steel using nanoindentation and electron backscatter diffraction. *J. Mater. Sci.* **2020**, *55*, 1738–1752. [CrossRef]

83. Kim, H.N.; Mandal, S.; Basu, B.; Kalidindi, S.R. Probing Local Mechanical Properties in Polymer-Ceramic Hybrid Acetabular Sockets Using Spherical Indentation Stress-Strain Protocols. *Integr. Mater. Manuf. Innov.* **2019**, *8*, 257–272. [CrossRef]

84. Pathak, S.; Kalidindi, S.R.; Klemenz, C.; Orlovskaya, N. Analyzing indentation stress–strain response of LaGaO3 single crystals using spherical indenters. *J. Eur. Ceram. Soc.* **2008**, *28*, 2213–2220. [CrossRef]

85. Kalidindi, S.R.; Vachhani, S.J. Mechanical characterization of grain boundaries using nanoindentation. *Curr. Opin. Solid State Mater. Sci.* **2014**, *18*, 196–204. [CrossRef]

86. Chang, C.; Verdi, D.; Garrido, M.A.; Ruiz-Hervias, J. Micro-scale mechanical characterization of Inconel cermet coatings deposited by laser cladding. *Boletín Sociedad Española Cerámica Vidrio* **2016**, *55*, 136–142. [CrossRef]

87. Weaver, J.; Aydogan, E.; Mara, N.A.; Maloy, S.A. *Nanoindentation of Electropolished FeCrAl Alloy Welds*; Los Alamos National Lab. (LANL): Los Alamos, NM, USA, 2017.

88. Mohanty, D.P.; Wang, H.; Okuniewski, M.; Tomar, V. A nanomechanical Raman spectroscopy based assessment of stress distribution in irradiated and corroded SiC. *J. Nucl. Mater.* **2017**, *497*, 128–138. [CrossRef]

89. Zhang, Q.; Li, X.; Yang, Q.J.A.A. Extracting the isotropic uniaxial stress-strain relationship of hyperelastic soft materials based on new nonlinear indentation strain and stress measure. *AIP Adv.* **2018**, *8*, 115013. [CrossRef]

90. Fernandez-Zelaia, P.; Melkote, S.N. Validation of Material Models for Machining Simulation Using Mechanical Properties of the Deformed Chip. *Procedia CIRP* **2017**, *58*, 535–538. [CrossRef]

91. Weaver, J.; Nunez, U.C.; Krumwiede, D.; Saleh, T.A.; Hosemann, P.; Nelson, A.T.; Maloy, S.A.; Mara, N.A. *Spherical Nanoindentation Stress-Strain Measurements of BOR-60 14YWT-NFA1 Irradiated Tubes*; Los Alamos National Lab. (LANL): Los Alamos, NM, USA, 2017.

92. Rossi, A. Mechanical Response of Polymer Matrix Composites Using Indentation Stress-Strain Protocols. Level of Thesis, Georgia Institute of Technology, Atlanta, Georgia, 2018.

93. Khosravani, A. High Throughput Prototyping and Multiscale Indentation Characterization of Metallic Alloys. Level of Thesis, Georgia Institute of Technology, Atlanta, Georgia, 2019.

94. Vachhani, S.J.; Trujillo, C.; Mara, N.; Livescu, V.; Bronkhorst, C.; Gray, G.T.; Cerreta, E. Local Mechanical Property Evolution During High Strain-Rate Deformation of Tantalum. *J. Dyn. Behav. Mater.* **2016**, *2*, 511–520. [CrossRef]

95. Weaver, J.S.; Khosravani, A.; Castillo, A.; Kalidindi, S.R. High Throughput Exploration of Process-Property Linkages in Al-6061 using Instrumented Spherical Microindentation and Microstructurally Graded Samples. *Integr. Mater. Manuf. Innov.* **2016**, *5*, 192–211. [CrossRef]

96. Donohue, B.R.; Ambrus, A.; Kalidindi, S.R. Critical evaluation of the indentation data analyses methods for the extraction of isotropic uniaxial mechanical properties using finite element models. *Acta Mater.* **2012**, *60*, 3943–3952. [CrossRef]

97. Patel, D.K.; Kalidindi, S.R. Correlation of spherical nanoindentation stress-strain curves to simple compression stress-strain curves for elastic-plastic isotropic materials using finite element models. *Acta Mater.* **2016**, *112*, 295–302. [CrossRef]

98. Patel, D.K.; Al-Harbi, H.F.; Kalidindi, S.R. Extracting single-crystal elastic constants from polycrystalline samples using spherical nanoindentation and orientation measurements. *Acta Mater.* **2014**, *79*, 108–116. [CrossRef]

99. Patel, D.K.; Kalidindi, S.R. Estimating the slip resistance from spherical nanoindentation and orientation measurements in polycrystalline samples of cubic metals. *Int. J. Plast.* **2017**, *92*, 19–30. [CrossRef]

100. Ohmura, T.; Tsuzaki, K.; Matsuoka, S. Nanohardness measurement of high-purity Fe–C martensite. *Scr. Mater.* **2001**, *45*, 889–894. [CrossRef]

101. Hutchinson, B.; Hagström, J.; Karlsson, O.; Lindell, D.; Tornberg, M.; Lindberg, F.; Thuvander, M. Microstructures and hardness of as-quenched martensites (0.1–0.5% C). *Acta Mater.* **2011**, *59*, 5845–5858. [CrossRef]

102. Molina, R.C.; Pla, M.; Rossi, R.H.; Páramo, J.A.B. Analysis of the decrease of the apparent young's modulus of advanced high strength steels and its effect in bending simulations. In Proceedings of the IDDRG 2009 International Conference, Golden, CO, USA, 1–3 June 2009; pp. 109–117.

103. Chen, Z.; Gandhi, U.; Lee, J.; Wagoner, R.H. Variation and consistency of Young's modulus in steel. *J. Mater. Process. Technol.* **2016**, *227*, 227–243. [CrossRef]

104. Rana, R.; Liu, C.; Ray, R.K. Low-density low-carbon Fe–Al ferritic steels. *Scr. Mater.* **2013**, *68*, 354–359. [CrossRef]

105. Furukawa, T.; Morikawa, H.; Endo, M.; Takechi, H.; Koyama, K.; Akisue, O.; Yamada, T. Process Factors for Cold-rolled Dual-phase Sheet Steels. *Trans. Iron Steel Inst. Jpn.* **1981**, *21*, 812–819. [CrossRef]

106. Tomita, Y. Effect of morphology of second-phase martensite on tensile properties of Fe-0.1C dual phase steels. *J. Mater. Sci.* **1990**, *25*, 5179–5184.

107. Byun, T.S.; Kim, I.S. Tensile properties and inhomogeneous deformation of ferrite-martensite dual-phase steels. *J. Mater. Sci.* **1993**, *28*, 2923–2932. [CrossRef]

108. Chang, C. Correlation between the Microstructure of Dual Phase Steel and Industrial Tube Bending Performance. Master's Thesis, University of Windsor, Windsor, ON, Canada, 2010.

109. Sarwar, M.; Priestner, R. Influence of ferrite-martensite microstructural morphology on tensile properties of dual-phase steel. *J. Mater. Sci.* **1996**, *31*, 2091–2095. [CrossRef]

110. Tavares, S.; Pedroza, P.; Teodosio, J.; Gurova, T.J.S.M. Mechanical properties of a quenched and tempered dual phase steel. *Scr. Mater.* **1999**, *40*, 887–892. [CrossRef]

111. Bag, A.; Ray, K.; Dwarakadasa, E.J. Influence of martensite content and morphology on tensile and impact properties of high-martensite dual-phase steels. *Metall. Mater. Trans.* **1999**, *30*, 1193–1202. [CrossRef]

112. Hayat, F.; Uzun, H. Effect of Heat Treatment on Microstructure, Mechanical Properties and Fracture Behaviour of Ship and Dual Phase Steels. *J. Iron Steel Res. Int.* **2011**, *18*, 65–72. [CrossRef]

113. Kocatepe, K.; Cerah, M.; Erdogan, M. Effect of martensite volume fraction and its morphology on the tensile properties of ferritic ductile iron with dual matrix structures. *J. Mater. Process. Technol.* **2006**, *178*, 44–51. [CrossRef]

114. Morsdorf, L.; Tasan, C.C.; Ponge, D.; Raabe, D. 3D structural and atomic-scale analysis of lath martensite: Effect of the transformation sequence. *Acta Mater.* **2015**, *95*, 366–377. [CrossRef]

115. Bhadeshia, H.K.D.H.; Keehan, E.; Karlsson, L.; Andrén, H.O. Coalesced bainite. *Trans. Indian Inst. Met.* **2006**, *59*, 689–694.

116. Rice, J.R. Inelastic constitutive relations for solids: An internal-variable theory and its application to metal plasticity. *J. Mech. Phys. Solids* **1971**, *19*, 433–455. [CrossRef]

117. Pedregosa, F.; Varoquaux, G.; Gramfort, A.; Michel, V.; Thirion, B.; Grisel, O.; Blondel, M.; Prettenhofer, P.; Weiss, R.; Dubourg, V.J.J. Scikit-learn: Machine learning in Python. *J. Mach. Learn. Res.* **2011**, *12*, 2825–2830.

118. Bergstr, Y.; Granbom, Y.; Sterkenburg, D. A Dislocation-Based Theory for the Deformation Hardening Behavior of DP Steels: Impact of Martensite Content and Ferrite Grain Size. *J. Metall.* **2010**. [CrossRef]

119. Liedl, U.; Traint, S.; Werner, E.A. An unexpected feature of the stress–strain diagram of dual-phase steel. *Comput. Mater. Sci.* **2002**, *25*, 122–128. [CrossRef]

120. Lorenz, D.; Zeckzer, A.; Hilpert, U.; Grau, P.; Johansen, H.; Leipner, H.S. Pop-in effect as homogeneous nucleation of dislocations during nanoindentation. *Phys. Rev.* **2003**, *67*, 172101. [CrossRef]

121. Ahn, T.-H.; Oh, C.-S.; Lee, K.; George, E.P.; Han, H.N. Relationship between yield point phenomena and the nanoindentation pop-in behavior of steel. *J. Mater. Res.* **2012**, *27*, 39–44. [CrossRef]

122. Pathak, S.; Riesterer, J.L.; Kalidindi, S.R.; Michler, J. Understanding pop-ins in spherical nanoindentation. *Appl. Phys. Lett.* **2014**, *105*, 161913. [CrossRef]

123. Wang, M.G.; Ngan, A.H.W. Indentation strain burst phenomenon induced by grain boundaries in niobium. *J. Mater. Res.* **2004**, *19*, 2478–2486. [CrossRef]

124. Morris, J.R.; Bei, H.; Pharr, G.M.; George, E.P. Size effects and stochastic behavior of nanoindentation pop in. *Phys. Rev. Lett.* **2011**, *106*, 165502. [CrossRef] [PubMed]

125. Phani, P.S.; Johanns, K.E.; George, E.P.; Pharr, G.M. A stochastic model for the size dependence of spherical indentation pop-in. *J. Mater. Res.* **2013**, *28*, 2728–2739. [CrossRef]

126. Ohmura, T.; Hara, T.; Tsuzaki, K. Evaluation of temper softening behavior of Fe–C binary martensitic steels by nanoindentation. *Scr. Mater.* **2003**, *49*, 1157–1162. [CrossRef]

127. Speich, G.R.; Leslie, W.C. Tempering of steel. *Metall. Trans.* **1972**, *3*, 1043–1054. [CrossRef]

128. Caron, R.N.; Krauss, G. The tempering of Fe-C lath martensite. *Metall. Trans.* **1972**, *3*, 2381–2389. [CrossRef]

129. Tkalcec, I.; Azcoitia, C.; Crevoiserat, S.; Mari, D. Tempering effects on a martensitic high carbon steel. *Mater. Sci. Eng.* **2004**, *387*, 352–356. [CrossRef]

130. Ohmura, T.; Tsuzaki, K.; Matsuoka, S. Evaluation of the matrix strength of Fe-0.4 wt% C tempered martensite using nanoindentation techniques. *Philos. Mag.* **2002**, *82*, 1903–1910.

131. Waseda, O.; Veiga, R.G.A.; Morthomas, J.; Chantrenne, P.; Becquart, C.S.; Ribeiro, F.; Jelea, A.; Goldenstein, H.; Perez, M. Formation of carbon Cottrell atmospheres and their effect on the stress field around an edge dislocation. *Scr. Mater.* **2017**, *129*, 16–19. [CrossRef]

132. Pereloma, E.V.; Miller, M.K.; Timokhina, I.B. On the decomposition of martensite during bake hardening of thermomechanically processed transformation-induced plasticity steels. *Metall. Mater. Trans.* **2008**, *39*, 3210–3216. [CrossRef]

133. Waterschoot, T.; de Cooman, B.C.; De, A.K.; Vandeputte, S. Static strain aging phenomena in cold-rolled dual-phase steels. *Metall. Mater. Trans.* **2003**, *34*, 781–791. [CrossRef]

134. Suresh, S.; Nieh, T.G.; Choi, B.W. Nano-indentation of copper thin films on silicon substrates. *Scr. Mater.* **1999**, *41*, 951–957. [CrossRef]

135. Michalske, T.A.; Houston, J.E. Dislocation nucleation at nano-scale mechanical contacts. *Acta Mater.* **1998**, *46*, 391–396. [CrossRef]

136. Pathak, S.; Stojakovic, D.; Doherty, R.; Kalidindi, S.R. Importance of surface preparation on the nano-indentation stress-strain curves measured in metals. *J. Mater. Res.* **2009**, *24*, 1142–1155. [CrossRef]

 *metals*

Article

Strength and Fracture Toughness of Hardox-400 Steel

Ihor Dzioba * and **Robert Pała**

Department of Machine Design, Faculty of Mechatronics and Mechanical Engineering,
Kielce University of Technology, Al. 1000-lecia PP 7, 25-314 Kielce, Poland; rpala@tu.kielce.pl
* Correspondence: pkmid@tu.kielce.pl; Tel.: +48-41-3424-303

Received: 1 April 2019; Accepted: 25 April 2019; Published: 30 April 2019

Abstract: This paper presents results of strength and fracture toughness properties of low-carbon high-strength Hardox-400 steel. Experimental tests were carried out for specimens of different thickness at wide temperature range from −100 to 20 °C. The dependences of the characteristic of material strength and fracture toughness on temperature based on experimental data are shown. Numerical calculation of the stress and strain distributions in area before crack tip using the finite element method (FEM) was done. Based on results of numerical calculation and observation of the fracture surfaces by scanning electron microscope (SEM), the critical local stress level at which brittle fracture takes place was assessed. Consideration of the levels of stress and strain in the analysis of the metal state at the tip of the crack allowed to justify the occurrence of the brittle-to-ductile fracture mechanism. On the basis of the results of stretch zone width measurements and stress components, the values of fracture toughness at the moment of crack initiation were calculated.

Keywords: Hardox-400 steel; strength properties; fracture toughness; critical local stress level

1. Introduction

During recent years, new low-carbon ferritic steels have been designed and produced which are characterized by their high yield stress values or high hardness with maintaining a good level of plasticity. Development of this type of steel and carrying out research to determine their mechanical properties is of great importance for their applications in various industries, including building construction, automotive, aerospace, and others.

To achieve such a high strength, these steels are subjected to thermomechanical processing (TMP) [1–5]. In these works, an accurate analysis of the microstructure, resulting from slightly different TMPs and various element contents, is also presented. Generally, it is a multiphase microstructure that consists of ferrite, bainite, retained austenite, and martensite. Changes in the microstructure lead to changes in material properties. The most frequently studied were the effect of microstructural changes on strength characteristics, plasticity and hardness [6–12], and less frequently—on impact resistance [13].

In a few publications, the results of fracture toughness test of high-strength steels (HSS) have been presented [14–21]. Lack of accurate information on the fracture toughness of HSS does not allow to properly assess the strength of structural elements made of these steels and assess the parameters of their safe use [22]. Therefore, this publication presents the results of investigations of mechanical properties of Hardox-400 steels in the temperature range of their use.

During production, the TMP technology aims to achieve uniform yield strengths (e.g., $\sigma_y \geq 960$ MPa for S960QC) or hardness (e.g., HV ≥ 400 for Hardox) for each plate thickness. As a result, negligible differences in the microstructure for different plate thicknesses and across plate thickness can be observed. These microstructural differences lead to small changes in strength properties across thickness, but fracture toughness values change considerably.

The influence of temperature in wide range from −100 to 20 °C on strength and fracture toughness characteristics were tested. Special attention allowed to determine fracture toughness characteristics. The change of critical value of fracture toughness with specimens' thickness was tested. Stress and strain distributions in front of the cracks were calculated by numerical modeling and FEM. Based on the results of numerical calculations, the values of critical local stress were calculated, the achievements of which enables a realization of brittle fracture in the specimens.

This paper presents a review of the results which were obtained in testing of high-strength steels: S960QC and Hardox-400 since 2009. Several papers published in this time period have reported the results obtained during investigation of these steels [16–21]. However, during next testing, the results were complemented and more exact. This paper includes an accurate data received within all test time. This paper mainly presents the results obtained during testing high-strength Hardox-400 steel. The results for strength and fracture toughness properties of the S960QC steel are qualitatively similar.

2. Materials and Methods

2.1. Microstructure and Hardness of Tested Plate of Hardox-400 Steel

The plates of Hardox-400 steel of 30 mm thickness was produced using a controlled thermomechanical treatment. The chemical composition of Hardox-400 steel is shown in Table 1. The microstructure of this steel is tempered martensite–bainite (Figure 1a). The grain sizes were within the range of 5–20 μm. A separate isolated large nonmetallic inclusions of oxides and sulphides (Figure 1b) and particles of titanium nitrides enriched with Nb (Figure 1c) of size 1.0–2.5 μm were present in this steel. Qualitative identification of the chemical composition of these particles was carried out by energy dispersive X-ray spectroscopy analysis (EDX) (JEOL, Peabody, MA, USA), (Figure 2). Additionally, numerous particles of carbides precipitates of sizes 50–300 nm (Figure 1d) were observed in areas of ferrite. The difference of microstructure between layers in the middle part of a plate and near surface is insignificant.

Table 1. The chemical composition of Hardox-400 steel according to [23].

C, %	Si, %	Mn, %	P, %	S, %	Cr, %	Ni, %	Mo, %	B, %
0.32	0.70	1.60	0.025	0.01	2.40	1.50	0.60	0.004

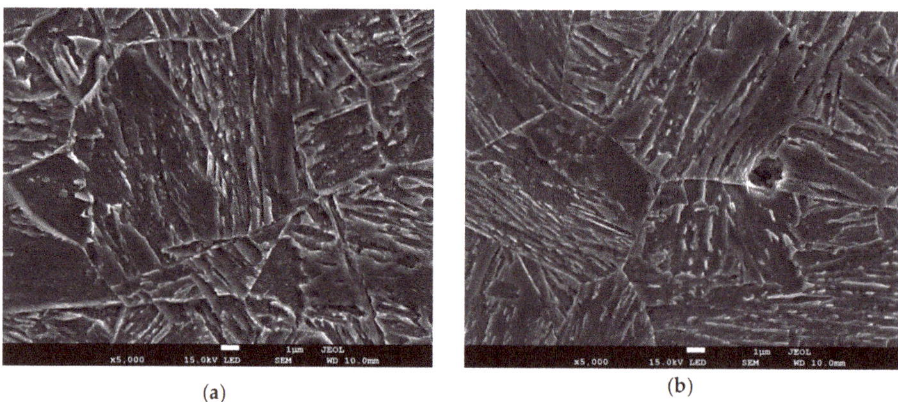

(a) (b)

Figure 1. *Cont.*

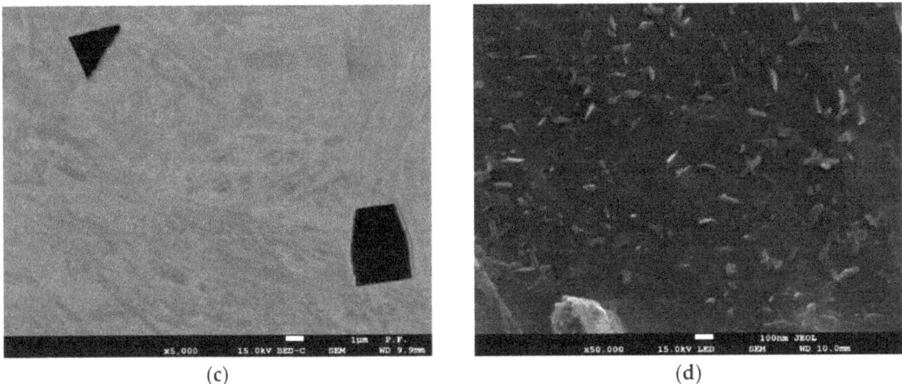

Figure 1. (**a**) Microstructure of Hardox-400 steel (**b**,**c**) with inclusions and (**d**) particles of carbides precipitates in ferritic regions.

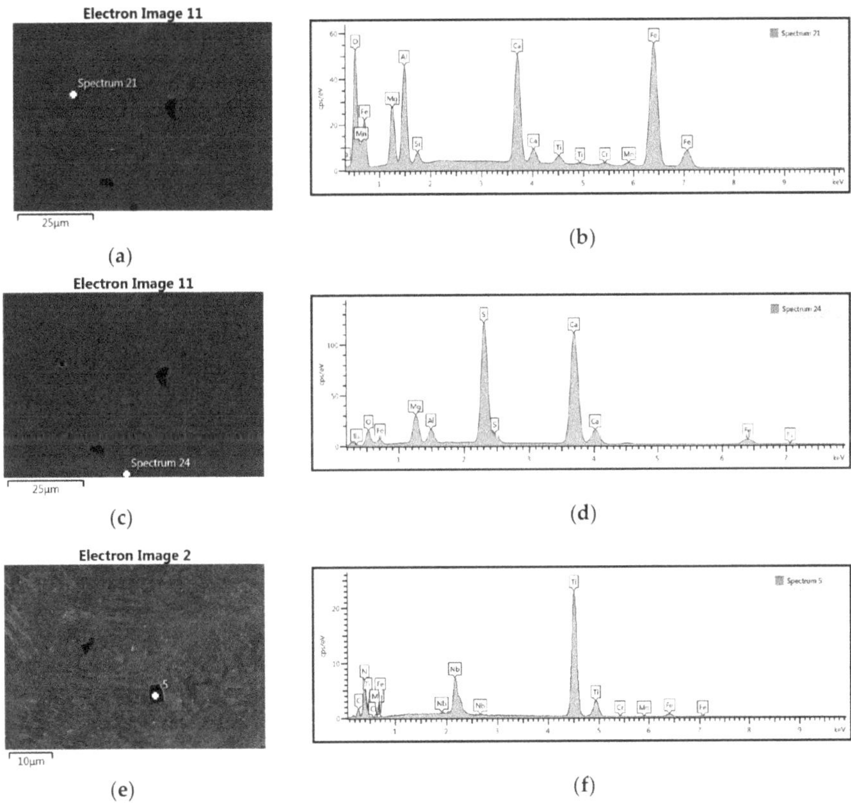

Figure 2. Large particles of inclusions and results of the energy dispersive X-ray spectroscopy (EDX) analysis of (**a**,**b**) oxides; (**c**,**d**) sulfides; (**e**,**f**) titanium nitrides enriched with Nb.

As a result of manufacturing by thermomechanical treatment, the hardness distributions changed along the plate thickness (see Figure 3a). In the middle part of a plate of 30 mm thick, the average level of hardness was about 350 HV10, but the near surface of a plate is about 400 HV10, and on a surface of a plate it exceeds 425 HV10. The significant and considerable scatter of hardness data along thickness was observed.

Tests were conducted on specimens that were cut out from a middle part of a plate of Hardox-400 steel, where hardness level was lower (Figure 3a,b). For determination of strength properties, cylindrical specimens were used with diameter of 5 mm and $L_0 = 25$ mm that were tested at tensile loading. Fracture toughness characteristics were obtained using Single Edge Notch Bend Specimens (SENB) ($B = 1.0$–24.0 mm; $W = 24$ mm; $S = 96$ mm; $a_0/W = 0.5$). Due to the frequent utilization of these types of steel in constructions in regions with low temperature, both strength and fracture toughness properties were determined in a wide temperature range: from −100 to 20 °C.

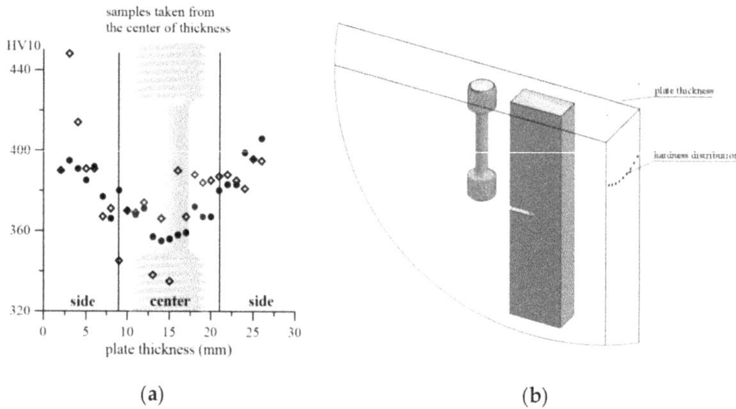

(a)   (b)

Figure 3. (**a**) Hardness distribution through the plate thickness; (**b**) the scheme of cutting out a specimen.

All experimental uniaxial tensile and fracture toughness tests were carried out using MTS (MTS Systems Corporation, Eden Prairie, MN, USA) and Zwick (Zwick Roell Group, Ulm, Germany) universal testing machine with system of automatic control of operating and data recording. The signals of load cell and specimen elongation extensometer were recorded during uniaxial tensile tests, and during fracture toughness tests the signals of load cell, extensometers of displacement of specimen in loading point and crack mouth opening, were recorded in real time and then used for material properties determination. The thermal chamber and vapor of liquid nitrogen was used for specimens tested at lowering temperatures. Temperature measurement made by thermocouple, which was placed immediately on the tested specimen, with accuracy of ±0.1 °C. All metallographic, fractographic tests and analyses were carried out by the scanning electron microscope (SEM) JSM-7100F (JEOL, Peabody, MA, USA).

2.2. Temperature Influence on Strength Properties of Hardox-400 Steel

The strength properties through the thickness were not uniform, but differences of these properties obtained on specimens cut from the middle layer and near surface were not large. The average values of strength characteristics for specimens from middle part of a plate were: $\sigma_y = 950$ MPa, $\sigma_{uts} = 1205$ MPa; and for specimens from materials near surface: $\sigma_y = 975$ MPa, $\sigma_{uts} = 1220$ MPa. However, during determining the strength properties of Hardox-400 steel, a wide scatter of data was observed for both localizations of specimens from the middle part and near the surface of plates. In the following part of the paper, we present the results according to the specimens cut out from the central layer of plates.

Based on the data recorded during uniaxial tensile tests, the nominal and real strength characteristics of Hardox-400 steel were received. Figure 4 shows the stress–strain (σ–ε) graphs for different test temperatures (−100–20 °C). While on scatter of data, the values of strength and plastic characteristics of Hardox-400 steel are increased with the decrease of temperature. These trends are well described by linear functions. The average values of the characteristics for test temperature are presented in the Table 2. There are also the equations obtained by fitting of linear function of all points received in all tested temperatures. The data of the true $\sigma–\varepsilon$ graphs are necessary to create material model for the numerical calculation, which will be presented in the next part of the paper.

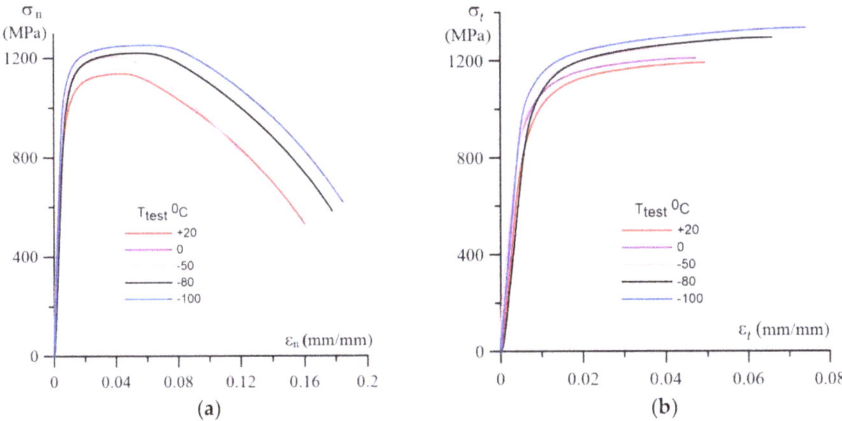

Figure 4. The diagrams of $\sigma–\varepsilon$ for the Hardox-400 steel: (**a**) Nominal and (**b**) true.

Table 2. The strength and plasticity properties of Hardox-400 steel.

Hardox-400	20 °C		0 °C		−50 °C		−80 °C		−100 °C		Equation
	Value	scat. %	Value	scat. %	Value	scat. %	Value	scat. %	Value	scat. %	
σ_{y_n}; σ_{y_t} (MPa)	947 955	1.1	958 966	0.4	990 999	0.5	1020 1030	0.1	1053 1062	1.4	$\sigma_{y\_t} = -0.8534 \cdot T_{test} + 961.91$
σ_{uts_n}; σ_{uts_t} (MPa)	1149 1205	1.1	1162 1212	0.2	1195 1254	0.7	1221 1297	0.3	1245 1327	1.1	$\sigma_{uts\_t} = -1.0138 \cdot T_{test} + 1212.39$
E_n; E_t (GPa)	180 181	2.1	179 180	0.8	176 177	3.1	173 174	5.9	202 203	8.3	$E\_t = -0.0767 \cdot T_{test} + 178.37$
n_n; n_t	10.9 9.8	4.6	10.2 9.3	2.8	10.96 9.76	4.7	12.42 10.31	1.0	14.36 11.72	7.3	$n\_t = -0.0131 \cdot T_{test} + 9.57$
ε_{uts_t}	0.05	6.4	0.045	9.0	0.052	13.7	0.066	1.1	0.071	2.7	$\varepsilon_{uts\_t} = -0.0002 \cdot T_{test} + 0.05$
ε_{c_n}	0.16	5.7	0.134	14.5	0.162	4.2	0.175	1.7	0.184	0.2	$\varepsilon_{cn} = -0.0003 \cdot T_{test} + 0.15$

2.3. The Influence of Temperature and Specimen Thickness on Fracture Toughness Characteristics

Fracture toughness characteristics were determined according to the recommendations of ASTM standards [24,25]. A drop potential and compliance change methods were used to obtain the critical values of J-integral, J_C, when ductile toughness at crack initiation and propagation took place. While, if crack initiation and propagation occurred as brittle, the critical value of J-integral, J_C, was calculated according to the formula: $J_C = \eta E_C/(B(W - a_0))$. If all requirements of ASTM standards according to specimen dimensions and procedures were satisfactory, the J_C are the material characteristic and denoted as J_{IC}. In the research program, SENB specimens with different thickness were tested. For some of them, the requirement on critical thickness was not observed, so the critical value of J-integral was not classified as J_{IC}, but as J_C. All results of the critical fracture toughness values are presented in J-integral units; for recalculation to stress intensity factor (SIF) units, the equation $K_C = (EJ_C/(1 - \nu^2))^{0.5}$ should be used.

A significant scatter range was typical for fracture toughness critical values J_C of Hardox-400 steel. A scatter range increased when fracture mechanisms of subcritical crack growth in SENB specimens are fully ductile, or firstly ductile and then brittle. When subcritical crack growth in specimens occurred according to the pure brittle-by-cleavage mechanism, the scatter of fracture toughness data was small. An example of data distributions of J_C of test temperature T for specimens of 2 mm and 12 mm are shown in Figure 5. The large data scatter was observed in the regions of ductile fracture for specimens of thickness 2 mm (see Figure 5a) and in the region of ductile-to-brittle change for specimens of thickness 2 mm and 12 mm (see Figure 5a,b). In specimens of both thicknesses in brittle fracture regions, the scatter of values were slight (see Figure 5a,b). Qualitatively, similar types of distributions results were observed for specimens of thickness 1.0, 2.0 mm, and from 8.0 to 24.0 mm.

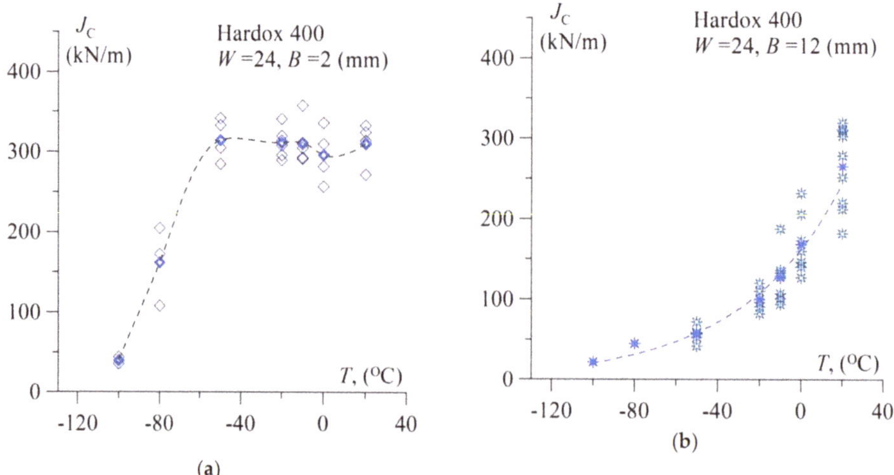

Figure 5. The critical values J_C with temperature changes for (**a**) specimens of 2 mm and (**b**) 12 mm thickness.

The distribution graphs of the average critical fracture toughness values J_C of test temperature T in intervals from −100 to 20 °C for different specimens thickness are presented in Figure 6a. In the distribution for specimens of 4 mm thickness, we can observe three regions of the fracture process: Ductile (0–20 °C), ductile-to-brittle (−60–0 °C), and brittle (−100–−80 °C). For thick specimens, which are less than 4 mm-thick, in J_C of T distributions were clearly present in two regions: Ductile fracture (−60–20 °C) and ductile-to-brittle (−100–−80 °C). The fracture toughness data level on higher plateau, where crack growth occurs according to full ductile mechanisms, decreases with reduction of thickness from 4 to 1 mm. The graphs J_C of T distributions for specimens of 8 mm and more were similar. Lower plateau, when the crack growth occurs to the brittle mechanism, and ductile-to-brittle regions were the two regions presented for specimens of thickness from 8 to 24 mm. For specimens of 12 mm thickness and more, the distributions of $J_C = f(T)$ were similar and practically covered.

The distributions graphs J_C of B for tested temperatures are shown in Figure 6b. For the test temperature of 20 °C, ductile fracture mechanism predominated in crack growth for all specimens' thickness. The critical fracture toughness values, J_C, for specimens of thickness from 4 to 24 mm were similar, and equal to about 340–360 kN/m. Only for thick specimens (1 mm and 2 mm), fracture toughness decreases with thickness. The similar critical fracture toughness data, J_C, were obtained for all specimens tested at $T = -100$ °C (about 33 kN/m). These data belong to lower plateau, where mechanism of crack extension is brittle. For specimens tested at temperature intervals from −80 to 0 °C, the critical values of fracture toughness changed with thickness. For specimens, which thickness corresponds to standard requirements, obtained data were of the similar level. If the thickness of specimens was a little less than critical standard thickness, the fracture toughness level increased. But for thick specimens' critical values of fracture toughness rapidly decreased. This tendency was most clearly observed for $T = 0$ °C and disappeared with test temperature lowering.

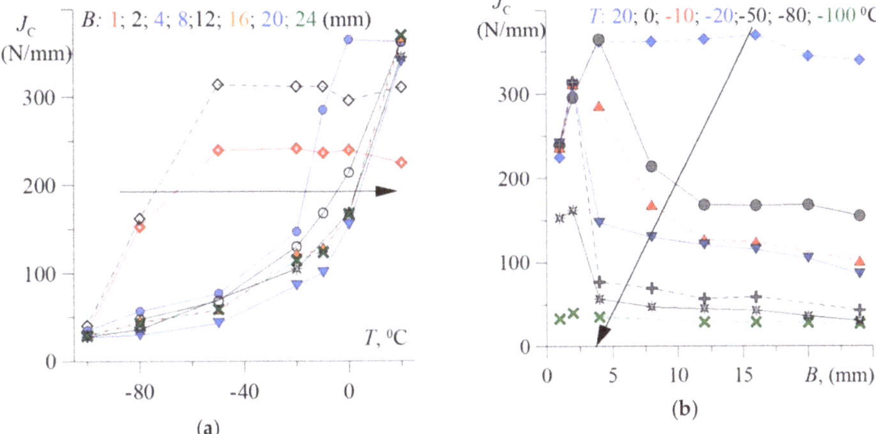

Figure 6. (**a**) The J_C with test temperature dependence for specimens of different thicknesses; (**b**) the J_C changes with specimens thickness for different test temperatures.

The differences in morphology of fracture surfaces of specimens of different thicknesses and tested at different temperatures are shown in Figure 7. For all thickness specimens tested at temperatures corresponding to the higher plateau on the fracture surfaces should show the full ductile fracture mechanism of cracks propagations. After wide zone of stretching crack growth by voids mechanisms (voids nucleation on particles of inclusions and precipitates), they rise and coalesce (see Figure 7a,b). At the region of ductile-to-brittle fracture, a stretch zone formed and thin zone with voids were created, then, crack propagated according to brittle mechanism by cleavage (see Figure 7c,d). On the lower plateau cracks grew by fully brittle cleavage cracking (see Figure 7e,f).

The width of stretch zones on fracture surfaces was used to determine the fracture toughness value at the crack initiation moment. From images of the fracture surfaces, it can be shown that a maximum of stretch zone width (*SZW*) corresponded to specimens, that fracture toughness concerns the higher plateau. In the transition region, the width of stretch zone reduced, as shown on fracture surfaces in Figure 7c,d, and when fracture toughness corresponded to lower plateau the *SZW* was inexpressive (weak) (see Figure 7e,f). The details of fracture toughness calculations in the crack initiation moment are presented further in the paper.

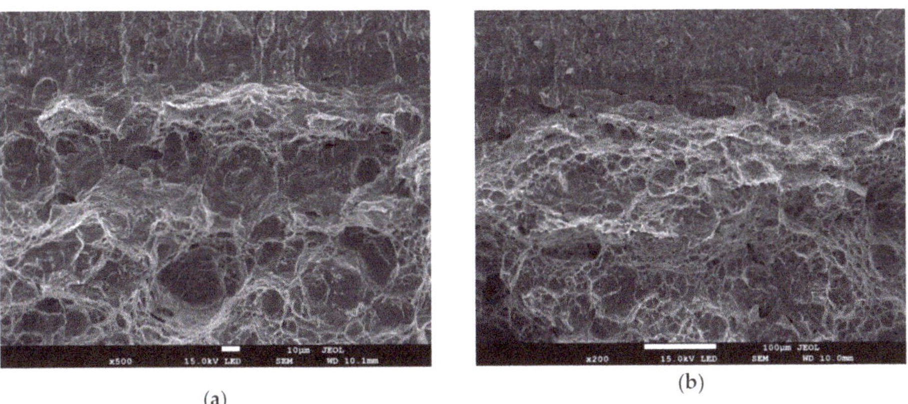

(a)

(b)

Figure 7. *Cont.*

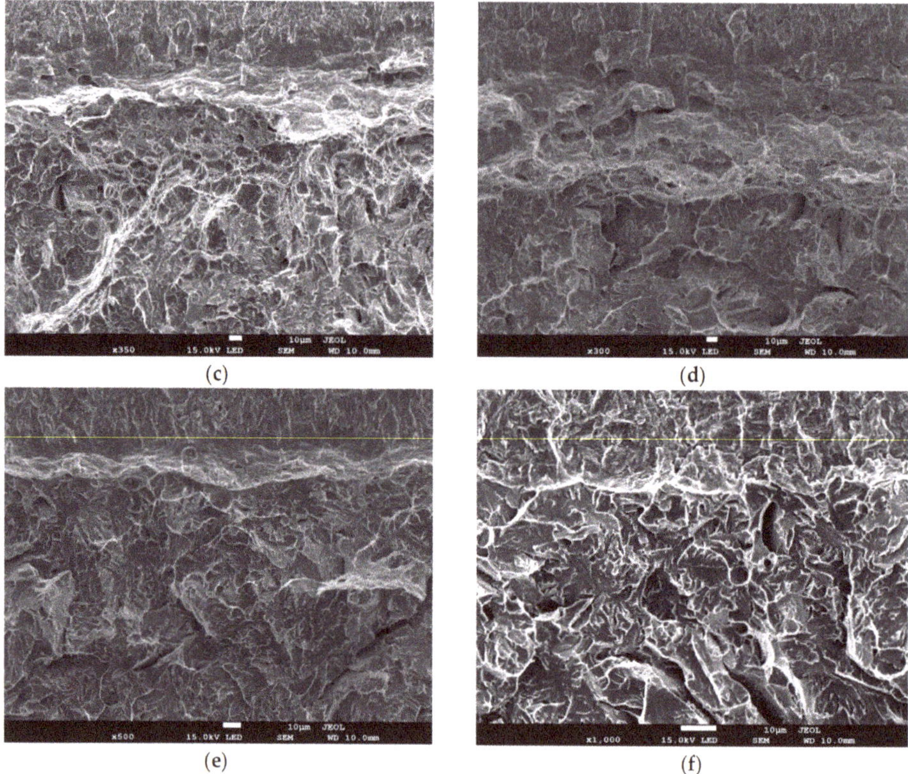

Figure 7. Microstructure of Hardox-400 steel (**a–c**) with inclusions and (**d**) particles of carbides precipitates in ferritic regions. The fracture surfaces of the specimens: From upper plateau (**a**) $B = 4$ mm, $T_{test} = 0$ °C and (**b**) $B = 12$ mm, $T_{test} = 20$ °C; from ductile-to-brittle region (**c**) $B = 8$ mm, $T_{test} = -50$ °C and (**d**) $B = 12$ mm, $T_{test} = -20$ °C; from lower plateau (**e**) $B = 2$ mm and (**f**) $B = 12$ mm, $T_{test} = -100$ °C.

For ferritic steels, the dependence of fracture toughness from temperature in brittle-to-ductile range is recommended to be describe by master curve (MC) [26]. The MC creation procedure is derived from the statistical analysis based on the Weibull model [27–31]. The application of MC is restricted to ferritic steels with a yield strength range between 275 MPa and 825 MPa. The values of yield strength of high-strength steels are higher. The attempts to describe the dependence of fracture toughness from temperature of Hardox-400 steel according to procedures recommended for master curves [26] were not successful. For high-strength steels, such as Hardox-400 and S960QC, other values of transition temperature and minimum value of fracture toughness should be introduced into the formula of master curve. Details of the investigations are presented in the papers [32,33].

2.4. Local Stress and Strain Distributions in Front of the Cracks

In the local criteria of fracture analysis, the local distributions of stress and strain before crack tip plays a great importance [34–38]. Based on these distributions, some values should be calculated as stress state triaxiality factors or Lode factor, which make it possible to provide the type of fracture mechanism that is realized during crack propagation. Knowledge of these distributions is required to establish the level of critical stress causing brittle fracture.

The stress and strain distributions before crack tip were calculated using the FEM. Material characteristics were defined based on the true σ–ε dependences obtained from a uniaxial tensile test. As a result of numerical calculations, stress distributions in front of crack were determined. The calculations were taken from the model of a three point bend specimen SENB, that has been used during fracture toughness tests (see Figure 8a). Calculations were performed in ADINA program. Length of the fatigue crack was equal to its average value calculated on the basis of standards [24,25]. The crack front was reflected as an arc of 0.01 mm radius (see Figure 8b). The specimen in thickness direction was divided into 11 layers, and distances between the layers decreased in the surface direction. The ratio of the thickness of the outer layer and the middle layer of the sample was 0.3. Size of the mesh elements decreased with approaching to the crack front. The zone surrounding the crack tip with a radius of 2 mm was divided in a radial direction into 30 sections with a density towards the tip of 14. In the model were applied 8-nodes, three-dimensional finite elements. In a definition of boundary conditions was a blocked possibility of a displacement of the specimen surface that is common with the crack front z, the surface in the specimen axis from x direction, and the supporting roll was completely immobilized. In the calculations was steered the displacement of the roll that loaded the tested specimen. The value of displacement corresponded to the initiation moment of the subcritical crack, determined as a result of the experimental research. The model of a large strain was adopted. The true σ–ε graphs were used as a material model for the numerical calculation.

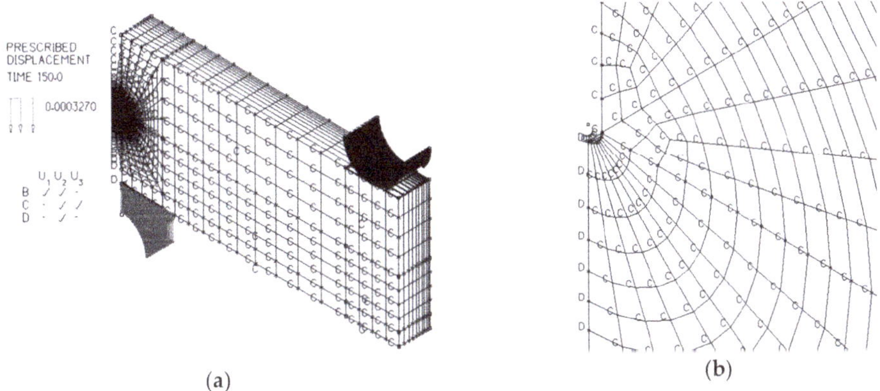

Figure 8. (a) The SENB specimen model and mesh used in numerical calculation; (b) crack tip zone details.

3. Results

The local stress and strain distributions before crack front in the critical moment of subcrack initiation are shown on graphs in Figure 9. Stress distributions for three perpendicular directions with distance from crack tip r at $T_{\text{test.}} = -100\ °C$ in the middle plane of specimen are presented in Figure 9a (where, σ_{11} is a stress in plane of crack in accordance with growth direction, σ_{22} is a stress perpendicular to crack plane in direction which opening crack, and σ_{33} is a stress in plane of crack in direction of thickness). The stress distributions at other test temperatures are qualitatively similar. The point of maximum stress distributions moves from crack tip with test temperature increase, as shown for the opening stress σ_{22} in Figure 9b. The opening stress σ_{22} is the highest and most important in the fracture analysis. In the next analysis, attention will be concentrated mainly on the opening stress σ_{22}.

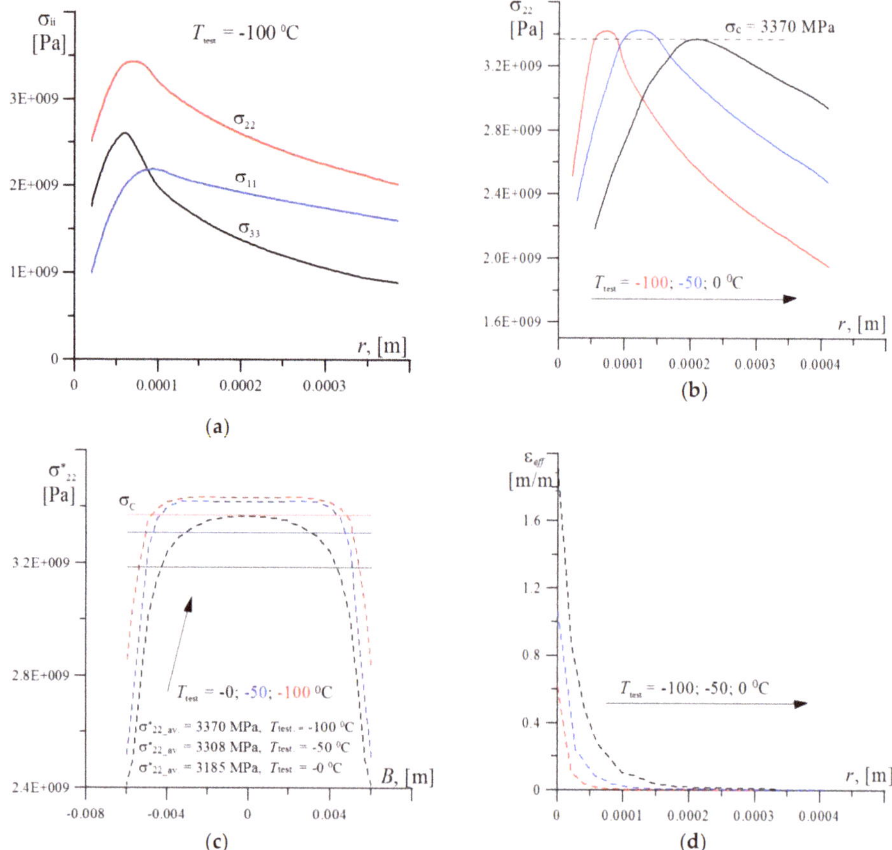

Figure 9. (**a**) The distributions of stress components before crack tip at $T_{\text{test.}} = -100\ °C$; (**b**) the distributions of the opening stress σ_{22} at different test temperature; (**c**) the distributions of max values σ^{*}_{22} through thickness; (**d**) the strain distributions at test temperatures.

4. Discussion

According to the model of a brittle fracture [34,37,38], it could happen if opening stress σ_{22} distribution exceeds the critical level σ_C on critical distance l_C. Full brittle fracture was observed during specimens tested at $T_{\text{test.}} = -100\ °C$ (see Figure 7e,f). At this $T_{\text{test.,}}$ the maximum value of the opening stress reached the level of more than $\sigma^{*}_{22} = 3440$ MPa (see Figure 9a). The distributions of maximum opening stress σ^{*}_{22} values before crack front through thickness of the tested specimens are presented in Figure 9c. Taking into account the fact that at $T_{\text{test.}} = -100\ °C$, fracture of tested specimens was fully brittle, the average level of σ^{*}_{22} through thickness can be considered as critical stress level, so that σ_C needs to realise brittle fracture. For Hardox-400 steel the critical stress level is determined as $\sigma_C = 3370$ MPa. The average levels of $\sigma^{*}_{22\_\text{av.}}$ through thickness distributions for different $T_{\text{test.}}$ shown in Figure 9c.

The critical stress level σ_C as straight line is presented in Figure 9b,c. The stress distributions σ_{22} and σ^{*}_{22} for specimens tested at $T_{\text{test.}} = -100\ °C$ and $-50\ °C$ exceed the critical stress σ_C on some regions, which are conditions for the realization of brittle fracture. While the stress distributions σ_{22} and σ^{*}_{22} for specimens tested at $T_{\text{test.}} = -0\ °C$ puts below the critical stress level σ_C, thus the fully ductile fracture mechanism must be realized during crack growth. The same ductile mechanism of fracture by growth and coalescence of voids had taken place for specimens tested at $T_{\text{test.}} = -0\,°C$ and $20\,°C$ (see Figure 9a,b).

The effective strain distributions presented in Figure 9d. Comparing the stress and strain distributions (see Figure 9b,d), we can observe that strain reached high values immediately before a crack tip, where stress was significantly decreased. In segments where the opening stress σ_{22} exceeded the critical level σ_C, the strain was low, less than 0.03 m/m. So, for the realization of mechanism by brittle fracture, there must be present high level of opening stress and insignificant strain level. Under the test temperature conditions $-100\,°C$, the low strain level of 0.4–0.6 m/m presented before crack tip and specimen breakthrough was characterized by nonsignificant plastic deformations of the stretch zone width about 5–15 μm (see Figure 7e,f).

With the rise of test temperature, the strain level before crack tip increased also and lead to more developed fracture by ductile mechanisms. On the photograph in Figure 7c, the width of stretch zone is equal about 20 μm and it was formed at strain level of 0.4–1.0 m/m for specimen tested at $T_{\text{test.}} = -50\,°C$. On the fracture surface of specimen tested at $T_{\text{test.}} = -20\,°C$, there can be observed a stretch zone of 20–25 μm width and fragments of the area of crack fracture surface by mechanisms of growth and coalescence of voids (see Figure 7c,d). For specimens tested at $T_{\text{test.}} = 0\,°C$, the *SZW* formed at strain level from range 0.40–1.8 m/m and it was equal about 40–45 μm (see Figure 7a).

Influence of specimen thickness on stress and strain distributions in the area before crack tip was also carried out. The results of calculation for the selected specimens of different thickness and tested at various temperature are presented in Figure 10.

According to numerical calculation results, brittle-by-cleavage fracture of all tested thickness can take place only at $T_{\text{test}} = -100\,°C$ or lower. In this case, the opening stress σ_{22} exceeded the critical level σ_C for specimens of all thickness (see Figure 10a,d). Investigation of fracture surfaces confirm this results (see Figure 7e,f). At $T_{\text{test.}} = -80\,°C$ in specimen of $B = 2$ mm sharply dropped the σ_{22} stresses to the level below σ_C (see Figure 10b,d), while effective strain ε_{eff} rapidly grew (see Figure 10e)—this means that ductile mechanism of subcrack propagation will be implemented. This suggestion is confirmed by the experimental results shown in Figure 6.

Temperature increase caused a reduction of the level of σ_{22} stresses and replacement of fracture mechanism from brittle on ductile, especially in thinner specimen. According to the data presented in Figure 10c, at test temperature $-20\,°C$ in specimens of thickness less than 4 mm should occur ductile fracture mechanisms of subcrack propagation. In specimens of thickness, more than 4 mm brittle fracture may occur. While in the area between crack tip and region, where $\sigma_{22} > \sigma_C$, current high level of strains (see Figure 10f), the two types of fracture will be present in crack propagation—ductile and brittle (see Figure 7c,d).

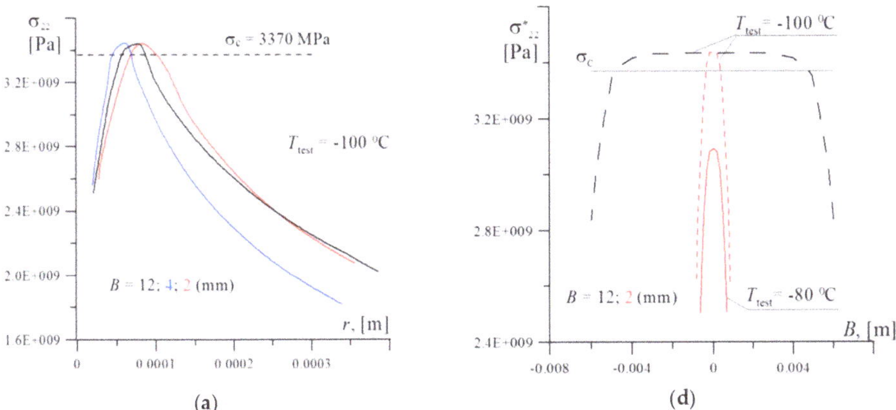

(**a**)　　　　(**d**)

Figure 10. *Cont.*

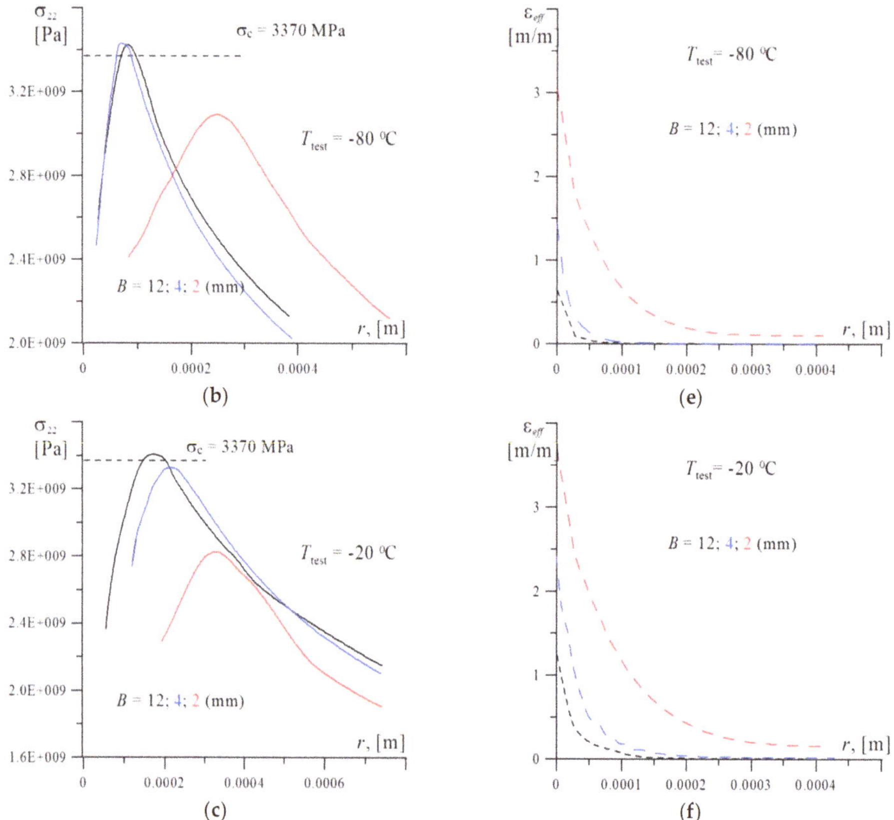

Figure 10. (**a**) The distributions of σ_{22} for specimens of B = 12. 0, 4.0, 2.0 (mm) at T_{test} = −100 °C; (**b**) −80 °C; (**c**) −20 °C. (**d**) The distributions of σ^{*}_{22} through thickness for specimens of B = 12. 0 and 2.0 (mm) at T_{test} = −100 °C and −80 °C. (**e**) The distributions of ε_{eff} for specimens of B = 12. 0, 4.0, 2.0 (mm) at T_{test} = −80 °C and (**f**) −20 °C.

The procedure for the determination of the critical fracture toughness value at the moment of start of subcrack, J_i, is based on the determination of the stretch zone width (Δa_{SZW}), which occurs prior to the subcritical crack initiation [39–41]. When a subcrack initiation is preceded by significant plastic strain, the formula proposed by Shih [42] is the most appropriate one:

$$J_i = (2\sigma_f/d_n)\Delta\bar{a}_{\text{SZW}}, \tag{1}$$

where, $\sigma_f = 0.5(\sigma_y + \sigma_{ut})$; d_n is a function dependent on the material hardening coefficient n and triaxiality coefficient $T_Z = \sigma_{33}/(\sigma_{11} + \sigma_{22})$ [43]. Values T_z were computed based on stress distribution data, which were determined in numerical methods. Coefficient d_n determined on the basis of formulas proposed by Guo [44]:

$$d_n = d_0 - (2T_z)^{a_3(n)}(d_0 - d_5), \tag{2a}$$

for $n > 5$:

$$
\begin{aligned}
d_0 &= 1 - 0.1240(1/n)^{1/2} + 0.8968(1/n) - 13.3941(1/n)^{3/2} + 15.3139(1/n)^2; \\
d_5 &= 0.78 + 0.0277(1/n)^{1/2} - 3.0791(1/n) + 2.4709(1/n)^{3/2}; \\
a_3(n) &= 11.4 - 45.8(1/n)
\end{aligned}
\tag{2b}
$$

The view of stretch zone obtained by SEM is shown in the upper part of the photo in Figure 11a. At the critical moment, the microcracks or voids of the damaged material before the blunted crack connects with the blunted crack and subcrack propagation starts. Based on the measurement of *SZW*, the critical values of fracture toughness at crack initiation, J_i, were estimated. The details of *SZW* measurement are presented in the papers [45,46]. The critical values of *J*-integral at the subcrack initiation, J_i, in the temperature range for which the growth proceeds according to the mixed mechanism (-20–20 °C), are located lower than J_{IC} (see Figure 11b). The difference increases with a rise of the test temperature. The reason why differences in J_{IC} and J_i values occur is the fact that for ductile growth of the subcrack, the values of J_{IC}, are characterized of fracture toughness for averaged extension of $\Delta a = 0.2$ mm length, while J_i—at the instant of initiation.

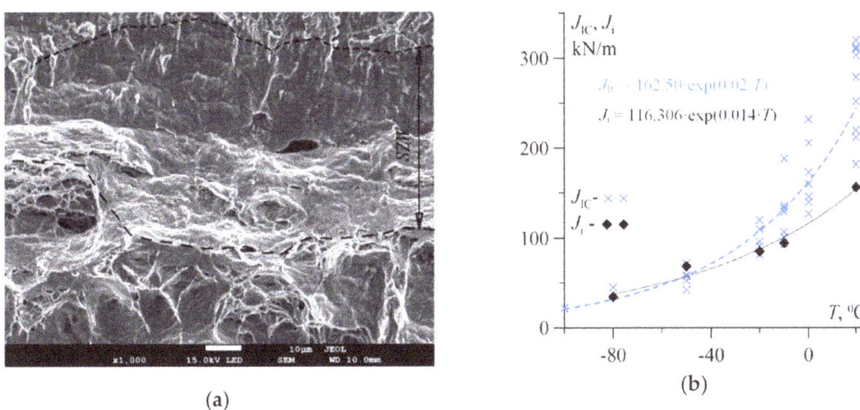

Figure 11. (**a**) A stretch zone view; (**b**) a comparison of the critical values of the J_{IC} and J_i.

5. Conclusions

As a result of the conducted research, dependencies of strength characteristics and fracture toughness in the range of operating temperatures (-100–20 °C) of Hardox-400 steels were determined. Also in this temperature range, the effect of thickness on the critical value of fracture toughness was investigated. The obtained results will allow to perform strength analysis of structural elements made of Hardox-400 steel according to FITNET [22] or other similar procedures. The results obtained during the examination of mechanical properties of Hardox-400 steels allowed to formulate several detailed conclusions presented below.

Hardox-400 steel is low-carbon steel which was produced using a controlled thermomechanical treatment that leads to formation of microstructure of tempered bainite–martensite, which contains numerous particles of carbides precipitates and separate isolated large nonmetallic inclusions (see Figure 1). This steel is characterized by high level of strength properties and relatively good level of plasticity (Table 2). Attention should be paid to the fact of increasing the plasticity with lowering test temperature. Also it is important to note that the hardness in the middle layers of the plate is significantly reduced. Lowering hardness of steel leads to a slight reduction of strength properties and slightly larger changes of fracture toughness.

The test temperature and thickness of specimens have a significant influence on the level of fracture toughness characteristics of Hardox-400 steel (see Figure 5). But this influence is not stable. High-strength steel of similar microstructure (S960QL) tested at room temperature was characterized of ductile fracture mechanism despite of the occurrence of the local areas of brittle fracture [47]. Generally, with test temperature decrease, fracture toughness characteristics of Hardox-400 steel decreases too. Such trends correspond to specimens, in which plane strain is dominant and the condition on the specimen thickness, $B \geq 25 J_{IC}/\sigma_y$, is fulfilled. But for specimens, in which plane strain is not dominant, the critical fracture toughness values show similar high values of J_C and located on higher plateau of brittle-to-ductile dependence. When more restrictive conditions on the specimen thickness were fulfilled, $B \geq 2.5(K_{IC}/\sigma_y)^2$, the fracture toughness values regardless of specimens' thickness were similar and located on the lower plateau. In this case, a full brittle-by-cleavage mechanism fracture takes place in tested specimens.

Brittle fracture is very dangerous, because it causes immediate destruction of elements. In order to assess the occurrence of a critical situation in which brittle fracture takes place, numerical FEM modelling and calculation of local stress distributions before the crack tip were performed. Based on the analysis of stress distributions and fractographic tests of specimens' fracture surfaces, the critical level of opening stress σ_{22} for Hardox-400 steel was determined as $\sigma_C = 3370$ MPa. If the stress σ_{22} exceeds the critical level σ_C, a brittle fracture occurs in the appropriate area and it grows in net cross section of the test specimens. When the stresses σ_{22} were lower of the critical level σ_C, a ductile fracture mechanism of crack growths by voids nucleation, growth and coalescence were observed. Hardox-400 steel contains large particles of inclusions, which additionally increases the level of local stresses. This is the reason why, in steel, there is a large data scatter of critical fracture toughness values J_C at test temperatures higher than $-50\,°C$.

Joint consideration of stress and strain levels in the specimens allows us to determine the subcrack growth mechanism: Full brittle, brittle-ductile, and full ductile. The results of numerical analysis were confirmed by the results of experimental research and the observation of fracture surfaces of specimens on the SEM.

Author Contributions: Conceptualized research, methodology and wrote article, I.D.; writing-review & editing, investigation, I.D. and R.P.; data curation, visualization R.P.

Funding: The research was financed by the National Science Centre, Poland (No. 2017/25/N/ST8/00179).

Acknowledgments: Authors would like to thank P. Furmanczyk for help in carrying out research by SEM.

Conflicts of Interest: The authors declare no conflicts of interest.

References

1. Porter, D. Developments in hot-rolled high strength structural steel. In Proceedings of the Nordic Welding Conference 2006, Tampere, Finland, 8–9 November 2006.
2. Kumar, A.; Singh, S.B.; Ray, K.K. Influence of bainit-martensite content on the tensile properties of low carbon dual phase steels. *Mater. Sci. Eng.* **2008**, *474*, 270–282. [CrossRef]
3. Somani, M.C.; Porter, D.A.; Karjalainen, L.P.; Suikkanen, P.P.; Misra, R.D.K. Process design for tough ductile martensitic steels through direct quenching and partitioning. *Mater. Today Proc.* **2015**, *2*, S631–S634. [CrossRef]
4. Kaijalainen, A.; Vahakuopus, N.; Somani, M.; Mehtonen, S.; Porter, D.; Komi, J. The effects of finish rolling temperature and niobium microalloying on the microstructure and properties of a direct quenched high-strength steel. *Arch. Metall. Mater.* **2017**, *62*, 619–626. [CrossRef]
5. Zhao, J.; Jiang, Z. Thermomechanical processing of advanced high strength steels. *Prog. Mater Sci.* **2018**, *94*, 174–242. [CrossRef]
6. Grajcar, A.; Kilarski, A.; Kozlowska, A. Microstructure—Properties relationships in termomechanically processed medium—Mn steels with high Al content. *Metals* **2018**, *8*, 929. [CrossRef]
7. Kim, S.J.; Lee, C.G.; Choi, I.; Lee, S. Effects of heat treatment and alloying elements on the microstructures and mechanical properties of 0.15% C transformation induced plasticity-aided cold-rolled steel sheets. *Metall. Mater. Trans. A* **2001**, *32*, 505–514. [CrossRef]
8. Soliman, M.; Palkowski, H. On factors affecting the phase transformation and mechanical properties of cold-rolled transformation-induced-plasticity-aided steel. *Metall. Mater. Trans. A* **2008**, *39*, 2513–2527. [CrossRef]
9. Grajcar, A.; Kwasny, W.; Zalecki, W. Microstructure–property relationships in TRIP aided medium-C bainitic steel with lamellar retained austenite. *Mater. Sci. Technol.* **2015**, *31*, 781–794. [CrossRef]

10. Hu, B.; Luo, H. Microstructures and mechanical properties of 7Mn steel manufactured by different rolling processes. *Metals* **2017**, *7*, 464. [CrossRef]

11. Lee, D.; Kim, J.K.; Lee, S.; Lee, K.; De Cooman, B.C. Microstructures and mechanical properties of Ti and Mo micro-alloyed medium Mn steel. *Mater. Sci. Eng. A* **2017**, *706*, 1–14. [CrossRef]

12. Kaar, S.; Krizan, S.; Schwabe, J.; Hofmann, H.; Hebesberger, T.; Commenda, C.; Samek, L. Influence of the Al and Mn content on the structure–property relationship in density reduced TRIP-assisted sheet steels. *Mater. Sci. Eng. A* **2018**, *735*, 475–486. [CrossRef]

13. Dudziński, W.; Konat, Ł.; Pękalska, L.; Pękalski, G. Structures and properties of Hardox 400 and Hardox 500 steels. *Mater. Eng.* **2006**, *3*, 139–142.

14. Wallin, K.; Karjalainen-Roikonen, P.; Pallaspuro, S. Low temperature fracture toughness estimates for Very High Strength Steel. *Int. J. Offshore Polar Eng.* **2016**, *26*, 333–338. [CrossRef]

15. Wallin, K.; Pallaspuro, S.; Valkonen, I.; Karjalainen-Roikonen, P.; Suikkanen, P. Fracture properties of high performance steels and their welds. *Eng. Fract. Mech.* **2015**, *135*, 219–231. [CrossRef]

16. Dzioba, I.; Pała, R.; Pała, T. Temperature dependency of fracture of high-strength ferritic Hardox-400 steel. *Mech. Autom. Rec.* **2013**, *7*, 222–225.

17. Neimitz, A.; Dzioba, I.; Pała, T. Master Curve of high–strength ferritic steel S960-QC. *Key Eng. Mater.* **2014**, *598*, 178–183. [CrossRef]

18. Neimitz, A.; Dzioba, I.; Janus, U. Cleavage Fracture of Ultra-High-Strength Steels, Microscopic Observations. Numerical Analysis. *Key Eng. Mater.* **2014**, *598*, 168–177. [CrossRef]

19. Neimitz, A.; Dzioba, I.; Pala, R.; Janus, U. The influence of the out-of-plane constraint on fracture toughness of high strength steel at low temperatures. *Solid State Phenom.* **2015**, *224*, 157–166. [CrossRef]

20. Neimitz, A.; Pala, T.; Dzioba, I. Fracture Toughness of Hardox-400 Steel at the Ductile-to-Brittle Transition Temperature Range. *Solid State Phenom.* **2015**, *224*, 167–172. [CrossRef]

21. Neimitz, A.; Dzioba, I. The influence of the out-of and in-plane constraint on fracture toughness of high strength steel in the ductile to brittle transition range. *Eng. Fract. Mech.* **2015**, *147*, 431–448. [CrossRef]

22. *FITNET: Fitness–for–Service. Fracture–Fatigue–Creep–Corrosion*; Koçak, M.; Webster, S.; Janosch, J.J.; Ainsworth, R.A.; Koerc, R., Eds.; GKSS Research Centre Geesthacht GmbH: Stuttgart, Germany, 2008.

23. Hardox Wearparts. Available online: https://www.hardoxwearparts.com/materials/hardox-400 (accessed on 29 April 2019).

24. ASTM E1737-96. *Standard Test Method for J-integral Characterization of Fracture Toughness*; ASTM International: West Conshohochen, PA, USA, 1996; pp. 1–24.

25. ASTM E1820-09. *Standard Test Method for Measurement of Fracture Toughness*; ASTM International: West Conshohochen, PA, USA, 2011.

26. ASTM E1921-05. *Standard test Method for Determination of Reference Temperature, T_0, for Ferritic Steels in the Transition Range*; ASTM International: West Conshohochen, PA, USA, 2011.

27. Wallin, K.; Saario, T.; Torronen, K. Statistical model for carbide induced brittle fracture in steel. *Met. Sci.* **1984**, *18*, 13–16. [CrossRef]

28. Ruggeri, C.; Dodds, R.H.; Wallin, K. Constraint Effects on Reference Temperature, T_0, for Ferritic Steels in the Transition Region. *Eng. Fract. Mech.* **1998**, *60*, 19–36. [CrossRef]

29. Gao, X.; Ruggieri, C.; Dodds, R.H. Calibration of Weibull stress parameters using fracture toughness data. *Int. J. Fract.* **1998**, *92*, 175–200. [CrossRef]

30. Gao, X.; Dodds, R.H. Constraint Effects on the Ductile-to-Cleavage Transition Temperature of Ferritic Steels: A Weibull Stress Model. *Int. J. Fract.* **2000**, *102*, 43–69. [CrossRef]

31. Gao, X.; Dodds, R.H. An engineering approach to assess constraint effects on cleavage fracture toughness. *Eng. Fract. Mech.* **2001**, *68*, 263–283. [CrossRef]

32. Neimitz, A.; Dzioba, I.; Limnell, T. Modified master curve of ultra high strength steel. *Int. J. Press. Vessels Pip.* **2012**, *92*, 19–26. [CrossRef]

33. Neimitz, A.; Dzioba, I.R. Fracture toughness of high-strength steels within the temperature range of ductile-to-cleavage transition. Master Curves. *Mater. Sci.* **2017**, *53*, 141–150. [CrossRef]

34. Ritchie, R.O.; Knott, J.F.; Rice, J.R. On the Relationship between Tensile Stress and Fracture Toughness in Mild Steels. *J. Mech. Phys. Solids* **1973**, *21*, 395–410. [CrossRef]

35. Beremin, N. A local criterion for cleavage fracture of a nuclear pressure vessel steel. *Met. Trans. A* **1983**, *14A*, 2277–2287. [CrossRef]

36. Berdin, C.; Besson, J.; Bugat, S.; Desmorat, R.; Feyel, F.; Forest, S.; Lorenz, E.; Maire, E.; Pardoen, T.; Pineau, A.; et al. *Local Approach to Fracture*; Besson, J., Ed.; Les Presses: Paris, France, 2004.
37. Neimitz, A.; Graba, M.; Gałkiewicz, J. An alternative formulation of the Ritchie-Knott-Rice local fracture criterion. *Eng. Fract. Mech.* **2007**, *74*, 1308–1322. [CrossRef]
38. Dzioba, I.; Gajewski, M.; Neimitz, A. Studies of fracture processes in Cr-Mo-V ferritic steel with various types of microstructure. *Int. J. Press. Vessels Pip.* **2010**, *87*, 575–586. [CrossRef]
39. Miyamoto, H.; Kobayashi, H.; Ohtsuka, N. Elastic-Plastic Fracture Toughness J_{IC} Test Method Recommended in Japan. In Proceedings of the ICF-IQ Conference, Beijing, China, 22–25 November 1983.
40. Schwalbe, K.H.; Landes, J.D.; Heerens, J. *Classical Fracture Mechanics Methods*; GKSS 2007/14.2007; Elsevier: Amsterdam, The Netherlands, 2007.
41. ISO 12135:2002. *Metallic Materials—Unified Method of Test for the Determination of Quasistatic Fracture Toughness*; International Organization for Standardization: Geneva, Switzerland, 2002.
42. Shih, C.F. Relationship between the J-integral and crack opening displacement for stationary and extending cracks. *J. Mech. Phys. Solids* **1981**, *29*, 305–329. [CrossRef]
43. Guo, W. Elastoplastic three dimensional crack border field—I. Singular structure of the field. *Eng. Fract. Mech.* **1993**, *46*, 93–104. [CrossRef]
44. Guo, W. Elasto-Plastic Three-Dimensional Crack Border Field—III. Fracture Parameters. *Eng. Fract. Mech.* **1995**, *51*, 51–71. [CrossRef]
45. Dzioba, I.; Furmanczyk, P.; Lipiec, S. Determination of the fracture toughness characteristics of the S355JR steel. *Arch. Metall. Mater.* **2018**, *63*, 497–503.
46. Furmanczyk, P.; Pała, R.; Dzioba, I. Strength properties and fracture toughness of high strength Hardox-400 steel in brittle-to-ductile temperature interval. In Proceedings of the 26th International Conference on Metallurgy and Materials, BRNO, Czech Republic, 24–26 May 2017; pp. 776–781.
47. Slezak, T.; Sniezek, L.; Torzewski, J.; Hutsaylyuk, V. Evaluation of Fatigue Failure of S960QL Steel in the Conditions of Plastic Strain. *Solid State Phenom.* **2016**, *250*, 175–181. [CrossRef]

Article

Influence of Mechanical Anisotropy on Micro-Voids and Ductile Fracture Onset and Evolution in High-Strength Low Alloyed Steels

Francesco Iob [1], Luca Cortese [2], Andrea Di Schino [3],* and Tommaso Coppola [1]

[1] Rina Consulting—Centro Sviluppo Materiali S.p.A, Via di Castel Romano 100, 00128 Roma, Italy; francesco.iob@rina.org (F.I.); tommaso.coppola@rina.org (T.C.)

[2] Department of Mechanical and Aerospace Engineering, Sapienza University of Rome, via Eudossiana 18, 00184 Roma, Italy; luca.cortese@uniroma1.it

[3] Dipartimento di Ingegneria, Università di Perugia, 06125 Perugia, Italy

* Correspondence: andrea.dischino@unipg.it

Received: 20 December 2018; Accepted: 12 February 2019; Published: 13 February 2019

Abstract: In this paper results of a wide and innovative mechanical assessment, that was performed on large diameter spiral line pipes for gas transportation, are reported. The anisotropic material hardening has been characterized by tensile (smooth and notched specimens), torsion, and compression tests. Tests were performed in the pipe of the pipe with specimens machined along several orientations, taking into account the pipe through thickness direction. The influence of different triaxiality stress states on anisotropic behavior of the material have also been analyzed by means of tensile tests on notched specimens. After the experiments, the material was assessed by measuring the void distribution on the material as is, and on many deformed and fractured specimens, including tensile tests at different triaxiality, and torsion tests. The results showed that in such a class of materials, the experimental void fraction is fully negligible and not related to the applied plastic strain, even at the fracture proximity. As a consequence it can be concluded that, the plastic softening hypothesis may be dropped and damage due to void evolution hypothesis is not adequate.

Keywords: mechanics; microstructure; fracture; anisotropy

---

1. Introduction

Oil and gas transmission pipeline steel, with high deformation capability, is required if environments, involving large strata movement, are considered. In fact, in such cases, the pipe is subjected to large plastic deformations, that could finally cause the pipeline failure. Pipelines that are used in such environments are designed according to the following strain-based approach, in order to prevent failure. This strain-based design strategy for transmission pipelines requires the capability to face high internal operating pressure and high deformation resistance. As a consequence, pipeline steel, designed for strain-based applications, will target high strength/toughness, good welding performance, and excellent plastic deformation capability. This implies that the pipeline steel transverse properties will meet the pipeline steel grade requiring technical specification in terms of strength, toughness, drop weight tear test, and hardness.

In addition to the transverse mechanical characteristic requirements, pipelines designed for such applications are also required to meet longitudinal targets. A higher degree of deformation-enhanced index (n), larger uniform elongation (A_g), low yield stress to tensile strength ratio ($R_{t0.5}/R_m$), continuous yielding (roundhouse) tensile stress-strain curve, with no yield point elongation, are needed to target an appropriate longitudinal deformation ability. The development of high-strength line pipes, aimed to target strain-based applications, has been carried out in recent years [1–4]. In this framework, the design and prediction methods have continuously improved [5]. Considering the plate material development, it is known that a proper design and control of micro-structure are key to success. Steel chemical composition and cooling processes are the most significant parameters affecting the plate micro-structure [6–11]. Also, small differences in cooling rates can strongly modify mechanical properties, such as uniform elongation and $R_{t0.5}/R_m$ [10].

Tensile properties of line pipes are strongly dependent on different positions along the pipe principal axes [12,13]. Anisotropic or orthotropic material properties strongly effect the pipeline performances, and large straining beyond the plastic instability limit (necking) needs to be properly described, taking into account of anisotropic plasticity constitutive law, aimed at capturing the material ductile behavior.

The scope of this paper is to describe a wide and innovative mechanical assessment, performed on large diameter spiral line pipes for gas transportation. The aim of such work is to validate a new plasticity model, developed at Rina Consulting–Centro Sviluppo Materiali [14–16]. Subsequent to mechanical tests, fracture surfaces and the section of selected specimens, were analyzed for voids- and inclusions-counting. Void counting was carried out on virgin material, and on areas next to the fracture surface, where the material reach elevated strain values. The results show that, in the considered material, the void fraction is fully negligible. Moreover, it is not related to the plastic strain applied, even at the fracture proximity. This result confirms those obtained in previous work on other materials [17], and the consequence is that, for such class of materials, the matrix softening hypothesis may be abandoned in the plastic behavior, and that damage evolution is not related do void nucleation and growth.

The Rina-CSM model is based on "Hill (48)" criterion, it provides a 3-D description of the anisotropic behavior to represent the ductile hardening in all the principal directions of the material. Moreover, the Lode angle effect is also introduced in the model, which allows a sound description of the effect of shear stress states. Such a model makes use of a series of coefficients which have a physical meaning, and that can be obtained by tensile and torsion testing, conducted in conventional mechanical laboratories.

The scope of this work is to measure differences in terms of initial yield stress and subsequent hardening of the material for different orientations under tensile, shear, and compressive stress states. The results should provide a useful complete characterization of the investigated materials, and will be used in a future work, to calibrate and implement the above-mentioned numerical model in a commercial finite element software.

2. Materials and Methods

The main information of the pipe used for specimen extraction is reported in Table 1.

Table 1. Investigated material.

Material Grade	Pipe Outer Diameter/mm	Pipe Wall Thickness/mm	Pipe Manufacturing Process
API X70	1219	19.3	Spiral

The chemical composition of the studied material is reported in Table 2.

Table 2. Material chemical composition.

Max, %							
C	Mn	Si	P	S	V	Nb	Ti
0.16	1.7	0.45	0.025	0.020	0.10	0.06	0.006

Material anisotropy has been assessed in terms of tensile and torsion tests performed in the base material. Tensile specimen were machined considering six different orientations. Each type of test was carried out on two specimen, in order to check the repeatability of the results. Three standard tensile tests, with geometry according to UNI EN ISO 6892-1:2009, are carried out on round bar, smooth specimens having 9 mm gauge section diameter, machined starting from the longitudinal (L), transversal (T), and at 45° degrees between L-T directions (45° LT) of the pipe. Three tensile tests are carried out on round specimens, having 2.5 mm gauge section diameter, extracted in the through thickness direction of the pipe (N), and oriented at 45° degrees from the longitudinal and through the thickness direction (45° LN), and at 45° degrees from the transversal and through the thickness direction (45° TN), (Figure 1).

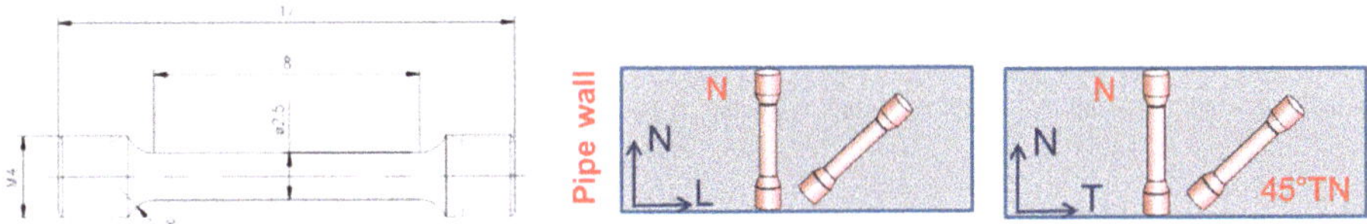

Figure 1. Mini round tensile specimen (mm) (N = normal; L = longitudinal).

Tensile specimen sizes are limited by the pipe wall thickness. To obtain experimental tensile stress strain curve from these specimens, the strain during the test was measured by a general purpose extensometer, with initial gauge length of 4 mm.

Additional tests are carried out on conventional round bar smooth specimens having 9 mm gauge section diameter, machined in the longitudinal (LC), transversal (TC) and 45° degrees between L-T directions (45° LTC) of the coil original rolling direction, (Figure 2).

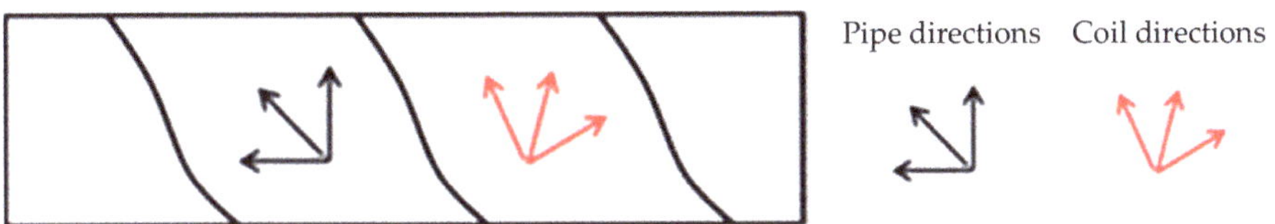

Figure 2. Principal orientations for in plane mechanical test.

Torsion tests were also performed in the longitudinal (L), and transversal direction (T), on round bar specimens having a gauge diameter of 8 mm and a gauge length of 17 mm (Figure 3).

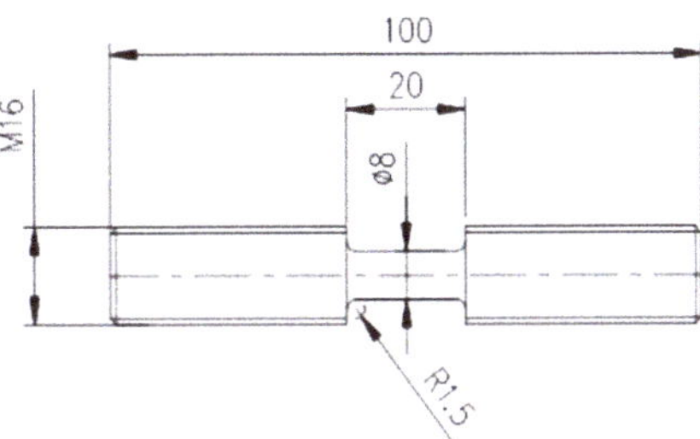

Figure 3. Torsion specimen (mm).

Custom tension-torsion equipment was used, that is capable of maximum axial and torsional loads of 100kN and 1000 Nm axial stroke up to 150 mm and unlimited rotation angle. Torque and rotation are measured by means of a bi-axial load cell and a digital encoder, embedded in the rotational actuator. The machine torsional stiffness is granted by a four-columns frame and by a bulky design of the grips. All tests have been executed in a free-end configuration (null axial load) to ensure a pure shear state of stress.

Further tensile tests are carried out on specimens with notched geometry (Figure 4), with two different notch radii of 10 and 2 mm. The specimens are extracted in the longitudinal (L), and transversal (T) of the pipe. The aim of these tension test is to assess the material anisotropic behavior under different stress states induced by the different notch radii [18].

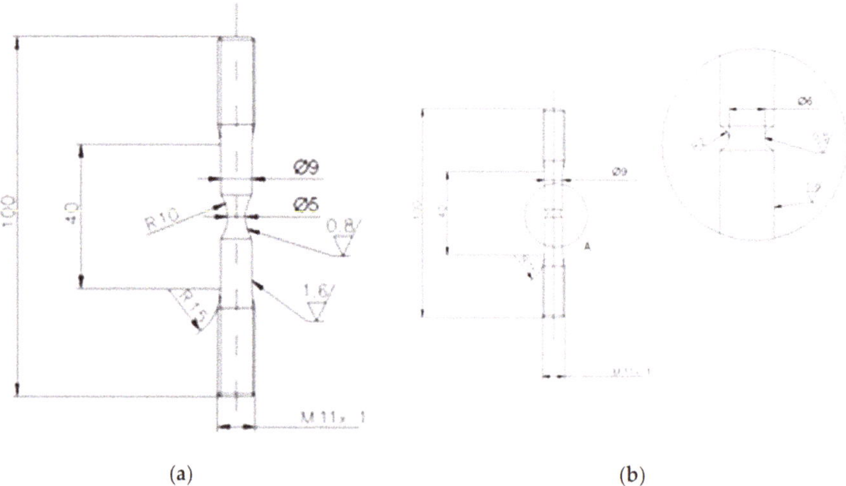

(a)   (b)

Figure 4. Round notched tensile specimen (mm): (**a**) Notch radius = 10 mm; (**b**) Notch radius = 2 mm.

In order to better assess the material behavior in the pipe through thickness direction (N), two compressive tests were performed on cylindrical specimen. As for the tensile specimens, the compression specimen size is limited by the pipe wall thickness. For current tests, a 12 mm height, with 6 mm diameter cylindrical geometry, was used.

After the execution of the mechanical tests, the fracture surfaces were analyzed by Scanning Electron Microscope Field Emission Gun (SEM FEG) LEO 1550 (Zeiss, Oberkochen, Germany) Selected specimens were sectioned and analyzed for voids and inclusions. Considered areas for such measure was far from fractured surface to assess the void fraction on as is material Next void counting was extended to areas taken along the longitudinal specimen axis, starting from the fractured surface (Figure 5).

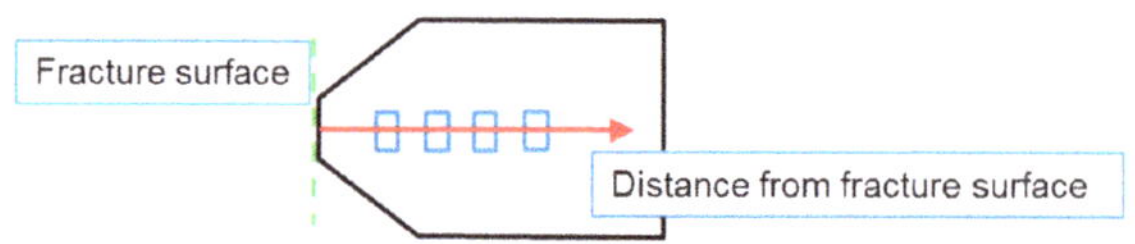

Figure 5. Void counting procedure.

3. Results

3.1. Experimental Tests Results

The results of the tensile tests on the round bar specimens, are shown in terms of true stress-strain curves for the strain, up to specimen necking. Standard tensile tests, performed on plane pipe directions, are reported in the subsequent Figure 6.

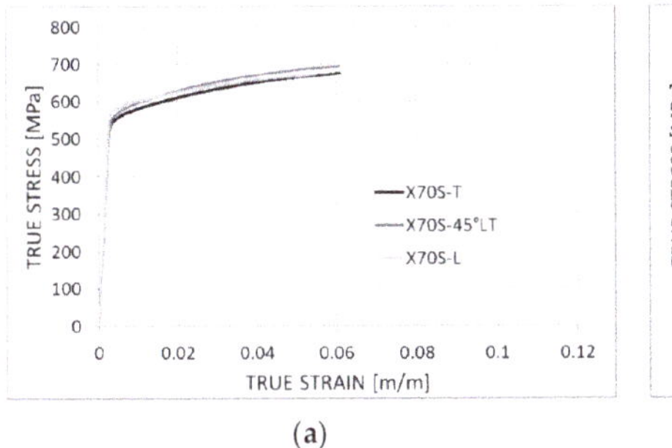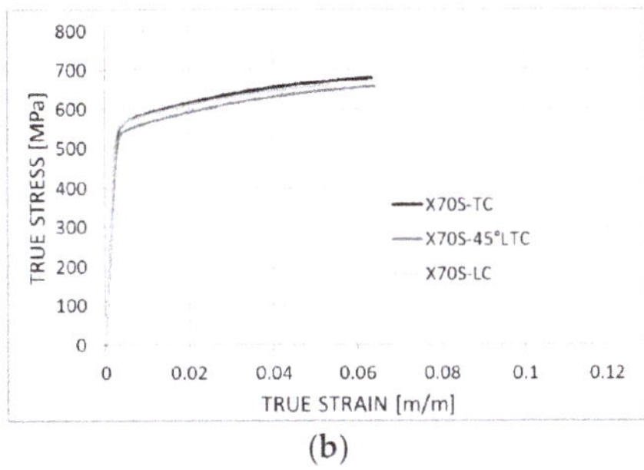

(a)   (b)

Figure 6. True stress-strain curves from tensile tests on X70 48″ Outer Diameter (OD) spiral Pipe: (**a**) Tests according to Pipe directions; (**b**) Tests according to Coil directions.

Considering a spiral pipe forming angle of about 25°, for the spiral pipe, 6 different orientations for the in plane tensile test are considered (Table 3).

Table 3. Orientation for tensile tests on X70 48″ OD Spiral Pipe.

Orientation ID	Orientation Direction
Longitudinal Pipe	0 and 180
Transversal Coil	25 and 205
45° LT Coil	70 and 250
Transversal Pipe	90 and 270
Longitudinal Coil	115 and 295
45° LT Pipe	135 and 315

The yield stress at different orientation are reported in Figure 7 below.

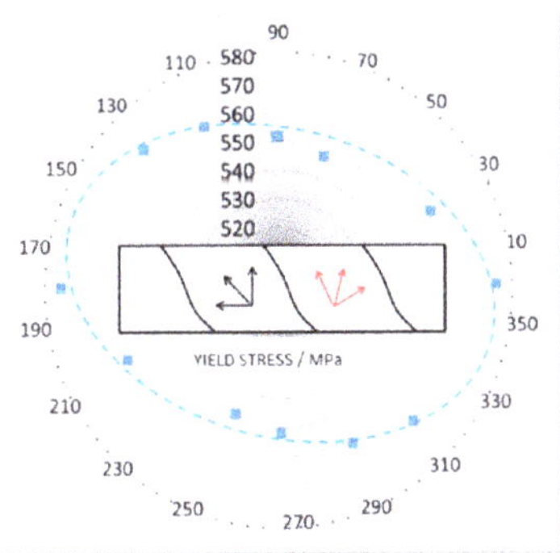

Figure 7. Yield stress derived from tensile tests on X70 48″ OD Spiral Pipe for different orientation. (Longitudinal pipe direction = 0°).

The tensile tests carried out on notched specimen are reported in terms of the engineering stress strain curve (Figure 8). The two specimen geometries have different values of notch radius and minimum diameter of the cross section (i.e., 5 mm diameter for notch radius = 10 mm and 6 mm diameter for notch radius = 2 mm) and this induces a different stress state (i.e., triaxiality) during the tests. The material shows very similar behavior in the longitudinal (L) and transverse (T) direction for the RNB10 specimens and identical behavior for the RNB2.

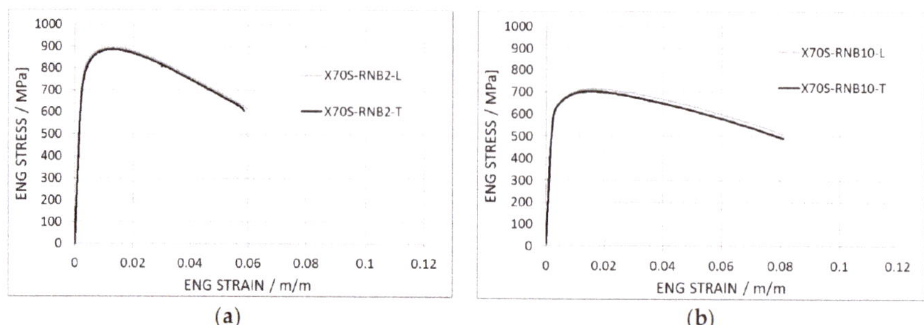

Figure 8. Engineering stress-strain curves derived from notched tensile tests on X70 48″ OD Spiral Pipe: (**a**) Notch radius = 2 mm; (**b**) Notch radius = 10 mm.

The results, obtained from tensile tests on mini round specimens, are summarized in Figure 9a. Specimens are extracted in the through thickness direction of the pipe (N), and oriented at 45° degrees from longitudinal, and through thickness direction (45° LN) and at 45° degrees from transversal, and through thickness direction (45° TN). In order to analyze a possible strength differential effect in the through thickness direction of the material, the tensile curves are compared with the compression curves in the Figure 9b.

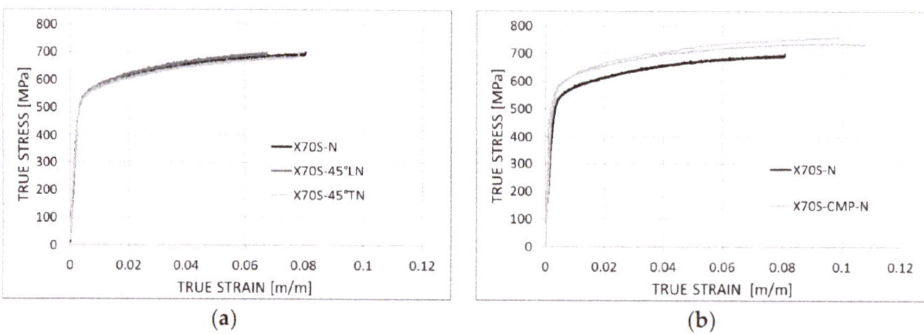

Figure 9. True stress-strain curves derived from tensile and compression tests on X70 48″ OD Spiral Pipe: (**a**) 2,5 mm OD round bar test; (**b**) 6 mm OD compression test.

True stress-strain curves of the material, for the in-plane pipe directions, can also be derived from torsion tests. True stress-strain curves are obtained by means of torsion tests up to large strain, exclusively from the elaboration of experimental data. This can be accomplished by relying on the equilibrium relations, or equivalently using the well-known Nadai's formula [19], which gives the material shear stress- shear strain ($\tau-\gamma$) curve, starting from the raw torque-displacement (M-θ) results:

$$\tau(\gamma_0) = \frac{1}{2\pi r_0^3}\left(\theta_N\frac{dM}{d\theta_N} + 3M\right)\gamma_0 = \gamma(r_0), \quad \theta_N = d\theta/dz \tag{1}$$

where r_0 and γ_0 are the outer radius and outer shear strain, and θ_n is the rotation per unit length. Then, using a standard J_2 (von Mises) plasticity equivalence, an average stress-strain relation can be finally calculated:

$$\sigma = \sqrt{3}\tau\sigma = \gamma/\sqrt{3} \tag{2}$$

In the previous equation an isotropic equivalence is used instead of an anisotropic formulation. This is justified by the consideration that, the measured M-θ global quantities just reflect an average material behavior, as a consequence of each point of the specimen being subjected to loading in a different anisotropic direction.

Figure 10 shows the obtained results for both longitudinal and transversal tests.

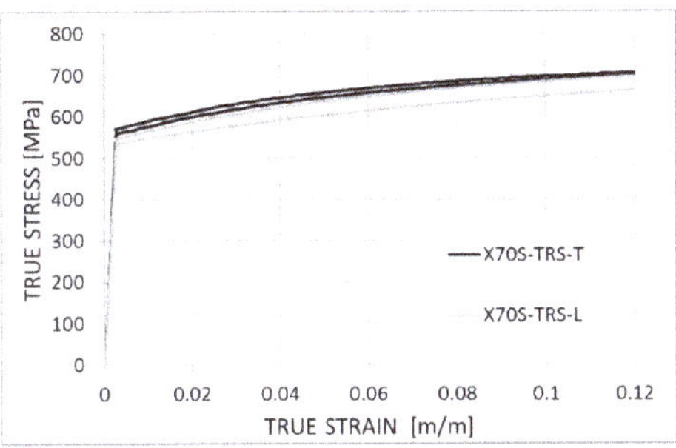

Figure 10. True stress-strain curves derived from torsion tests on X70 48″ OD Spiral Pipe.

In this case, material behavior seems only marginally affected by the test direction. This can be explained, considering that under torsion, the local state of stress is not directional. As a consequence, the overall response is somehow a mixture of the material behavior under all possible directions perpendicular to the specimen cut axis.

3.2. Fractographic Analysis Results

The X70 spiral pipe material has been assessed, measuring the void distribution on material as is, and on deformed and fractured tensile specimens, machined in the longitudinal (L), transversal (T) and 45° degrees, between L-T directions (45° LT) of the pipe. For every specimen, the SEM analysis of fractured surfaces, aimed at analyzing and measuring the dimple pattern, was performed (Figures 11–13).

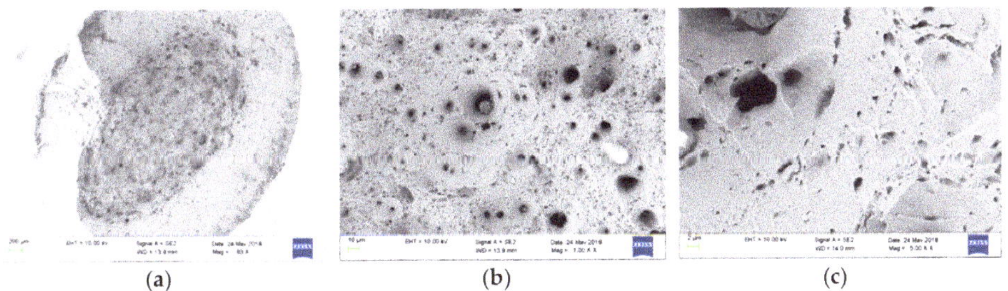

Figure 11. SEM Analysis of X70 Transversal pipe tensile specimen: (**a**) Mag 63 X; (**b**) Mag 1000 X; (**c**) Mag 5000 X.

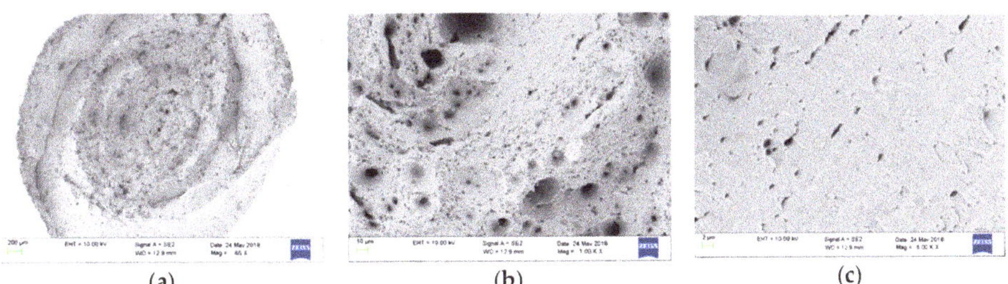

Figure 12. SEM Analysis of X70 Longitudinal pipe tensile specimen: (**a**) Mag 63 X; (**b**) Mag 1000 X; (**c**) Mag 5000 X.

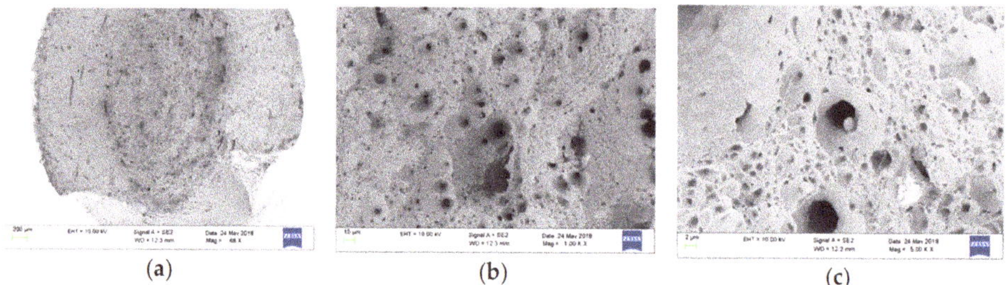

(a)   **(b)**   **(c)**

Figure 13. SEM Analysis of X70 45° LT pipe tensile specimen: (**a**) Mag 63 X; (**b**) Mag 1000 X; (**c**) Mag 5000 X.

3.3. Material Voids Count

After the tests, the specimens were sectioned along a radial plane, to consider the longitudinal section. The void counting results are reported in the following Figures 14 and 15.

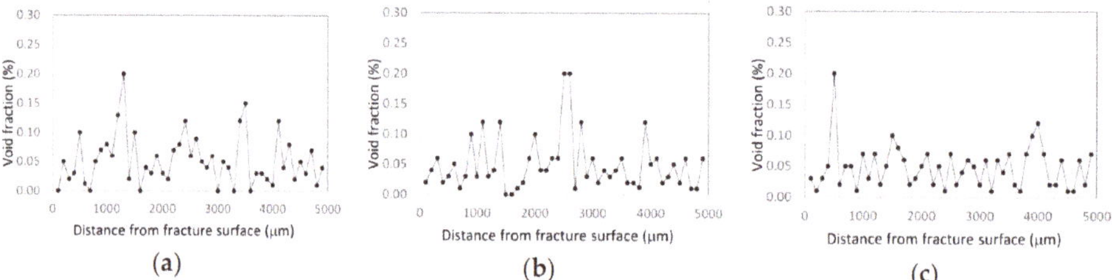

(a)   **(b)**   **(c)**

Figure 14. Void counting performed on tensile specimens after the test: (**a**) Transversal pipe tensile specimen; (**b**) Longitudinal pipe tensile specimen; (**c**) 45° LT pipe tensile specimen.

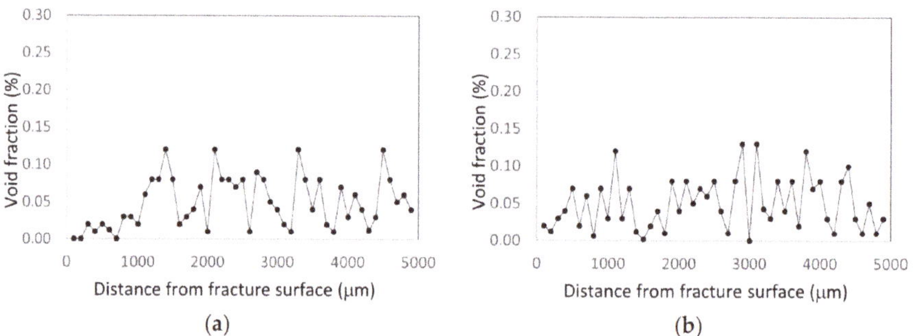

(a)   **(b)**

Figure 15. Void counting performed on torsion specimens after the test: (**a**) Transversal pipe tensile specimen; (**b**) Longitudinal pipe tensile specimen.

4. Discussion

4.1. Mechanical Analysis

The manufacturing rolling process of the coil, and the following process of spiral pipe shaping, induce anisotropy in the final product. This is visible in analyzing the yield stress values of the material obtained from tensile tests, carried out on in-plane pipe directions (Figure 7). Spiral pipes show an anisotropic behavior, with the principal anisotropic axis rotated with a respect to both the pipe principal directions (i.e., longitudinal and circumferential) and coil principal directions.

For a better understanding of the material behavior differences in the three orientations, longitudinal, transversal, and radial, the stress-strain curves are reported in Figure 16. The first outcomes are that the differences in the stress strain curves evolve with the strain, such that the anisotropy of the pipes changes during deformation. The differences in material properties are found if the longitudinal and transverse directions are considered. It is worth noting that the thickness and material properties affect the pipe response to bi-axial stress state, like pipe-bending with internal pressure. Pipes with the same longitudinal and transversal material properties can show a different response to the same loading, due to the differences in material, through thickness behavior.

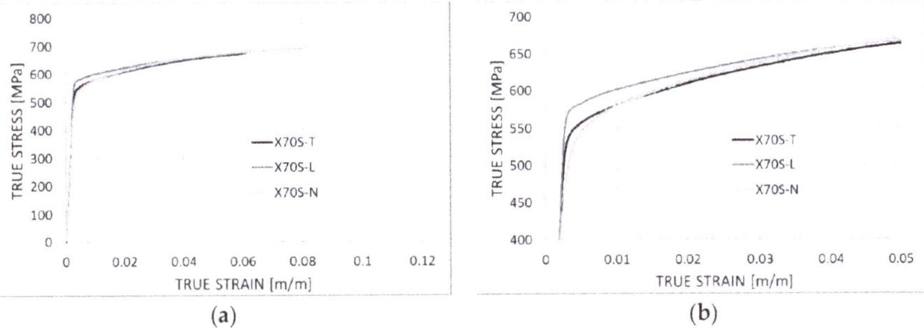

Figure 16. X70 true stress-strain curves derived from tensile tests on 48″ OD Spiral Pipe: (**a**) Standard view; (**b**) Magnified view.

A comparison of the true stress strain curves, obtained from tensile and torsion tests, shows a substantial match in the behavior of the material when subjected to tensile or shear stress states.

4.2. Fractographic Analysis

Voids nucleation from defects, like inclusions and the subsequent coalescence, is one of the most accepted theories to describe the damaged mechanisms of the alloy steels during deformation. Many criteria have been proposed, based on void evolution theories [20,21] and porous models, like the popular Gurson model [22] are used to describe the softening of the materials.

From the results reported in the previous paragraph, the main outcomes can be summarized by the following bullet points:

From the SEM analysis of fracture surfaces:

- No clear differences have been observed for the dimple geometries of the three tensile specimens oriented in different directions;
- The geometry of the dimples on the fracture surfaces is almost circular, even though the cross sectional area of the specimens presented an elliptical geometry, due to the anisotropic behavior of the material.

From material voids count:

- Voids fraction has an almost constant value (<0.5%) for the three specimens, also close to the fractured surfaces. Voids fractions are unrelated to the plastic strain reached from the material, and to the material orientation.
- Damage evolution seems therefore not governed by microvoids evolution.

5. Conclusions

A thorough characterization of the anisotropic plastic behavior, and damage evolution of a structural steel for pipelines, was carried out successfully. The main results showed that yielding is affected by material direction, proving its anisotropic behavior, while the local behavior at fracture seems to be quite insensitive to anisotropy for the material under investigation. The dimples geometry, observed on fractured specimens, is similar for different orientations. This could justify a different approach to the problem of modelling the plastic deformation and damage of these alloy steels. The damage evolution could be regarded as uncoupled from the plastic behavior, and even if an anisotropic plasticity criterion is needed to correctly describe the material, the damage model could be even based on a simple isotropic formulation as a first attempt.

Author Contributions: Data curation, L.C.; Investigation, F.I. and A.D.S.; Methodology, F.I.; Supervision, T.C.; Validation, F.I.

Funding: This research received no external funding.

Conflicts of Interest: Authors declare no conflict of interest.

References

1. Liessem, A.; Rueter, P.; Pant, M.; Schwinn, V. Production and development update of x100 for strain-based design applications. *J. Pipeline Eng.* **2010**, *9*, 9–27.
2. Hara, T.; Shinohara, Y.; Terada, Y.; Asahi, H.; Doi, H. Development of high-deformable high-strength line pipe suitable for strain-based design. In Proceedings of the 4th Pipeline Technology Conference, Hannover, Germany, 22–23 April 2009; pp. 154–160.
3. Muraoka, R.; Kondo, J.; Ji, L.; Chen, H.; Feng, Y.; Ishikawa, N.; Okatsu, M.; Igi, S.; Suzuki, N.; Masamura, K. Production of grade x80 high strain linepipes for seismic region application. *Proc. 8th Int. Pipeline Conf.* **2010**, 287–292. [CrossRef]
4. Seo, D.-H.; Yoo, J.-Y.; Soon, W.-H.; Kang, K.-B.; Cho, W.-Y. Development of X80/X100 linepipe steel plates and pipes for strain based design pipeline. In Proceedings of the Nineteenth International Offshore and Polar Engineering, Osaka, Japan, 21–26 July 2009; pp. 61–66.
5. Macia, M.L.; Kibey, S.A.; Arslan, H.; Bardi, F.; Ford, S.J.; Kan, W.C.; Cook, M.F.; Newbury, B. Approaches to qualify strain-based designed pipelines. In Proceedings of the 8th International Pipeline Conference, Calgary, AB, Canada, 27 September–1 October 2010; pp. 365–374.
6. Suikkanen, P.; Karjalainen, L.; DeArdo, A. Effect of carbon content on the phase transformation characteristics, microstructure and properties of 500 mpa grade microalloyed steels with non-polygonal ferrite microstructures. *La Metallurgia Italiana* **2008**, *6*, 41–54.
7. Regier, R.W.; Speer, R.W.; Matlock, D.K.; Choi, J.K. Effect of cooling rate on the crystallographic effective grain size of ferrite in low-carbon microalloyed steels. In Proceedings of the Association for Iron and Steel Technology International Symposium on the Recent Developments in Plate Steels, Winter Park, CO, USA, 19–22 June 2011; pp. 351–363.
8. Di Schino, A.; Di Nunzio, P.E.; Turconi, G.L. Microstructure evolution during tempering of martensite in medium C steel. *Mater. Sci. Forum* **2007**, *558–559*, 1435–1441. [CrossRef]
9. Di Nunzio, P.; Di Schino, A. Effect of Nb microalloying on the heat affected zone microstructure of girth welded joints. *Mater. Lett.* **2017**, *186*, 86–89.
10. Rufini, R.; Di Pietro, O.; Di Schino, A. Predictive simulation of plastic processing of welded stainless steel pipes. *Metals* **2018**, *8*, 519. [CrossRef]
11. Corradi, M.; Di Schino, A.; Borri, A.; Rufini, R. A review of the use of stainless steel for masonry repair and reinforcement. *Constr. Build. Mater.* **2018**, *181*, 335–346. [CrossRef]
12. Sun, H.; An, S.; Meng, D.; Xia, D.; Kang, Y. Influence of cooling condition on microstructure and tensile properties of x80 high deformability line pipe steel. In Proceedings of the 10th International Conference on Steel Rolling, Beijing, China, 15–18 September 2010; pp. 420–426.
13. Tanguy, B.; Luu, T.T.; Perrin, G.; Pineau, A.; Besson, J. Plastic and damage behaviour of a high strength X100 pipeline steel: Experiments and modelling. *Int. J. Pres. Ves. Pip.* **2008**, *85*, 322–335. [CrossRef]
14. Iob, F.; Campanelli, F.; Coppola, T. Modelling of anisotropic hardening behavior for the fracture prediction in high strength steel line pipes. *Eng. Fract. Mech.* **2018**, *148*, 363–382. [CrossRef]
15. Research Programme of the Research Fund for Coal and Steel, Contract N. RFSR-CT-2011-00029, project acronym ULCF. Available online: http://ec.europa.eu/research/participants/data/ref/other_eu_prog/rfcs/rfcs-call-summaries_en.pdf (accessed on 20 December 2018).

16. Research Programme of the Research Fund for Coal and Steel, Contract N. RFSR-CT-2013-00025, project acronym SBD-SPipe. Available online: https://ec.europa.eu/research/industrial_technologies/pdf/rfcs/summaries-rfcs_en.pdf (accessed on 20 December 2018).

17. Coppola, T.; Iob, F.; Campanelli, F. Critical review of ductile fracture criteria for steels. In Proceedings of the 20th European Conference on on Fracture (ECF20) Steel Rolling, Trondheim, Norway, 30 June–4 July 2014.

18. Coppola, T.; Demofonti, G.; Mannucci, G. Numerical-experimental procedures to identify the ductile fracture strain limits in pipeline steels. In Proceedings of the 19th International Offshore and Polar Engineering Conference, Osaka, Japan, 21–26 July 2009.

19. Nadai, A. *Theory of flow and fracture of solids*, 2nd ed.; McGraw-Hill: New York, NY, USA, 1950.

20. McClintock, F.A. A criterion for ductile fracture by growth of holes. *J. Appl. Mech.* **1968**, *35*, 363–371. [CrossRef]

21. Rice, J.R.; Tracey, D.M. On the ductile enlargement of voids in triaxial stress fields. *J. Mech. Phys. Solids* **1969**, *17*, 201–217. [CrossRef]

22. Gurson, A.L. Continuum theory of ductile rupture by void nucleation and growth. Part I. Yield criteria and flow rules for porous ductile media. *J. Eng. Mater. Technol.* **1977**, *99*, 2–15. [CrossRef]

 *metals*

Article

Mechanical Properties of Direct-Quenched Ultra-High-Strength Steel Alloyed with Molybdenum and Niobium

Jaakko Hannula *, David Porter, Antti Kaijalainen , Mahesh Somani and Jukka Kömi

Materials and Mechanical Engineering, Faculty of Technology, University of Oulu, P.O. Box 4200, 90014 Oulu, Finland; david.porter@oulu.fi (D.P.); antti.kaijalainen@oulu.fi (A.K.); mahesh.somani@oulu.fi (M.S.); jukka.komi@oulu.fi (J.K.)
* Correspondence: jaakko.hannula@oulu.fi; Tel.: +35-8408473448

Received: 6 March 2019; Accepted: 16 March 2019; Published: 19 March 2019

Abstract: The direct quenching process is an energy- and resource-efficient process for making high-strength structural steels with good toughness, weldability, and bendability. This paper presents the results of an investigation into the effect of molybdenum and niobium on the microstructures and mechanical properties of laboratory rolled and direct-quenched 11 mm thick steel plates containing 0.16 wt.% C. Three of the studied compositions were niobium-free, having molybdenum contents of 0 wt.%, 0.25 wt.%, and 0.5 wt.%. In addition, a composition containing 0.25 wt.% molybdenum and 0.04 wt.% niobium was studied. Prior to direct quenching, finish rolling temperatures (FRTs) of about 800 °C and 900 °C were used to obtain different levels of austenite pancaking. The final direct-quenched microstructures were martensitic and yield strengths varied in the range of 766–1119 MPa. Mo and Nb additions led to a refined martensitic microstructure that resulted in a good combination of strength and toughness. Furthermore, Mo and Nb alloying significantly reduced the amount of strain-induced ferrite in the microstructure at lower FRTs (800 °C). The steel with 0.5 wt.% Mo exhibited a high yield strength of 1119 MPa combined with very low 28 J transition temperature of −95 °C in the as-quenched condition. Improved mechanical properties of Mo and Mo–Nb steels can be attributed to the improved boron protection. Also, the crystallographic texture of the investigated steels showed that Nb and Nb–Mo alloying increased the amount of {112}<131> and {554}<225> texture components. The 0Mo steel also contained the texture components of {110}<110> and {011}<100>, which can be considered to be detrimental for impact toughness properties.

Keywords: high-strength steels; micro-alloying; martensite; microstructure; toughness

1. Introduction

Thermomechanically controlled processing (TMCP) combined with direct quenching (DQ) is a novel and effective processing route to produce ultrahigh-strength, high-performance steels [1–3]. Energy efficiency has become a critical issue in recent times as a result of environmental reasons; hence, the need for making lighter and more energy-efficient structures with high-strength steels has become increasingly important. Quenching directly after hot rolling is an interesting energy-efficient alternative to the conventional processing route, as the re-heating prior to quenching can be omitted.

Conventional offline re-heating and quenching (RQ) has been a well-established process to produce high-strength steel plates for quite some time. The use of microalloying in high-strength steels has become more common because of the possibility to further improve the steel properties. Metallurgically, it is well established that the use of Nb in thermomechanical processing is highly effective as it retards the static recrystallization (SRX) of austenite at high temperatures [4]. It is also known that while Mo alone does not have a very significant impact on recrystallization behaviour, Mo combined with Nb leads to a synergistic increase in the retardation of recrystallization process [5]. Mo reduces the activity of C and N, which retards the amount of Nb precipitation, thus leading to more solute Nb retarding SRX [6]. However, as direct-quenching is a relatively new process industrially, the metallurgical effects of Mo and Nb alloying on direct-quenched steels are not completely understood or established. The purpose of this study was to examine the effects of Mo and Nb additions on the mechanical properties and microstructures of thermomechanically processed, direct-quenched, high-strength low carbon steels.

2. Materials and Methods

The materials used in the present investigation were low carbon (~0.16 wt.%) steels with constant levels of manganese (1.1 wt.%), chromium (0.5 wt.%), and nickel (0.5 wt.%) alloying. Boron contents were also kept constant at ~15 ppm. The compositions studied covered three molybdenum levels without Nb, that is, 0 wt.%, 0.25 wt.%, and 0.5 wt.% Mo. A fourth composition bearing 0.25 wt.% Mo was also alloyed with 0.04 wt.% Nb. Each steel was named after the contents of Mo and Nb, as shown in Table 1. The chosen compositions were vacuum cast into approximately 70 kg slabs at the Tornio Research Centre of Outokumpu Oyj, Finland. Then, 180 mm × 80 mm × 55 mm pieces of the castings were homogenized at 1100 °C for 2 h and thermomechanically rolled into approximately 11 mm thick plates according to the rolling schedule given in Table 2. The temperature of the samples during rolling and direct quenching was monitored by thermocouples placed in holes drilled in the edges of the samples to the mid-width at mid-length. The effective cooling rate at the centre of the plates was ~40 °C/s. Two finish rolling temperatures (FRTs) of 800 and 900 °C (accuracy ± 20 °C) were used prior to quenching in a water bath.

Table 1. Chemical compositions of investigated steels in wt.%.

Steel	C	Si	Mn	Cr	Ni	Mo	Nb	Al	B	N
0Mo	0.16	0.2	1.0	0.5	0.5	-	-	0.03	0.0014	0.0050
0.25Mo	0.16	0.2	1.1	0.5	0.5	0.25	-	0.03	0.0015	0.0043
0.5Mo	0.16	0.2	1.1	0.5	0.5	0.5	-	0.025	0.0016	0.0051
0.25Mo–Nb	0.16	0.2	1.1	0.5	0.5	0.25	0.04	0.02	0.0016	0.0047

Table 2. Hot rolling pass schedules giving finish rolling temperatures (FRTs) of 800 and 900 °C.

Pass	Thickness (mm)	Temperature (°C)	Reduction per Pass (%)	Total Reduction (%)	Reduction after Pass 3 (%)
-	52	1100	-	-	-
1	42	1100	19	19	-
2	33	1050/1080	21	37	-
3	26	1000/1060	21	50	-
4	20	910/1030	23	62	23
5	15	850/960	25	71	42
6	11.2	800/900	25	78	57

Tensile tests and Charpy-V impact tests were performed to evaluate the strength and toughness properties using specimens with their long axes parallel to the rolling direction, that is, longitudinal specimens. Tensile tests were performed at room temperature using three round-bar specimens/rolled samples with a diameter of 6 mm and parallel length of 40 mm, following the standard ISO 6892-1:2009 [7]. Charpy V-transition curves were obtained using $10 \times 10 \times 55$ mm^3 specimens with through-thickness notches tested at various temperatures with two specimens/temperatures in the range 20 to −140 °C, following the European standard EN 10 045-1:1990 [8].

Microstructural characterization was executed using field-emission scanning electron microscopy (Carl Zeiss AG, Oberkochen, Germany) combined with electron backscatter diffraction (FESEM-EBSD).

The linear intercept method applied to laser scanning confocal microscopy (LSCM) images from specimens etched with saturated picral in soap solution was used to determine the prior austenite grain sizes in three principal directions: the rolling direction (RD), transverse to the rolling direction (TD), and the plate normal direction (ND) at the quarter-thickness of the specimen. On the basis of these measurements, the aspect ratio (r), total reduction below the recrystallization temperature (R_{tot}), grain boundary surface area per unit volume (S_v), and average grain size (d) were determined using the equations shown in Table 3 [9].

EBSD measurements and analyses were performed using the EDAX-OIM acquisition and analysis software (7.1.0, Amatek Inc., Berwyn, PA, USA). The FESEM for the EBSD measurements was operated at 15 kV using a step size of 0.15 µm. The scanned area was approximately 90 µm × 90 µm. Lath and effective grain sizes were determined as equivalent circle diameter (ECD) values with low-angle (2.5–15°) and high-angle boundary misorientations (15–63°), respectively. Minimum grain size was determined as three pixels (at least three points to be defined as grains), which corresponds to a grain diameter over 0.27 µm. The confidence index value (CI) was set to be more than 0.1, and clean-up of the data was executed using grain dilatation procedure, which will modify the orientations of points that do not belong to any grains.

X-Ray diffraction (XRD) analyses were carried out using a Rigaku SmartLab 9 kW X-ray diffractometer (Rigaku Corporation, Tokyo, Japan) with Cu Kα radiation. Data analysis was performed using PDXL2 analysis software (PDXL 2.6.1.2, Rigaku Corporation, Tokyo, Japan) to estimate the lattice parameters, microstrains, and crystallite sizes of the investigated steels. Furthermore, dislocation densities were estimated using the Williamson–Hall method (Equation (1)) [10,11].

$$\rho = \sqrt{\rho_s \rho_p}, \tag{1}$$

where ρ_s is dislocation density calculated from strain broadening and ρ_p is dislocation density calculated from particle, that is, crystallite size, see Equations (2) and (3). Furthermore, according to Williamson et al. [10,11]:

$$\rho_s = \frac{k\,\varepsilon^2}{F\,b^2}, \tag{2}$$

and

$$\rho_p = \frac{3n}{D^2}, \tag{3}$$

where ε is microstrain, b is the burgers vector, F is an interaction factor assumed to be 1, factor k is assumed as 14.4 for body-centred cubic metals, and D is crystallite size. In the equation, n is dislocations per block face, assumed to be 1 [11].

Table 3. Austenite grain structure parameters [9].

Parameter	Equation
r	d_{RD}/d_{ND}
R_{tot}	$1 - \sqrt{(1/r)}$
S_v	$0.429 \times (1/d_{RD}) + 0.571 \times (1/d_{TD}) + (1/d_{ND})$
d	$(d_{RD} \times d_{TD} \times d_{ND})^{1/3}$

3. Results and Discussion

3.1. Prior Austenite Grain Structure after Hot Rolling

Table 4 shows the measured austenite grain sizes, calculated S_v values, and corresponding R_{tot} values for FRTs of 800 and 900 °C. Generally, the S_v values and reduction percentages below the recrystallization temperature were higher when both Mo and Nb were present. The lower FRT of 800 °C led to higher austenite pancaking, that is, higher R_{tot} values, although calculated S_v values were generally lower, which is probably caused by grain boundary ferrite complicating the grain size calculations. Mo alloying did retard the recrystallization process without the presence of Nb, but the combination of Nb and Mo had a greater effect in delaying recrystallization. In the case of the 0.25Mo–Nb composition, the same level of austenite reduction below the recrystallization temperature was achieved with both FRTs, indicating that the no-recrystallization temperature (T_{NR}) of this composition is higher than 900 °C. Comparing the values of R_{tot} with the reductions after pass three in Table 2 indicates that the T_{NR} temperature lies close to ~1030 °C in the case of 0.25Mo–Nb steel. Similarly, the T_{NR} temperature of 0.5Mo steel seems to be about 910 °C, considering the case of FRT at 800 °C. Prior austenite grain size measurements confirmed that Mo too affected the recrystallization kinetics as expected and increased the T_{NR} temperature.

Table 4. Average austenite grain sizes in three principal directions and corresponding d, S_v, and R_{tot} estimated for different finish rolling temperatures (FRTs).

FRT	Steel	d_{RD} (µm)	d_{ND} (µm)	d_{TD} (µm)	d (µm)	S_v (mm²/mm³)	R_{tot} (%)
900 °C	0Mo	19.2	14.9	19.5	17.7	119	12.0
	0.25Mo	19.2	8.4	15.9	13.7	177	33.8
	0.5Mo	16.1	7.4	12.2	11.3	210	32.4
	0.25Mo–Nb	23.5	5.8	14.5	12.6	229	50.2
800 °C	0Mo	26.7	10.7	22.2	18.5	135	36.7
	0.25Mo	36.9	11.3	23.5	21.4	125	44.7
	0.5Mo	30.8	7.3	21.8	16.9	178	51.4
	0.25Mo–Nb	27.8	6.0	19.4	14.8	212	53.7

The differences in prior austenite grain structures and the influence of Mo and Nb in increasing the amount of austenite pancaking are evident from Figures 1 and 2. In addition, at the low FRT of 800 °C, a significant amount of strain-induced ferrite formed at the prior austenite grain boundaries, which can be seen as bands of white grains along the prior austenite grain boundaries. However, the presence of Mo and Nb significantly reduced the occurrence of strain-induced ferrite, although a small amount of ferrite did also form in the Mo and Nb alloyed steels (Figure 2a–d). Therefore, a combination of Nb–Mo alloying had a stronger effect on hindering ferrite formation compared with only Mo-alloyed steel, as seen when comparing Figure 2b,d. However, an increase in Mo content to 0.5% (0.5Mo steel; Figure 2c) showed a pronounced effect on hindering ferrite formation, if compared with 0.25Mo steel (Figure 2b).

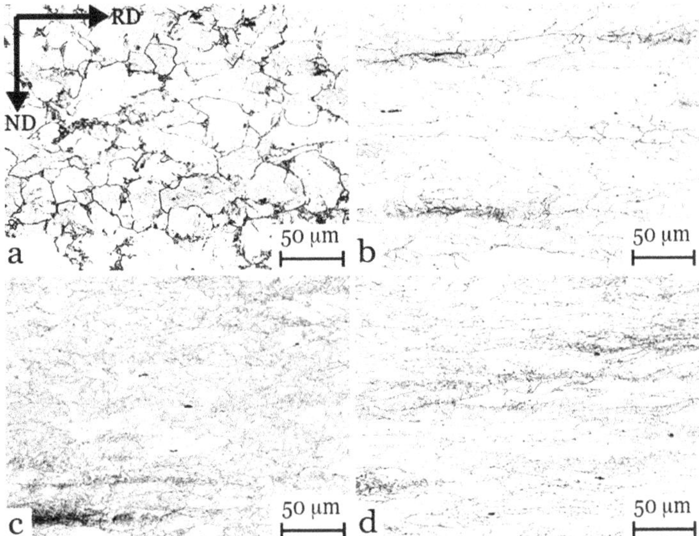

Figure 1. Prior austenite grain boundaries of the investigated steels with finish rolling temperature (FRT) of 900 °C: (**a**) 0Mo, (**b**) 0.25Mo, (**c**) 0.5Mo, and (**d**) 0.25Mo–Nb. RD = rolling direction, ND = normal direction.

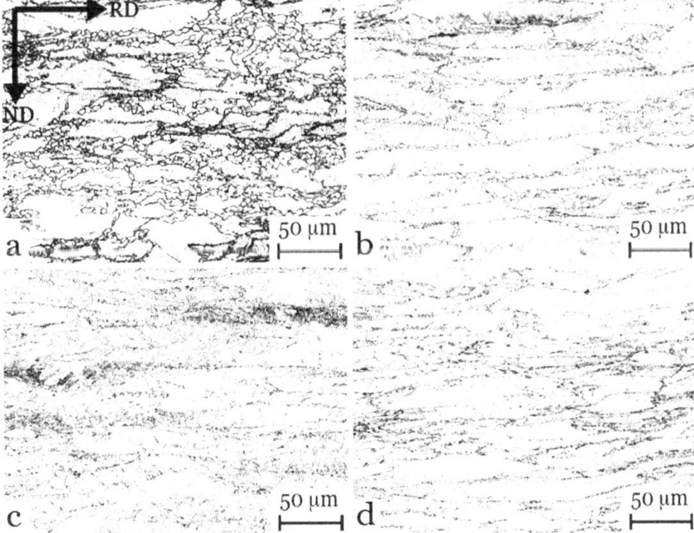

Figure 2. Prior austenite grain boundaries of the investigated steels with FRT of 800 °C: (**a**) 0Mo, (**b**) 0.25Mo, (**c**) 0.5Mo, and (**d**) 0.25Mo–Nb. Grain boundary ferrite can also be discerned. RD = rolling direction, ND = normal direction.

3.2. Transformed Microstructure

On the basis of the chemical compositions of the steels, continuous cooling transformation (CCT) diagrams were constructed using JMatPro 6.0® software to help interpret the results of the hot rolling experiments (Figure 3). This software can be used to predict CCT diagrams only for recrystallized, that is, undeformed, austenite. It can be seen that for such conditions, the addition of Mo is predicted to delay ferrite formation to lower cooling rates. As shown in Figure 3, the predicted equilibrium A_3 temperatures for the studied compositions are in the range of 812–821 °C. On the basis of the JMatPro calculations, Nb is predicted to have no effect on the transformation behaviour. At the cooling rates that follow the actual hot rolling experiments, that is, ~40–50 °C/s, microstructures should be fully martensitic with a Vickers hardness of ~420 HV10. However, it is clear from Figure 2 that the low-temperature deformation with an FRT of 800 °C led to the formation of strain-induced ferrite at the prior austenite grain boundaries, either during rolling or subsequent quenching even, to some extent, in the 0.5% Mo composition. Such a deformation in the vicinity of A_3 temperature, therefore, leads to accelerated ferrite formation compared with the predictions of Figure 3.

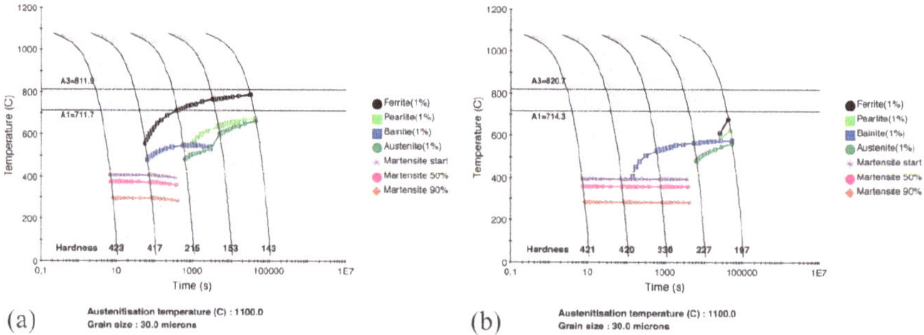

Figure 3. CCT diagrams constructed using JMatPro software. (**a**) 0Mo, (**b**) 0.5Mo, no deformation, austenite grain size 30 μm.

Figure 4 shows typical microstructures of the investigated steels with an FRT of 900 °C in the as-quenched condition at the $\frac{1}{4}$ depth position. Generally, the microstructures consisted of packets and blocks of martensitic laths refined and randomized in different directions. Auto-tempering of the martensite also occurred in some regions of the microstructures, as is evident from very fine carbides within the martensite laths. On the basis of the microstructural characterizations, the higher FRT of 900 °C produced mainly martensitic microstructures in all cases (Figure 4), however, a slightly coarser lath structure can be seen in 0Mo steel.

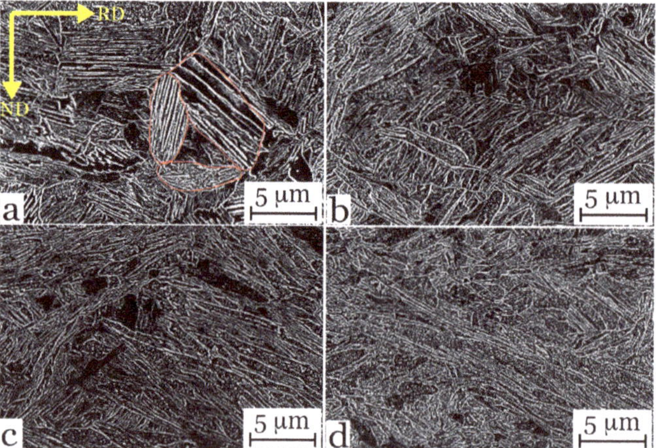

Figure 4. Typical microstructures (field-emission scanning electron microscopy (FESEM) after etching in 2% nital) of investigated steels in as-quenched condition, FRT 900 °C. (**a**) 0Mo, (**b**) 0.25Mo, (**c**) 0.5Mo, (**d**) 0.25Mo–Nb. Red lines show examples of martensite packets in the microstructure.

The microstructures with an FRT of 900 °C were further characterized using EBSD. The mean effective grain and lath sizes were determined as equivalent circle diameter (ECD) values with low-angle boundary misorientations (2.5–15°), corresponding to lath size and high-angle boundary misorientations (>15°) corresponding to effective grain size. The effective grain size is used to describe the packet/block size of martensite divided by high-angle boundaries Also, the effective coarse high-angle grain sizes at the 90th percentile in the cumulative grain area distribution ($d90\%$) were determined. The results, presented in Figure 5a, revealed no great differences in the various grain sizes of the transformed microstructures. Mo and Mo–Nb microalloying led to a small decrease in the mean effective grain and lath sizes, presumably because of the more pancaked austenite structures, corroborating the results presented in Table 2. However, no significant differences between $d90\%$ grain sizes were observed, although the 0.25Mo–Nb steel shows a somewhat higher $d90\%$ value. Furthermore, on the basis of the EBSD data, ECD effective high-angle grain sizes at different percentiles in the cumulative grain area distribution of the investigated steels after direct quenching from an FRT of 900 °C are given in Figure 6. Differences can only be seen in the 80th and 90th percentile grain diameters ($d80\%$ and $d90\%$), indicating that there are only minor differences in the grain size distributions.

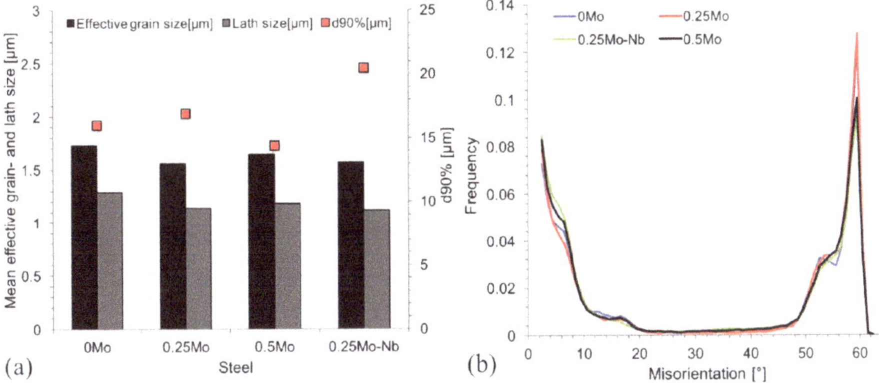

Figure 5. Results of the electron backscatter diffraction (EBSD) analysis in the as-quenched condition (FRT 900 °C). (**a**) Mean effective equivalent circle diameter (ECD) grain and lath sizes, and the ECD size of the coarsest grains ($d90\%$), (**b**) grain boundary misorientation distributions (>2.5°).

Also, grain misorientation angle distributions were determined using the EBSD data, and the results are presented in Figure 5b. Misorientation peaks at ~7.5° (sub-block boundaries), 16°, 52.5°, and 59° (packet or/and block boundaries) were detected, and they are a result of the different variants of the Kurdjumov–Sachs orientation relationship. Clear peaks at ~7.5° and ~59° are typical peaks for martensite or lower bainite microstructures [12,13]. On the basis of the misorientation distributions, no clear differences between the studied steels are apparent, which further supports the conclusion that all the investigated steels comprised mainly martensite in the direct-quenched condition when an FRT of 900 °C was used, as was also observed in other microstructural investigations. The defined low-angle and high-angle boundaries of investigated steels with an FRT of 900 °C can be seen from Figure 7. The lath-type structure with high density of low-angle boundaries (red lines) are clearly visible in Figure 7. Also slightly more elongated grain structure of 0.25Mo-Nb steel can be seen from Figure 7d.

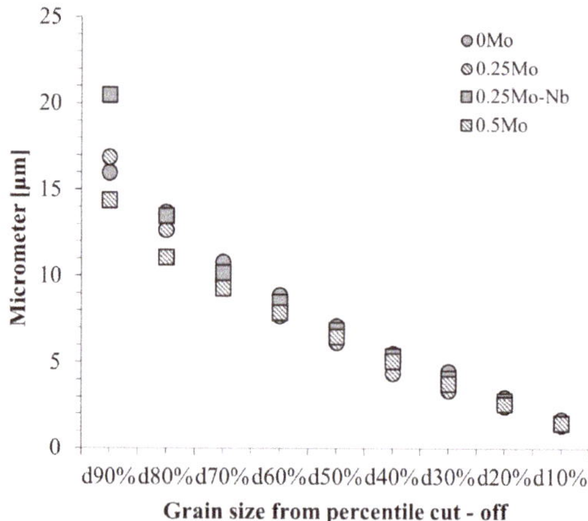

Figure 6. ECD effective high-angle grain sizes at different percentiles in the cumulative grain area distribution of the investigated steels after direct quenching for an FRT of 900 °C.

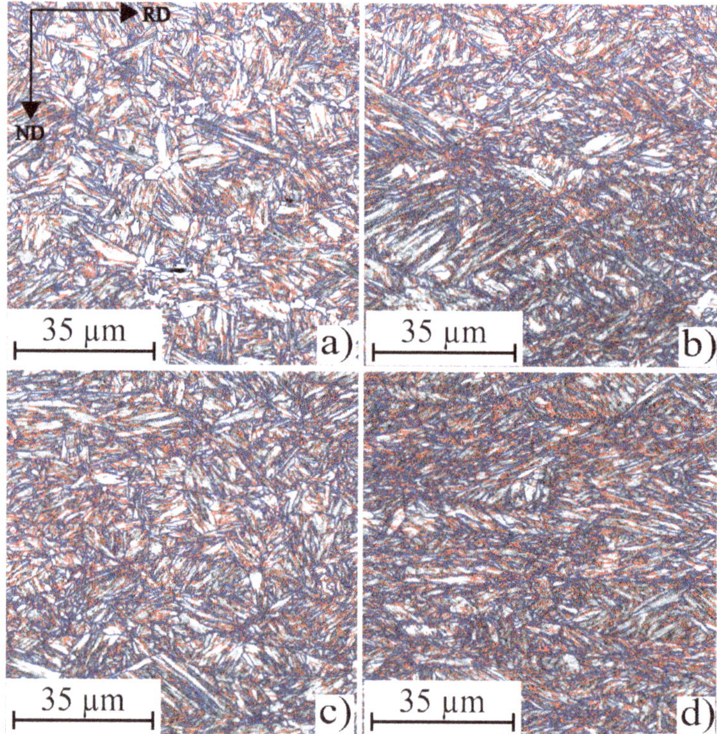

Figure 7. EBSD maps of investigated steels with an FRT of 900 °C showing low-angle boundaries (2.5–15°, red lines) and high-angle boundaries (>15°, blue lines). (**a**) 0Mo, (**b**) 0.25Mo, (**c**) 0.5Mo, (**d**) 0.25Mo–Nb.

3.3. Mechanical Properties after Hot Rolling and Direct Quenching

The tensile properties of the investigated steels in the as-quenched condition are presented in Table 5. The FRT of 900 °C resulted in a yield stress ($R_{p0.2}$, i.e., 0.2% offset proof stress) in the range of

950–1119 MPa and total elongation to fracture values in the narrow range of 8.6%–10.7%. With the FRT of 900 °C, the 0Mo steel showed rather low yield and tensile strengths, but an addition of 0.25 wt.% Mo enhanced the strength by more than 100 MPa (Table 5). The influence of ferrite can be discerned by comparing the results obtained at the FRT of 800 °C (Table 6). Strength values dropped in the case of steels with 0 wt.%–0.25 wt.% Mo, where significant amounts of ferrite formed at the lower FRT. In the case of the 0.5Mo and 0.25Mo–Nb steels, however, nearly identical yield and tensile strengths were achieved at both FRTs, which indicates that the presence of the very small amounts of strain-induced ferrite (corresponding to 800 °C FRT) had practically no appreciable effect on the yield and tensile strengths. In the case of the two leanest compositions, the presence of larger fractions of strain-induced ferrite seemed to have a negative effect on ductility, thus reducing the total elongations to fracture (A) in comparison with the values obtained at higher FRTs (without ferrite), presumably because of strain concentration and fracture in the softer phase. A corresponding effect is observed in the $R_m \times A$ values shown in Table 5.

Table 5. Tensile properties and hardness of the investigated steels in the as-quenched condition. (0.2% offset proof stress, tensile strength, plastic component of the uniform elongation, and total elongation to fracture with a gauge length of 5× the specimen diameter).

FRT	Steel	$R_{p0.2}$ (MPa)	R_m (MPa)	A_g (%)	A (%)	$R_m \times A$ (MPa.%)	HV10
900 °C	0Mo	950	1310	3.5	10.7	14,661	400
	0.25Mo	1078	1436	3.2	10.0	15,096	440
	0.5Mo	1119	1485	3.2	8.8	13,915	445
	0.25Mo–Nb	1100	1473	3.3	8.6	13,497	440
800 °C	0Mo	766	1204	5.0	7.4	9488	336
	0.25Mo	1003	1400	3.4	6.6	10,011	390
	0.5Mo	1107	1513	3.0	7.7	12,515	440
	0.25Mo–Nb	1086	1496	3.2	8.2	13,087	440

Transition curves were constructed based on the Charpy-V impact test results obtained in the temperature range +20 to −140 °C, see Figure 8. It can be seen that additions of Mo and Nb significantly improved the impact toughness, even though both the yield and tensile strengths were higher compared with those of the unalloyed 0Mo steel. Figure 9 shows how, for both FRTs, excellent combinations of strength and 28 J transition temperatures are obtained with the highest level of Mo or a lower Mo content in combination with Nb.

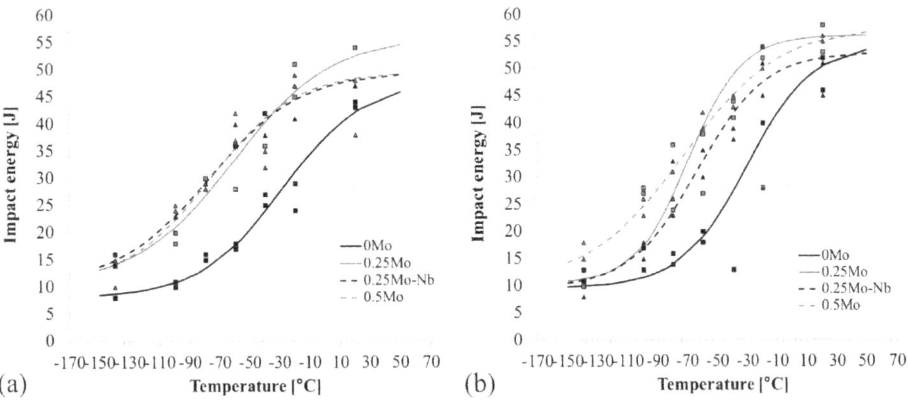

Figure 8. Transition curves in longitudinal direction for the as-quenched specimens: (**a**) FRT 800 °C, and (**b**) FRT 900 °C.

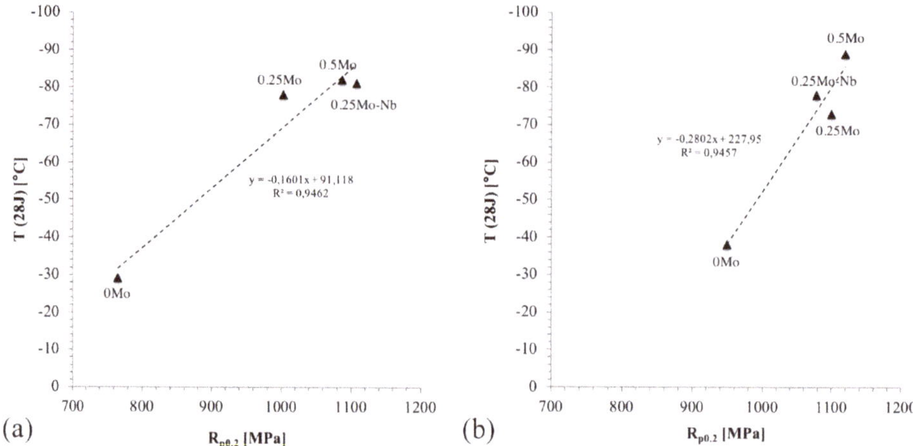

Figure 9. Correlation of the 28 J transition temperatures and the yield strength: (**a**) FRT 800 °C and (**b**) FRT 900 °C.

3.4. Correlation between Mechanical Properties and Microstructural Features

Crystallite sizes, microstrains, and dislocation densities measured using XRD are presented in Table 6 for the materials rolled with the FRT of 900 °C. According to the XRD measurements of the studied steels in the quenched conditions, crystallite sizes and microstrains were almost equal, producing dislocation densities varying in the narrow range 3.52–4.04 $\times$ 10^{15} m^{-2}. On the basis of earlier studies, it has been reported that dislocation density of lath martensite can be between 1 $\times$ 10^{15} m^{-2} and 1 $\times$ 10^{16} m^{-2} [14,15]. Also, Saastamoinen [16] et al. have reported similar dislocation densities for martensitic steels with corresponding carbon levels to those in the present study. Slight differences can be noticed when comparing the dislocation densities of the direct-quenched samples, but no statistically acceptable conclusions can be made.

Table 6. Crystallite sizes, microstrains, and dislocation densities of investigated steels, FRT 900 °C.

Material	Crystallite Size (Å)	Microstrain (%)	Dislocation Density ($\times$ 10^{15}(m^{-2}))
0Mo	341	0.377	3.6
0.25Mo	308	0.380	4.0
0.5Mo	312	0.389	4.0
0.25Mo–Nb	316	0.344	3.5

Figure 10 presents the transformation texture of the investigated steels measured by using X-ray diffraction. In the case of the Nb and Nb–Mo alloyed steels, stronger {112}<131> and {554}<225> components were discovered as a result of martensitic ferrite formation from unrecrystallized pancaked austenite. The 0Mo steel showed clear {110}<110> and {011}<100> components, but exhibited only weak {112}<131> and {554}<225> components, see Figure 10a. In previous studies, it has been noticed that {110}<110> texture components can enhance the propagation of cleavage fracture and thus decrease low-temperature toughness [17].

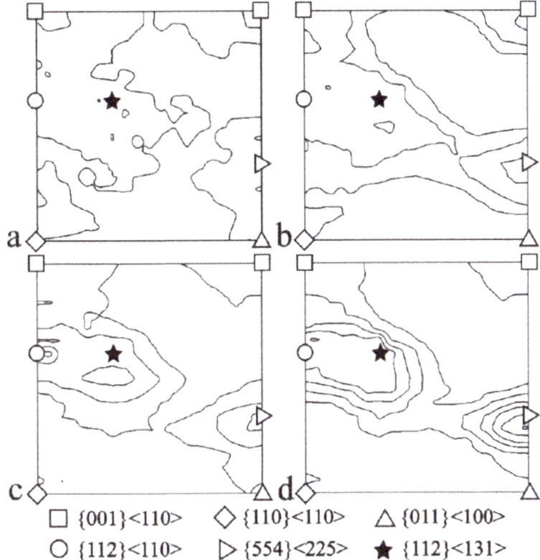

Figure 10. $\varphi2 = 45°$ sections of ODFs for specimen (**a**) 0Mo, (**b**) 0.25Mo, (**c**) 0.5Mo, (**d**) 0.25Mo–Nb. (Levels: 0.5,1.0,1.5,2.0 …).

The better properties of Mo alloyed steels can also be a result of boron protection. It is commonly known that soluble boron increases the hardenability by segregating at the prior austenite grain boundaries and thereby reducing the tendency to the formation of grain boundary ferrite [9,10]. In the present study, the addition of Mo and Mo–Nb efficiently reduced the amount of grain boundary ferrite, which can be the result of the segregation of Mo and Nb at the austenite grain boundaries, and at the same time, preventing formation of boron-containing precipitates. The formation of such precipitates at grain boundaries can provide nucleation sites for ferrite, thereby reducing the hardenability and resulting in lower strength and hardness. Further, the precipitates can deteriorate the impact toughness by acting as nucleation sites for fracture [18,19]. Table 7 presents the calculated hardness values using JMatPro® software of 0Mo steel with and without boron alloying to see the effect of boron on the hardenability of this steel. The boron alloying is predicted to increase the hardness by 50 HV, which corresponds to an approximately 150 MPa increase in tensile strength. When comparing the tensile strengths of 0Mo and 0.25Mo steels with FRT of 900 °C to exclude the effect of grain boundary ferrite, the difference between tensile strength values was 126 MPa, which is relatively close to predicted difference based on JMatPro calculations (Table 7). The higher than predicted tensile strengths of the hot-rolled and direct-quenched steels when compared with the predictions of JMatPro can be explained by the finer martensitic microstructure achieved with austenite pancaking during controlled deformation and rapid direct quenching.

Table 7. Calculated hardness values using JMatPro® software with and without B alloying compared to measured hardness values from hot-rolled and direct-quenched (DQ) steels.

Steel	Cooling Rate 50 °C/s (JMatPro)		Steel	Hot-rolled and DQ	
	Hardness (HV)	R_m [1] (MPa)		*Hardness* (HV)	R_m (MPa)
0Mo,B-free	374	1234	0Mo	400	1310
0Mo	424	1386	0.25Mo	440	1436

[1] Predicted R_m values are calculated using the same tensile strength/hardness ratio of 3.3 as achieved in actual tensile tests.

Fracture surfaces of Charpy V-notch test samples were characterized to evaluate the crack propagation through the material. Figure 11 presents the brittle fracture surfaces of 0Mo and 0.25Mo steels analyzed with scanning electron microscope (SEM). In the case of 0Mo steel, the fracture type was mainly transgranular fracture, also showing some indications of quasi-cleavage fracture (Figure 11c). However, in the case of 0.25Mo steel, the fracture type was more quasi-cleavage showing more areas with features of ductile fracture, such as tear ridges and dimpled rupture fracture (Figure 11d). No features of intergranular fracture were detected in the case of both steels, indicating that boron was not deteriorating the grain boundary toughness by precipitating at prior austenite grain boundaries. It can be concluded that for 0Mo steel, crack propagation through the specific crystallographic planes is easier than for 0.25Mo steel, which can be attributed to differences in texture components, shown in Figure 10.

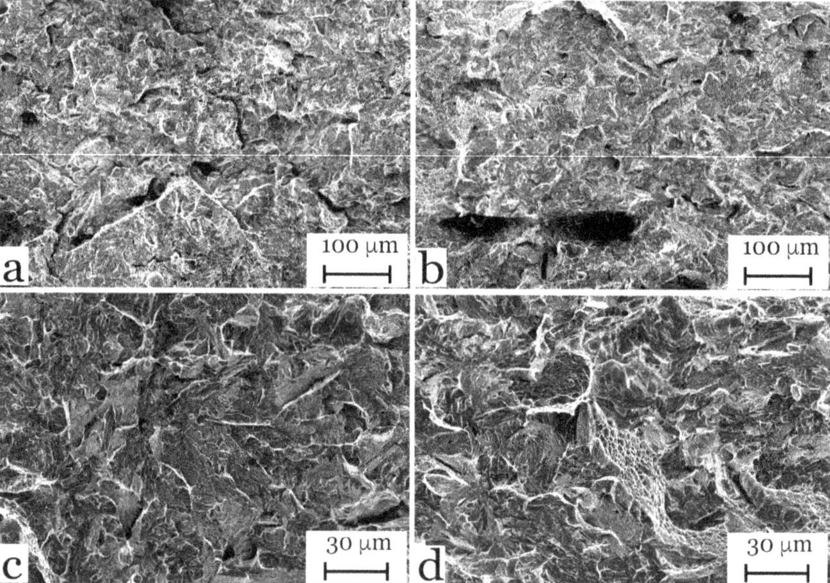

Figure 11. Brittle fracture surfaces (scanning electron microscope, SEM) of (**a**,**c**) 0Mo steel and (**b**,**d**) 0.25Mo steel.

4. Summary and Conclusions

The effect of molybdenum and niobium on the microstructures and mechanical properties of laboratory control rolled and direct-quenched 11 mm thick steel plates containing 0.16 wt.% C was studied. Two finish rolling temperatures of 800 and 900 °C were used. The plates were direct quenched to room temperature following thermomechanical processing at a cooling rate of ~40–50 °C/s. On the basis of the results, the following conclusions can be drawn:

- Mo and Nb microalloying raise the no-recrystallization temperature, leading to a more pancaked austenite and higher S_v values. There is a strong synergy between Nb and Mo.
- On the basis of microstructural and SEM-EBSD analyses, microstructures were essentially martensitic when the finish rolling temperature was 900 °C. There were no significant differences in the lath sizes, mean effective grain sizes, or the 10–90th percentile effective grain sizes among the different compositions studied. Also, grain boundary misorientation distributions were identical and typical for martensite.
- The finish rolling temperature of 800 °C led to the formation of strain-induced ferrite at the austenite grain boundaries, which deteriorated yield and tensile strengths, but the addition of Mo

and Mo–Nb significantly enhanced hardenability and decreased the amount of ferrite formation, and thereby increased the strength.

- For FRTs of 800 and 900 °C, Mo and Mo–Nb microalloying increased both the strength and impact toughness of the direct-quenched state.
- For the FRT of 900 °C, where the incidence of ferrite is very limited, there is a positive correlation between yield strength and transition temperature and the specific prior austenite grain boundary area S_v. However, EBSD analysis did not show any significant differences in the various martensite grain sizes (lath and effective grain size), thus indicating that a finer grain structure is not the reason for the higher strength. Nor were there significant differences in the dislocation densities of the steels.
- JMatPro calculations indicated that, for the present steel compositions, the increase in strength caused by the addition of Mo is partly explained by an additional hardenability increase caused by boron protection.
- The crystallographic texture of the investigated steels with an FRT of 900 °C showed that Nb and Nb–Mo alloying increased the amount of {112}<131> and {554}<225> texture components, whereas in the absence of Mo and Nb, the texture components {110}<110> and {011}<100> appear, which are detrimental to impact transition temperature. Brittle fracture surfaces of the Charpy V-notch test samples showed that for 0Mo steel, crack propagation through the crystallographic planes was easier, which can be the result of differences in texture components.
- With the addition of Mo, and Mo–Nb microalloying and direct quenching, martensitic steel with over 1400 MPa tensile strength combined with excellent impact toughness properties was produced, which can be used for demanding structural applications.

Author Contributions: Writing—original draft, J.H.; writing—review and editing, A.K., D.P., and M.S.; supervision, J.K.

Funding: This research was funded by IMOA (International Molybdenum Association).

Acknowledgments: Financial support of the IMOA (International Molybdenum Association) is gratefully acknowledged.

References

1.   Endo, S.; Nakata, N. Development of Thermo-Mechanical Control Process (TMCP) and high performance steel in JFE Steel. *JFE Tech. Rep.* **2015**, *20*, 1–7.

2.   Nishioka, K.; Ichikawa, K. Progress in thermomechanical control of steel plates and their commercialization. *Sci. Technol. Adv. Mater.* **2012**, *13*, 023001. [CrossRef] [PubMed]

3.   Kaijalainen, A.J.; Suikkanen, P.P.; Limnell, T.J.; Karjalainen, L.P.; Kömi, J.I.; Porter, D.A. Effect of austenite grain structure on the strength and toughness of direct-quenched martensite. *J. Alloy. Compd.* **2013**, *577*, S642–S648. [CrossRef]

4.   Palmiere, E.J.; Garcia, C.I.; DeArdo, A.J. Supression of static recrystallization in microalloyed steels by strain-induced precipitation. In Proceedings of the International Symposium on Low-Carbon Steels for the 90s, Warrendale, PA, USA, 17–21 October 1993; Asfahani, R., Tither, G., Eds.; TMS: Pittsburgh, PA, USA, 1993; pp. 121–130.

5.   Pereda, B.; Fernández, A.I.; López, B.; Rodriguez-Ibabe, J.M. Effect of Mo on Dynamic Recrystallization Behavior of Nb-Mo Microalloyed Steels. *ISIJ Int.* **2007**, *47*, 860–868. [CrossRef]

6.   Akben, M.G.; Weiss, I.; Jonas, J.J. Dynamic precipitation and solute hardening in A V microalloyed steel and two Nb steels containing high levels of Mn. *Acta Metall.* **1981**, *29*, 111–121. [CrossRef]

7.   ISO 6892-12016. *Metallic materials—Tensile testing—Part 1: Method of test at room temperature*; International Organization for Standardization: Geneva, Switzerland, 2016.

8.   ISO 148-12016. *Metallic materials—Charpy pendulum impact test—Part 1: Test method*; International Organization for Standardization: Geneva, Switzerland, 2016.

9. Higginson, R.L.; Sellars, C.M. *Worked Examples in Quantitative Metallography*; Maney: London, UK, 2003; pp. 45–47. ISBN 1902653807.

10. Williamson, G.K.; Hall, W.H. X-Ray broadening from filed aluminium and tungsten. *Acta Metall.* **1953**, *1*, 22–31. [CrossRef]

11. Williamson, G.K.; Smallman, R.E. III. Dislocation densities in some annealed and cold-worked metals from measurements on the X-ray debye-scherrer spectrum. *Philos. Mag.* **1956**, *1*, 34–46. [CrossRef]

12. Dong, H.; Sun, X.; Yong, Q.; Li, Z.; Weng, Y.; Yang, Z. Third generation high strength low alloy steels with improved toughness. *Sci. China Technol. Sci.* **2012**, *55*, 1797–1805.

13. Zajac, S.; Schwinn, V.; Tacke, K.H. Characterisation and Quantification of Complex Bainitic Microstructures in High and Ultra-High Strength Linepipe Steels. *Mater. Sci. Forum* **2005**, *500–501*, 387–394. [CrossRef]

14. Kennett, S.C.; Krauss, G.; Findley, K.O. Prior austenite grain size and tempering effects on the dislocation density of low-C Nb-Ti microalloyed lath martensite. *Scr. Mater.* **2015**, *107*, 123–126. [CrossRef]

15. Morito, S.; Nishikawa, J.; Maki, T. Dislocation Density within Lath Martensite in Fe–C and Fe–Ni Alloys. *ISIJ Int.* **2003**, *43*, 1475–1477. [CrossRef]

16. Saastamoinen, A.; Kaijalainen, A.; Porter, D.; Suikkanen, P.; Yang, J.-R.; Tsai, Y.-T. The effect of finish rolling temperature and tempering on the microstructure, mechanical properties and dislocation density of direct-quenched steel. *Mater. Character.* **2018**, *139*, 1–10. [CrossRef]

17. Pallaspuro, S.; Kaijalainen, A.; Mehtonen, S.; Kömi, J.; Zhang, Z.; Porter, D. Effect of microstructure on the impact toughness transition temperature of direct-quenched steels. *Mater. Sci. Eng. A* **2018**, *712*, 671–680. [CrossRef]

18. Taylor, K.A. Grain-boundary segregation and precipitation of boron in 0.2 percent carbon steels. *Metall. Trans. A* **1992**, *23*, 107–119. [CrossRef]

19. Antunes, J.P.G.; Nunes, C.A. Characterization of Impact Toughness Properties of DIN39MnCrB6-2 Steel Grade. *Mater. Res.* **2017**, *21*, 2–6. [CrossRef]

 *metals*

Article

Modelling and Microstructural Aspects of Ultra-Thin Sheet Metal Bundle Cutting

Jarosław Kaczmarczyk [1,*], Aleksandra Kozłowska [2], Adam Grajcar [2] and Sebastian Sławski [1]

[1] Institute of Theoretical and Applied Mechanics, Silesian University of Technology, 44-100 Gliwice, Poland; sebastian.slawski@polsl.pl

[2] Institute of Engineering Materials and Biomaterials, Silesian University of Technology, 44-100 Gliwice, Poland; aleksandra.kozlowska@polsl.pl (A.K.); adam.grajcar@polsl.pl (A.G.)

[*] Correspondence: jaroslaw.kaczmarczyk@polsl.pl; Tel.: +483-2237-2790

Received: 31 December 2018; Accepted: 29 January 2019; Published: 1 February 2019

Abstract: The results of numerical simulations of the cutting process obtained by means of the finite element method were studied in this work. The physical model of a bundle consisting of ultra-thin metal sheets was elaborated and then submitted to numerical calculations using the computer system LS-DYNA. Experimental investigations rely on observation of metallographic specimens of the surfaces being cut under a scanning electron microscope. The experimental data showing the microstructure of an ultra-thin metal bundle were the basis for the verification of the numerical results. It was found that the fracture area consists of two distinct zones. Morphological features of the brittle and ductile zones were identified. There are distinct differences between the front and back sides of the knife. The experimental investigations are in good agreement with the simulation results.

Keywords: steel sheet; cutting; finite element method; scanning electron microscope; plastic zone; brittle area; fracture

1. Introduction

In this paper the cutting problem of ultra-thin steel sheets arranged in bundles is considered. The sheets are made of C75S (carbon steel) cold rolled steel. The thickness of a single plate is 100 μm. The process of preparation of such thin sheets is complex and the cutting technology is also sophisticated. Cutting of bundles is more efficient compared to cutting individual sheets, because it enables the cutting of many sheets with a single cutting tool passage. The most frequently occurring defects are the bends of the cut sheet edges, burrs as well as defects in the form of vertical scratches. In order to improve the cutting process, numerical investigations were carried out concerning mainly the use of the finite element method for numerical simulation of cut sheet separation together with experimental research aimed at verifying the obtained results.

2. State of the Art

The cutting process of metals is the subject of interest of many researchers [1–4] as well as entrepreneurs and engineers working in the industrial sector. There are many interesting items in the literature regarding mainly machining [5–8], but much less attention is given to cover mechanical cutting of guillotines [9,10]. Different approaches are used for simulating failure and metal separation [11–13]. The most important issue is to select an appropriate FEM (finite element method) [14]. Machine deformation can also affect significantly the cutting process [15,16].

The cutting of a single sheet [9,17,18] is much simpler in comparison with cutting a stack of sheets. The smaller the number of sheets in the bundle the better the quality of the sheets cut in cross section [17]. By the good quality of sheets being cut, one means: low roughness of the sheets in cross section, small number of defects in the shape of vertical craters, burrs that occur very rarely and insignificant edge bending. The best solution to achieve these goals is to avoid any kind of possible defects, but from the practical point of view it seems to be impossible because by nature the mentioned problems are often met in industry [19]. The only thing one can do is to improve the cutting process by changing the selected parameters that affect the process. Nowadays, it is rather difficult to eliminate entirely the negative causes of defects during the cutting process. It can be achieved by optimisation taking into account one criterion or many criteria [20–23] to minimise the chosen entity e.g., roughness, edge bending, burrs, vertical scratches in the shape of vertical craters, etc. Before optimisation one should be familiar with the mechanisms of the cutting process. Then one can influence many parameters of the cutting process in order to achieve the minimum roughness, minimum edge bending, the rare probability of occurrence of burrs and as well as vertical scratches. Hence, the main goal of this paper is better understanding of the mechanism of the cutting process. The physical models and the corresponding mathematical ones were elaborated in order to visualise the stresses and strains in elastic and plastic zones at the chosen time instances during cutting of the bundle of ultra-thin sheets made of C75S cold rolled high-strength steel. Next, the experimental investigations were taken into consideration to show the shape of separate sheets after cutting under large magnification. The problem with observation of the cutting process under a scanning electron microscope is because of watching the sheets only after cutting. It would be perfect to see the whole cutting process, but up to now there is no such sufficiently fast apparatus with the appropriate large chambers in which the whole industrial cutting machine could be placed. That is why numerical simulations using the finite element method applying the computer program LS-DYNA were adopted [24]. It is worth mentioning that using the finite element method one should remember that it is only an approximation of a small portion of reality. To obtain good results one needs to assume the proper number of elements. The number of elements cannot be too small or too large because the obtained results will suffer an error. The smaller the element the smaller the digits are describing its volume, mass, inertia etc. During numerical simulations the small digits are multiplied, divided etc. That is why serious problems with numerical round off to a certain number of significant digits can occur in the case of too small elements. Too large elements can produce an error also. So, the optimum size of elements cannot be either too large or too small [25–30].

It was found out that in practice solution difficulties usually arise only when the finite element discretization is very fine, and for this reason the condition of infinite stresses under concentrated load is frequently ignored. Much finer discretization would lead to stress singularities at the point where the tip of the blade touches the sheet being cut. Additional errors might be caused also by computer approximation of the very small floating-point digits because the numbers are processed by round off. However, in general the finer the mesh the better the results (small errors) but the time of calculations increases. Therefore the balance between accuracy and time of computations should be established. If the size of the elements in the model is too large the time of computations decreases but the value of errors increases. This is caused usually by assumed linear shape function and the number of nodes. In the current work the proper size of the mesh was established on the basis of experiment and the numerical calculations were verified with experimental data.

The cutting of a bundle of metal sheets is the field of interest of such authors as Gasiorek et al. [4]. Their paper presents numerical modelling of the guillotine cutting process of sheet aluminium bundles. A finite element method with a smoothed particle hydrodynamics approach was coupled to simulate the cutting process. Experimental results of the cutting were presented for the validation purposes. The measured force characteristics in all numerical simulations and experiment were compared. The mentioned researchers focused on investigations of deformations and plastic strains.

It can be concluded that cutting on guillotines is a niche subject appearing in the literature rather sporadically when the problems related to machining processes have been thoroughly discussed. Research concerning cutting of single sheets using a guillotine shear can be found in Ref. [10]. The authors formulated physical models and corresponding to them mathematical ones using the computer program LS-DYNA [24].

The authors of the current paper covered such topics as mechanisms of the cutting process, plastic strains, reduced Huber–Mises stresses as well as the microstructure of the surfaces of sheets being cut.

In a previous work [9] the authors investigated the relationships between the cutting depth and the values of reduced Huber–Mises stresses for one sheet. In the current study, the model of mechanism of sheet separation using two thin sheets forming a bundle, is presented. The microstructural aspects of the cutting process are also addressed. Moreover, the differences in cutting behaviour between the front part of the cutting tool and the back part are identified.

3. Methodology

To achieve the goal of the study the following methodology was applied:

- to conduct fast changing dynamical calculations, the finite element method and the computer system LS-DYNA (LSTC, Livermore, CA, USA) were deployed;
- to obtain the experimental verification of the cutting, microstructural observations of fracture surfaces were performed, the SEM images were obtained using a Zeiss SUPRA 25 (Carl Zeiss AG, Jena, Germany) scanning microscope operating at 20 kV under two different magnifications: $400\times$ and $800\times$;
- the final stage concerns the comparison of the numerical simulations with the experimental results.

3.1. Material

To conduct the numerical investigations of the cutting process, a bilinear material model for the bundle of sheets made of C75S high-strength steel was assumed. The detailed data are collected together and shown in Table 1. Such a material model is very popular in the literature because on one hand it describes the elastic-plastic behaviour and on the other hand it is relatively simple. The bilinear material model (Figure 1) consists of two lines which describe the double relationship, respectively, the elastic one which is modelled by a straight line that starts from the origin of the Cartesian coordinate system of the stress–strain relation and the plastic one modelled by the second line which starts at the yield point and describes the plastic reinforcement. Despite the assumed bilinear material model being simple it describes the nonlinear behaviour between stress and strain.

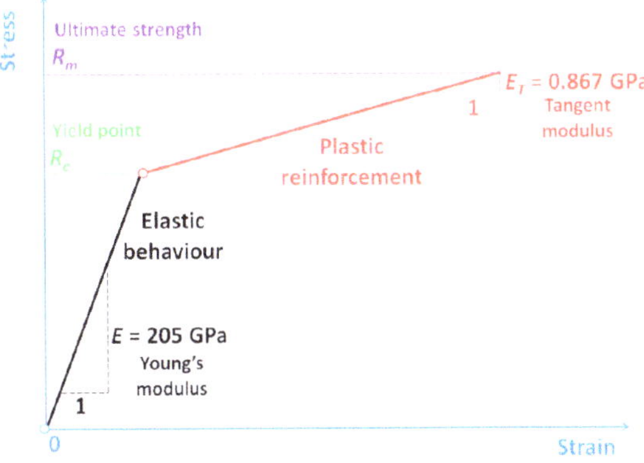

Figure 1. Bilinear material model.

Table 1. Material properties for steel sheets being cut.

No	Name of the Material Properties	Symbol	Value
1.	Young's modulus	E	205 GPa
2.	Poison's ratio	ν	0.28
3.	Kirchhoff's modulus	G	80 GPa
4.	Tangent modulus	E_T	0.867 GPa
5.	Failure strain	ε_f	0.15
6.	Yield stress	R_e	0.51 GPa
7.	Ultimate tensile strength	R_m	0.64 GPa

3.2. Physical Model of the Bundle of Sheets

In order to model the numerical cutting process, the physical model of the process was elaborated. It contains such parts as:

- cutting tool,
- pressure beam,
- bundle of sheets which consists of two separate metal sheets being cut,
- worktable.

All mentioned elements were modelled as rigid except the bundle of sheets which was modelled as deformable. This problem is focused on sheets being cut and on the quality of the cut surfaces understood as the smallest edge bending possible, burrs and vertical scratches in the shape of craters. The contact between the earlier mentioned working surfaces of all parts are taken into consideration. The unilateral constraints are imposed on the nodes that are in mutual contact during cutting i.e., between (Figures 2 and 3):

- the rigid cutting tool and the first deformable sheet,
- the rigid cutting tool and the second deformable sheet,
- the rigid pressure beam and the first deformable sheet,
- the first deformable sheet and the second deformable sheet,
- the second deformable sheet and the rigid worktable.

The models of Coulomb's and Moren's friction laws are taken into consideration. The static and kinetic coefficients of friction for all working surfaces in contact are assumed as for steel, respectively:

- static coefficient of friction $\mu_s = 0.22$,
- kinetic coefficient of friction $\mu_k = 0.11$.

The presented physical model (Figure 2) is divided into finite coplanar rectangular elements with four nodes (Figure 3). One node has two degrees of freedom (x and y translation). During modelling of the cutting process, the plane state of strain was applied [31]. Individual parts were divided into finite elements and nodes (Table 2). The size of the finite element of the sheet being cut is 0.02 mm horizontally and 0.01 mm vertically. The proposed element size was chosen on the basis of experiment as a compromise between the number of elements and the time consumption needed for the geometrically as well as the materially nonlinear fast changing dynamic cutting problem with cracking.

Table 2. Information about parts discretized into finite elements and nodes.

No	Name of the Part	Kind of Part	Number of Nodes	Number of Elements
1.	Cutting tool (knife)	Rigid	112	90
2.	First metal sheet being cut	Deformable	572	500
3.	Second metal sheet being cut	Deformable	572	500
4.	Pressure beam	Rigid	546	500
5.	Worktable	Rigid	1071	1000
	Total number		2873	2590

The physical model of the sheet metal cutting process is shown in Figure 2. On a motionless perfectly rigid table the sheets being cut are placed and then they are pressed from the top using a perfectly rigid pressure beam made in the form of a wedge with convergence 1:30. In this work it is assumed that the cutting tool blade has an apex angle $\alpha = 30°$ and the gap between the cutting tool and the pressure beam is 0.1 mm. The unilateral constraints in the form of so-called contact were imposed on the surfaces of the sheets being cut, worktable, pressure beam and cutting tool. The contact is based on the condition of impenetrability, namely the condition that two bodies cannot interpenetrate.

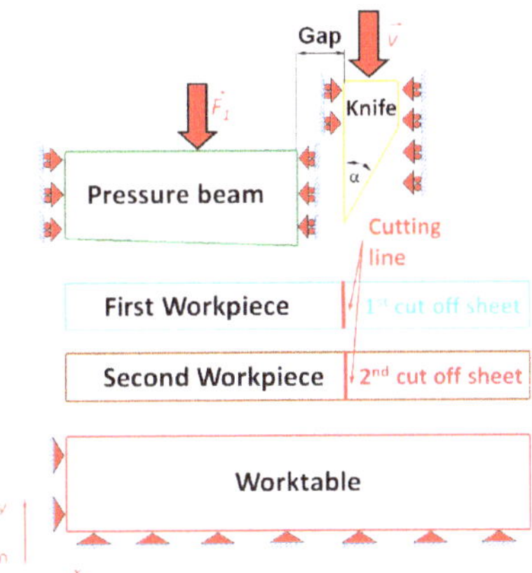

Figure 2. Physical model of a bundle consisting of two sheets being cut.

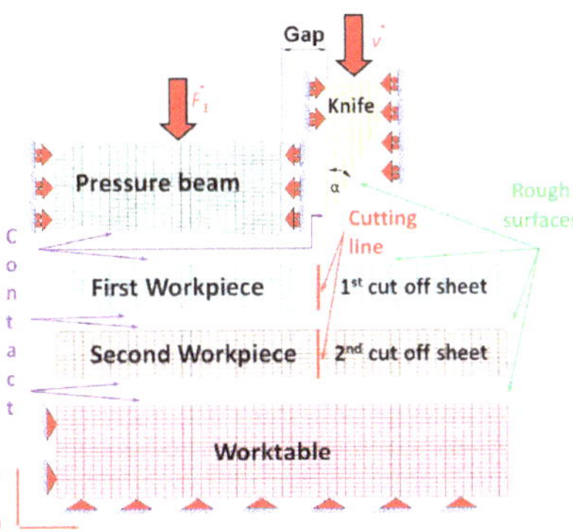

Figure 3. Mesh of the physical model of a bundle being cut.

To simulate the cutting process the failure model was applied. It consists of the separation of the nodes belonging to the cutting line (Figure 3). The nodes are defined as separable in the case when the reduced Huber–Mises strain at the closest node to the tip of the blade of a cutting tool approaches the equivalent value of strain equal to 0.15. The mentioned value of strain corresponds to the rupture during the experimental uniaxial tensile test. If the reduced Huber–Mises strain is higher than 0.15 the node which is closest to the tip of the blade is separated into two independent nodes allowing the cutting tool to penetrate into the sheet being cut. In the case when the reduced Huber–Mises strain in the sheet is less than the equivalent value of strain (0.15), then the node which is closest to the tip of the blade is not separated. So it is left as a single node which means that further penetration is impossible because the mentioned node is not separated into two nodes and the cutting process stops at this point (node).

In order to carry out the numerical simulation of a cutting process, the nonlinear incremental differential equation of dynamical equilibrium taking into account the phenomena occurring in mutual contact between bodies should be applied [32]:

$$\begin{bmatrix} M & 0 \\ 0 & 0 \end{bmatrix}\begin{bmatrix} {}^{t+\Delta t}\Delta\ddot{u}^{(i)} \\ 0 \end{bmatrix} + \begin{bmatrix} C & 0 \\ 0 & 0 \end{bmatrix}\begin{bmatrix} {}^{t+\Delta t}\Delta\dot{u}^{(i)} \\ 0 \end{bmatrix} + \left\{ \begin{bmatrix} {}^{t+\Delta t}K^{(i-1)} & 0 \\ 0 & 0 \end{bmatrix} + \begin{bmatrix} {}^{t+\Delta t}K_c^{(i-1)} \end{bmatrix} \right\}\begin{bmatrix} {}^{t+\Delta t}\Delta u^{(i)} \\ {}^{t+\Delta t}\Delta\lambda_k^{(i)} \end{bmatrix} =$$
$$\begin{bmatrix} {}^{t+\Delta t}R \\ 0 \end{bmatrix} - \begin{bmatrix} M & 0 \\ 0 & 0 \end{bmatrix}\begin{bmatrix} {}^{t+\Delta t}\ddot{u}^{(i-1)} \\ 0 \end{bmatrix} - \begin{bmatrix} C & 0 \\ 0 & 0 \end{bmatrix}\begin{bmatrix} {}^{t+\Delta t}\dot{u}^{(i-1)} \\ 0 \end{bmatrix} - \begin{bmatrix} {}^{t+\Delta t}F^{(i-1)} \\ 0 \end{bmatrix} + \begin{bmatrix} {}^{t+\Delta t}R_c^{(i-1)} \\ {}^{t+\Delta t}\Delta_k^{(i-1)} \end{bmatrix} \qquad (1)$$

where:

M, C—mass and damping matrices, respectively,

${}^{t+\Delta t}K^{(i-1)}$—tangent stiffness matrix including the material and geometric nonlinearities after iteration $(i-1)$,

${}^{t+\Delta t}K_c^{(i-1)}$—contact matrix after iteration $(i-1)$,

${}^{t+\Delta t}R$—vector of externally applied forces at time $t+\Delta t$,

${}^{t+\Delta t}F^{(i-1)}$—vector of nodal point forces after iteration $(i-1)$,

${}^{t+\Delta t}R_c^{(i-1)}$—vector of contact forces after iteration $(i-1)$,

${}^{t+\Delta t}\Delta_k^{(i-1)}$—vector of material overlaps at contactor nodes after iteration $(i-1)$,

${}^{t+\Delta t}\Delta u^{(i)}$, ${}^{t+\Delta t}\Delta\dot{u}^{(i)}$, ${}^{t+\Delta t}\Delta\ddot{u}^{(i)}$—vectors of incremental displacements, velocities and accelerations respectively in iteration (i),

${}^{t+\Delta t}\Delta\lambda_k^{(i)}$—vector of Lagrange multipliers in iteration (i),

${}^{t+\Delta t}\dot{u}^{(i-1)}$, ${}^{t+\Delta t}\ddot{u}^{(i-1)}$—vector of velocities and accelerations respectively after iteration $(i-1)$.

In the successive stage Equation (1) is converted into a formula enabling direct integration of the considered dynamic problem using the computer system LS-DYNA.

The rupture in metal sheets being cut is described by the following equation:

$$\text{if } \varepsilon \geq \varepsilon_f\text{—node separation belonging to the cutting line takes place,} \qquad (2)$$

$$\text{if } \varepsilon < \varepsilon_f\text{—no separation possible in nodes belonging to the cutting line,} \qquad (3)$$

where: ε—reduced Huber–Mises strain, ε_f—failure strain.

3.3. Microscopic Investigations

Observations of fracture surface of 1st (top) and 2nd (bottom) sheets located in the bundle were carried out to determinate the rupture mode. This part of research was done on the basis of SEM images obtained by using Zeiss SUPRA 25 (Carl Zeiss AG, Jena, Germany) scanning microscope operating at 20 kV under two different magnifications: 400× and 800×. The details of the fracture were observed at the front and the back of the cutting tool.

4. Results and Discussion

The research is composed of the three following stages:

- the numerical simulations concerning the cutting process of a bundle of sheets,
- the experimental investigations consisting of a series of microstructural images obtained by means of a scanning electron microscope (SEM),
- the comparison of the numerical and experimental results.

4.1. Numerical Simulations

The bundle consisting of two sheets was analysed. The numerical calculations were carried out using the finite element method and computer system LS-DYNA which is a general-purpose finite element program capable of simulating complex real world problems. It is used by the automobile, aerospace, construction, military, manufacturing, and bioengineering industries. This program is optimized for shared and distributed memory Unix, Linux, and Windows based platforms. The code's origins lie in highly nonlinear, transient dynamic finite element analysis using explicit time integration. LS-DYNA capabilities include [24]: nonlinear dynamics, rigid body dynamics, normal modes, thermal analysis, fluid analysis, etc. The simulations were performed at the Silesian University of Technology using the computer cluster Ziemowit (http://www.ziemowit.hpc.polsl.pl). The cutting process begins at the moment at which the pressure beam starts moving slowly from zero to maximum speed ($\vartheta_{max} = 0.012$ mm/s), then moves with constant maximum speed and next slows down imperceptibly until its speed approaches zero. Following this, the pressure beam touches the top of the bundle (Figures 4–7) and compresses the sheets with a small force a priori assumed and treated as negligibly small compared with the force produced by the knife during cutting. The small force guarantees the cutting of the bundle without any shifting of the sheets being cut. At the moment when the pressure beam stops, the cutting tool starts traveling from zero to maximum speed ($\vartheta_{max} = 0.022$ mm/s) and then moves with constant maximum speed and starts cutting the bundle of sheets. It should be mentioned that the tip of the blade of a cutting tool travels parallel to the worktable. The cutting tool first cuts the top sheet in the bundle and after that it starts cutting the final sheet, then it stops at the worktable and goes back to the original position and the cutting process repeats itself.

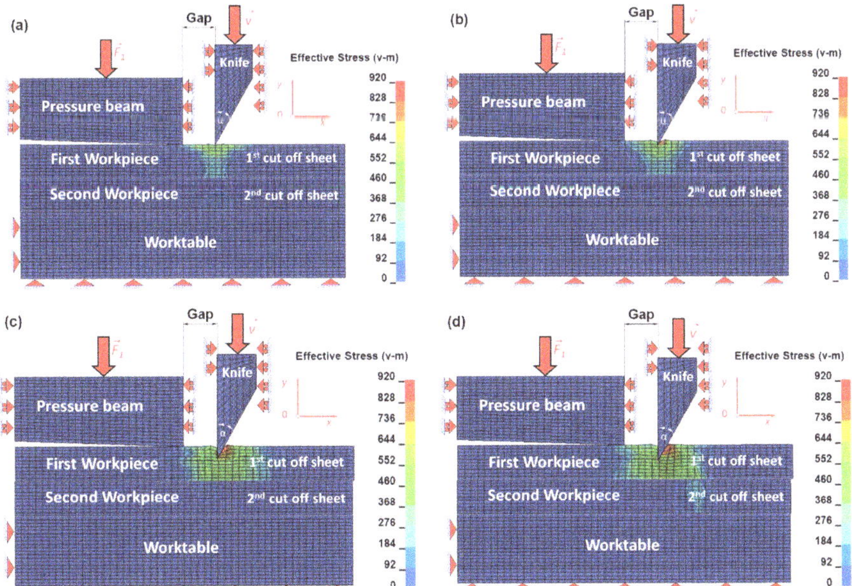

Figure 4. *Cont.*

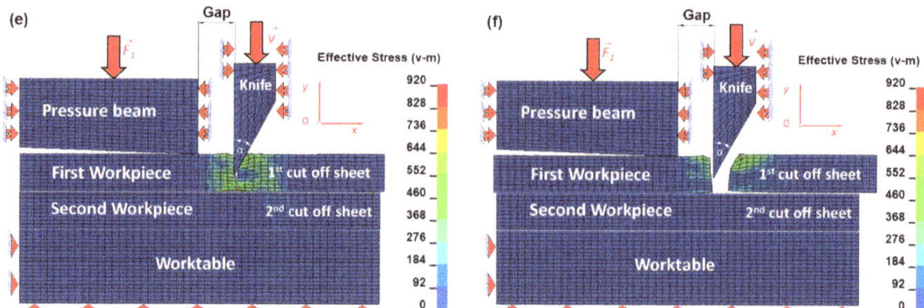

Figure 4. Huber–Mises stresses in the first sheet being cut for the following time instances: (**a**) t = 3 s, (**b**) t = 3.5 s, (**c**) t = 4.5 s, (**d**) t = 5 s, (**e**) t = 5.55 s, (**f**) t = 5.56 s. The visualisation of stresses for the next sheet being cut is continued in the next figure.

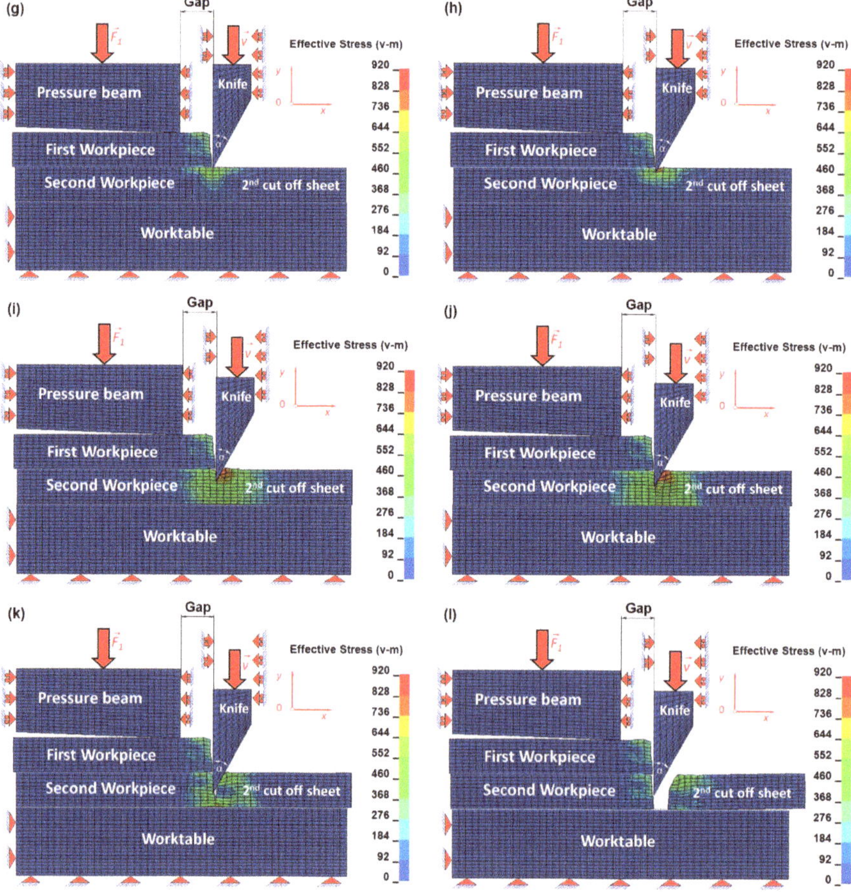

Figure 5. Huber–Mises stresses in the second sheet being cut for the following time instances: (**g**) t = 7.5 s, (**h**) t = 8 s, (**i**) t = 9 s, (**j**) t = 9.5 s, (**k**) t = 10.1 s, (**l**) t = 10.11 s (continuation of Figure 4). The visualisation of stresses for the previous sheet being cut is presented in the previous figure.

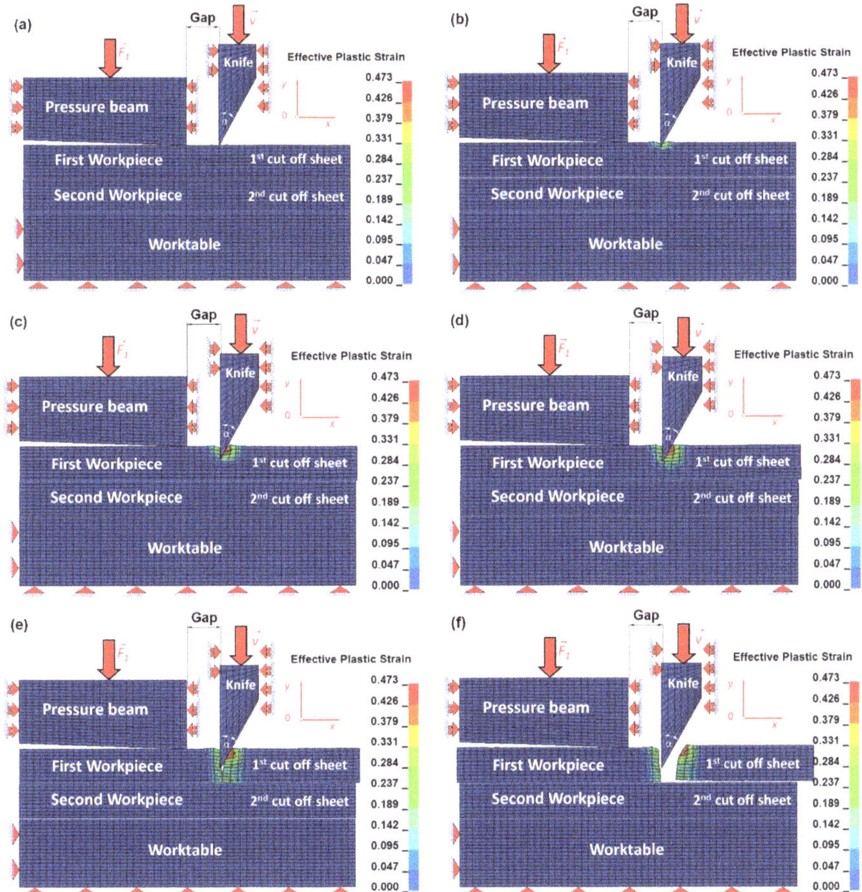

Figure 6. Effective plastic strains in the first sheet being cut for the following time instances: (**a**) $t = 3$ s, (**b**) $t = 3.5$ s, (**c**) $t = 4.5$ s, (**d**) $t = 5$ s, (**e**) $t = 5.55$ s, (**f**) $t = 5.56$ s. The effective plastic strains in the next sheet being cut are shown in the next figure.

During cutting in the first stage, the knife travels with constant speed downwards and touches the first top sheet being cut and starts pressing it and forming the elastic zone of the reduced Huber–Mises stresses (Figure 4a) and corresponding to it the equivalent Huber–Mises plastic strains (Figure 6a). So the reduced Huber–Mises stresses and strains grow from zero to the maximum values which correspond to the elastic limit. Since the cutting tool moves farther into the first sheet being cut, the stresses and strains exceed the elastic limit and start creating a plastic region close to the tip of the blade of the cutting tool (Figures 4b and 6b). The C75S steel begins to reinforce and finally the stresses and strains are so high that the material cannot withstand it anymore and so loses its continuity. Then the tip of the blade of the cutting tool penetrates into the first top sheet being cut and begins to shear it causing a ductile fracture as consequence. The sheet is being cut by separating it plastically (Figures 4c and 6c). The separation is possible since the early mentioned failure model was additionally adopted to enable the cutting process to be continued which consisted of the nodes being split up. The numerical simulations show that from the very beginning of cutting which starts at the top of each sheet until circa 1/3 of the height of the sheet being cut measured from its top downwards, the ductile fracture caused by shearing suddenly changes into brittle fracture caused by the tensile state of the stresses which begins to dominate instead of shearing after exceeding the early mentioned 1/3 of the height of each sheet (Figure 4e,f and Figure 6e,f).

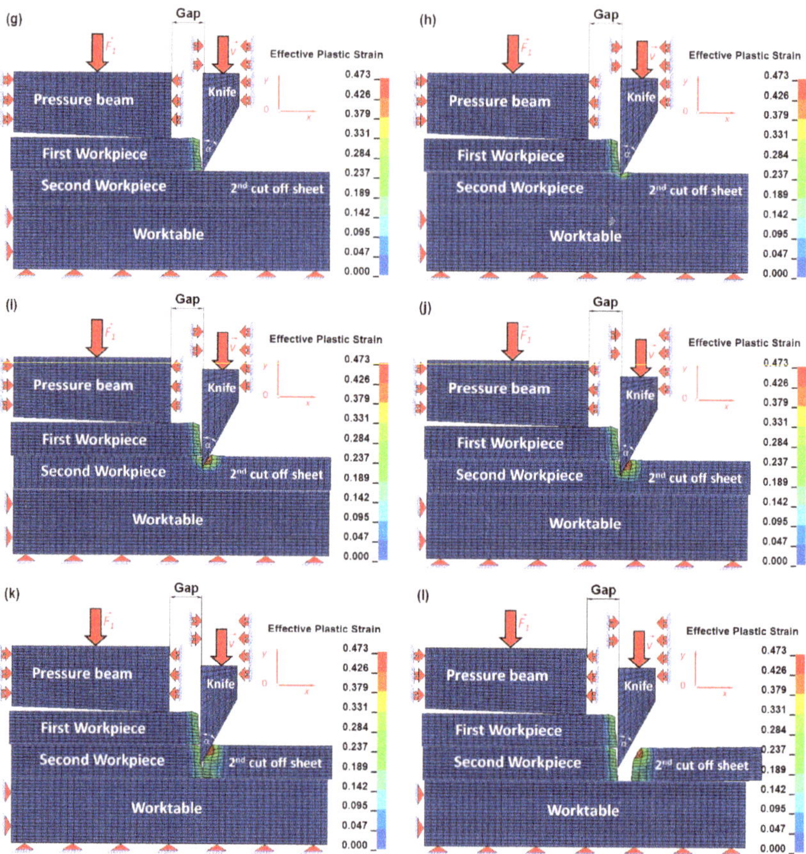

Figure 7. Effective plastic strains in the second sheet being cut for the following time instances: (**g**) t = 7.5 s, (**h**) t = 8 s, (**i**) t = 9 s, (**j**) t = 9.5 s, (**k**) t = 10.1 s, (**l**) t = 10.11 s (continuation of Figure 6). The effective plastic strains in the previous sheet being cut are shown in the previous figure.

The final stage of cutting is very interesting because it is connected with brittle cracking produced by tensile stresses. The described phenomenon is very similar to the experimental uniaxial tensile test. At the end of the stretching test, a neck is formed and cracking appears. This experimental tensile test is similar to the cracking observed during cutting of the sheet in the case where the tip of the blade penetrates into the sheet being cut at a height bigger that 1/3 of the height of the sheet. The cut surface is smooth, without burrs and vertical scratches. Such cut surfaces are required because they are fine unlike those induced by shearing accompanied by many defects. So, the only problem is the first stage concerning the shearing. It takes place from zero to circa 1/3 of the height of the sheet measured from its top downwards. During the shearing edge bending, scratches and large plastic zones are formed. On such cut surfaces there are many defects which are not desired (Figure 4a–c, Figure 5g–i, Figures 6a–c and 7g–i). The more sheets in a bundle, the more numerous are the defects. The sheet with the most defects is always the last one at the bottom. This work deals with only two sheets in a bundle but it is enough to prove numerically as well as experimentally that plastic deformations are more severe during the cutting of the last bottom sheet in a bundle (Figure 4c,d, Figure 5i,j, Figure 6c,d and Figure 7i,j). The elastic and plastic zones during cutting of the second sheet are similar to the first one.

The only difference is that the bottom sheet is more severely plastically deformed. Comparing these two plastic zones in the two sheets being cut one can state that the second sheet is more deformed so the plastic strains are slightly higher. The second stage concerning the sheet being cut from 1/3 of the height to the whole height of the second sheet (Figure 5k,l and Figure 7k,l) is very similar to the same stage concerning the cutting of the first sheet (Figure 4e,f and Figure 6e,f). These stages are highly desired from a theoretical as well as practical point of view since a good quality of sheets being cut is highly sought after. On these cut surfaces where the brittle fracture occurs there are no burrs or vertical scratches in the shape of craters, etc.

4.2. Microscopic Details

The sheet metal bundle cutting process leads to the occurrence of two types of fracture. There is a substantial difference in the fracture character when comparing the back side and front side of the cutting tool (Figure 8). The presence of a fracture consisting of plastic and brittle parts in the zone located at the back of the cutting tool was observed. The ductile zone is always located in the upper part of the observed cross section, whereas the brittle zone occurs in its lower part (Figure 8a,c). The boundary between the ductile and brittle zones is located at a 1/3 of the sheet height measured from the top downwards independent of the place in the bundle.

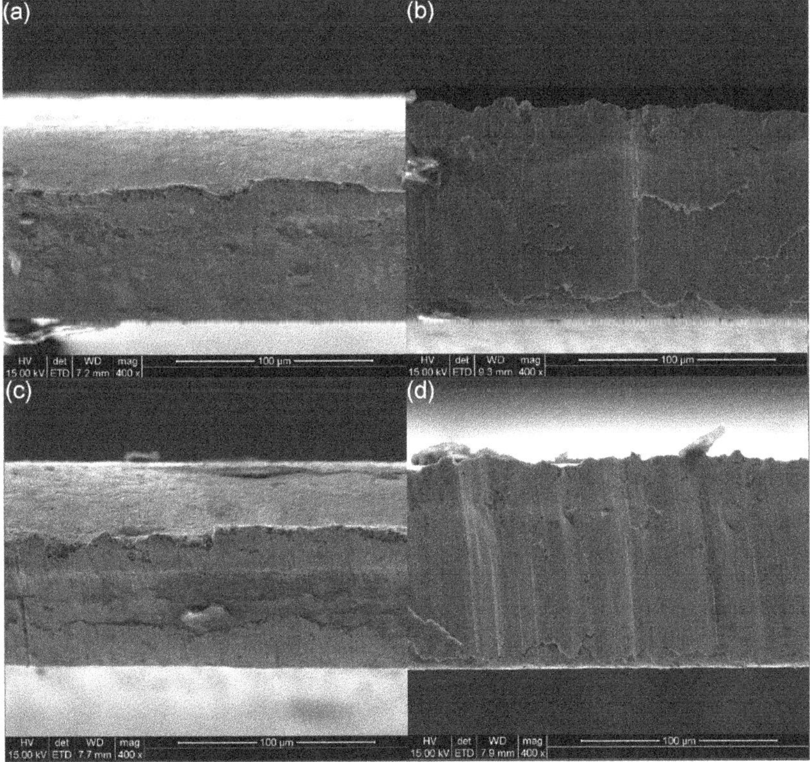

Figure 8. Cross sections of the steel sheets being cut depending on the localization of the observed surface: top sheet (**a**,**b**) and bottom sheet (**c**,**d**) at the back side of the cutting tool (**a**,**c**) and at the front side of the cutting tool (**b**,**d**) mag. 400×.

The plastic zone is characterized by a dimple mode of rupture. Most of the observed dimples are characterized by an oval shape. Moreover, some of them have a spherical shape. The dimples are characterized by various sizes. Most of them are relativity small due to the presence of numerous nucleating sites, which are activated during the deformation process. The micro-voids are closely located and they coalesce before they have an opportunity to grow to a larger size [33]. The lower part of the samples shows typical brittle fracture. Numerous brittle walls can be observed (Figure 9a,c) but no dimples were detected.

Figure 9. Details of the surface fracture of the steel sheets being cut depending on the localization of the observed surface: top sheet (**a,b**) and bottom sheet (**c,d**) at the back side of the cutting tool (**a,c**) and at the front side of the cutting tool (**b,d**) mag. 800×.

The cut surfaces located at the front of the cutting tool are characterized by the different nature of fracture in comparison to the surfaces located at the back side of the cutting tool. In this case, the observed cross sections are homogenous (Figure 8b,d). The typical features of brittle fracture are observed. The decohesion of all the steel sheets occurred without plastic deformation. The presence of irregular shear planes and numerous scratches are visible (Figure 9b,d) as well as some cracks.

There are no big differences when comparing the top and bottom sheets (Figure 8). However, one can see that the sheet located at the bottom of the bundle is more heavily affected by the cutting process.

4.3. Comparison of the Numerical and Experimental Investigations

The elaborated physical and mathematical models of the cutting process were the basis to carry out numerical simulations taking into account the failure model which corresponds to all nodes that belong to the cutting line (Figure 3). Experimental verification using a scanning electron microscope was carried out. At the final stage the numerical as well as the experimental results were juxtaposed. This comparison is presented in Figure 10 and shows that the plastic region of sheets being cut begins from zero to 1/3 of the sheet height measured from the top of each sheet downwards. The plastic zone is slightly larger in the bottom sheet (Figure 10c,d). It might be explained by taking into account the friction which resists the impending motion of the cutting tool. During cutting of the first sheet, the friction force occurs only between the top sheet being cut and the cutting tool (Figures 4 and 6) whereas during cutting of the second sheet the friction force occurs not only between the bottom sheet being cut and the cutting tool but also between the mentioned cutting tool and the cut off top sheet which is on the left side between the pressure beam and the bottom sheet (Figures 5 and 7). That is why the conditions of cutting of the bottom sheet are slightly more severe than the conditions of cutting of the top sheet. Evidence that the second sheet being cut works in slightly heavier conditions comes from the graphs presenting the internal energy accumulated during cutting in both sheets (Figure 11) and for each sheet successively (Figure 12).

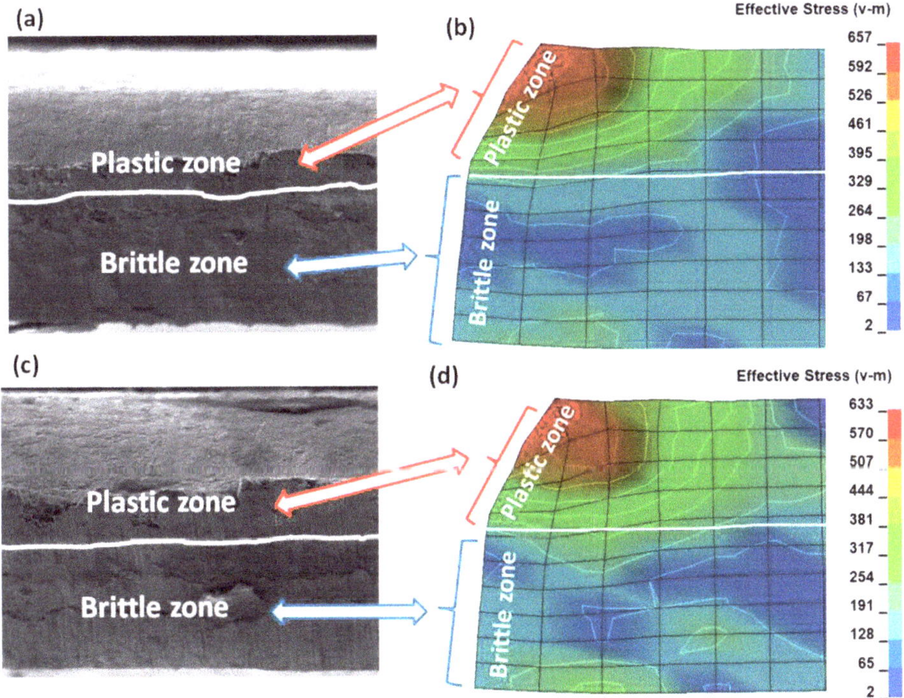

Figure 10. Comparison of the experimental and numerical results.

A similar comparison analysis was conducted by Bohdal [10], who investigated the cutting process of a single sheet being cut using a bench shear. The cutting process was simulated by means of the FEM-SPH model and was compared with the experimental data using recorded images from a high-speed camera. The experiment shows good agreement in the depth angle and the burr height with the numerical simulations. From the obtained results it can be seen that the fracture process becomes less steady and progresses in a non-uniform manner in some locations along the shearing line. An analysis of the state of stress, strain and fracture mechanisms of the material was presented. This approach of the author concerns shearing using bench shears however, in the current paper the cutting process of ultra-thin metal sheets arranged in a bundle deals mainly with shearing but stretching as well.

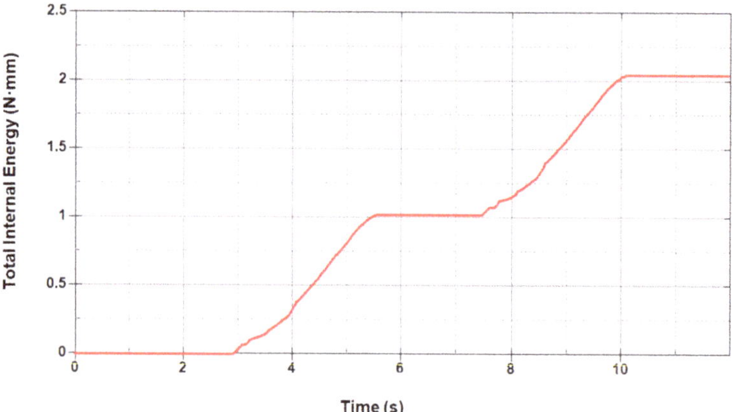

Figure 11. Total internal energy in both sheets being cut versus time.

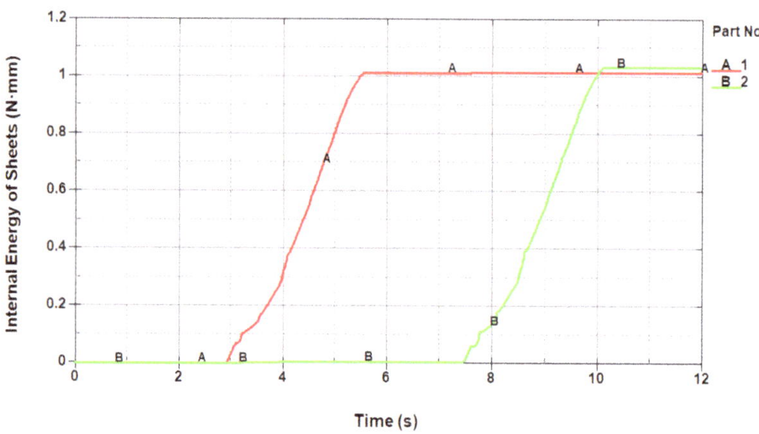

Figure 12. Internal energy for separated sheets being cut versus time: (**A 1**)—in the first sheet, (**B 2**)—in the second sheet.

5. Conclusions

In this paper the cutting problem was analysed for better understanding of the mechanisms involved. The cutting process is characterised by fast change of physical phenomena such as displacements, strains, stresses, etc. versus time and an additional problem constitutes the excessive speed of the process which prevents observation in detail by the unaided eye. That is why the physical model and corresponding to it the mathematical model of the cutting process were elaborated here. Next, numerical simulation using the finite element method and adopting the computer system LS-DYNA was performed in order to present the mechanism of the cutting process of sheets in a bundle. The elastic and plastic zones as well as the ductile and brittle fractures that occur during cutting at selected time intervals are presented. Experimental investigation was also carried out by observation of the steel sheets after cutting under a scanning electron microscope. The numerical and experimental results were compared with each other.

During study of the fast changing cutting process the following conclusions were drawn:

- the maximum stresses and the plastic strains obtained in the numerical simulations are mainly concentrated in sheets being cut close to the tip of the blade of a cutting tool,
- the maximum reduced Huber–Mises stresses in the first stage of cutting are higher than the yield point and that is why sheets being cut cannot withstand it with the result that ductile fracture occurs at a height from zero to circa 1/3 of the height of the sheet being cut measured from the top of the sheets downwards. In this particular ductile zone shearing dominates, justified on the base of numerical as well as experimental data,
- the maximum reduced Huber–Mises stresses in the final stage of cutting are higher even than the ultimate strength which occurs at a height larger than circa 1/3 of the height of the sheets being cut, measured from the top of the sheets downwards. The steel sheet being cut in this particular zone cannot withstand it and brittle fracture occurs because instead of shearing, stretching dominates implied on the basis of numerical simulations and verified by experiment,
- on the basis of SEM imaging it can be stated that ductile fracture is accompanied by many defects like burrs, vertical scratches, higher roughness, larger edge bending, etc.,
- brittle fracture has no significant defects and is characterised by a smooth surface in the cross section of the sheets being cut, observed under the scanning electron microscope.

Author Contributions: Conceptualization, J.K.; Data curation, J.K.; Formal analysis, J.K.; Funding acquisition, J.K.; Investigation, A.K., A.G. and S.S.; Methodology, J.K.; Software, J.K. and S.S.; Validation, J.K. and A.G.; Visualization, A.K.; Writing—original draft, J.K.; Writing—review and editing, A.K., A.G. and S.S.

Funding: This work was financially supported by statutory funds from the Faculty of Mechanical Engineering of Silesian University of Technology in 2018.

Acknowledgments: Numerical calculations were carried out using the computer cluster Ziemowit (http://www.ziemowit.hpc.polsl.pl) funded by the Silesian BIO-FARMA project No. POIG.02.01.00-00-166/08 in the Computational Biology and Bioinformatics Laboratory of the Biotechnology Centre in the Silesian University of Technology.

References

1. Arslanov, M.Z. A polynomial algorithm for one problem of guillotine cutting. *Oper. Res. Lett.* **2007**, *35*, 636–644. [CrossRef]
2. Alvarez-Valdes, R.; Parajon, A.; Tamarit, J.M. A tabu search algorithm for large-scale guillotine (un)constrained two-d imensional cutting problems. *Comput. Oper. Res.* **2002**, *29*, 925–947. [CrossRef]
3. Tiwari, S.; Chakraborti, N. Multi-objective optimization of a two-dimensional cutting problem using genetic algorithms. *J. Mater. Process. Technol.* **2006**, *173*, 384–393. [CrossRef]
4. Gasiorek, D.; Baranowski, P.; Malachowski, J.; Mazurkiewicz, L.; Wiercigroch, M. Modelling of guillotine cutting of multi-layered aluminum sheets. *J. Manuf. Process.* **2018**, *34*, 374–388. [CrossRef]
5. Nouari, M.; Makich, H. On the Physics of Machining Titanium Alloys: Interactions between Cutting Parameters, Microstructure and Tool Wear. *Metals* **2014**, *4*, 335–338. [CrossRef]
6. Razak, N.H.; Chen, Z.W.; Pasang, T. Effects of Increasing Feed Rate on Tool Deterioration and Cutting Force during and End Milling of 718Plus Superalloy Using Cemented Tungsten Carbide Tool. *Metals* **2017**, *7*, 441. [CrossRef]
7. Koklu, U.; Basmaci, G. Evaluation of Tool Path Strategy and Cooling Condition Effects on the Cutting Force and Surface Quality in Micromilling Operations. *Metals* **2017**, *7*, 426. [CrossRef]
8. Haddag, B.; Atlati, S.; Nouari, M.; Moufki, A. Dry Machining Aeronautical Aluminum Alloy AA2024-T351: Analysis of Cutting Forces, Chip Segmentation and Built-Up Edge Formation. *Metals* **2016**, *6*, 197. [CrossRef]
9. Kaczmarczyk, J.; Grajcar, A. Numerical Simulation and Experimental Investigation of Cold-Rolled Steel Cutting. *Materials* **2018**, *11*, 1263. [CrossRef] [PubMed]
10. Bohdal, Ł. Application of a SPH Coupled FEM Method for Simulation of Trimming of Aluminum Autobody Sheet. *Acta Mech. Autom.* **2016**, *10*, 56–61. [CrossRef]
11. Fedeliński, P.; Górski, R.; Czyż, T.; Dziatkiewicz, G.; Ptaszny, J. Analysis of Effective Properties of Materials by Using the Boundary Element Method. *Arch. Mech.* **2017**, *66*, 19–35.

12. Paggi, M. Crack Propagation in Honeycomb Cellular Materials: A Computational Approach. *Metals* **2012**, *2*, 65–78. [CrossRef]

13. Perez, N. *Fracture Mechanics*; Kluwer Academic Publishers: Boston, MA, USA, 2004.

14. González, H.; Pereira, O.; Fernández-Valdivielso, A.; López de Lacalle, L.N.; Calleja, A. Comparison of Flank Super Abrasive Machining vs. Flank Milling on Inconel® 718 Surfaces. *Materials* **2018**, *11*, 1638.

15. Del Pozo, D.; López de Lacalle, L.N.; López, J.M.; Hernández, A. Prediction of Press/Die Deformation for an Accurate Manufacturing of Drawing Dies. *Int. J. Adv. Manuf. Technol.* **2008**, *37*, 649–656. [CrossRef]

16. Lamikiz, A.; López de Lacalle, L.N.; Sanchez, J.A.; Bravo, U. Calculation of the Specific Cutting Coefficients and Geometrical Aspects in Sculptured Surface Machining. *Mach. Sci. Technol.* **2005**, *9*, 411–436. [CrossRef]

17. Kaczmarczyk, J. Numerical Simulations of Preliminary State of Stress in Bundles of Metal Sheets on the Guillotine. *Arch. Mater. Sci. Eng.* **2017**, *85*, 14–23. [CrossRef]

18. Kaczmarczyk, J.; Gąsiorek, D.; Mężyk, A.; Skibniewski, A. Connection Between the Defect Shape and Stresses which Cause it in the Bundle of Sheets Being Cut on Guillotines. *Model. Optim. Phys. Syst.* **2007**, *6*, 81–84.

19. Show, M.C. *Metal Cutting Principles*; Oxford University Press: New York, NY, USA, 2005.

20. Bhatti, M.A. *Practical Optimization Methods*; Springer-Verlag Inc.: New York, NY, USA, 2000.

21. Rothwell, A. *Optimisation Methods in Structural Design*; Springer International Publishing AG: Delft, The Netherlands, 2017.

22. Goldberg, D.E. *Genetic Algorithms in Search, Optimization, and Machine Learning*; Addison-Wesley Publishing Company, Inc.: New York, NY, USA, 1989.

23. Michalewicz, Z. *Genetic Algorithms + Data Structures = Evolution Programs*; Springer: Berlin/Heidelberg, Germany, 1996.

24. *LS-DYNA Keyword User's Manual*; Livermore Software Technology Corporation: Livermore, CA, USA, 2017.

25. Bathe, K.J.; Chaudhary, A. A Solution Method for Planar and Axisymmetric Contact Problems. *Int. J. Numer. Method Eng.* **1985**, *21*, 65–88. [CrossRef]

26. Belytschko, T.; Liu, W.K.; Moran, B.; Elkhodary, K.I. *Nonlinear Finite Elements for Continua and Structure*; John Wiley & Sons, Ltd.: Southern Gate, CA, USA, 2014.

27. Fish, J.; Belytschko, T.A. *First Course in Finite Elements*; John Wiley & Sons, Ltd.: Southern Gate, CA, USA, 2007.

28. Mohammadi, S. *Discontinuum Mechanics Using Finite and Discrete Elements*; WIT Press Southampton: Boston, MA, USA, 2003.

29. Hughes, T.J.R. *The Finite Element Method. Linear Static and Dynamic Finite Element Analysis*; Manufactured in the United States by RR Donnelley: Chicago, IL, USA, 2016.

30. Zienkiewicz, O.C.; Taylor, R.L. *The Finite Element Method, Solid Mechanics*; Butterworth-Heinemann: Oxford, UK, 2000.

31. Timoshenko, S.P.; Goodier, J.N. *Theory of Elasticity*; McGraw-Hill Education: New York, NY, USA, 2017.

32. Chaudhary, A.B.; Bathe, K.J. A Solution Method for Static and Dynamic Analysis of Three—Dimensional Contact Problems with Friction. *Comput. Struct.* **1986**, *24*, 855–873. [CrossRef]

33. ASM Handbook. *Volume 12: Fractography*; ASM International: Cleveland, OH, USA, 1987.

 *metals* 

Article

Effect of the $t_{8/5}$ Cooling Time on the Properties of S960MC Steel in the HAZ of Welded Joints Evaluated by Thermal Physical Simulation

Miloš Mičian [1],*, Daniel Harmaniak [1], František Nový [1], Jerzy Winczek [2], Jaromír Moravec [3] and Libor Trško [4]

[1] Faculty of Mechanical Engineering, University of Žilina, Univerzitná 8215/1, 010 26 Žilina, Slovak Republic; daniel.harmaniak@fstroj.uniza.sk (D.H.); frantisek.novy@fstroj.uniza.sk (F.N.)

[2] Faculty of Mechanical Engineering and Computer Science, Czestochowa University of Technology, Dabrowskiego 69, 42-201 Czestochowa, Poland; winczek@imipkm.pcz.czest.pl

[3] Faculty of Mechanical Engineering, Technical University of Liberec, Studentská 1402/2, 461 17 Liberec I, Czech Republic; jaromir.moravec@tul.cz

[4] Research Centre of the University of Žilina, University of Žilina, Univerzitná 8215/1, 010 26 Žilina, Slovak Republic; libor.trsko@rc.uniza.sk

* Correspondence: milos.mician@fstroj.uniza.sk; Tel.: +421-41-513-2768

Received: 29 December 2019; Accepted: 4 February 2020; Published: 7 February 2020

Abstract: The heat input into the material during welding significantly affects the properties of high-strength steels in the near-weld zone. A zone of hardness decrease forms, which is called the soft zone. The width of the soft zone also depends on the cooling time $t_{8/5}$. An investigation of the influence of welding parameters on the resulting properties of welded joints can be performed by thermal physical simulation. In this study, the effect of the cooling rate on the mechanical properties of the heat-affected zone of the steel S960MC with a thickness of 3 mm was investigated. Thermal physical simulation was performed on a Gleeble 3500. Three levels of cooling time were used, which were determined from the reference temperature cycle obtained by metal active gas welding (MAG). A tensile test, hardness measurement, impact test with fracture surface evaluation, and microstructural evaluation were performed to investigate the modified specimen thickness. The shortest time $t_{8/5} = 7$ s did not provide tensile and yield strength at the minimum required value. The absorbed energy after recalculation to the standard sample size of 10×10 mm was above the 27 J limit at −40 °C. The hardness profile also depended on the cooling rate and always had a softening zone.

Keywords: $t_{8/5}$ cooling time; S960MC steel; mechanical properties; heat-affected zone (HAZ); physical simulation

1. Introduction

High-strength low-alloy (HSLA) steels provide design engineers with new opportunities in the design of mobile vehicles. These include the use of smaller cross-sections and thinner sheets. This has beneficial effects such as weight reduction, production cost savings, simplified transport, lower fuel consumption, and consequently reduced emissions. One method of producing high-strength steels is thermo-mechanical processing, which is a technological process of controlled rolling in the area of transformation temperatures of austenite to bainite and martensite. High-strength steels are achieved by plastic deformation, grain refinement, adjustment of the chemical composition, precipitation hardening by micro-alloying elements (NB, V, and Ti), and austenite-to-martensite and -bainite phase transformation. The strength obtained by these factors allows the carbon content and amount of alloying elements added to be reduced and compared to standard steel of the same properties. These generally known properties of HSLA steels are described in a series of research papers by Schaupp et al. [1], Branco et al. [2], and others [3–5]. Since the carbon equivalent of these steels is at the level of standard steels with a lower strength, they therefore have excellent weldability in terms of the chemical composition. Compared to mild steels, structures can achieve a comparable or higher fatigue life. This is described in Ślęzak's [6] and Lago's [7] research.

Research in the field of the welding of high-strength steels is concentrated in several areas. The formation of a soft zone in the heat-affected zone (HAZ) where the hardness decreases compared to the base material, is one of the problems in HSLA steel welding. A soft zone does not always mean a decrease in the static strength of the weld joint in the transverse tensile test. Hochhauser et al. [8] and Björk [9] concluded from their research that the decrease in strength mainly depends on the width of the soft zone. It is possible to determine the ratio of the width of the soft zone to the thickness of the material to be welded. This ratio is called the relative thickness of the soft zone—X_{SZ}. A reduction in the relative thickness of the soft zone results in an increased strength of the weld joint during tensile tests. Nowacki et al. [10] also expressed the view that a small width of the tempered heat-affected zones (often having lower strength properties than required) does not significantly affect the strength of the whole joints. This can be explained by the occurrence of the phenomenon of the so-called "contact strengthening", generated by plane stresses in the less hard areas of joints.

Welding procedures with a limited heat input result in the formation of very narrow, softened areas located within the heat-affected zones of joints. Due to this, they are suitable for welding technologies with a concentrated heat source such as laser welding or electron beam welding. These phenomena are described in research papers by Guo et al. [11,12] and Kopas et al. [13]. Combining multiple techniques into one—the use of so-called hybrid technologies—is also one of the ways to achieve HSLA welded joints with satisfactory properties. Górka [14] examined hybrid welding technologies (laser beam + MAG) of S700MC steel T-joints with a thickness of 10 mm. Sajek and Nowacki [15] applied the same technology to S960QL.

Nevertheless, MAG welding in the field of the production of steel structures is still the most commonly used technique. MAG welding conditions can be improved by means of additional cooling of the welds (cupper backing pieces and heat sinks). Laitila and Larkiola [3] described that the use of heat sinks results in a beneficial increase of the cross-weld yield strength; however, at the expense of the yield-to-tensile strength ratio.

Generally, two mechanisms of "softening" are described. If the maximum temperature of the thermal cycle is below A_1, it is tempering softening. Tempering processes occurring in the HAZ of welds involve a process similar to that when manufacturing the base material. They include carbon rejection of the supersaturated martensite, the transformation of metastable carbides to stable ones, or the spheroidization of carbides. This phenomenon is typical of structural steels in a quenched and tempered condition. Tempering softening is less pronounced for thermo-mechanically-controlled processed (TMCP) steels due to a beneficial precipitation hardening effect and less extensive transformation of the hardened microstructure. However, a thermo-mechanically-treated microstructure changes irreversibly if the microstructure is exposed to temperatures above the A_1 temperature (representing another phenomenon—transformation softening). A longer period of heating to higher temperatures leads to grain growth in the HAZ. Such a structure is sensitive to the reduction of toughness. Samardžić et al. [16] and Lahtinen et al. [17] analyzed the effect of the temperature input and cooling rate on the toughness of individual HAZs.

The filler metal also has a considerable effect on the resulting properties of the weld joint. With respect to the strength of the base material and the strength of the filler metal, several cases may occur. With undermatching weld metal, the yield strength is smaller than the yield strength of the base material. The matching weld metal has the same properties as the base material. This category includes MAG welding wires with a classification of G 89, according to the EN ISO 16834-A standard. Sefcikova et al. [18], Pisarski et al. [19], and other authors [4,20,21] used this type of add-on material in their experiments. They compared the results of the mechanical properties of welded joints, when applying undermatching and matching welding wire. The lower-strength additive material was shown to be suitable for reducing the need for preheating. Such welds have lower residual stresses than those using higher-strength filler metal and are less susceptible to crack initiation. In this case, not the soft zones, but the strength of the weld metal, may be the limitation of the overall strength of the weld joint.

A physical simulation of material involves the exact reproduction of the thermal and mechanical processes in the laboratory that the material is subjected to in real-life welding. In the case of the physical simulation of the welding process, the most common task is to investigate the loading of the material by the temperature cycle. This is important due to its effect on the resulting structure and mechanical properties of the HAZ. These experiments have been described in several articles. The impact of temperature and holding time for S700MC steel on grain growth in HAZs was investigated by Moravec et al. [22]. Laitila and Larkiola [3] used physical simulation to evaluate the properties of S960 steel with multi-layer MAG welding. Gáspár [23] and Dunder et al. [24] evaluated the effect of the cooling time $t_{8/5}$ on the hardness and toughness of the S960QL steel HAZ.

The research carried out so far and the published results cited in this work relate to only the welding of high-strength steel in the range of sheet thicknesses from 8 to 20 mm. S960MC steel is manufactured with a minimum thickness of 3 mm and is suitable for cold bending. The production of various bend profiles also requires single-layer butt MAG welds. The behavior of HLSA thin sheets when subjected to a welding cycle load is very interesting, but not featured in the literature so far.

2. Materials and Methods

2.1. Experimental S960MC Material

In this experiment, the S960MC (Strenx 960MC—SSAB) steel was supplied, according to the EN 10149-2 Standard. The required chemical composition according to this standard and the chemical composition according to the inspection certificate of investigated steel are shown in Table 1.

Table 1. Chemical composition of investigated steel.

According to	Chemical Composition [%] wt.—S960MC Steel										
	C	Si	Mn	P	S	Al	Nb	V	Ti	Mo	B
EN 10149-2 *	0.20	0.60	2.20	0.025	0.010	0.015	0.090	0.20	0.250	1.000	0.005
Inspection ** certificate	0.085	0.18	1.06	0.01	0.003	0.036	0.002	0.007	0.026	0.109	0.001

* Maximum values of alloying elements except aluminum. At a total of aluminum, it is a minimum value. The sum of Nb, V, and Ti shall be max. 0.22%. ** According to the EN 10204 3.1 material certificate provided by the steel producer

The SSAB manufacturer (SSAB AB, Stockholm, Sweden) for this steel reduces the maximum required values of C = 0.12%, Si = 0.25%, and Mn = 1.30% wt. The total content of Nb + V + Ti micro alloys may be 0.18% wt. at the most. The base material was subjected to tensile testing according to EN ISO 6892-1. Tests were performed on the device INSTRON Series 5985. All samples for this test were cut by a fiber laser. Their orientation was in the rolling direction (sample designation 0°), perpendicular to the rolling direction (sample designation 90°), and at a 45° angle to the rolling direction (sample designation 45°) for an evaluation of planar anisotropy. The mechanical properties of the investigated steel, according to the supply standard and our own measurements, are shown in Table 2.

The values of the anisotropy coefficient show that the base material does not show any signs of planar anisotropy in all investigated parameters (R_m, $R_{p0.2}$, and A). The microstructure of the base material was evaluated by the optical microscopy ZEISS LSM 700 device.

Table 2. Mechanical properties of investigated steel.

According to	Angle to the Rolling Direction	Mechanical Properties S960MC, Thickness 3 mm				Planar Anisotropy Coefficient P [%]		
		$R_{p0.2}$ [MPa]	R_m [MPa]	$R_{p0.2}/R_m$	A [%]	PR_m	$PR_{p0.2}$	PA
EN 10149-2	-	min. 960	980–1250	-	min. 7	-	-	-
Experimental measurements	0°	1007	1092	0.92	7.9	-	-	-
	45°	1018	1106	0.92	6.7	1.2	1.1	−14.4
	90°	1044	1124	0.93	6.5	2.9	3.6	−17.0

In general, the S960MC is a microalloyed, thermo-mechanically processed high-strength structural steel, with a microstructure consisting of tempered martensite and rest austenite. Only in the case of thick plates can a small amount of bainite occur. As shown in Figure 1, the microstructure of the base experimental steel consists of tempered martensite and rest austenite. The amount of rest austenite was not exactly determined in this study. This kind of microstructure provides a good combination of a high tensile strength and high fracture toughness.

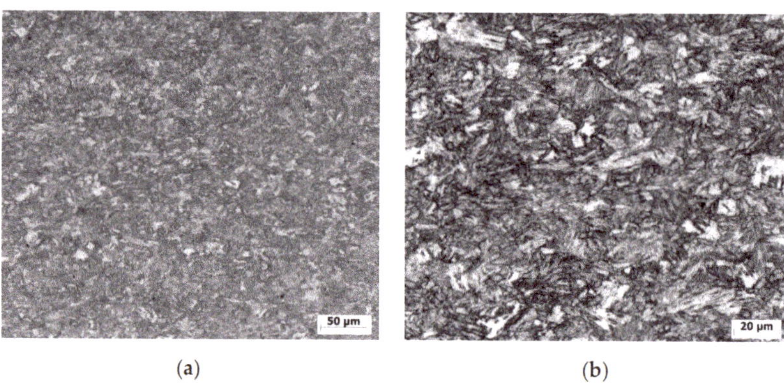

(a)   (b)

Figure 1. Microstructure of investigated steel S960MC: (**a**) magnification 200×; (**b**) magnification 500×.

2.2. Experimental Welding

The test sample was welded to set the parameters of the physical simulation. Real courses of the temperature cycle were obtained from welding. The joint was designed as a butt weld with a square groove and root gap of 1.5 mm on 3-mm thick plates. Welding was done by means of the linear automatic machine FVD 15 MF (Fronius, Wels, Austria) with a Fronius TransPuls Synergic 2700 power source. The welding parameters were monitored with Fronius Explorer software at a frequency of 10 Hz. The following actual welding parameters were obtained by means of monitoring: mean current I_z = 102 A, wire feed speed v_d = 3.8 m·s^{-1}, mean voltage U_z = 16.6 V, and welding speed v_z = 3.7 mm·s^{-1}. Using the energy efficiency factor for MAG welding h = 0.8, the effective heat input was Q_{ef} = 0.37 kJ·mm^{-1}.

The copper-coated solid wire Union X 96 (classified as G 89 5 M21 Mn4Ni2.5CrMo, according to the EN ISO 16834-A standard; manufactured by Böhler Welding) was used for welding. This wire belongs to undermatching filler materials, where the yield strength of the weld metal is less than the yield strength of the base material. The manufacturer's minimum guaranteed yield strength value is $R_{p0.2}$ = 930 MPa, and the tensile strength is R_m = 980 MPa for M21 shielding gas. Filler material with the same classification for welding steels of the strength grade 960 MPa was also used in studies by Guo et al. [11], Jambor et al. [25,26], Schneider et al. [20], and Taavitsainen et al. [5]. Type MIX 18 (classification M21, manufactured by Linde) was used as the shielding gas at a gas flow of 15 L/min when welded without preheating.

During the welding process, temperature cycles in the HAZ were recorded by the Temperature Input Module NI-9212 (manufactured by National Instruments) with an application prepared in the LabView software environment. For each channel, the temperature scan rate was 95 Hz. T-type thermocouples with a diameter of 0.32 mm were condenser-welded on the bottom of the plate.

Figure 2 shows a view of the bottom of the sheet in the area of the weld root and welded thermocouples. The distance of the TC1 thermocouple from the root edge was 0.96 mm. The distance between the TC1 and TC2 thermocouples was 3.18 mm, and 4.69 mm between TC3 and TC2. The thermal effect from welding the thermocouples on the material created local craters.

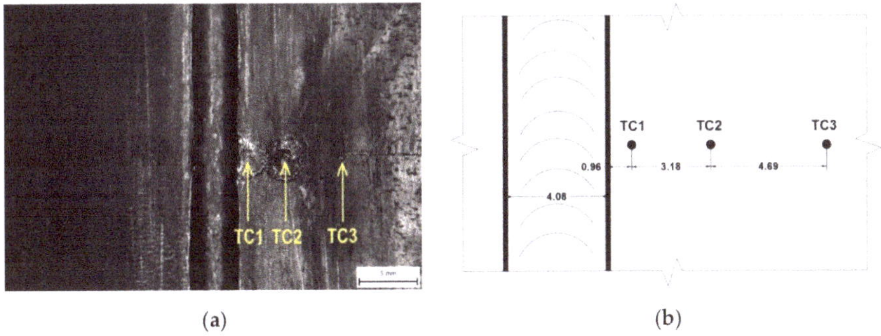

(**a**)　　　　　　　　　　　　　　　　(**b**)

Figure 2. Position of thermocouples on the root side of the weld: (**a**) marking the locations of the welded thermocouples; (**b**) geometry of the thermocouple position from the edge of the weld root.

Geometrical evaluation of the butt weld joint in the cross-section through an area of welded thermocouples is shown in Figure 3. The weld reinforcement was 1.71 mm at the weld width of 6.79 mm. The root reinforcement was 1.43 mm at the root weld width of 4.08 mm. The weld showed a little linear misalignment between plates.

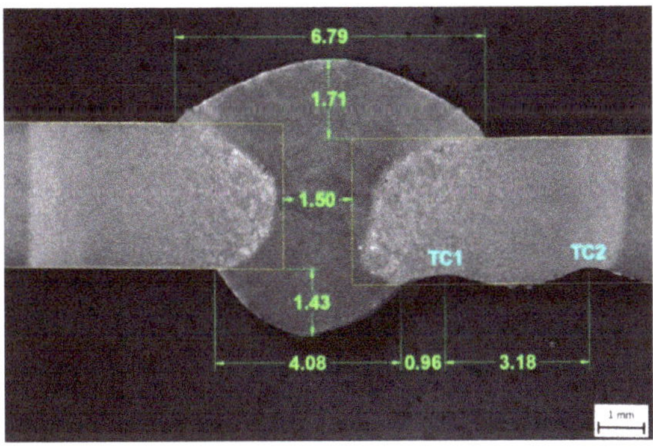

Figure 3. Geometrical evaluation of the experimental butt weld joint with marked points of thermal cycle measurement. (unit: mm)

The thermal cycles obtained from the welding are shown in Figure 4. To determine the cooling time $t_{8/5}$, the cycle temperature must be higher than 800 °C. Only the TC1 thermocouple reached a temperature above 800 °C, and its maximum temperature was 1105 °C. The cooling time $t_{8/5}$ = 17.5 s was determined from the thermal cycle of the TC1 thermocouple. The maximum temperature of the TC2 thermocouple was 688 °C. This temperature indicated that the TC2 thermocouple was on the border of the HAZ and the base material. The maximum temperature of the TC3 thermocouple was

283 °C. At this temperature, there was no change in the material properties, so the thermocouple was positioned in the base material.

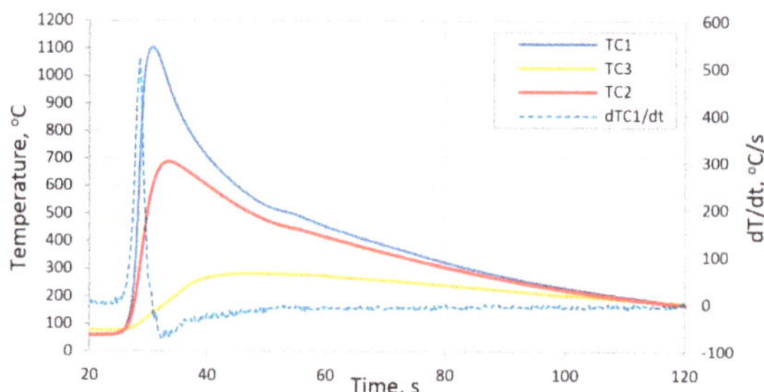

Figure 4. Thermal cycle obtained during the welding of a S960MC steel butt weld joint.

2.3. Preparation of Input Data for Physical Simulation

Physical simulation provides a unique opportunity to describe a HAZ in steels and other metal materials for various fusion welding methods. It includes processes with a high energy density such as electron beam welding (EBW) and laser beam welding (LBW) as well as processes with a medium to low energy density such as arc welding. In this paper, HAZ physical simulations were carried out on a Gleeble 3500 thermo-mechanical physical simulator (manufactured by Dynamic Systems Inc., Poestenkill, NY, USA), which is installed at the Technical University of Liberec, Liberec, Czech Republic.

The heating and cooling of the test samples in this test was controlled according to the specified temperature cycle set to the Gleeble system—the so-called "TC$_{prog}$ Program Cycle". The above-mentioned cycle can be generated based on the calculation model in QuikSIM2 software, according to the welding input conditions such as the type of material, heat input, metal sheet thickness, etc. Six functions can be used [27]. The first function is based on a real measured cycle. We can also use four equations derived from certain heat transfer equations such as the Hannerz equation, Rykalin 2D and 3D equations, and the Rosenthal equation. In the last one above-mentioned—the exponential cooling method—the sample is heated to a maximum temperature given by the rate. Then, the sample is exponentially cooled at a predetermined rate from 800 to 500 °C.

In the experiment, the method based on the description of the TC$_{prog}$ control cycle was used according to the measured data from the actual TC1 temperature cycle depicted in Figure 4, which was termed as the "TC$_{ref}$ Reference Cycle" for further description. The TC$_{ref}$ cycle consists of a heating phase, where the heating rate can be calculated as its first derivative, dT/dt. The maximum heating rate value was 522.5 °C/s and was used to determine the TC$_{prog}$ control cycle heating rate. The maximum heating cycle rate value was T_{max} = 1105 °C. The cooling phase was described by the cooling time $t_{8/5}$. The Strenx 960MC steel manufacturer recommends a $t_{8/5}$ time interval of 1–15 s to ensure satisfactory mechanical properties of the weld joint [28]. For the experiment using the Gleeble instrument, three program cycles were designed with different levels of the $t_{8/5}$ cooling time. These program cycles were designated as TC1$_{prog}$, TC2$_{prog}$, and TC3$_{prog}$. The cooling time $t_{8/5}$ was programmed at the following three levels: $t_{8/5}$ = 7 s (for TC1$_{prog}$), $t_{8/5}$ = 10 s (for TC2$_{prog}$), and $t_{8/5}$ = 17 s (for TC3$_{prog}$). At the same time and for the given test conditions, the time $t_{8/5}$ = 7 s was the shortest that could be achieved using Gleeble. This was done when the control system turned off the power supply as the maximum cycle temperature was reached, and the sample was cooled through heat transfer by water-cooled jaws that clamped the sample. In the TC1$_{prog}$ program cycle, this phase represented the sample temperature requirement of 20 °C. Figure 5 shows the program temperature cycles used in the experiment.

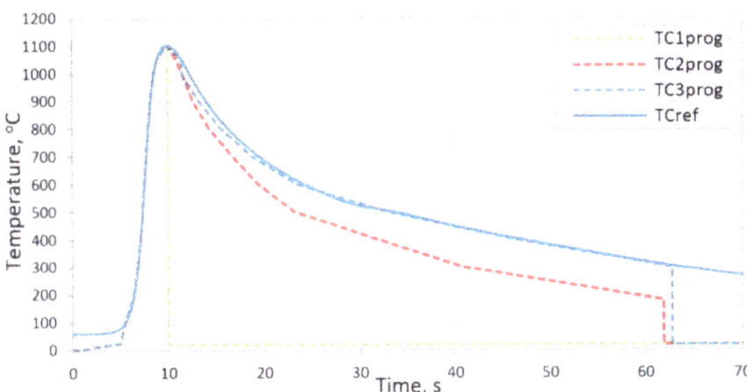

Figure 5. Program temperature cycles for the control Gleeble.

2.4. Preparation of Test Samples and Their Physical Simulation on Gleeble

Samples are normalized when testing materials because they follow standards such as EN ISO. However, samples in a physical simulation can often be arbitrary in order to achieve a satisfactory goal of the simulation. For HAZ simulation, round bars are most often recommended for tensile tests. Samples of a square cross-section are recommended for Charpy impact tests. A higher cooling rate can be achieved in the case of round samples and shortening the free length of the sample also speeds up cooling.

Since a material of thickness $t = 3$ mm was used in the experiment, the samples for simulation were flat and their dimensions were modified to fit the clamping jaws. As the samples could not be clamped at their original thickness, they were reduced by grinding on both sides to a thickness of 2 mm after laser cutting. The sampling was carried out in the direction of the metal sheet rolling. The overall dimensions of the test sample for the Gleeble simulation are shown in Figure 6.

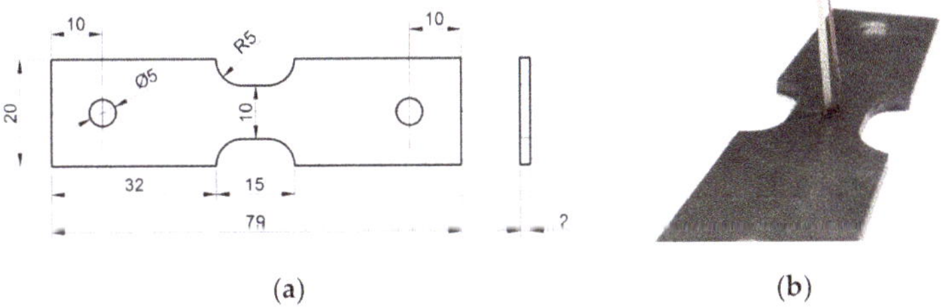

Figure 6. Samples for testing on a Gleeble 3500 device and a tensile test after simulation: (**a**) dimensions of the test sample (unit: mm); (**b**) view of the test sample with a thermocouple welded on it.

A thermocouple was welded onto the sample—in the middle of its length—using a condenser welding machine. This thermocouple sensed the real temperature during heating and cooling. At the same time, this cycle was compared with the program cycle, and the control system adjusted the power to achieve matching of the two cycles. The sample clamping during thermal loading in the Gleeble device is shown in Figure 7.

For three control cycles, three sets of test samples were prepared for the Gleeble for subsequent mechanical testing and microstructure assessment. The test samples were labeled according to their cooling time $t_{8/5}$, namely 7s, 10s, and 17s. The device was set to "Force-Control" when the force $F = 0$. This means that the distance between the jaws changed automatically with respect to the thermal expansion of the samples. As a result, no internal stresses occurred in the samples. The whole process took place in a vacuum. A record of the real temperature cycles measured during the simulation for different times $t_{8/5}$ is shown in Figure 8. In a temperature cycle corresponding to the cooling time $t_{8/5} = 7$ s, an increase in temperature can be observed at 480 °C. This increase is associated with the latent heat release during austenite transformation. A similar phenomenon can be observed in the cycle for $t_{8/5}$ = 10 s. In this case, the austenite phase transformation beginning temperature shifted to 500 °C as the austenite decay temperature curve in the CCT diagram increased with time. This phenomenon was not observed in the last temperature cycle with a $t_{8/5}$ = 17 s cooling time. This is because the Gleeble device was able to transfer the latent heat generated in the sample by cooling according to the program cycle already at slower cooling.

Figure 7. Position of the sample in the Gleeble 3500 device during the temperature cycle application.

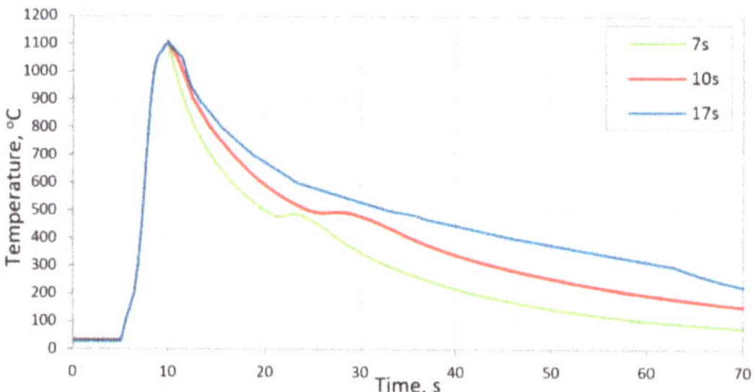

Figure 8. Thermal cycle from the Gleeble physical simulation with different cooling times $t_{8/5}$.

Samples directly from the Gleeble device were used for the static tensile test, without further treatment or modification. The test was performed on an INSTRON Model 5985 at room temperature. Figure 9 shows the samples after physical simulation milled to the required dimensions for the Charpy pendulum impact test, with a maximum impact energy of 150 J. At the same time, shims were adhered to the sample in the same way as recommended by the EN ISO 148-1 standard for low energies. The fracture surfaces after the impact test were shot by a scanning electron microscope (SEM), model TESLA VEGA II. Samples for the assessment of the microstructure and microhardness were taken by the means of a longitudinal section using precise water-cooled cutting equipment to prevent heat affecting the samples. Sample dimensions are shown in Figure 10. Samples were prepared by a standard procedure for light metallographic microscopy and etched with 2% Nital. Samples were assessed using a ZEISS LSM 700 optical microscope (Carl Zeiss Microscopy GmbH, Jena, Germany). Microhardness was measured in a line drawn through the thickness center using an Innovatest NOVA 130 device (INNOVATEST Europe BV, Maastricht, The Netherlands). This measurement utilized a load force of 9.807 N for 10 s. The distance between indentations was 0.5 mm.

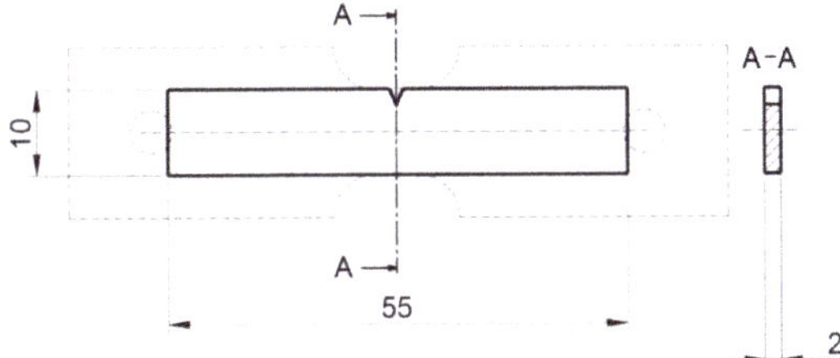

Figure 9. Sample with a reduced thickness of 2 mm used to measure the Charpy pendulum impact test (unit: mm).

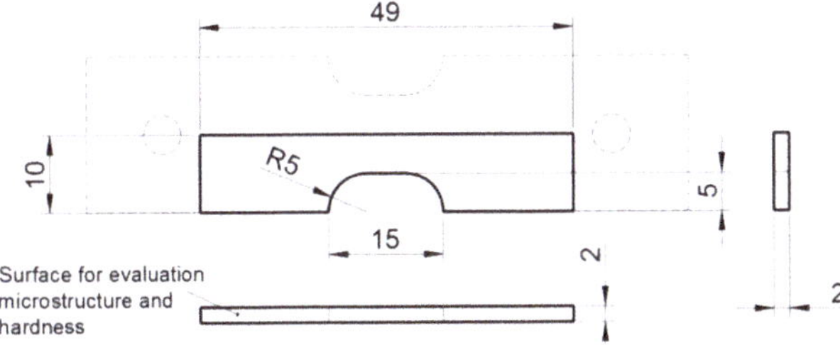

Figure 10. Sample with a reduced thickness of 2 mm used for the microstructure and hardness evaluation (unit: mm).

3. Results and Discussion

3.1. Tensile Test and Charpy Pendulum Test

In the static tensile test, we assessed the yield strength and tensile strength. Subsequently, we used these values to calculate their mutual ratio. The Charpy pendulum impact test was carried out at the temperature of −40 °C. The sample width was 2 mm, which is not a normalized dimension. Samples of the same width were also made for the base material (BM). The average value was calculated from five measurements. The criterion value of absorbed energy for S960MC steel at −40 °C is KV = 27 J in the case of samples with a standard 10 mm width. To compare the results of 2 mm wide samples with the criterion value, the results were recalculated using a coefficient given by the ratio of reduced width and standard width (i.e., 10 mm/2 mm = 5.) The same approach was also used by Schneider et al. [20] and Guo [11] in their works.

The results of tensile testing showed that the load on the base material by a thermal cycle with a maximum temperature of 1105 °C causes a decrease in the tensile and yield strength. The cooling rate expressed by time $t_{8/5}$ had a decisive influence on the above-mentioned phenomenon. The tensile test samples marked "7s" reached an average tensile strength of 1027.2 MPa, which is 94% of the value of the base material. This value is above the minimum required value of 980 MPa for the S960MC steel. The yield strength was 917.4 MPa, which is 91% of the base material value. However, it was below the minimum required yield strength of 960 MPa for the S960MC steel. In the case of the 10s and 17s samples, the tensile strength and yield strength did not reach the minimum required values for the steel that was investigated. The ratio of yield strength to tensile strength rose with an increasing cooling time $t_{8/5}$ value from 0.89 in the case of the 7s sample up to 0.91 in the 17s sample. The above ratio for the base material reached 0.92. The results of the tensile test are presented in Table 3, and are graphically shown in Figure 11.

Table 3. Results of the tensile test (at a room temperature of 23 ± 2 °C) for the base material and samples affected by the thermal cycle under different cooling times $t_{8/5}$.

Sample No.	$R_{p0.2}$ [MPa]	Average Value of $R_{p0.2}$ [MPa]	R_m [MPa]	Average Value of R_m [MPa]	$R_{p0.2}/R_m$
BM	1007.2	1007.2	1092.2	1092.2	0.92
7s-01	923.1	917.4	1031.5	1027.2	0.89
7s-02	911.7		1022.9		
10s-01	873.8	875.1	974.4	972.7	0.90
10s-02	876.5		970.9		
17s-01	836.3	852.2	920.7	936.6	0.91
17s-02	868.2		952.6		

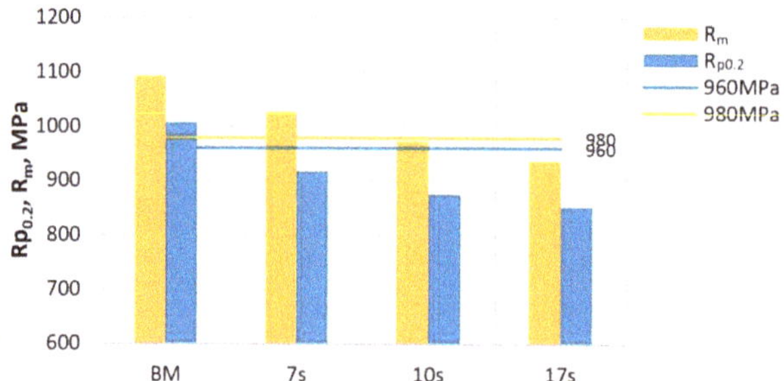

Figure 11. Yield strength and tensile strength for the base material and samples with different cooling times $t_{8/5}$.

As a result of the above-mentioned testing, the tensile strength and yield strength decrease as the $t_{8/5}$ time value increases. This is a phenomenon described by other authors in their work [10,19,20]. However, even the minimum cooling time of $t_{8/5}$ = 7 s that could be simulated on the Gleeble device did not achieve the required mechanical properties.

The afore-mentioned reduction in tensile strength can be explained by several aspects. In the investigated HAZ region, grain coarsening, and a higher degree of martensite tempering (lowering the degree of tetragonality) occurred, resulting in a decreased hardness and hence tensile strength. At the same time, we expect the precipitate dissolution of micro-alloying elements that lose their ability to prevent the growth of austenite grains.

In the Charpy pendulum impact tests of the assessed samples with a reduced width of 2 mm, the absorbed energy values in all simulated samples were higher than those of the base material. The test was carried out at −40 °C. This value was 20 J in the 7s sample. These values were higher in the 10s and 17s samples. The base material reached an average absorbed energy value of 17 J. At the same time, there was a trend of an increasing absorbed energy value with an increasing $t_{8/5}$ cooling time. This increase can be explained by a more pronounced effect of tempering the martensitic structure in repeated heating. On the contrary, the influence of grain growth had a negative effect on the absorbed energy. The results of the Charpy pendulum impact tests are shown in Table 4 and are graphically shown in Figure 12. At the same time, this table also shows the recalculated absorbed energy values based on the dimensions of a standard 10-mm wide sample. These values in all samples were higher than the minimum required value of 27 J for the S960MC steel at −40 °C.

Table 4. Results of the Charpy pendulum impact test (at the temperature of −40 °C) for the base material and samples affected by the thermal cycle under different cooling times $t_{8/5}$.

Sample No.	KV [J], 2 × 10 mm	Average Value of KV [J], 2 × 10 mm	KVC [J·cm⁻²]	KV [J], 10 × 10 mm (Calculated)
BM	17	17	106.3	85.0
7s-01	19	20	125.0	100.0
7s-02	21			
10s-01	25	23,5	146.9	117.5
10s-02	22			
17s-01	26	28	175.0	140.0
17s-02	30			

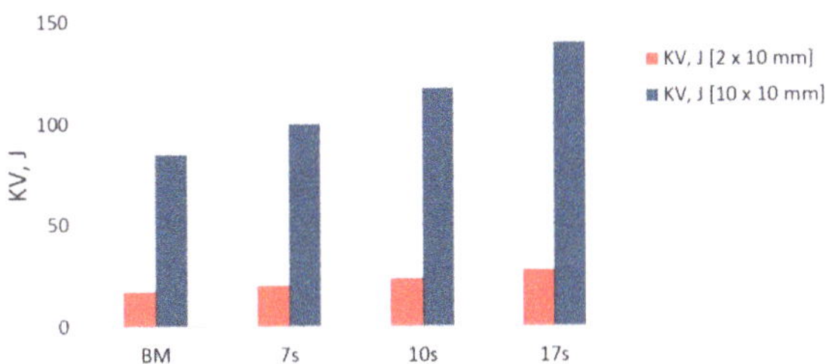

Figure 12. Absorbed energy KV for the base material and samples with different cooling times $t_{8/5}$.

3.2. Charpy Fracture Surfaces

The observation of the fracture surface topography of the broken samples after the Charpy impact tests using scanning electron microscopy (SEM) only revealed the ductile fracture. In Figure 13, it can be seen that the ductile fracture is accompanied by plastic deformation, which is manifested macroscopically by the cross-section distortion of all broken samples. Microscopically, pure transcrystalline ductile fractures with a typical dimple morphology, without any signatures of intercrystalline or brittle fracture, can be observed. The variation of dimple size is affected by the microstructural features of the HAZ. As can be seen in Figure 14, the different cooling times have a negligible influence on the fracture changes in this case. The ductile dimples are generated around carbide particles (Fe_3C). Therefore, the size, shape, and distribution of dimples depend on the distribution of carbide particles and grain size.

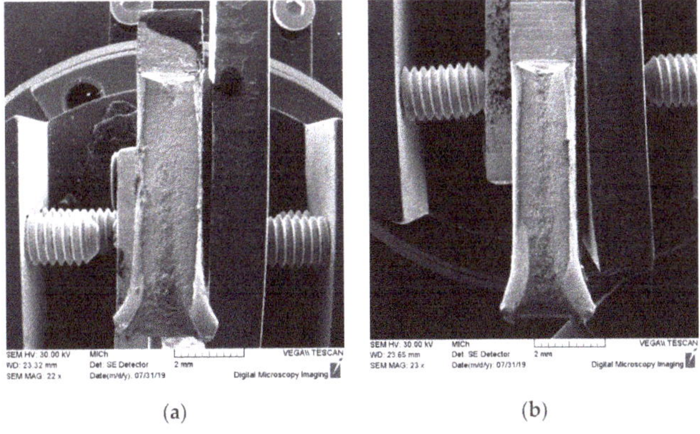

(a)   (b)

Figure 13. SEM images of the whole fracture surface after the Charpy pendulum impact test with different cooling times $t_{8/5}$: (**a**) sample 7s; (**b**) sample 17s.

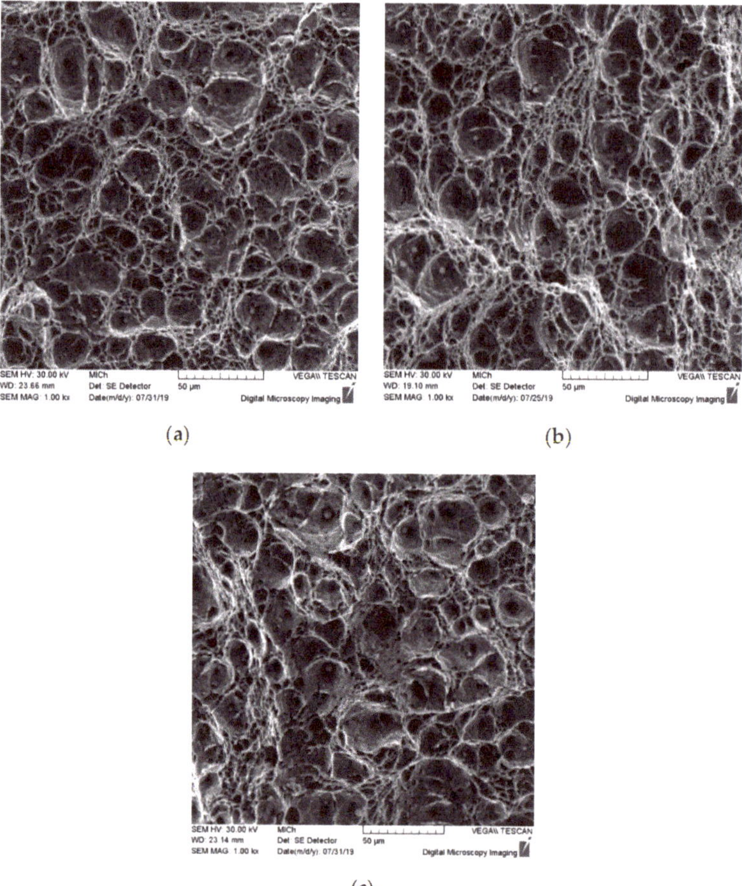

Figure 14. SEM micrographs of the fracture surface after the Charpy pendulum impact test with different cooling times: (**a**) sample 7s; (**b**) sample 10s; (**c**) sample 17s.

3.3. Microstructure Analysis

The main emphasis in evaluating the microstructure was placed on comparing the structural changes in the central part of the simulated samples that were exposed to the same maximum temperature of 1105 °C, but cooled at different rates, expressed by time $t_{8/5}$. Heating the base material to 1105 °C represents the transformation of the original structure into austenite, which grew at the same time. After rapid cooling, the enlarged austenitic grains were converted back to coarse martensite. This region (Figures 15a–c and 16—area f) is called the coarse grain heat-affected zone (CGHAZ). Moravec et al. [22] used the temperature of 1350 °C to evaluate this zone. It is obvious that even at 1105 °C—which is highly in the austenitic region—this results in the growth of austenitic grains, a phenomenon that strongly depends on temperature and time. This growth will increase as the increasing temperature approaches the solidus temperature. Figure 15a–c show the microstructure in the center of the samples heated to the temperature of 1105 °C and cooled at different rates. The microstructure had a similar character in light microscopy evaluation. The grain size was also determined using the linear method according to EN ISO 2624, with an average grain diameter of 18 μm for the 7s sample, 17 μm for the 10s sample, and 14 μm for the 17s sample. The results show that the cooling rate had no significant effect on the grain size. A similar statement was also presented by Laitila et al. [3] in their work.

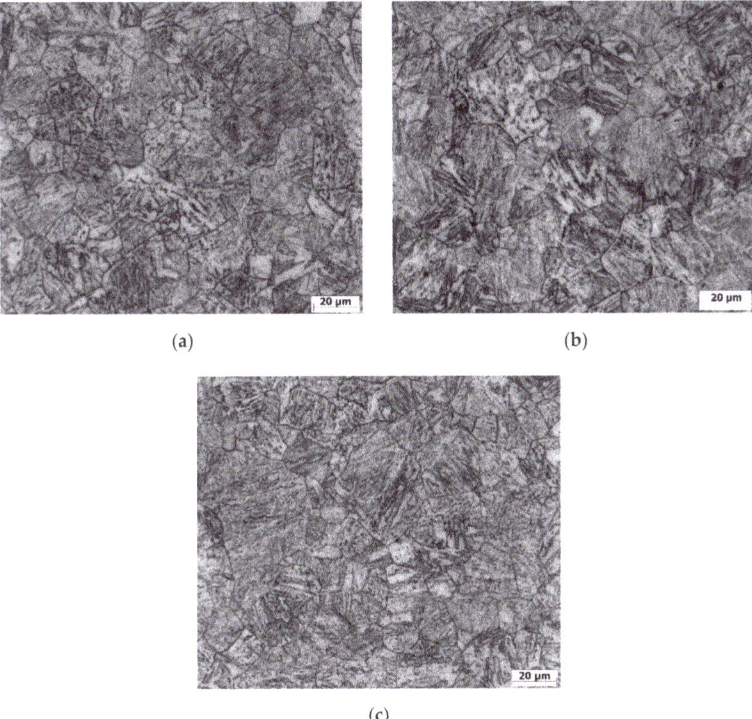

Figure 15. Microstructure in the center of the sample with different cooling times: (a) $t_{8/5} = 7$ s; (b) $t_{8/5} = 10$ s; (c) $t_{8/5} = 17$ s.

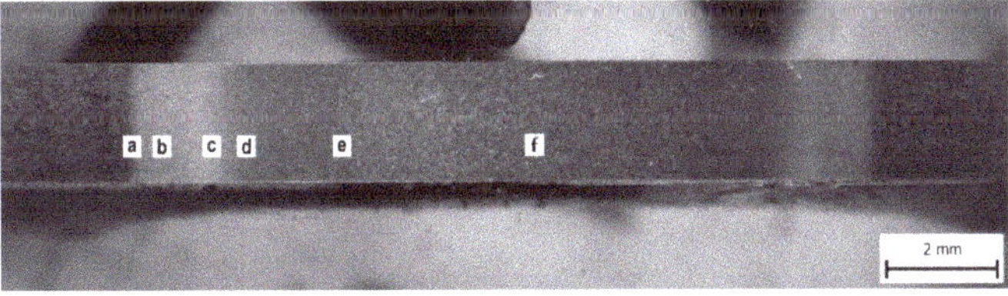

Figure 16. Macrostructure of the 7s sample with marked spots of microstructure evaluation: (**a**) Base material (BM)—inter-critical heat-affected zone (ICHAZ) transition area; (**b**) ICHAZ area; (**c**) ICHAZ—fine-grained area of the heat-affected zone (FGHAZ) transition area; (**d**) FGHAZ area; (**e**) FGHAZ—coarse grain heat-affected zone (CGHAZ) transition area; (**f**) CGHAZ area.

In the HAZ prepared by our physical simulation, except for the coarse-grained microstructure located in the central part of the specimen, additional HAZ areas with different microstructures were visible. Similar observations in real welded joints have also been reported in works published by Jambor et al. [25,26], Guo [11], Ferreira [29], and Bayock [4]. Next to the central coarse-grained area of the HAZ, a fine-grained area of the heat-affected zone (FGHAZ) occurs. Material in the FGHAZ was heated slightly above the A_{c3} temperature, but only for a very short time. This caused the transformation of the base martensitic microstructure to austenite. However, due to the relatively low temperature and a very short time of rapid cooling, which resulted in a refinement of the austenitic microstructure and, later, also of the resulting martensitic microstructure (Figure 16, area d). The last observed area of the HAZ with a changed microstructure is the region exposed to temperatures ranging from A_{c1} to A_{c3}, where martensite was partially transformed into austenite. This temperature exposure resulted in the formation of a mixture of martensite and austenite, which, due to the slower cooling, transformed martensite into ferrite, while the untransformed martensite was tempered. The above-mentioned region is called the inter-critical heat-affected zone (ICHAZ) (Figure 16, area b). As stated in the continuous cooling transformation (CCT) diagram of S960QL steel [15,16], ferritic transformation can occur after 50–90 s in the 500–700 °C temperature range. In fact, there is a small difference in the chemical composition between S960QL and S960MC steel. The S960QL steel has a higher content of Cr and Ni, and the cooling time to the beginning of ferritic transformation will be shorter in the case of S960MC steel. Of course, some transition areas of the HAZ are also present (Figure 16a,c,e). Detailed microstructures of typical subzones of the HAZ that are formed in the S960MC steel due to different maximum temperatures of the thermal cycle and cooling rates can be seen in Figure 17.

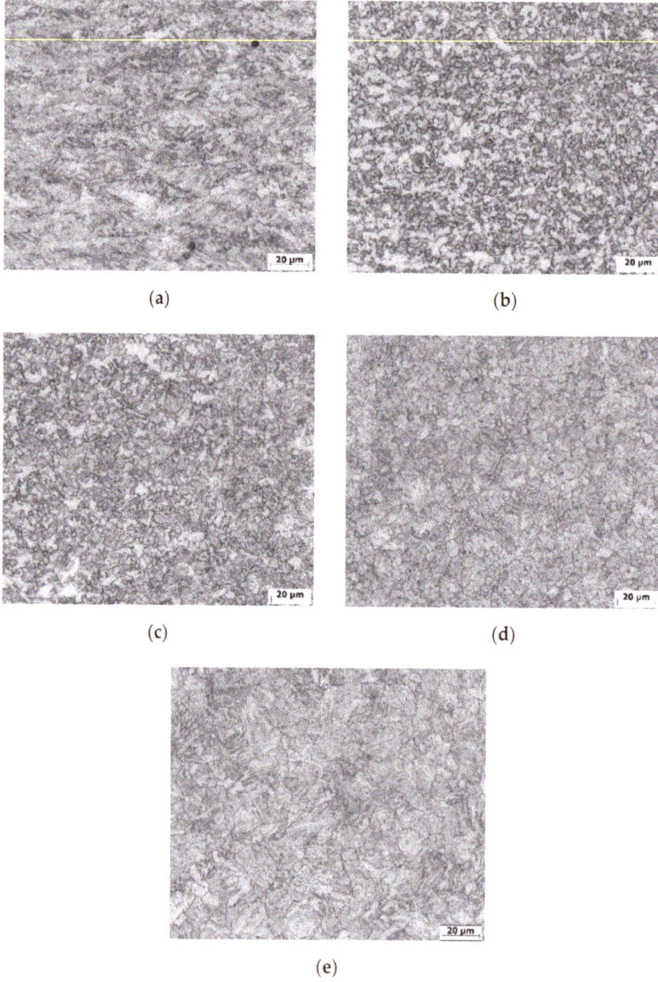

Figure 17. Microstructure of individual subzones from the center of the 7s sample to its edge: (**a**) BM—ICHAZ transition area; (**b**) ICHAZ; (**c**) ICHAZ—FGHAZ transition area; (**d**) FGHAZ; (**e**) FGHAZ—CGHAZ transition area.

3.4. Hardness Test

The HV1 hardness was measured in a line on the longitudinal section surface of a Gleeble sample according to Figure 10, in the central part of the material thickness. For the evaluation of the effect of $t_{8/5}$ on the resulting structure hardness, we calculated the average value using the six measurements in the range of −1.5 to 1.5 mm from the sample center. A decrease in hardness in this region compared to the base material was observed in all test samples. The base material had a hardness of 361 HV1. This value was determined as the average one obtained from five measurements. In the 7s sample, the value was lower by 33 HV1 than the base material hardness, reaching 328 HV1. In the 10s and 17s samples, the decrease was even greater at 309 HV1 and 302 HV1, respectively. The results showed that altering the thermal load of the base material to 1105 °C caused a decrease in hardness, and this is associated with a change in structure. As the cooling time $t_{8/5}$ increases, a higher decrease in hardness is generated. A summary of the results is shown in Table 5, and graphically in Figure 18.

Table 5. Result of the hardness test for the base material and the samples affected by the thermal cycle under different cooling times $t_{8/5}$.

Sample No.	Hardness in the Centre of Sample, HV1
BM	361
7s	328
10s	309
17s	302

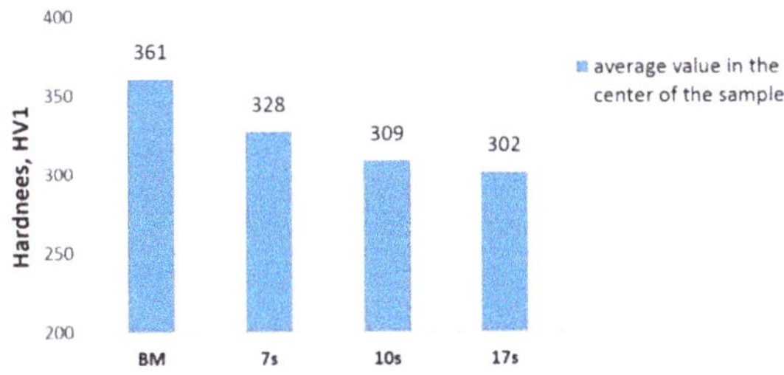

Figure 18. Hardness values of the base material and samples with different cooling times $t_{8/5}$.

4. Conclusions

Assessing the impact of welding on the change in mechanical properties, especially in HAZ high-strength steels, is a current issue and, therefore has been addressed by many researchers. The work mainly focused on steels with a thickness of 5 mm and more, where it is easier to carry out all of the required tests (especially the Charpy pendulum test). However, thin metal sheets of the S960 strength class are very often used in the manufacture of structures in order to reduce the weight. As part of our research, we performed an analysis of tensile testing, Charpy pendulum tests, fracture surfaces, and microhardness and microstructure tests in order to assess the cooling rate effect—expressed by the $t_{8/5}$ time—on the change of mechanical properties of the S960MC steel HAZ. This steel was supplied at a 3 mm thickness by physical simulation using a Gleeble device. Since the Gleeble device did not allow the clamping of flat samples of such a thickness, this had to be reduced to 2 mm. The main conclusions of the above research are as follows:

- The results of the tensile test show that the loading of the base material by a temperature cycle with a maximum temperature of 1105 °C and cooling time $t_{8/5} = 7$ s caused a decrease in the tensile strength to 94% and in the yield strength to 91% of the base material values. The value of tensile strength was above the minimum required tensile strength value, but the yield strength was already below the minimum yield strength value required for S960MC steel. For the samples

10s and 17s with longer $t_{8/5}$ times, this decrease was even greater and the tensile strength and yield strength did not reach the minimum values for S960MC steel. This phenomenon is associated with grain coarsening and a higher degree of martensite tempering and results in a decrease in hardness and hence strength;

- For the Charpy pendulum impact tests of samples with a reduced width of 2 mm, the absorbed energy values at the test temperature of −40 °C of all the simulated samples were higher than in the base material. At the same time, there was a trend of an increasing absorbed energy value with an increasing $t_{8/5}$ cooling time. This increase can be explained—just like in strength tests—by a more pronounced effect of tempering the martensitic structure during reheating. However, the effect of grain growth in this case did not affect the absorbed energy value. After converting the absorbed energy to a standard sample width of 10 mm, all samples exceeded the KV = 27 J limit for S960MC steel. Fracture surfaces in all tested samples including the base material had the same characteristics of transcrystalline ductile fracture with a typical dimple morphology;

- Microstructure analysis was carried out in the central part of the simulated samples that were exposed to the same maximum temperature of 1105 °C, but cooled at different rates expressed by $t_{8/5}$ time. The character of the microstructure was similar in all samples, consisting of coarse grains of the original austenite transformed back into martensite and tempered martensite. The grain size determined by the linear method was 18 μm for the 7s sample. A slight increase was observed for the 10s and 17s samples. The results indicate that the cooling rate had no significant effect on the grain size;

- A decrease in microhardness in the middle part of the Gleeble sample corresponding to the CGHAZ zone compared to the base material was observed in all test samples. For the 7s sample, this value was lower by 33 HV1 than the base material hardness, reaching 328 HV1. For the 10s and 17s samples, the value dropped even lower, to 309 HV1 and 302 HV1, respectively. The base material hardness was 361 HV1. The results show that thermal loading of the base material to a temperature of 1105 °C causes a decrease in hardness and is associated with a change in structure. A greater decrease in hardness is generated as the $t_{8/5}$ cooling time increases;

- When MAG welding 3 mm thick S960MC steel, it is necessary to ensure that appropriate welding conditions are used so that the temperature cycle in the HAZ reaches $t_{8/5} < 7$ s. It is possible to use an external method of heat removal from the HAZ by using coolers in combination with a welding mode that generates the lowest possible heat input while maintaining weld integrity.

Author Contributions: Conceptualization, writing—original draft, and project administration, M.M.; Funding acquisition, M.M. and L.T; Data curation and formal analysis, D.H.; Investigation, D.H., F.N., and J.M; Methodology, F.N. and L.T.; Validation, J.W and J.M.; Writing—review and editing, J.W. All authors have read and agreed to the published version of the manuscript.

Funding: This research was funded by APVV, grant number APVV-16-0276; KEGA, grant number KEGA 009ŽU-4/2019; and VEGA, grant number VEGA 1/0951/17.

Conflicts of Interest: The authors declare no conflicts of interest.

References

1. Schaupp, S.; Schroepfer, D.; Kromm, A.; Kannengiesser, T. Welding residual stresses in 960 MPa grade QT and TMCP high-strength steels. *J. Manuf. Process.* **2017**, *27*, 226–232. [CrossRef]
2. Branco, R.; Berto, F. Mechanical Behavior of High-Strength, Low-Alloy Steels. *Metals* **2018**, *8*, 610. [CrossRef]
3. Laitila, J.; Larkiola, J. Effect of enhanced cooling on mechanical properties of a multipass welded martensitic steel. *Weld. World* **2019**, *63*, 637–646. [CrossRef]
4. Njock Bayock, F.; Kah, P.; Mvola, B.; Layus, P. Effect of heat input and undermatched filler wire on the microstructure and mechanical properties of dissimilar S700MC/S960QC high-strength steels. *Metals* **2019**, *9*, 883. [CrossRef]

5. Taavitsainen, J.; Santos Vilaca da Silva, P.; Porter, D.; Suikkanen, P. Weldability of Direct-Quenched Steel with a Yield Stress of 960 MPa. In Proceedings of the 10th International Conference on Trends in Welding Research and 9th International Welding Symposium of Japan Welding Society (9WS), Tokyo, Japan, 11–14 October 2017; pp. 52–55.

6. Ślęzaka, S.; Śnieżeka, L. Comparative LCF study of S960QL high strength steel and S355J2 mild steel. *Procedia Eng.* **2015**, *114*, 78–85. [CrossRef]

7. Lago, J.; Trško, L.; Jambor, M.; Nový, F.; Bokůvka, O.; Mičian, M.; Pastorek, F. Fatigue life improvement of the high strength steel welded joints by ultrasonic impact peening. *Metals* **2019**, *9*, 619. [CrossRef]

8. Hochhauser, F.; Ernst, W.; Rauch, R.; Vallant, R.; Enzinger, N. Influence of the soft zone on the strength of welded modern hsla steels. *Weld. World* **2012**, *56*, 77–85. [CrossRef]

9. Björk, T.; Toivonen, J.; Nykänen, T. Capacity of fillet welded joints made of ultra high-strength steel. *Weld. World* **2012**, *56*, 71. [CrossRef]

10. Nowacki, J.; Sajek, A.; Matkowski, P. The influence of welding heat input on the microstructure of joints of S1100QL steel in one-pass welding. *Arch. Civ. Mech. Eng.* **2016**, *16*, 777–783. [CrossRef]

11. Guo, W.; Crowther, D.; Francis, J.A.; Thompson, A.; Liuc, Z.; Li, L. Microstructure and mechanical properties of laser welded S960 high strength steel. *Mater. Des.* **2015**, *85*, 534–548. [CrossRef]

12. Guo, W.; Li, L.; Dong, S.; Crowther, D.; Thompson, A. Comparison of microstructure and mechanical properties of ultra-narrow gap laser and gas-metal-arc welded S960 high strength steel. *Opt. Lasers Eng.* **2017**, *91*, 1–15. [CrossRef]

13. Kopas, P.; Sága, M.; Jambor, M.; Nový, F.; Trško, L.; Jakubovičová, L. Comparison of the mechanical properties and microstructural evolution in the HAZ of HSLA DOMEX 700MC welded by gas metal arc welding and electron beam welding. *MATEC Web Conf.* **2018**, *244*, 01009. [CrossRef]

14. Górka, J. Structure and Properties of Hybrid Laser Arc Welded T-joints (Laser Beam–MAG) in Thermo-mechanical Control Processed Steel S700MC of 10 mm thickness. *Arch. Metall. Mater.* **2018**, *63*, 1125–1131.

15. Sajek, A.; Nowacki, J. Comparative evaluation of various experimental and numerical simulation methods for determination of $t_{8/5}$ cooling times in HPAW process weldments. *Arch. Civ. Mech. Eng.* **2018**, *18*, 583–591. [CrossRef]

16. Samardžić, I.; Dunđer, M.; Katinić, M.; Krnić, N. Weldability investigation on real welded plates of fine-grained high-strength steel S960QL. *Metalurgija* **2017**, *56*, 207–210.

17. Lahtinen, T.; Vilaça, P.; Peura, P.; Mehtonen, S. MAG welding tests of modern high strength steels with minimum yield strength of 700 MPa. *Appl. Sci.* **2019**, *9*, 1031. [CrossRef]

18. Sefcikova, K.; Brtnik, T.; Dolejs, J.; Keltamaki, K.; Topilla, R. Mechanical properties of heat affected zone of high strength steels. *IOP Conf. Ser. Mater. Sci. Eng.* **2015**, *96*, 012053. [CrossRef]

19. Pisarski, H.G.; Dolby, R.E. The significance of softened HAZs in high strength structural steels. *Weld. World* **2003**, *47*, 32. [CrossRef]

20. Schneider, C.; Ernst, W.; Schnitzer, R.; Staufer, H.; Vallant, R.; Enzinger, N. Welding of S960MC with undermatching filler material. *Weld. World* **2018**, *62*, 801. [CrossRef]

21. Barsoum, Z.; Khurshid, M. Ultimate strength capacity of welded joints in high strength steels. *Procedia Struct. Integrity* **2017**, *5*, 1401–1408. [CrossRef]

22. Moravec, J.; Novakova, I.; Sobotka, J.; Neumann, H. Determination of grain growth kinetics and assessment of welding effect on properties of S700MC steel in the HAZ of welded joints. *Metals* **2019**, *9*, 707. [CrossRef]

23. Gáspár, M. Effect of welding heat input on simulated HAZ areas in S960QL high strength steel. *Metals* **2019**, *9*, 1226. [CrossRef]

24. Dunđer, M.; Vuherer, T.; Samardžić, I. Weldability prediction of high strength steel S960QL after weld thermal cycle simulation. *Metalurgija* **2014**, *53*, 627–630.

25. Jambor, M.; Ulewicz, R.; Nový, F.; Bokůvka, O.; Trško, L.; Mičian, M.; Harmaniak, D. Evolution of microstructure in the heat affected zone of S960MC GMAW weld. *Mater. Res. Proc.* **2018**, *5*, 78–83.

26. Jambor, M.; Novy, F.; Mician, M.; Trsko, L.; Bokuvka, O.; Pastorek, F.; Harmaniak, D. Gas metal arc welding of thermo-mechanically controlled processed S960MC steel thin sheets with different welding parameters. *Commun. Sci. Lett. Univ. Zilina* **2018**, *20*, 29–35.

27. Gleeble User Training 2012. *Gleeble Systems and Applications*; Dynamic Systems Inc.: Poestenkill, NY, USA, 2013.

28. Strenx. *Welding of Strenx*; SSAB AB: Stockholm, Sweden, 2017.
29. Ferreira, D.N.B.; Alves, A.N.S.; Neto, R.M.A.C.; Martins, T.F.; Brandi, S.D. A New Approach to simulate HSLA steel multipass welding through distributed point heat sources model. *Metals* **2018**, *8*, 951. [CrossRef]

 *metals*

Article

Effect of Welding Heat Input on Simulated HAZ Areas in S960QL High Strength Steel

Marcell Gáspár

Institute of Materials Science and Technology, University of Miskolc, H-3515 Miskolc, Hungary; gasparm@uni-miskolc.hu

Received: 15 October 2019; Accepted: 13 November 2019; Published: 16 November 2019

Abstract: When the weldability of high strength steels is analyzed, it is the softening in the heat-affected zone (HAZ) that is mostly investigated, and the reduction of toughness properties is generally less considered. The outstanding toughness properties of quenched and tempered high strength steels cannot be adequately preserved during the welding due to the unfavorable microstructural changes in the HAZ. Relevant technological variants ($t_{8/5}$ = 2.5–100 s) for arc welding technologies were applied during the HAZ simulation of S960QL steel (EN 10025-6) in a Gleeble 3500 physical simulator, and the effect of cooling time on the critical HAZ areas of single and multipass welded joints was analyzed. Thermal cycles were determined according to the Rykalin 3D model. The properties of the selected coarse-grained (CGHAZ), intercritical (ICHAZ) and intercritically reheated coarse-grained (ICCGHAZ) zones were investigated by scanning electron microscope, macro and micro hardness tests and instrumented Charpy V-notch pendulum impact tests. The examined HAZ subzones indicated higher sensitivity to the welding heat input compared to conventional structural steels. Due to the observed brittle behavior of all subzones in the whole $t_{8/5}$ range, the possible lowest welding heat input should be applied in order to minimize the volume of HAZ that does not put fulfillment of the allowed maximal (450 HV10) hardness at risk and does not lead to the formation of cold cracks.

Keywords: high strength steel; heat-affected zone (HAZ); physical simulation; instrumented Charpy V-notch pendulum impact test; toughness; welding heat input

1. Introduction

High strength steels have an increasingly important role in engineering applications, especially in the vehicle and transportation industry, where the producers are faced with the challenges of strict requirements for lower CO_2 emission, and there is customer demand for lower operational costs. By the combination of alloying elements, rolling and heat-treating processes, various high strength steel grades were developed in the past few decades, and therefore, nowadays high strength steels are also available in thin sheets and thicker plates. Besides increasing the strength and toughness properties, the development of weldability characteristics also plays an important role in steel producers. However, the determination of the right welding technology, including the optimal process window, may still cause difficulties for welding engineers [1–6].

Quenched and tempered (Q + T) steels are considered the highest strength structural steels [2,5]. Due to their outstanding mechanical properties, especially strength, significant weight reduction can be achieved by their application. Besides their high yield strength, they have also good toughness properties thanks to the fine-grained tempered martensitic and bainitic microstructure [7]. In the present paper, the weldability of S960QL according to EN 10025-6 is investigated in the aspect of microstructural changes in the heat-affected zone (HAZ). Due to water cooling during quenching and to the application of high temperature tempering (HTT), this grade generally has a non-equilibrium tempered martensitic microstructure. For improved hardenability properties in the whole plate thickness, alloying elements (chromium (Cr), molybdenum (Mo)) are added to the steel, which moves the continuous cooling transformation (CCT) diagram to the right. Microalloying components (niobium (Nb), vanadium (V) and titanium (Ti)) are often used in order to ensure a fine grain microstructure and to prevent grain coarsening. High strength steels are generally more sensitive to the impurity level and inclusions; therefore, strict limitations are set for phosphorus (P) and sulphur (S) content [8]. The excellent strength of Q + T high strength steels originates primarily from the fine tempered lath-like martensitic microstructure and the dislocation density. The effect of the dissolved carbon is generally lower [9].

During the welding of quenched and tempered high strength steels, the biggest weldability problems can be caused by the unfavorable changes of the original fine-grained microstructure. In the HAZ, both hardened and softened areas can be found [10,11]. Cold cracking in weld and HAZ can be simply identified by non-destructive materials tests, but the reduction of strength and toughness cannot be examined on the welded structure. Quality expectations and the continuous training of the welders are indispensable to guarantee the required mechanical properties of the welded joint. As the yield strength grows, the area of the optimal welding lobe generally decreases. According to welding experiments, mechanical requirements can be fulfilled with the application of low welding heat input, although the risk of cold cracking increases. Since mobile structures welded from high strength steels often operate at low temperature when occasionally even overload may occur, the fulfillment of the guaranteed minimum impact energy is critical in terms of safe operation. Consequently, the influence of welding parameters on HAZ characteristics is examined, and sufficient toughness should be guaranteed across the whole welded joint. Welding parameters are often described by the $t_{85/5}$ cooling time in order to be able to compare the different technological variants (heat input, preheating/interpass temperature) and welding processes. Within the weldability challenges, the effect of the limited filler metal choice and the mismatch effect on the static and dynamic behavior of the welded joint cannot be neglected. Due to the higher crack sensitivity of the weld and HAZ microstructure and the possible cyclic loading conditions in mobile structures, fracture mechanics approach is recommended to be used during the structural and technological design [12,13].

When the HAZ of structural steels is examined, a coarse-grained zone is considered as the most critical one in terms of the toughness properties. In the case of quenched and tempered high-strength steels, the intercritical zone can be also similarly disadvantageous due to the higher carbon equivalent. In multipass welded joints, the unfavorable microstructural changes of both abovementioned zones (grain growth and brittle microstructure at grain boundaries) can occur in the intercritically reheated coarse-grained zone, which is supposed to be the region of the whole welded joint with the lowest toughness [14,15].

The aims of the present research are to provide a detailed characterization (microstructure, hardness, toughness) of the HAZ of S960QL high strength steel and to analyze the effect of welding parameters during the application of the most common arc welding technologies. The present study raises also the question whether the requirements of the governing standards are sufficient to guarantee the avoidance of brittle fracture in the HAZ. The properties of HAZ subzones are investigated as a function of $t_{8/5}$ cooling time for the applicable arc welding technologies by the application of physical simulation.

2. Materials and Methods

2.1. The Investigated Base Material

The chemical composition of the investigated S960QL base material according to the material certificate is summarized in Table 1. Based on the chemical composition the carbon equivalents according to EN 1011-2 were determined: CEV = 0.54%, CET = 0.36%. The required and the measured mechanical properties of the examined S960QL steel according to the material certification are summarized in Table 2. The examined steel has a guaranteed 27 J Charpy-V energy (CVN) at −40 °C thanks to its fine-grained microstructure (approx. 10 µm grain size) and low impurity level (minimized S and P content).

Table 1. Chemical composition and carbon equivalents of the investigated S960QL steel in wt. %.

Chemical Composition (wt. %)													
C	Si	Mn	P	S	Cr	Ni	Mo	V	Ti	Al	Nb	B	N
0.17	0.23	1.23	0.011	0.001	0.20	0.04	0.588	0.041	0.004	0.061	0.017	0.001	0.002

According to the EN 10204 3.1 material certificate provided by the steel producer.

Table 2. Mechanical characteristics of the investigated S960QL steel.

Mechanical Characteristics					
S960QL	$R_{p0.2}$ MPa	R_m MPa	$R_{p0.2}/R_m$	A_5 %	CVN (at −40 °C) J
EN 10025-6	≥960	980–1150	-	≥10	≥27
Material certificate	1014	1053	0.96	14	75

According to EN 10204 3.1 material certificate provided by the steel producer.

2.2. Physical Simulation of Critical HAZ Areas

Physical simulation provides unique opportunities for HAZ characterization in metals during different kind of fusion welding technologies (from high energy density processes such as laser beam welding to the medium or low energy density processes as gas metal arc welding). In the present paper, HAZ tests were performed by a Gleeble 3500 thermomechanical physical simulator (Dynamic Systems Inc., Poestenkill, NY, USA), installed in. With the HAZ test, the desired part of the HAZ can be precisely and homogeneously created in a volume sufficient for the further material tests, e.g., the Charpy V-notch pendulum impact test [15,16].

The time and temperature points in the heating and cooling part of HAZ thermal cycles were determined based on the Rykalin 3D model, and the GSL programs were manually written [17]. This model describes the temperature field generated by a moving point-like heat source on the surface of a semi infinity body. In this case, 3D thermal conductivity is dominant while surface heat transfer (convection) is negligible. This model was selected since the investigated S960QL is generally used in medium and heavy plate thickness (>15 mm), where a 3D model may give more precise results. Furthermore, this equation is independent from the plate thickness; consequently, a reduced number of variables need to be considered during the evaluation.

Thermophysical properties were determined for the whole relevant temperature range (between 20 and 1400 °C) by the application of JMatPro software (v7.0, Sente Software Ltd., Guildford, UK), and then, their average values were used in the Rykalin equation (λ = 37.8 W/(m °C), c_p = 690.2 J/(kg °C), ρ = 7614.7 kg/m^3). Preheating and interpass temperatures (T_0) were set to 150 °C based on previous welding experiments [18].

The aim was to simulate the CGHAZ, ICHAZ and ICCGHAZ of the investigated S960QL for a $t_{8/5}$ cooling range, typical for the arc welding processes: 2.5–30 s. In the case of CGHAZ, $t_{8/5}$ = 100 s was also examined, since in the 2.5–30 s interval basically martensitic transformation was expected, and the presence of a bainite microstructure was only expected after 60 s (based on the performed JMatPro calculations 27% bainite was prognosed at 100 s). Considering the selection of peak temperatures, the motivation was to generate the most critical parts of the selected HAZ areas, those having the lowest toughness. Hence, for CGHAZ the peak temperature of 1350 °C was selected in order to produce the occurring largest grains. This value is safely below than the nil-strength temperature of the investigated steel (NST = 1408 °C) [19].

Determination of the peak temperature for ICHAZ was based on the literature [7,9,20] and preliminary experiments. Three simulations were performed with 760, 780, 815 and 850 °C peak temperatures at $t_{8/5} = 5$ s. Then Charpy V-notch impact tests were performed. The lowest energy values were measured in case of 760 °C and 780 °C peak temperatures. Finally, 775 °C was selected for the further simulation by ensuring the required transformation ratio in the examined $t_{8/5}$ interval. For the simulation of ICCGHAZ the two peak temperatures (1350 and 775 °C) were applied together. Specimens with a 10 mm × 10 mm square cross section and 70 mm length were machined from a $s = 15$ mm plate, and they were applied in the $t_{8/5} = 5$–100 s interval with a 10 mm free span. For achieving the 2.5 s cooling time a special, previously tested drilled specimen was applied with a 7.5 mm free span and external air cooling [21].

The applied HAZ cycles for the different cooling time values (indicating the effect of welding heat input and preheating) between 2.5 and 30 s are illustrated in Figures 1–3. The thermal cycle for the $t_{8/5} = 100$ s is not shown in Figure 1 due to its wide extent. For the simulation of ICCGHAZ, due to practical reasons, a forced cooling (10 °C/s) was applied after 200 °C, which was followed by a 5 s holding time before the second thermal cycle started.

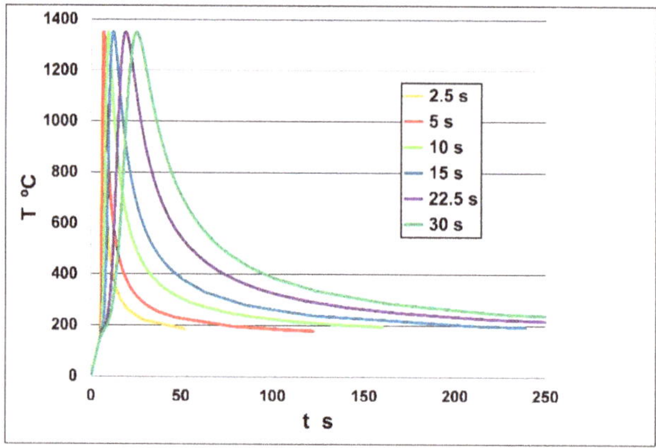

Figure 1. Coarse-grained heat-affected zone (CGHAZ) thermal cycles.

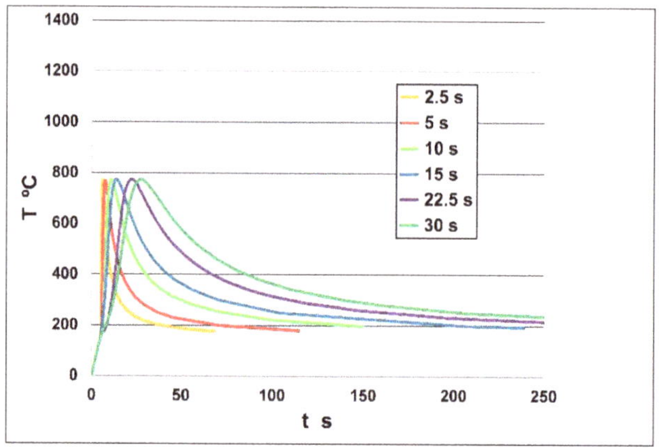

Figure 2. Intercritical heat-affected zone (ICHAZ) thermal cycles.

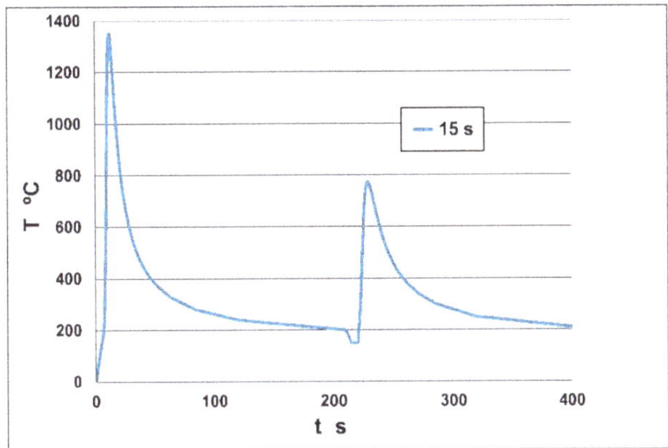

Figure 3. Intercritically reheated coarse-grained heat-affected zone (ICCGHAZ) thermal cycle ($t_{8/5}$ = 15 s).

3. Results and Discussion

After the simulations the HAZ subzones and the effect of welding parameters were examined by scanning electron microscopic (SEM), macro and micro hardness tests and instrumented Charpy V-notch pendulum impact tests.

3.1. SEM Tests

A ZEISS EVO MA10 (Carl Zeiss AG, Oberkochen, Germany) scanning electron microscope was applied for the examination illustrated in Figures 4–7. Samples were covered with a golden layer for improved picture quality. The tempered martensitic microstructure of the base material is presented in Figure 4, where the formation of carbides refers to the tempering after quenching [2].

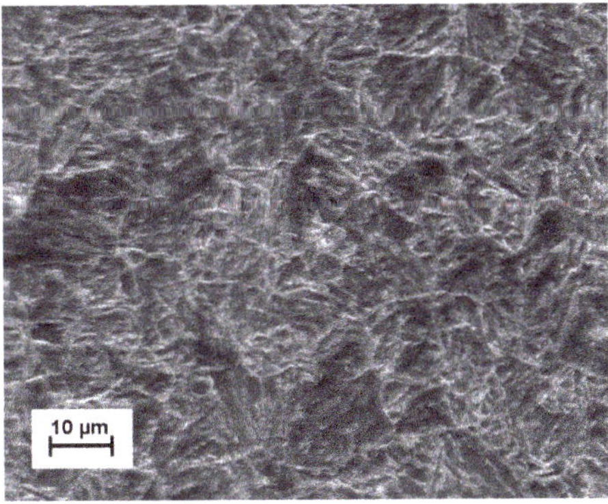

Figure 4. Tempered martensitic microstructure of the base material (M = 1000×, 2% HNO_3).

Based on the microscopic tests the HAZ subzones were adequately simulated in all cases for the arc welding of S960QL grade, and the results were correlated with the JMatPro calculations. In CGHAZ martensitic microstructure with large (>100 µm), prior austenite grain size (Figure 5) was observed in the 2.5–30 s interval. Occasionally the measured grain size was even as large as 200 µm, although in physical simulations the grain size of CGHAZ is generally larger than in the real welded joints [16]. Furthermore, at this high (1200–1400 °C) temperature range the nature of grain growth is unstable, and unexpected large grains may appear [7,9]. The type of microstructure did not change significantly between $t_{8/5}$ = 2.5 and 30 s; however, the self-tempering of martensite was identified at longer cooling times (Figure 5b). The reduced toughness properties can be explained by the large grains, the presence of unstable blocky austenite and the relatively high carbon equivalent. The presence of upper bainite was observed only at $t_{8/5}$ = 100 s (Figure 5d). The upper-bainitic microstructure may increase the risk of the formation of elongated martensite-austenite (M-A) constituents, which can cause the further reduction of impact energy values. In CGHAZ the M-A constituents generally appear between the ferrite laths of bainite and has a long and thin morphology, different than in ICHAZ or in ICCGHAZ [15]. The cold cracking sensitivity of CGHAZ should be emphasized, since hydrogen can diffuse from the fusion zone to this brittle microstructure [2,7,22].

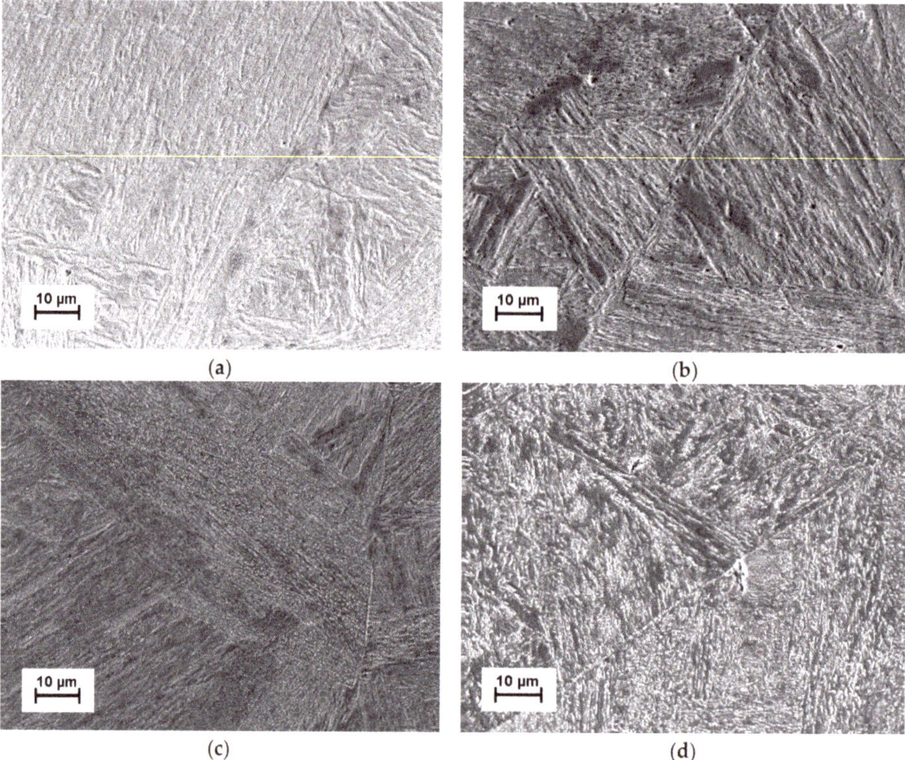

Figure 5. CGHAZ microstructure: (**a**) $t_{8/5}$ = 2.5 s; (**b**) $t_{8/5}$ = 15 s; (**c**) self-tempering of martensite at $t_{8/5}$ = 30 s; (**d**) upper bainitic microstructure at $t_{8/5}$ = 100 s (M = 1000×, 2% HNO_3).

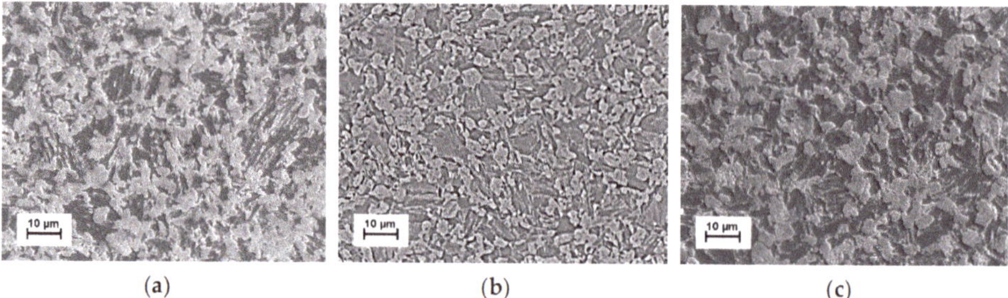

Figure 6. ICHAZ microstructure: (**a**) $t_{8/5}$ = 2.5 s; (**b**) $t_{8/5}$ = 15 s; (**c**) $t_{8/5}$ = 30 s (M = 1000×, 2% HNO_3).

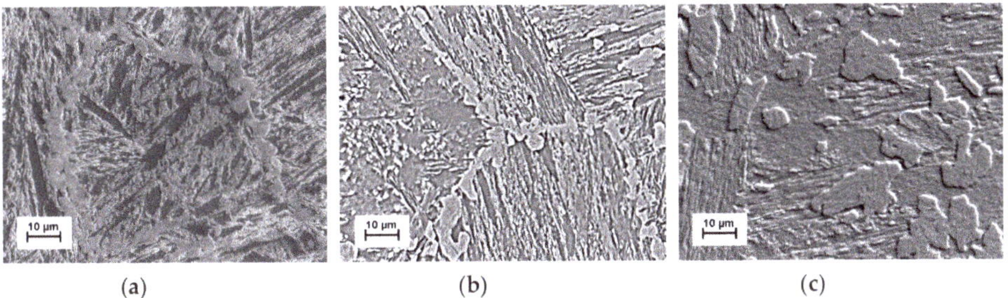

Figure 7. ICCGHAZ microstructure: (**a**) $t_{8/5}$ = 2.5 s; (**b**) $t_{8/5}$ = 15 s; (**c**) $t_{8/5}$ = 30 s (M = 1000×, 2% HNO$_3$).

The microstructures of ICHAZ and ICCGHAZ are presented in Figures 6 and 7. The SEM images in Figures 6b and 7b were prepared in backscatter mode for a more complex evaluation of the transformed parts. In ICHAZ, transformed parts at the boundaries of original grains generally have a higher carbon content, since austenite has a higher carbon solubility in this temperature range. In the case of high heat input, when this area stays in this intercritical range for a long period, the carbon content of austenitized parts further increases. Then, in the cooling part, these austenitic parts transform to a brittle, generally martensitic or bainitic microstructure as a function of cooling speed and the strength category. The transformed parts may have higher hardness than CGHAZ, while the original parts are tempered; furthermore, their carbon content reduces, and they soften. Retained austenite can be often noticed near the brittle martensitic islands; therefore, these areas are called together as M-A constituents. In terms of the simulated ICHAZ, fine M-A islands formed as the result of partial austenitic transformation, while the middle of the original grains became tempered. By the increase of cooling time, the transformed parts are less homogeneous and less brittle (Figure 6c) compared to this zone at $t_{8/5}$ = 2.5 s (Figure 6a). In ICCGHAZ similar microstructural changes can be identified with grains essentially ten times larger than in ICHAZ. In many cases, this local zone has the lowest toughness in the welded joint, since the disadvantageous properties of CGHAZ and ICHAZ meet here. The toughness of ICCGHAZ is determined by the tempered coarse-grained martensite and the amount, distribution, type and hardness of the austenitized parts. The properties of M-A constituents are basically influenced by the peak temperature and cooling time. In the case of short cooling times, the quantity of M-A parts has an increased role on impact energy, while the hardness difference between the original (but tempered) coarse grains and the M-A parts is more relevant in long cooling times. In real welded joints, the abovementioned unfavorable properties of ICCGHAZ are less harmful, since this zone just forms locally while ICHAZ can be found in the whole plate thickness [14,15].

3.2. Hardness Tests

A Reicherter UH250 universal macro-hardness (BUEHLER Worldwide Headquarters, Lake Bluff, IL, USA) and a Mitutoyo micro-hardness tester (Mitutoyo Hungária Ltd., Budapest, Hungary) were used for the investigations. Evaluation was performed according to the EN 15614-1 standard, which permits HV$_{max}$ = 450 HV10 for the non-heat-treated welded joints (including HAZ) of quenched and tempered high strength steels belonging to the 3rd group of CR ISO 15608. Specimens were perpendicularly cut to their longitudinal size at the thermocouples. Five measurements were done on the cut surface. The average macro hardness values of the simulated subzones are presented in Figure 8.

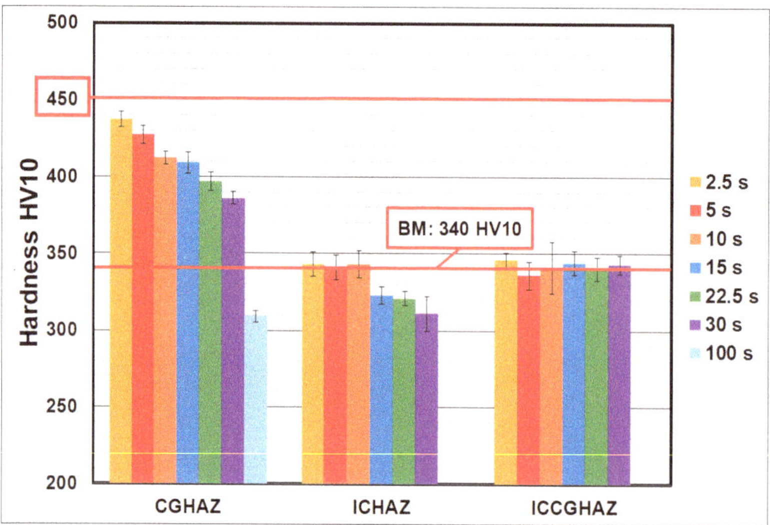

Figure 8. Macro hardness test results in HAZ subzones.

All HAZ subzones at each cooling time fulfilled the requirements of the governing standard. This was true even given the fact that in the case of high energy density processes (e.g., electron beam welding) the maximum hardness may even exceed 500 HV in the HAZ of S960QL. Furthermore, even in the same steel category, the difference in the type and the number of alloying elements can have an effect on HAZ hardenability. By considering all subzones 437 HV10, maximum hardness was measured in CGHAZ when $t_{8/5}$ = 2.5 s was simulated. With the increase in cooling time, the hardness proportionately decreased in this zone, but it was still above the base material hardness between 2.5 and 30 s. Although the hardness limit of the standard was fulfilled in this zone, the hardness of the fine-grained zone (FGHAZ) can occasionally become even higher [10]. Compared to the base material, softening was only noticed at $t_{8/5}$ = 100 s. As can be seen in Figure 8, the ICHAZ tends to soften the most in the investigated cooling time interval, where the lowest hardness was 311 HV10 at $t_{8/5}$ = 30 s. Regarding the effect of a softened zone on the behavior of the whole welded joint, it is important to note that HAZ softening depends mainly on the relative extension of the soft zone and has only a small effect on the transverse tensile strength of the joint [11].

In terms of ICCGHAZ, the average hardness was similar to that of the base material, and the cooling time did not influence the macro hardness. The hardness of the coarse grains of the first thermal cycle significantly decreased due to the tempering effect of the intercritical (second) thermal cycle. The hardness of primer coarse grains did not have any effect on the final hardness of ICCGHAZ.

Microhardness tests were also performed for the more-in-depth examination of ICHAZ and ICCGHAZ. The possible lowest mechanical load (HVM0.1) was set on the equipment due to the limited area of the transformed islands. The highest local hardness values (often much above 450 HVM0.1) were measured in the M-A parts of ICCGHAZ. Generally, lower microhardness was noticed in ICHAZ, although all values were above the hardness of the original microstructure. The effect of welding parameters could be clearly seen in the hardness of grain centers due to the tempering effect of the second heat cycle. In ICHAZ, the average microhardness decreased to 265 HVM0.1 at $t_{8/5}$ = 30 s. A similar tendency was observed in ICCGHAZ, where the minimum hardness was 311 in the middle of the grains at this cooling time.

In Table 3, the hardness values of gas metal arc (GMA) welded joints from the investigated S960QL material are presented as a function of the cooling time in order to provide a connection between the simulation and the real welding experiments [18]. The plates were 15 mm thick and V-groove welds were prepared. During these previously performed welding experiments, the welds were manually welded; therefore, in most cases, an interval is given for the cooling time, which was calculated and measured at each welding pass. In general, based on the hardness results a correlation can be found between the physical simulation and the real welding experiments. At shorter cooling times, higher values than 400 HV10 can be measured, and the increasing heat input reduced the maximal hardness in the HAZ. During the hardness tests of real welded joints, five measurements of each HAZ were performed. The correlation might be improved by a more detailed hardness distribution of the real HAZ.

Table 3. Measured maximal hardness values of gas metal arc (GMA) welded joints from S960QL steel.

Measured Maximal Hardness Values in Real HAZ of GMA Welded Joints				
$t_{8/5}$ s	6–10	10–15	15–20	20–30
HV10	405	390	385	345

3.3. Instrumented Charpy V-Notch Pendulum Impact Tests

Five specimens from each thermal cycle were used for the Charpy V-notch pendulum impact tests performed with PSD 300/150 instrumented equipment (WPM Werkstoffprüfsysteme GmbH, Markkleeberg, Germany). According to EN 10025-6 the required minimum impact energy is 27 J at −40 °C for S960QL steel. According to the material certificate, the investigated steel plate has 75 J CVN, although 167 J was measured during the performed impact test on the base material. The average CVN of the specimens of the performed impact tests is given in Figure 9.

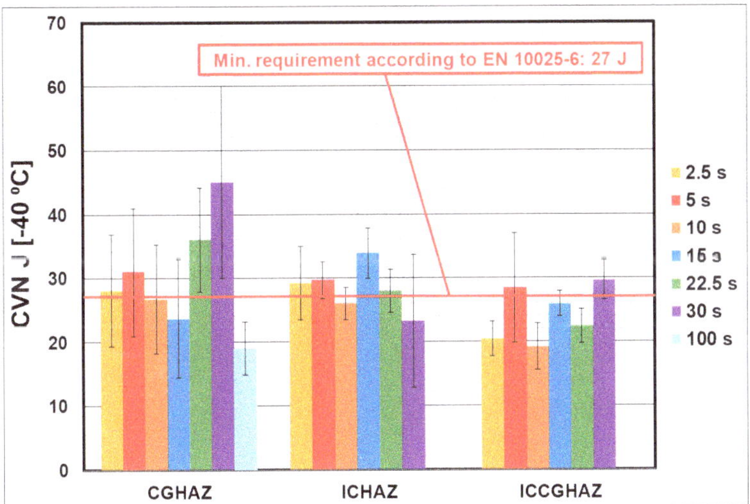

Figure 9. Charpy V-notch pendulum impact tests on HAZ subzones.

All of the investigated subzones in the examined $t_{8/5}$ cooling time interval had significantly lower toughness than the base material. The cooling time does not affect equally the different HAZ areas. The minimum and the maximum impact energy values of the given subzones did not appear at the same cooling time. Cooling times had the largest effect on the toughness of CGHAZ, although the highest scatter was noticed here. By considering the relatively high standard deviations of the measured values, the results were statistically evaluated by the ANOVA method. According to the calculations, there is a statistically significant difference among the measured CVN values in CGHAZ at the different cooling times ($F = 4.65 > F_{crit} = 2.29$, $p = 0.000834$). Then, post hoc tests were performed in order to determine which results are different from each other. Based on the calculations (Dunnett's T3) only the CVN values of $t_{8/5}$ = 22.5 s and 100 s are statistically different (p = 0.009). However, the increase of sample size may also indicate statistical difference between other values. Therefore, it can be recommended to test more samples in CGHAZ for similar physical simulation investigations. The occasionally measured higher impact energy values in CGHAZ might be explained by the observed self-tempering effect (Figure 5c) [21]. When upper bainitic microstructure appeared at $t_{8/5}$ = 100 s in CGHAZ (Figure 5d), the toughness significantly reduced. In the examined S960QL steel, ICHAZ has similar impact energy to CGHAZ between 2.5–10 s, and significantly lower in the 22.5–30 s interval. In terms of the toughness properties ICCGHAZ seems to be the most critical part, where more values, including the standard deviations, were under the required value.

In Table 4 the CVN values of the real GMA welded joints from the investigated S960QL material are presented in terms of the calculated and measured cooling time values, similarly to the evaluation of the hardness results [18].

Table 4. Measured CVN values of GMA welded joints from S960QL steel.

Measured CVN Values of GMA Welded Joints from S960QL Steel				
$t_{8/5}$ s	6–10	10–15	15–20	20–30
VWT * 1	43, 46, 49	38, 45, 47	40, 41, 42	37, 38, 39
VHT * 1/1	19, 33, 42	36, 42, 54	28, 33, 53	22, 27, 30

* V: V-notch; W: notch in weld zone; H: notch in HAZ; T: notch through thickness.

In the examined $t_{8/5}$ cooling time range, higher CVN values were measured in the weld than in the HAZ. Therefore, HAZ can be considered as the most critical part concerning the toughness distribution in the welded joint. The low CVN values at the different cooling time range indicate the brittle behavior of the HAZ, in accordance with the physical simulation results. In some cases, the higher impact energy values can be connected with the uncertain notch location and crack propagation due to the limited extension of critical HAZ areas. With the increasing cooling time, lower CVN values were measured, which might be connected with the wider HAZ and the lower number of welding passes (5 instead of 9).

Comparing to the conventional Charpy V-notch impact test where the whole energy absorbed during the fracture is determined, the instrumented impact testing can provide more detailed information about the fracture process and the ductile/brittle behavior of the material. Using strain gauge measurement technology, the load-time diagram can be determined, and the characteristic points of the fracture process can be identified (the start of the plastic strain, maximal force, start of the unstable crack propagation, end of the unstable crack propagation). In the registered diagram, the unstable crack propagation stage is correlated with the amount of brittle fracture on the fracture surface. From the load-time diagram, the force-displacement diagram can be calculated. Assuming that the crack initiation occurs at the maximal force, the registered diagram can be divided into two parts according to the maximal force. Until the maximum force, the area under the curve is considered as the absorbed energy for crack initiation (W_i), while the remaining area is for crack propagation (W_p). As the ratio of the absorbed energy for the crack initiation increases, the toughness of the examined material reduces [23]. The detailed method of the instrumented Charpy V-notch pendulum impact test is described in EN ISO 14556.

During the instrumented Charpy V-notch impact tests force-displacement diagrams were determined for the base material and the HAZ areas (Figure 10). It can be seen that characteristics of the diagrams drastically changed compared to the base material. Although only minor unstable crack initiation was noticed in the base material, the behavior of the investigated subzones was totally brittle, almost independently from the set $t_{8/5}$ cooling time.

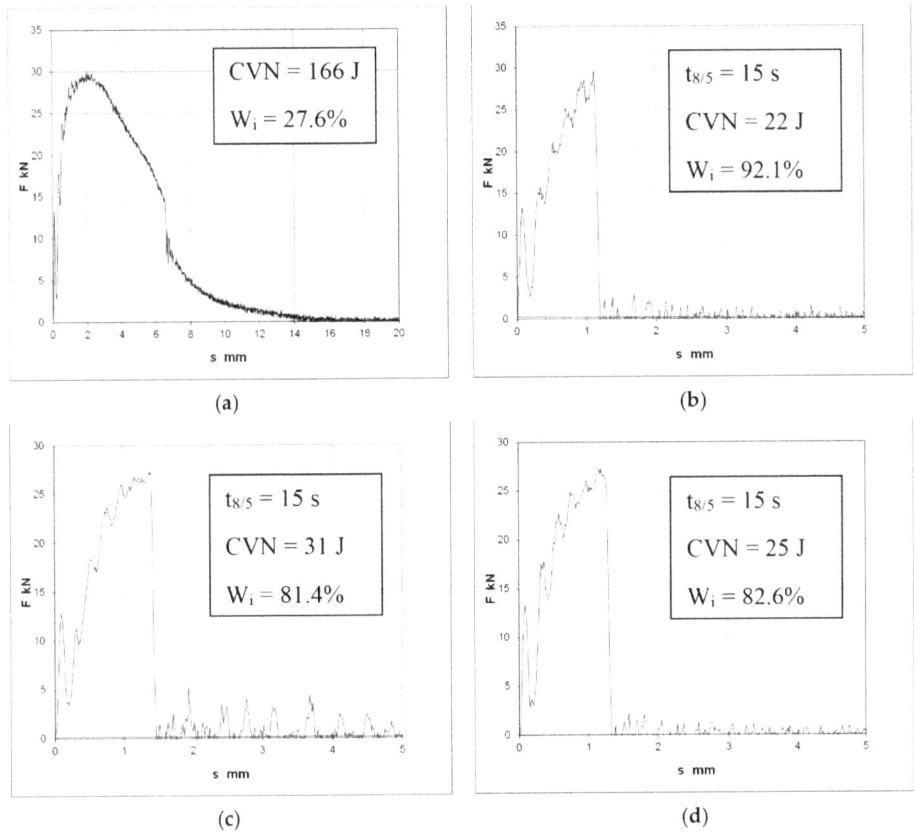

Figure 10. Registered force-displacement diagrams: (**a**) base material, (**b**) CGHAZ, (**c**) ICHAZ, (**d**) ICCGHAZ.

In Table 5 the ratio of the absorbed energy for the crack initiation is summarized for the base material and the HAZ. The values indicate the brittle behavior of CGHAZ, ICHAZ and ICCGHAZ where mostly unstable crack propagation can be seen in all diagrams. According to the values in Table 5, CGHAZ seems more brittle than ICHAZ, although the difference is minimal. The $t_{8/5}$ cooling time did not affect significantly the ratio of W_i in the examined interval (similarly to CVN). It is important to mention that the behavior of the investigated zones was brittle even though in some cases they met the 27 J requirement at −40 °C. On the basis of the presented instrumented Charpy V-notch impact tests, it can be concluded that the critical, local toughness reduction in the HAZ of the investigated S960QL cannot be avoided, and its extension cannot be significantly influenced by the $t_{8/5}$ cooling time. The effect of cooling time primarily appears in the variation of the width of the critical HAZ areas.

Table 5. Percentage of the absorbed energy for crack initiation compared to CVN.

$t_{8/5}$ s	2.5	5	10	15	22.5	30	Average
BM				-			25.9
CGHAZ	91.9	90.1	87.0	89.9	89.2	87.3	89.2
ICHAZ	90.7	84.2	80.7	77.4	81.5	79.2	82.3
ICCGHAZ	93.1	85.6	88.0	87.7	88.5	84.2	87.8

The presented considerations were supplemented by the fractographic analysis of the fracture surfaces. The backscatter SEM images about the fracture surfaces of the ductile base material and the brittle HAZ subzones, at $t_{8/5} = 15$ s, are presented in Figure 11. The sharp edges indicate cleavage cracking in all subzones, and the difference of the grain size can be also observed. However, in HAZ the qualitative improvement of fracture surfaces, with the presence of micro-voids separated by plastic areas, can be obtained by post-weld treatment in high strength steels [24].

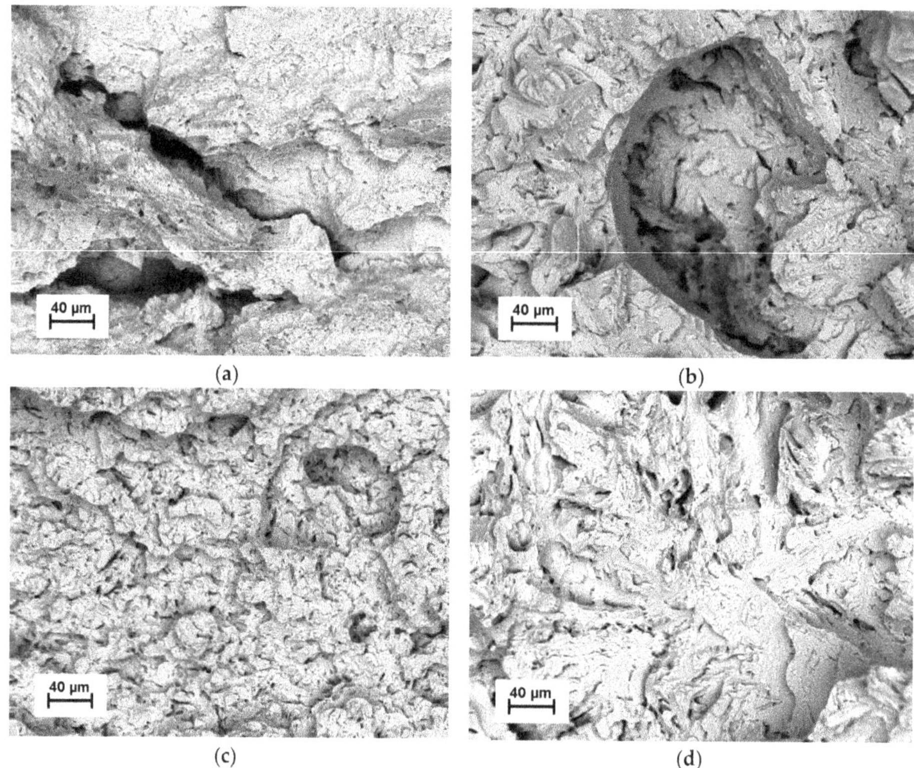

Figure 11. Fracture surface of the impact test specimens: (**a**) base material; (**b**) CGHAZ at $t_{8/5} = 15$ s; (**c**) ICHAZ at $t_{8/5} = 15$ s; (**d**) ICCGHAZ at $t_{8/5} = 15$ s (M = 250×).

4. Conclusions

The toughness properties of the HAZ in the welded joint of the investigated S960QL steel in the critical subzones (CGHAZ, ICHAZ and ICCGHAZ) were investigated by the Gleeble 3500 physical simulator for a $t_{8/5} = 2.5$–30 s cooling time interval (in case of CGHAZ $t_{8/5} = 100$ s was also simulated).

The HAZ indicated higher sensitivity to the welding heat input compared to conventional structural steels. While in low or medium strength steels the HAZ toughness and hardness can be significantly affected by the $t_{8/5}$ cooling time, in S960QL significant hardening and toughness reduction was observed in the whole cooling time range of the most common arc welding processes $t_{8/5} = 2.5$–30 s. In case of $t_{8/5} = 100$ s softening and extremely low CVN values were identified in CGHAZ.

The reduction of CVN was significant compared to the base material in all HAZ areas in the investigated cooling time interval. In most cases, the required 27 J at −40 °C was in the deviation interval of the CVN values of the subzones. Whenever the requirement was fulfilled, basically a brittle fracture was detected in the registered force-displacement diagrams, since the absorbed energy mostly consisted of the crack initiation energy. The fact that the toughness requirement is fulfilled may be

Metals **2019**, 9, 1226

connected to the higher strength of the material, despite the observed brittle behavior in HAZ. Besides CGHAZ, the ICHAZ should be similarly considered when the ductile-brittle behavior of the HAZ in Q + T steels is investigated. In multipass welded joints, ICCGHAZ seems to be the most critical part of the HAZ, with mostly lower CVN than the required value.

The critical, local toughness reduction in the HAZ of the investigated S960QL cannot be avoided by the modification of the welding parameters in the examined $t_{8/5}$ cooling time range; however, the application of post weld heat treatment (PWHT) can be an opportunity for the improvement of toughness properties. The effect of cooling time primarily appears in the variation of the width of HAZ areas. Therefore, the possible lowest welding heat input that does not risk the fulfillment of the allowed maximal (450 HV10) hardness and the formation of cold cracks should be applied in order to minimize the volume of HAZ.

Funding: This research was supported by the European Union and the Hungarian State, co-financed by the European Regional Development Fund in the framework of the GINOP-2.3.4-15-2016-00004 project, aimed to promote the cooperation between the higher education and the industry.

Acknowledgments: The author is grateful to the Institute of Physical Metallurgy, Metalforming and Nanotechnology at the University of Miskolc for ensuring the scanning electron microscope during the evaluation of HAZ tests. The author is also grateful for the opportunity to use JMatPro software at the University of Oulu during his internship. He also thanks the Department of Business Statistics and Economic Forecasting at the University of Miskolc for providing the SPSS software used during the statistical calculations.

Conflicts of Interest: The author declares no conflict of interest.

References

1. Branco, R.; Berto, F. Mechanical Behavior of High-Strength, Low-Alloy Steels. *Metals* **2018**, *8*, 610. [CrossRef]
2. Węglowski, M.S. High Strength Quenched and Tempered Steels: Weldability and Welding. In *High-Strength Steels: New Trends in Production and Applications*; Branco, R., Ed.; Nova Science Publishers: New York, NY, USA, 2018; pp. 143–224.
3. Bayock, F.N.; Kah, P.; Layus, P.; Karhin, V. Numerical and experimental investigation of the heat input effect on the mechanical properties and microstructure of dissimilar weld joints of 690-MPa QT and TMCP steel. *Metals* **2019**, *9*, 355. [CrossRef]
4. Spena, P.R.; Rossi, S.; Wurzer, R. Effects of Welding parameters on strength and corrosion behavior of dissimilar galvanized Q&P and TRIP spot welds. *Metals* **2017**, *7*, 1–14.
5. Kurc-Lisiecka, A.; Piwnik, J.; Lisiecki, A. Laser welding of new grade of advanced high strength steel STRENX 1100 MC. *Arch. Metal. Mater.* **2017**, *62*, 1651–1657. [CrossRef]
6. Varbai, B.; Sommer, C.; Szabó, M.; Tóth, T.; Májlinger, K. Shear tension strength of resistant spot welded ultra high strength steels. *Thin-Walled Struct.* **2019**, *142*, 64–73. [CrossRef]
7. Bhadesia, H.K.D.H.; Honeycombe, R.W.K. *Steels: Microstructure and Properties*, 3rd ed.; Elsevier: Oxford, UK, 2006; pp. 95–258.
8. Tervo, H.; Kaijalainen, A.; Pikkarainen, T.; Mehtonen, S.; Porter, D. Effect of impurity level and inclusions on the ductility and toughness of an ultra-high-strength steel. *Mater. Sci. Eng. A* **2017**, *697*, 184–193. [CrossRef]
9. Porter, D.A.; Easterling, K.E. *Phase Transformations in Metals and Alloys*, 2nd ed.; Chapman and Hall: London, UK, 1996; pp. 387–441.
10. Dunne, D.P.; Pang, W. Displaced hardness peak phenomenon in heat-affected zone of welded quenched and tempered EM812 steel. *Weld. World* **2017**, *61*, 57–67. [CrossRef]
11. Hochauser, F.; Ernst, W.; Rauch, R.; Vallant, R.; Enzinger, N. Influence of the soft zone on the strength of welded modern HSLA steels. *Weld. World* **2012**, *56*, 77–85. [CrossRef]
12. Lukács, J.; Dobosy, Á. Matching effect on fatigue crack growth behaviour of high-strength steels GMA welded joints. *Weld. World.* **2019**, *63*, 1315–1327. [CrossRef]
13. Hassanifard, S.; Mohtadi-Bonab, M.A.; Jabbari, Gh. Investigation of fatigue crack propagation in spot-welded joints based on fracture mechanics approach. *J. Mater. Eng. Perform.* **2013**, *22*, 245–250. [CrossRef]
14. Leroy Olscon, D.; Siewert, T.A.; Liu, S.; Edwards, G.R. *ASM Handbook: Welding Brazing and Soldering*; ASM International: Novelty, OH, USA, 1995; pp. 191–212.

15. Nevasmaa, P. *Evaluation of HAZ Toughness Properties in Modern Low Carbon Low Impurity 420, 550 and 700 MPa Yield Strength Thermomechanically Processed Steels with Emphasis on Local Brittle Zones*; Lisensiaatintyö, University of Oulu: Oulu, Finland, 1996.
16. Adony, Y. Heat-affected zone characterization by physical simulations. *Weld. J.* **2006**, *85*, 42–47.
17. Rykalin, N.N.; Nikolaev, A.V. Welding arc heat flow. *Weld. World* **1971**, *9*, 112–133.
18. Gáspár, M.; Balogh, A. GMAW experiments for advanced (Q + T) high strength steels. *Prod. Proc. Syst.* **2013**, *6*, 9–24.
19. Kuzsella, L.; Lukács, J.; Szűcs, K. Nil-strength temperature and hot tensile tests on S960QL high-strength low-alloy steel. *Prod. Proc. Syst.* **2013**, *6*, 67–78.
20. Laitinen, R.; Porter, D.A.; Karjalainen, L.P.; Leiviskä, P.; Kömi, J. Physical simulation for evaluating heat-affected zone toughness of high and ultra-high strength steels. *Mater. Sci. For.* **2013**, *762*, 711–716. [CrossRef]
21. Gáspár, M.; Balogh, A.; Lukács, J. Toughness examination of physically simulated S960QL HAZ by a special drilled specimen. In *Vehicle and Automotive Engineering*; Springer: Berlin, Germany, 2017; pp. 469–481.
22. Mohtadi-Bonab, M.A.; Ghesmati-Kucheki, H. Important factors on the failure of pipeline steels with focus on hydrogen induced cracks and improvement of their resistance: Review paper. *Met. Mater. Int.* **2019**, *25*, 1109–1134. [CrossRef]
23. Lenkeyné, B.Gy.; Winkler, S.; Tóth, L.; Blauel, J.G. Investigations on the brittle to ductile fracture behaviour of base metal, weld metal and HAZ material by instrumented impact testing. In Proceedings of the 1st International Conference on Welding Technology, Materials and Material Testing, Fracture Mechanics and Quality Management, Wien, Austria, 22–24 September 1997; pp. 423–432.
24. Konat, L. Structural aspects of execution and thermal treatment of welded joints of hardox extreme steel. *Metals* **2019**, *9*, 915. [CrossRef]

metals

Article

Determination of Grain Growth Kinetics and Assessment of Welding Effect on Properties of S700MC Steel in the HAZ of Welded Joints

Jaromir Moravec *, Iva Novakova, Jiri Sobotka and Heinz Neumann

Faculty of Mechanical Engineering, Department of Engineering Technology, Technical University of Liberec, Studentská 1402/2, 461 17 Liberec, Czech Republic; iva.novakova@tul.cz (I.N.); jiri.sobotka@tul.cz (J.S.); heinz.neumann@tul.cz (H.N.)
* Correspondence: jaromir.moravec@tul.cz; Tel.: +420-485-353-341

Received: 30 May 2019; Accepted: 18 June 2019; Published: 24 June 2019

Abstract: The welding of fine-grained steels is a very specific technology because of the requirement for the heat input limit value. Applying temperature cycles results in an intense grain growth in a high-temperature heat-affected zone (HAZ). This has a significant effect on the changing of strength properties and impact values. The intensity of grain coarsening in the HAZ can be predicted based on the experimentally determined activation energy and material constant, both of which define grain growth kinetics. These quantities, together with real measured welding cycles, can be subsequently used during experiments to determine mechanical properties in a high-temperature HAZ. This paper shows a methodical procedure leading to the obtainment of the material quantities mentioned above that define the grain growth, both at fast and slow temperature cycles. These data were used to define the exposure temperature and the soaking time in a vacuum furnace to prepare test samples with grain sizes corresponding to the high-temperature HAZ of welded joints for the testing procedures. Simultaneously, by means of the thermo-mechanical simulator Gleeble 3500, testing samples were prepared which, due to a temperature gradient, created conditions comparable to those in the HAZ. The experiments were both carried out with the possibility of free sample dilatation and under a condition of zero dilation, which happens when the thermal expansion of a material is compensated by plastic deformation. It has been found that shape of the temperature cycle, maximal achieved cycle temperature, cooling rate, and, particularly, the time in which the sample is in the austenite region have significant effects on the resulting change of properties.

Keywords: grain grow kinetics; high-strength steels; gleeble 3500; welding cycles; mechanical properties; heat affected zone

1. Introduction

The effort to obtain steels with higher mechanical properties is also associated with attempts to reduce grain size, as the yield strength (the Hall-Petch equation) and other mechanical properties are dependent on grain size. In addition to other changes caused by welding processes, a change in grain size, especially in a high-temperature heat-affected zone (HAZ) area, is a very negative phenomenon. For thermomechanically processed steels, these phenomena are unfavorable and, unfortunately, irreversible. Considerable attention therefore needs to be paid to the issue of grain growth in the HAZ of welded joints. Grain size is important, not only for the material mechanical properties but also for the course of transformation processes. An austenitic transformation leads to longer times and increases the risk of martensite formation.

Changes in grain size occur during transformation processes, recrystallization processes, and long-term temperature exposures. The thermodynamic driving force of grain growth is the reduction of surface Gibbs free energy. Grain growth leads to a reduction in the grain boundary area and, thus, to a reduction in the amount of free energy. The change of grain size is related to changes in temperature and is also dependent on the proportional amount of the relevant phase. The following basic rules apply to grain growth [1]:

- Grain growth occurs by moving grain boundaries rather than coalescing.
- The motion of the grain boundary is interrupted, and the direction of motion may suddenly change.
- Grain can grow into another at the expense of its volume.
- Grain consumption rates at the expense of others are often increased when grains are almost consumed.
- The curved border usually migrates to its center of curvature.
- If the grain boundaries of one phase meet at angles other than 120° grains with an acute angle are consumed so that the angle approaches 120°.

The first models based on these physical foundations were developed in the early 1950s (Smith. 1952; Burke and Turnbul, 1952) [1]. However, there was a big difference between theoretically and experimentally determined values. In 1980, a new approach based on computer simulations was applied, and processes that were difficult to observe experimentally (such as the volume change rate of individual grains) could now be observed on the basis of simulation calculations. Several simulation methods have been developed over the years, with the "Monte Carlo Potts model" being the most used. A description of this method is given in [2,3]. The general relationship for grain growth is given by (1), and the constant K is expressed by (2) [4].

$$D^m - D_0^m = K \cdot t, \tag{1}$$

$$K = K_0 \cdot \exp\left(-\frac{Q}{R \cdot T}\right), \tag{2}$$

In Equations (1) and (2), D is the current grain size, D_0 is the initial grain size, K is the proportionality constant that depends on the thermodynamic temperature T and the activation energy Q required for grain growth, R is a universal gas constant, t is a soaking time at a given temperature, and m is a material exponential coefficient.

It has been shown experimentally that the values of the coefficient m are in the range of 2–5. The value of $m = 2$ is valid if the grain growth process is exclusively controlled by diffusion. The value of $m = 4$ is used in the case of combination precipitation and diffusion along the grain boundary. As is shown below, other factors also affect grain growth [4,5].

As was already mentioned above, grain growth is dependent on the exposure temperature and the soaking time at this temperature, but it is also affected by other phenomena such as secondary precipitation. It has been found experimentally that the partial growth of grains already occurs after the transformation temperature has been reached. However, a noticeable grain growth in steels is only evident at temperatures above 900 °C [6,7]. At the same time, as temperature rises, grain growth rate increases, but there are a number of factors that slow down its growth kinetics. Most often, there is growth slowdown due to the presence of other particles that prevent the grain boundaries from moving. These are mainly very small oxides, sulfides, nitrides, carbides, or silicate particles [4]. These particles may already be present in the material, or they may precipitate along grain boundaries at the exposure temperature. Thanks to them, restoring forces (Zener drag forces) act against the grain growth direction and arise at the borders [4]. As a result, the grain size limit, for which the grain growth driving force is in balance with Zener drag forces, can be determined. This dimension is defined by Equation (3) as following [8]:

$$R_{crit} = A \cdot r \cdot f^{-1}, \tag{3}$$

In Equation (3), R_{crit} defines the grain size limit expressed as the critical grain radius, A is the material constant for a given steel type, r is the particle mean radius preventing grain motion, and f is the volume fraction of these particles. The influence of Zener forces on the fixation of grain boundaries can be observed, especially at lower temperatures when all particles preventing the motion of grain boundaries are present. At higher temperatures, these particles gradually dissolve, and the Zener drag forces decrease [8].

In his work, Górka described the influence of temperature cycles on the mechanical properties of the S700MC steel [9]. The maximum value of a temperature cycle was 1250 °C. An important conclusion of this work was the fact that, due to performed temperature cycles, all realized cooling conditions ($t_{8/5}$—the time required to cool the sample from 800 to 500 °C for the given temperature cycle) revealed a significant decrease in the impact strength at the test temperature of −30 °C. In another publication [10], the same author studied the effect of a simulated temperature cycle on the properties and structure of the HAZ of S700MC 10 mm thick steel plates. The simulation was prepared for both a simple and complex temperature cycle. The heat treatment results were studied by a Charpy pendulum impact test, a static tensile test, a hardness measurement, and metallographic analysis.

An analysis of the results proved that the temperature cycles during welding strongly influence the structural and phase changes in a HAZ. A rapid decrease of toughness in the affected material is associated with the separation processes of MX-type phases and the dissolution of Nb carbides and V carbonitrides in austenite during heating. Checking the amount of heat input into the joint area during welding makes it possible to reduce the unfavorable precipitation processes in the weld and HAZ, which ensures adequate joint strength. Similar conclusions were made by Górka in his work [11].

In [12], Górka dealt with the assessment of the microstructure and properties of the S700MC heat-affected zone after heating to 1250 °C and cooling by specified cooling rates as following: $t_{8/5} = 3$, 5, 10, 15, 30, 60, and 120 s. The influence of temperature cycle parameters was evaluated by a metallographic analysis, a hardness evaluation, a Charpy impact test, and a static tensile test. It was determined that for all $t_{8/5}$ values, the impact strength values were very low, and the austenite transformation conditions were not a reliable basis to evaluate the weldability of this group of steels. Similar results were provided by study [13], where the author focused on the evaluation of S700MC weldability by various welding methods. Lahtinen et al. [14] dealt with the weldability of 8 mm sheets from two types of high-strength fine-grained steels—namely, S700MC and S690QL. Two-layer welded joints made by MAG technology with four different heat input values from 0.7 to 1.4 kJ·mm^{-1}, which corresponded to values $t_{8/5} = 5$, 10, 15, and 20 s. Tests carried out on all materials included the determination of ultimate tensile strength, hardness profiles (HV5), Charpy impact tests, and microstructure analysis by scanning electron microscopy (SEM). The most notable differences in the mechanical properties of welded joints among tested materials were observed in Charpy impact tests, mostly close to the melting boundary, whereas the heat-treated steel was more vulnerable to heat affected zone (HAZ) embrittlement than TMCP steel. Rahman et al., in their work [15], dealt with S700MC-steel austenite-grain-size, which has a direct influence on the course of austenite transformation during cooling. They described a model of non-isothermal grain growth which includes the influence of the Zener effect. The study confirmed that the grain growth kinetics of a HAZ strongly depend on precipitates dissolution kinetics.

Numerical simulations have provided a significant contribution to the optimization of welding conditions. Kik et al. [16] dealt with the application of simulation computations for a T-weld made by laser and hybrid methods for sheets of 10 mm thickness. The simulation program SYSWELD was used in these computations, and possibilities of its application are now presented.

The SYSWELD simulation program is based on Equations (1) and (2), and computations are based on Equation (4), this expressing the grain growth rate.

$$\dot{D}^{m} = K_0 \cdot \exp\left(-\frac{Q}{R \cdot T}\right), \tag{4}$$

This computational relationship is intended for cases where the amount of austenite in the structure is constant or decreases. In cases where the austenite content of the structure increases, two situations may arise:

- The size of existing grains increases;
- there is the formation of new grains with an initial zero size or with a much smaller initial size than the existing grains in the structure.

In the second case, the grain size computation is carried out in accordance with Equation (5). In Equation (5), the λ value expresses the current ratio of austenite in the structure and the $\dot{\lambda}$ value of the transformation rate of this phase. In cases where austenite is no longer formed ($\lambda \leq 0$), Equation (4) is applied [17].

$$\dot{D}^m = K_0 \cdot \exp\left(-\frac{Q}{R \cdot T}\right) - \frac{\dot{\lambda}}{\lambda} \cdot D^m, \tag{5}$$

For the purpose of simulation computations, input data are necessary to predict the finite states of the structure and properties in the welded joint area. Therefore, it is not necessary to determine and describe all phenomena and processes during individual temperature cycles. However, it is necessary to obtain data for the prediction of finite states on the basis of realized experiments. This is also the background for the method for determining HAZ grain size values for S700MC steel.

2. Materials and Methods

2.1. Experimental S700MC Material

S700MC steel is one of the fine-grained high-strength steels. It is a thermomechanically rolled steel with a minimum guaranteed yield strength of 700 MPa, a low carbon content, an increased manganese content, and a reduced sulfur content. It is designed for cold forming, can be welded, and is micro-alloyed by Al, Ti, Nb, and V, where the sum of Nb, V, and Ti should not exceed 0.22 wt%. Micro-alloying, together with thermomechanical processing, leads to grain refinement and thus to increasing strength and toughness. S700MC steel is used for statically and dynamically stressed structures, especially in the field of transport. It can be used in the manufacture of axles and the undercarriage frames of cars and trucks, sub-assemblies of cranes, and possibly also mining machines. The reason for its use is to reduce weight while maintaining sufficient strength properties.

The steel tested in this work was delivered in the form of a 10 mm sheet. It has a ferritic-bainitic structure (Figure 1, left) and its grain size was measured by electron backscatter diffraction (EBSD) analysis (Figure 1, right). The resulting mean grain size of the supplied material was 3.35 μm. The chemical composition of the S700MC steel found by a Q4 Tasman spectrometer is shown in Table 1. The EN 10149-2 standard specifies a minimum allowable yield strength of $R_e = 700$ MPa for S700MC steel, a tensile strength R_m in the range of 750–950 MPa, and a ductility greater than 12%.

Table 1. Chemical composition (wt%) of the S700MC steel.

Elements	C	Si	Mn	P	S	Al	Nb	V
EN 10149-2	max. 0.12	max. 0.6	max. 2.2	max. 0.025	max. 0.010	min. 0.015	max. 0.09	max. 0.2
Experiment	0.051	0.197	1.916	0.006	0.006	0.038	0.063	0.072
	Ti	Mo	B	N	Ni	Cr	W	
EN 10149-2	max. 0.25	max. 1	max. 0.005	-	-	-	-	
Experiment	0.056	0.113	0	0.013	0.154	0.036	0.035	

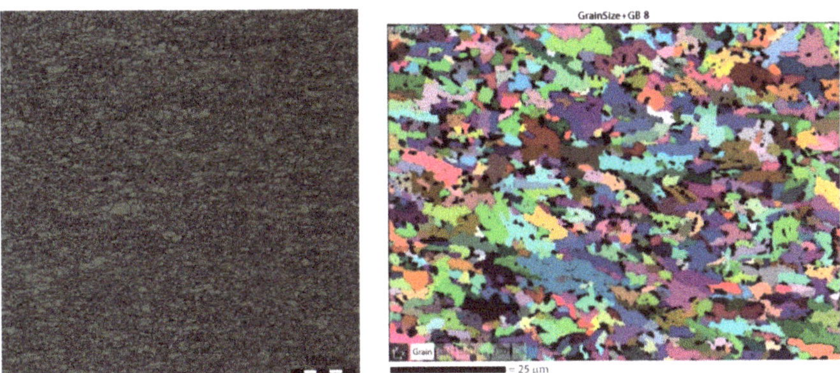

Figure 1. Ferritic-bainitic structure (**left**) and EBSD grain size analysis (**right**) of tested S700MC steel.

2.2. Methodological Procedure and Experimental Determination of Activation Energy Q Values and Kinetic Constant of Grain Growth K

For the prediction of grain coarsening intensity during welding, it is necessary to start from Equations (1) and (2). For this, however, it is necessary to determine the value of the proportionality constant K as well as the value of the activation energy Q needed to determine the total exponential constant K_0. Both of these values are temperature dependent and can be experimentally determined from samples exposed to at least two different temperatures and at least three different soaking times at these temperatures. Nevertheless, only a very approximate activation energy value of Q can be obtained by this method. It is more accurate to use at least 9, but preferably 12, samples exposed to heat under different conditions [6].

To determine the Q and K values, S700MC steel samples were subjected to a set of temperature-loadings as following: Exposure temperatures of 900, 1000, 1100, and 1200 °C; and soaking times of 0.5, 2, 4, and 8 h. Samples of $12 \times 12 \times 20$ mm^3 were exposed to temperatures in a Reetz vacuum furnace at a pressure about 6×10^{-4} Pa. Both the heating and cooling rate were $7\,°\text{C·min}^{-1}$. The reason for using the vacuum furnace was to prevent surface oxidation and the possible contamination of the sample by the diffusion of gases from the surrounding atmosphere.

After temperature exposure, the samples were cut in the middle and metallographically prepared for EBSD analysis to evaluate their grain size with an electron microscope (HV 30 kV; beam intensity 18; WD 12.9 mm). A total of 3 areas of 450×450 μm^2 were scanned on each sample. Figure 2 show the results of EBSD analysis on samples with temperatures of 900, 1000, 1100, and 1200 °C and a soaking time of 2 h. The mean grain size in μm (determined as the average from three measured areas) is given in Table 2 for all types of samples.

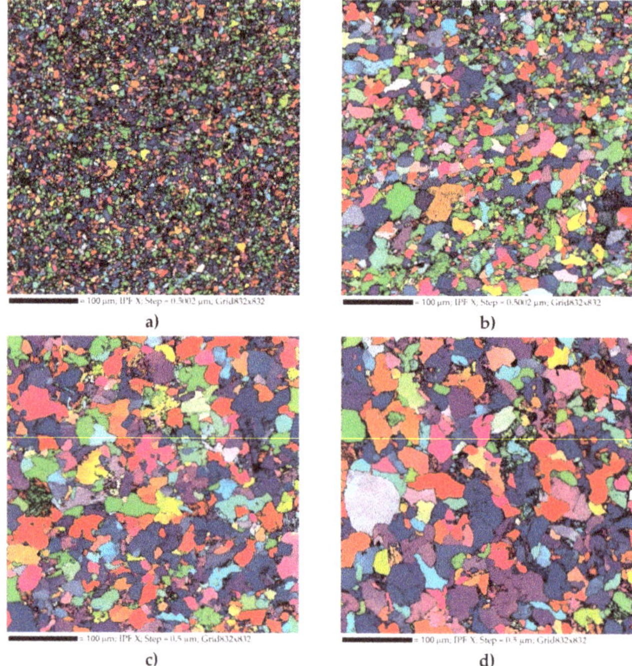

Figure 2. Results of EBSD analysis of grain size determination after 2 h at exposure temperature:
(**a**) 900 °C; (**b**) 1000 °C; (**c**) 1100 °C; and (**d**) 1200 °C.

Table 2. EBSD grain size analysis results expressed as mean grain size (μm) for the indicated exposure
temperatures and soaking times on these temperatures.

S700MC		Soaking Time at Relevant Temperature			
		0.5 h	2 h	4 h	8 h
	900 °C	3.99	4.29	4.87	5.91
Exposure	1000 °C	5.15	7.06	7.98	10.53
temperature	1100 °C	6.81	8.62	10.78	13.82
	1200 °C	8.53	10.02	12.78	16.97

Based upon the data in Table 2, it is already possible to determine the activation energy for the
grain growth Q of the S700MC steel. However, in the first phase, the partial exponential constants K_T
characteristic for each of the exposure temperatures and the initial grain size values D_0 corresponding
to individual temperatures at a soaking time of $t = 0$ s must be determined first. The initial grain size
D_0 should approximately correspond to the HAZ grain coarsening values when applying a welding
cycle, with a maximum that corresponds to the exposure temperature.

The way to obtain values D_0 and K_T is shown in Figure 3. The graph shows, for the individual
temperatures of the experiment, the average mean values of grain size (Table 2) obtained experimentally
or, more precisely, their second squares representing the mean grain areas. Each of the exposure
temperatures in the graph is expressed by a trend line and its equation. From the slope of the trend
line, it is then possible to deduct the value of each partial exponential constant K_T as well as the initial
grain size D_0.

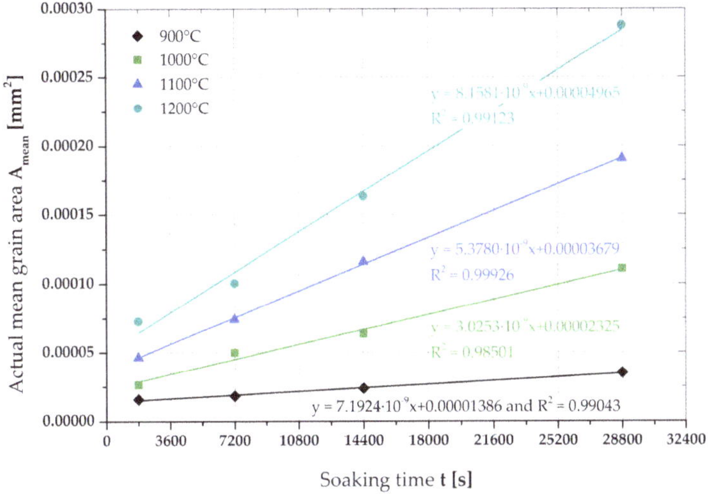

Figure 3. Actual mean grain area for S700MC steel vs. soaking time at temperatures of 900 °C, 1000 °C, 1100 °C, and 1200 °C (according to ISO 643).

Table 3. shows the values of the determined partial exponential constant K_T and the initial grain size D_0.

Table 3. Values of partial exponential constants K_T and initial grain size D_0 for a given temperature.

S700MC	Exposure Temperature (°C)			
	900	1000	1100	1200
K_T (mm² s⁻¹)	7.1924×10^{-10}	3.0253×10^{-9}	5.3780×10^{-9}	8.1581×10^{-9}
D_0 (mm)	0.00001386	0.00002325	0.00003679	0.00004965

The method of determining the grain growth activation energy Q as well as the total exponential constant K_0 is apparent from Figure 4. The graph shows the values of the natural logarithm for determined partial exponential constants K_T in dependence on the reciprocal values of exposure temperatures expressed in kelvin $[T^{-1}]$. As a result, a trend line can be created, and its equation is expressed in the following format (6).

$$y = C{\cdot}x + B, \tag{6}$$

The constant C in Equation (6) defines the slope of trend line and, together with the universal gas constant R, is used to calculate the value of activation energy Q in accordance with Equation (7). The constant B in Equation (6) is then used to calculate the total exponential constant K_0, in accordance with Equation (8).

$$Q = -(2.3{\cdot}R{\cdot}C), \tag{7}$$

$$K_0 = e^B, \tag{8}$$

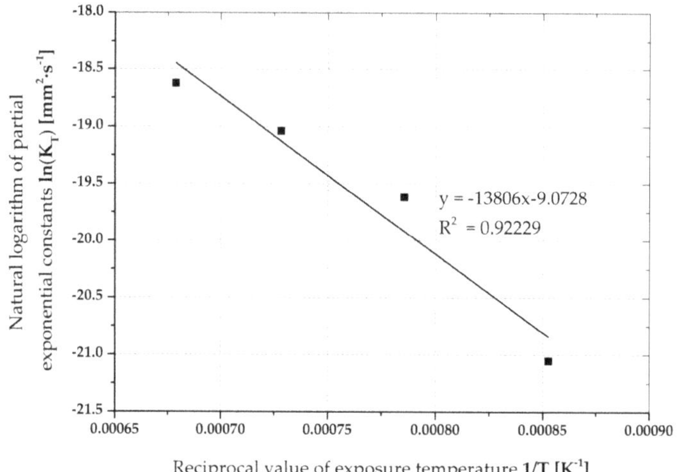

Figure 4. Determination the total exponential constant K_0 and the grain growth activation energy Q.

Using the trend line data (Figure 4) and Equations (7) and (8), the activation energy of grain growth $Q = 264.02$ kJ·mol^{-1} and proportionality constants $K_0 = 11.47 \times 10^{-5}$ (mm^2·s^{-1}) were obtained for the S700MC steel. Using these variables, the temperature dependence of the proportionality constant K, also known as the grain growth kinetics curve, can be defined in accordance with Equation (2). Table 4 shows the computed values of the proportionality constant K for a temperature range of 20–1500 °C. The data can be used as input data for numerical simulations of grain growth in any computational program. They can also be used for the simplified prediction of a grain growth rate made manually based on the knowledge of a real welding cycle.

Table 4. Computed proportional constant values K defining grain growth kinetics for S700MC steel.

Temperature (°C)	20	100	200	300	400	500
K (mm^2·s^{-1})	9.839×10^{-52}	1.224×10^{-41}	8.019×10^{-34}	9.825×10^{-29}	3.702×10^{-25}	1.657×10^{-22}
Temperature (°C)	600	700	800	900	1000	1100
K (mm^2·s^{-1})	1.832×10^{-20}	7.698×10^{-19}	1.612×10^{-17}	2.009×10^{-16}	1.685×10^{-15}	1.035×10^{-14}
Temperature (°C)	1200	1300	1400	1500		
K (mm^2·s^{-1})	4.982×10^{-14}	1.961×10^{-13}	6.555×10^{-13}	1.912×10^{-12}		

2.3. Welding Experiments

To obtain the real temperature welding cycles and to assess the effect of grain coarsening in a HAZ on the change of strength and brittle-fracture properties, welding experiments (according to ISO 4063-135) were performed for S700MC steel. A joint was designed as a double-sided fillet weld on 10 mm thick plates. Welding was done by means of linear automatic machine with a BDH Puls Syn 550 power source in a synergistic mode with an adjusted current of 315 A and a welding speed of 0.5 m·min^{-1}. All process parameters were monitored by Weld Monitor with a 25 kHz recording frequency. The following actual welding parameters were obtained by means of monitoring: The effective current $I_{ef} = 353.2$ A, the effective arc voltage $U_{ef} = 29.7$ V, the welding speed $v_s = 0.503$ m·min^{-1}, the wire feed speed $v_d = 10.03$ m·min^{-1}, and the gas flow of 15 dm^3·min^{-1}. A Boehler UNION NiMoCr with a diameter of 1.2 mm was used as filler material, and the mixed gas M21 was used as a shielding gas, according to ISO 14175.

During the welding process, temperature cycles in the HAZ were recorded by the DiagWeld system. This is a special device developed on TUL that allows for temperature recording by various types of thermocouples with a recording frequency of up to 100 Hz (depending on the number of thermocouples). The device has a galvanically isolated control panel, thus eliminating the influence of magnetic fields and high voltages during the welding and high-frequency heating of a material. The thermocouples were condenser welded into 4 mm diameter holes drilled into the weld flange at different distances from the material surface and, thus, at different distances from the fusion boundary. In total, 6 thermocouples of the S-type (90%Pt/10%Rh–Pt) were used. The temperature cycles obtained are shown in Figure 5. The only thermocouple TC2 (located 0.2 mm from the fusion boundary) measured a temperature cycle with a maximum temperature of 1383 °C—a temperature that was already part of the high-temperature HAZ area. This temperature cycle was then used bothto predict grain coarsening results in the HAZ by using the experimentally measured proportionality constant K and to apply these results to the thermal-mechanical simulator Gleeble 3500; this cycle was also used to prepare test specimens to determine mechanical properties and KV_2 in the HAZ area of the tested fillet weld joint of S700MC steel.

Figure 6 shows a metallographic scratch pattern, including the basic geometric evaluation of the individual runs. An EBSD analysis was performed on the weld on the Run 1 to determine the grain size during the transition from the weld through the HAZ to the parent material. In the EBSD analysis, a total area of 0.6×4 mm^2 was scanned, and this is shown in Figure 6 by a green rectangle and a green arrow. A detail of the grain coarsening in the high-temperature HAZ is shown in Figure 7.

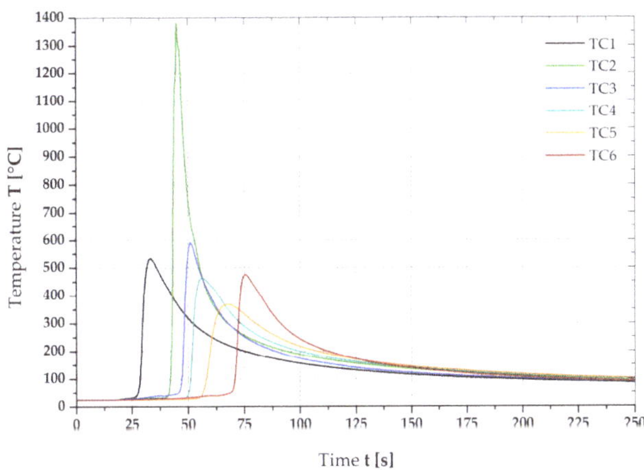

Figure 5. Thermal cycles obtained during welding of a S700MC steel fillet weld joint.

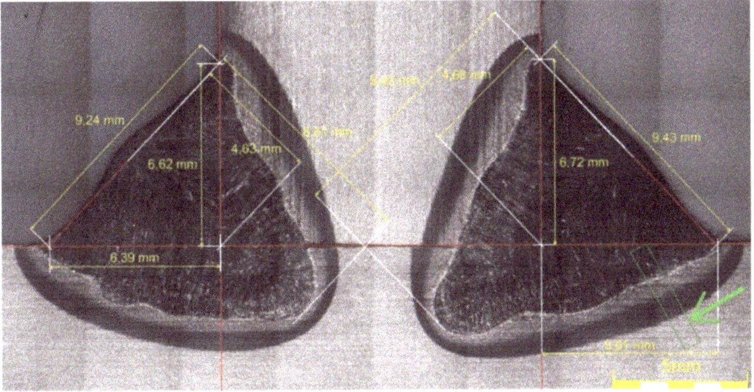

Figure 6. Geometrical evaluation of the weld and the grain size measurement area (green rectangle).

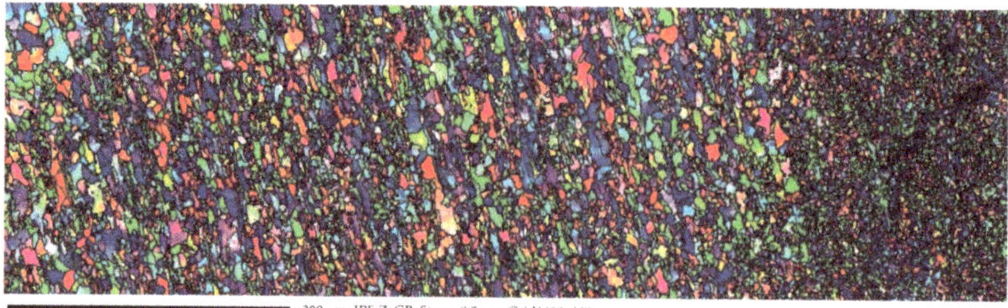

Figure 7. Detail of grain coarsening in the high-temperature heat-affected zone (HAZ) as it is shown in Figure 6.

2.4. Preparation of Test Samples for Measurement of Mechanical Properties and KV_2 Values in HAZ

From our knowledge of the actual temperature cycle measured during welding, the proportionality constant K and the actual grain size in the high-temperature HAZ, samples could be prepared to determine the mechanical properties of the KV_2 values in the HAZ of a S700MC steel fillet weld. Two aspects, local and global, were used to prepare the sample. The local aspect allowed for the determination of the mechanical properties of a particular grain size applied to the entire sample volume. The final machined testing samples were placed in a Reetz vacuum furnace and, based upon the knowing of proportionality constant K, the exponent m, the actual experimentally determined grain size in the high-temperature HAZ, the exposure temperature, and the soaking time at this temperature could be determined. Specifically, the exposure temperature was 1100 °C, and the soaking time was 418 min. In this way, the samples shown in Figure 8 were used for a static tensile test, Figure 9 shows samples used for static tensile test in the Gleeble 3500 and the samples in Figure 10 were used for a Charpy impact test, according to ISO 148-1.

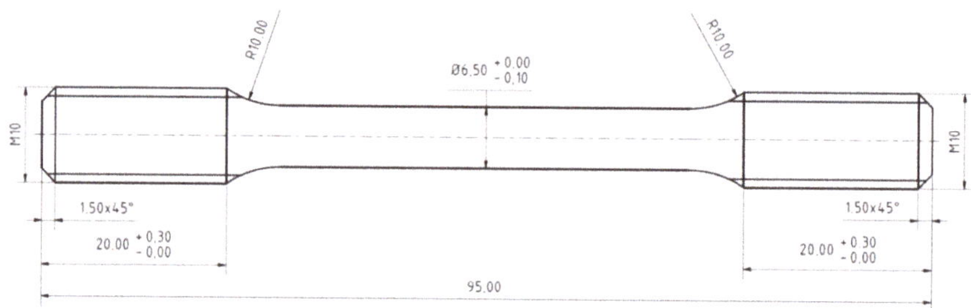

Figure 8. Samples used to measure mechanical properties as exposed in a vacuum furnace (unit: mm).

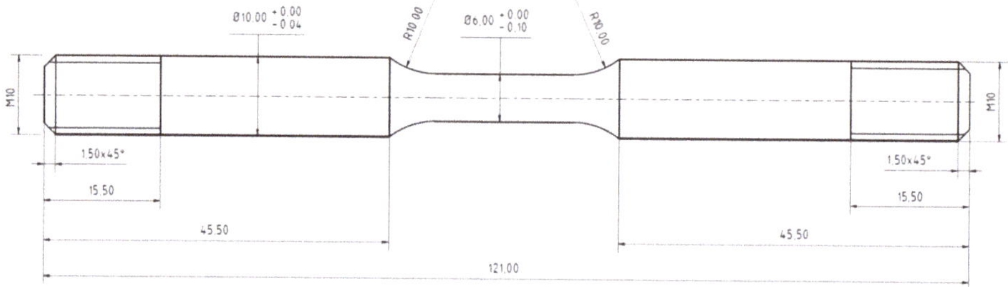

Figure 9. Samples used to measure mechanical properties after applying a temperature cycle in the Gleeble 3500 (unit: mm).

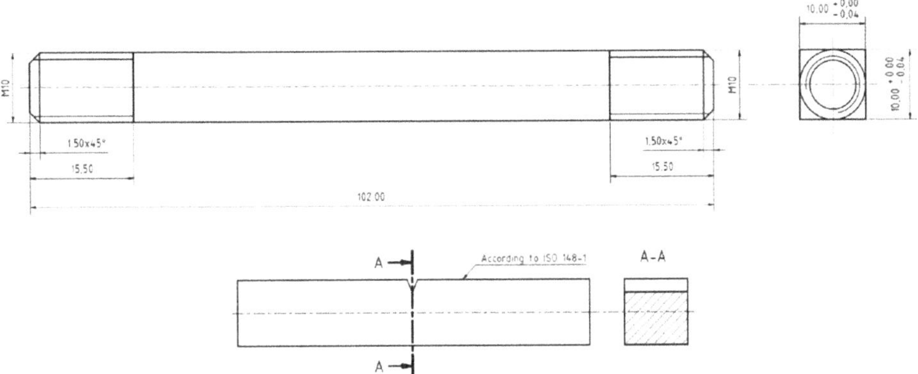

Figure 10. Samples used to measure the Charpy impact test after applying a temperature cycle in the Gleeble 3500 (unit: mm).

Other test specimens were designed and manufactured is such way that it was (in the Gleeble 3500 thermomechanical simulator, Dynamic Systems Inc., Poestenkill, NY, USA) possible to apply them for the experimentally measured temperature cycle with a maximum temperature of 1383 °C. This was a sample for a static tensile test (Figure 9) and a Charpy impact test (Figure 10).

Two methods of applying temperature cycles can be performed on testing samples (Figure 9; Figure 10). The first method rests in the application a temperature cycle in a force-control mode where force is equal to 0. This means that the clamping jaws move during heating and cooling according to the direction of sample dilatation so that no tension is created in the specimen, which could cause plastic deformation. The second method utilizes the application of a temperature cycle in an extensometer-control mode (L-gauge control mode) without dilatation or, more precisely, where dilatation is equal to 0. This method simulates the rigid clamping of the sample or the rigidity of the weldment itself. Increasing stress in the sample is firstl compensated by elastic and then by plastic deformation.

The temperature vs. stress course for the zero dilation sample designed for the static tensile test is shown in Figure 11. It is evident that, from a temperature of about 300 °C, the sample was already deformed plastically. Then, in the temperature range of 750–900 °C, the internal stress in the sample was significantly reduced due to decrease of yield strength, as well as phase transformation. The thermal expansion of the sample was here accompanied by further plastic deformation. When the maximum temperature of 1383 °C was reached and the cooling of the sample began, the compressive stresses were transferred to the tensile ones. During cooling of the sample, one significant drop in the sample stresses appeared in the temperature range corresponding to the phase transformation. This was already visible from the course of real temperature cycle.

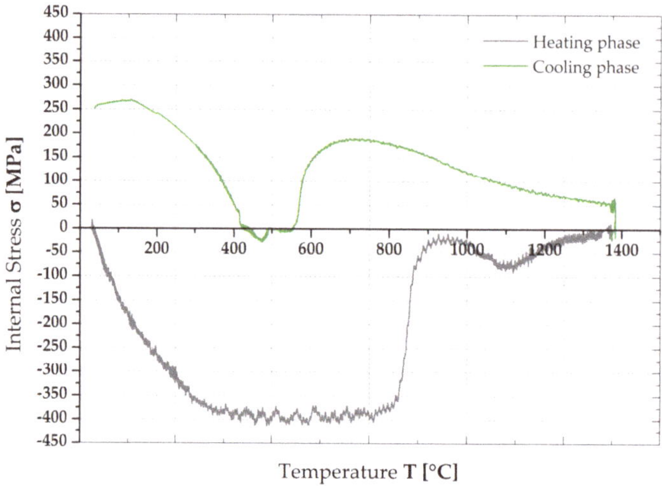

Figure 11. Graph of the dependence of the stresses in the sample on the temperature when applying a temperature cycle with zero dilatation, with a recording frequency of 500 Hz.

The overall setting of the sample in the Gleeble 3500 before the application of the temperature cycle is shown in Figure 12. The detail of the programmed (black dashed line) and real (red line) applied temperature cycle is shown In Figure 13. From Figure 13, is clear that the real temperature curve had a slight delay behind the program in the heating section. At the beginning of the cooling phase, the real sample temperature copied the program up to 600 °C, when, due to the latent heat release during the austenite transformation, it was not possible to cool the sample sufficiently enough with just Cu high temperature jaws. At a temperature of 400 °C, the cooling rate continued to follow the programmed temperature cycle.

Figure 12. Position of sample in the Gleeble 3500 device before the temperature cycle application.

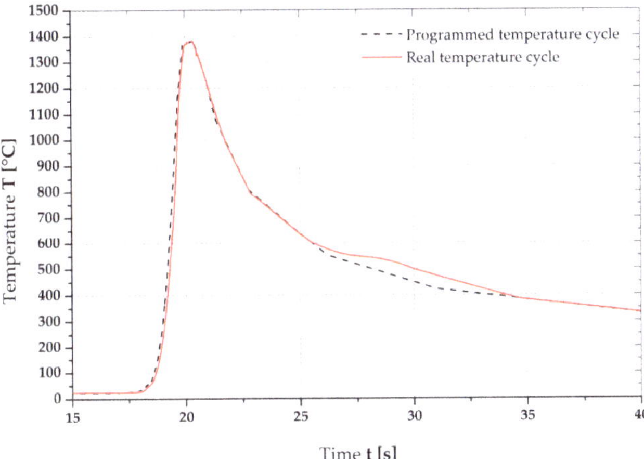

Figure 13. Comparison of the programmed (black dashed line) and the real (red line) applied temperature cycle.

3. Results

3.1. Evaluation of Welding Experiments and Comparison with Prediction Using the Proportionality Constant K

The first step was to compare the HAZ grain coarsening found in the welding experiments with the grain size predicted numerically based on the temperature dependence of the proportionality constant K and the temperature cycle measured by TC2 (Figure 5) at welding the fillet weld. The grain coarsening in the HAZ (Figure 7) of the weld was evaluated in three different, partially overlapping, areas. Thus, mean grain sizes of d_1 = 13.24 µm, d_2 = 13.09 µm, and d_3 = 13.28 µm were obtained. The average mean grain size in the high-temperature HAZ area was d = 13.21 µm.

The grain size prediction was performed in the Sysweld simulation program for the given temperature cycle. As input data, the temperature dependence of the proportionality constant K, the real measured temperature cycle (T_{max} = 1383 °C), and the exponential coefficient m = 2.42 were used [4,5,15]. The simulation was based on Equations (1), (2) and (4)

The predicted mean grain size obtained by the simulation was d_{str} = 12.04 µm, which is lower by about 9% compared to the real one. However, it should be noted that the prediction was made for one specific temperature cycle with a maximum temperature of 1383 °C, while visible grain coarsening in a HAZ area occurs during welding in a temperature range of 1250–1500 °C. Therefore, the grain growth simulation in the given temperature range was performed, and the predicted value of the mean grain size was d_{st} = 12.54 µm, which differs from the real one by 5.5%.

3.2. Evaluation of Welding Influence on Mechanical Properties and Impact Value in a HAZ Area

To assess the effect of welding on the change of properties in the HAZ area, a temperature cycle corresponding to that of Figure 13 was applied to the samples in Gleeble 3500. Multiple test variants were performed, and these differed only in boundary conditions. First, the temperature cycle was applied to the static tensile test specimens in a controlled force mode, allowing for the free dilatation of sample without plastic deformation. Samples were designated as GT-FD-0X. In the second phase, the samples were controlled by an L-gauge with zero dilatation during the application of the temperature cycle. Samples that were internally compensated by plastic deformation were designated as GT-D0-0X. The static tensile test results are shown in Table 5 (R_e is yield strength, R_m is ultimate tensile strength, Ag is uniform ductility, and A_{20} is total ductility).

Table 5. Mechanical properties for temperature cycles (temperature 23 °C, strain rate 10^{-2} s^{-1}).

Sample No.	Diameter (mm)	R_e (MPa)	R_m (MPa)	A_g (%)	A_{20} (%)
BMT-01	6.52	749	851	10.97	23.91
BMT-02	6.53	748	852	12.33	25.22
BMT-03	6.50	731	849	10.72	23.62
Average value		743	851	11.01	24.25
GT-FD-01	6.48	708	834	8.64	23.85
GT-FD-02	6.51	712	826	8.36	22.97
GT-FD-03	6.50	699	829	9.02	23.68
GT-FD-04	6.51	704	822	8.58	23.33
Average value		706	828	8.65	23.46
GT-D0-01	6.50	617	785	5.49	23.38
GT-D0-02	6.48	626	797	5.32	23.63
GT-D0-03	6.51	621	783	5.12	21.78
GT-D0-04	6.50	634	790	4.94	24.21
Average value		625	789	5.22	23.25
VFT-FD-01	6.51	301	448	13.24	37.65
VFT-FD-02	6.49	306	452	13.45	38.63
VFT-FD-03	6.48	311	446	13.52	38.16
VFT-FD-04	6.51	304	457	13.09	37.93
Average value		306	451	13.33	38.09

For an easier assessment of the welding cycle influence on the HAZ, the tensile test results of the parent material (BMT-0X samples) are also in Table 5. The last phase of the experiment used specimens where grain coarsening was achieved over the entire sample volume by the exposure to temperature of 1100 °C for 418 min. The mean grain size obtained from these samples was 13.39 μm. These ones were designated as VFT-FD-0X. Samples labeled as GT had a shape identical to Figure 9, and the samples labeled as BMT and VFT had the shapes shown in Figure 8. Samples were also prepared in the same manner for the Charpy pendulum impact test. The samples were geometry identical to Figure 10, and there was an applied temperature cycle in the Gleeble 3500 device by means of a force-control mode (thus with free dilation of sample). These were designated as GCH-FD-0X. The L-gauge controlled samples (which had zero dilatation during the application of the temperature cycle) where the internal stress was compensated by plastic deformation, were designated as GCH-D0-0X. Samples from the parent material that were produced in accordance with ISO 148-1 were designated as BMCH-0X. Samples for the Charpy impact test (prepared in the vacuum furnace) revealed a mean grain size of 13.43 μm and were designated as VFCH-FD-0X. The results of the Charpy impact test are shown in Table 6 (evaluated according to ISO 14556). All tests were performed on the 450 J Charpy impact machine with an instrumented impact edge and, in addition to KCV$_2$ and KV$_2$, the maximum achieved force F_{max} and the real impact velocity v_r were evaluated. Moreover, the percentage ratio of the ductile fracture from the photos of fracture surfaces was also evaluated by using an image analysis in NIS Elements 3.2. AR.

Table 6. Results of the Charpy impact test for the parent material and samples affected by the temperature cycle under different conditions (a temperature of 23 °C).

Sample No.	KCV$_2$ (J/cm^2)	Absorbed Energy KV$_2$ (J)	F_{max} (N)	Actual Impact Speed [m·s^{-1}]	Ductile Fracture [%]
BMCH-01	294.7	235.8	25119	5.529	100
BMCH-02	367.1	293.7	25536	5.478	100
BMCH-03	336.1	268.9	25343	5.426	100
BMCH-04	327.3	261.8	25413	5.429	100
BMCH-05	306.4	245.1	25323	5.426	100
Average value	326.3	261.1			
GCH-FD-01	254.7	187.8	25714	5.439	62.5
GCH-FD-02	237.5	174.0	25649	5.426	59.8
GCH-FD-03	263.2	194.6	25803	5.511	63.3
GCH-FD-04	271.8	201.4	25852	5.442	63.9
Average value	256.8	205.4			
GCH-D0-01	166.3	133.0	26094	5.512	35.3
GCH-D0-02	174.1	139.3	25927	5.426	33.6
GCH-D0-03	187.6	150.1	25973	5.431	36.1
GCH-D0-04	170.9	136.7	26011	5.473	37.0
Average value	174.7	139.8			
VFCH-FD-01	466.6	373.3	18694	5.426	100
VFCH-FD-02	473.9	379.1	18670	5.426	100
VFCH-FD-03	432.3	345.8	18434	5.684	100
VFCH-FD-04	458.9	367.2	18030	5.383	100
Average value	457.9	366.3			

4. Discussion

In general, the welding process has a considerable influence on the changes of entire joint properties. While changes in the weld metal properties can be partially compensated for by an appropriate choice of filler material, the properties in the HAZ can be influenced only by the welding process parameters. This paper was focused on the changes that occur in a high-temperature HAZ area. This area is characterized by intense grain growth, which has a significant impact on the changes of strength properties. This phenomenon is particularly significant in the case of fine-grained steels, and that is why it is highly-recommended to weld these materials with limited heat input.

4.1. Verification of the Grain Size Determination Method Applicability for Welding Cycles

It is very difficult to carry out HAZ grain size prediction based on data obtained from highly-dynamic temperature cycles due to a high measurement error. Real welding thermal cycles achieve a temperature range above A$_{c3}$ for just a single (or, maximally tens of) second. Within such a time interval, the change in grain size at temperatures up to 1200 °C is very small (see Table 2). Above the indicated temperature, it can be tough to determine the time dependence of grain size within such a short time interval. Therefore, much longer exposure times at temperatures not exceeding 1200 °C are used to determine grain growth intensity [4,5]. As mentioned in Section 3.1, by obtaining the temperature dependence of the proportionality constant K (Table 4) and on the basis of known temperature cycles, it is possible to predict the grain size in a high-temperature HAZ area with sufficient accuracy. Most of the calculations and simulation programs are based on grain size prediction via Equation (1), and this equation can be modified based on, in the given predicted location and at the given temperature, the amount of austenite increases, decreases, or remains constant; see Equations (4) and (5).

From the prediction accuracy point of view, the appropriate exponential coefficient m should be selected. According to [4,5], this coefficient should be in the range of 2–5. According to work of Hu et al. [15], which was devoted to predicting the grain growth of S700MC steel by using tests on samples heated in a quenching dilatometer, the coefficient m should move within the interval of 2–3.3. For S700MC steel (described in Section 2.1) which was welded by the method described in Section 2.4, optimized results were obtained with an exponential coefficient of $m = 2.42$.

Due to the values mentioned above (K, m), a grain size of 12.047 μm was predicted in the simulation for a real temperature cycle with a maximum temperature of 1383 °C, which was lower by about 9% compared to the grain size of the HAZ found in real welding. By including a temperature range of 1250–1500 °C, even better results were obtained, and the deviation in grain size was less than 6%. The same proportional constant values K and the exponential coefficient m were also used to determine the soaking time at an exposure temperature of 1100 °C to obtain samples with a defined grain size throughout whole volume. In this case, the difference between measured grain size on the samples was less than 2%. From the results mentioned above, is clear that the experimentally determined temperature dependence of the proportionality constant K can be used with sufficient precision, not only to predict grain at long-term exposures at chosen temperatures but also for rapid processes that occur during welding.

4.2. Discussion of the Influence of Welding on Mechanical Properties and Impact Value of a HAZ Area

Information about changing the mechanical properties and impact value in a HAZ is very important in terms of changing the properties of a welded joint. Especially for fillet welds, it is impossible to measure KV_2 values in a HAZ, and measuring mechanical properties using the static tensile testing also has its limits. Therefore, for testing fillet welds, a breaking test is prescribed. However, this can be realized only with the one-sided fillet welds. Another possibility is to measure the hardness values in a HAZ and in weld metals. Changes that occur in a HAZ and, especially, the coarsening of the grain are very important for the properties of welded joints, mainly in the case of fine-grained steels.

The methodology described in Section 2.4 provides the possibility to prepare samples where both the values of the mechanical and impact properties in a HAZ can be determined. Various authors applied temperature cycles to [9–16] fine-grained test specimens, but different limits appeared in their works. As a first, there is the T_{max} of the temperature cycle. The authors of [15,18] used temperature cycles with a maximum temperature generally not exceeding $T_{max} = 1000$ °C and 1200 °C, respectively. In his work, Gorka [9–13] used a temperature cycle with maximum temperatures of 1250 °C and 1299 °C. Only Spanos et al. [19] applied a T_{max} of 1400 °C in the dilatometer temperature cycle; however they used very low cooling rate that corresponded to the time $t_{8/5} = 60$ s. All of these temperature cycles will undoubtedly have an impact on the structural changes in the material and, thus, on the changes of mechanical properties. However, as is shown in Table 2 and Figure 2, for short times and temperatures up to 1200 °C, grain size change was very small. This is confirmed by the temperature dependence of the proportionality constant K, from which it is evident that intense grain growth occurs only at temperatures above 1300 °C. The second limit of testing is the possibility of including the influence of the rigidity of the welded clamping part and, eventually, the rigidity of the weldment itself. When comparing the GT_FD_0X temperature cycle samples that could freely dilate with GT_D0_0X, where the temperature cycle was applied and which had zero dilatation, differences in yield strength R_e and in the uniform ductility A_g are evident (see Table 5). The difference is even more pronounced when comparing KV_2 values. When applying the same temperature cycle to the freely dilated samples of GCH_FD_0X and samples that couldn't freely dilate, as with GCH_F0_0X, the difference in the value of KV_2 was 66 J; see Table 6.

The discussed problem is also evident from the dependence of the force on the displacement during the Charpy impact test (Figure 14) and the corresponding types of fracture surfaces. Figure 15 shows these fracture surfaces of relevant samples.

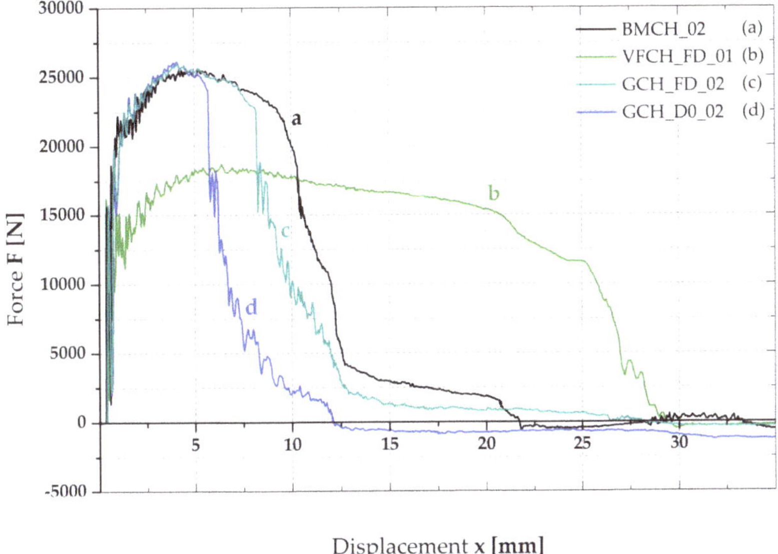

Displacement **x** [mm]

Figure 14. Force vs. displacement for the Charpy impact test for following samples (**a**) BMCH_02; (**b**) VFCH_FD_01; (**c**) GCH_FD_02; and (**d**) GCH_D0_02.

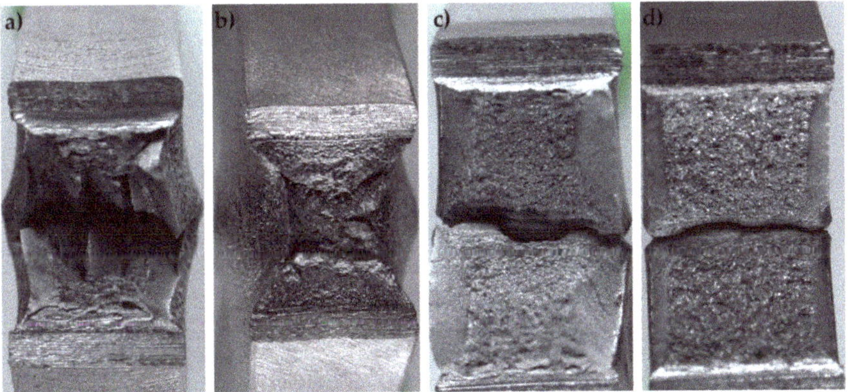

Figure 15. Fracture surfaces of samples after the Charpy impact test for following samples: (**a**) BMCH_02; (**b**) VFCH_FD_01; (**c**) GCH_FD_02; and (**d**) GCH_D0_02.

The results obtained for samples VFT_FD_0X and VFCH_FD_0X are then completely outside from the expected values both for mechanical properties and impact values. The aim of the experiments with these samples was to assess the situation where the grain size intensity would be the same in the entire sample volume to avoid the effect of heterogeneity. However, the long-term temperature exposure resulted not only in a grain growth to the desired size but also the removal of hardening after thermomechanical processing by the restoration mechanism. This caused a half-reduction in the mechanical properties and an increase in the KV_2 by 105 J. As a solution of this problem, there could be a shape inductor capable of rapidly heating the test sample throughout its whole cross-section area.

The last thing that should be mentioned is the influence of the heating and cooling rate that generally defines the time of the temperature cycle in the area above A_{c3}. Based on the real temperature cycles for the given process parameters and the welding method, the instruments used for applying this cycle to the sample were able to maintain the required heating and cooling rates; the results will correspond to the weld joint behavior. The producer [20] set for the S700MC steel recommended a range $t_{8/5}$ from 1 to 20 s. The real temperature cycle (Figure 13) used in this work had a cooling time of $t_{8/5}$ = 7.2 s, and the total cycle time at 500 °C was t_{500} = 10.8 s. When using a temperature cycle with half this heating rate and twice this cooling rate ($t_{8/5}$ = 14.4 s, the total cycle time at 500 °C was t_{500} = 21.6 s), the KV$_2$ values and mechanical properties will be significantly affected. In addition, a double-long temperature cycle was applied to samples with zero dilation for GT_D0_0X and GCH_D0_0X. The measurement was always performed on three samples under an ambient temperature of 23 °C. The average values obtained from the three measurements are as follows: KV_2 = 37 J, R_e = 586 MPa, R_m = 712 MPa, A_g = 5.88%, and A_{20} = 23.38%. The two-fold prolongation of the temperature cycle resulted mainly in the reduction of KV$_2$ by 103 J. The fracture was evaluated as the brittle one and is shown in Figure 16.

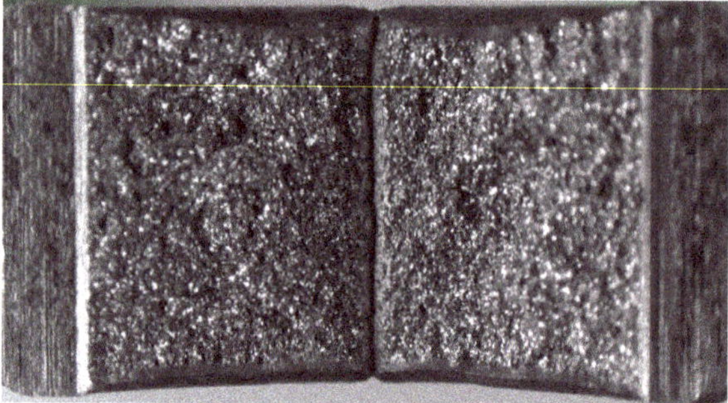

Figure 16. Fracture surfaces of GCH_D0_01_after the Charpy impact test (with a twice as long temperature cycle).

5. Conclusions

The influence of the welding process on the changes in a HAZ for fine-grained welded joints is a topical issue, and, thus, many researchers are dedicated to it. These works are most often included in the field of assessment of the welding influence on the structural changes in a HAZ and the associated changes of mechanical and brittle-fracture properties. Another large group is focused on grain size prediction—but mainly from the perspective of mathematical solutions. The presented work is focused on assessing the effect of grain coarsening in a high-temperature HAZ on the changes of mechanical properties and KV$_2$. Grain coarsening is an important material parameter, and sufficient attention should be given to it, especially in the case of thermomechanically processed steels that are hardened by grain boundary strengthening. From the measured values, it is clear that the concept of validation experiments should be based on the methodological procedures shown in Section 2 under precisely defined boundary conditions. Otherwise, results with considerable errors can be obtained. Therefore, the following recommendations should be taken into account when conducting verification experiments:

- Before the verification experiment, the type of welded joint, the possibilities of its expansion during welding, the method of its clamping, and the rigidity of the parts to be joined should all be defined.
- Consider the T_{max} for the temperature cycle applied to the test specimen, especially when it is a thermomechanically fine-grained steel. The effect of grain coarsening is more pronounced in welding cycles at temperatures higher than 1300 °C.
- Consider the shape of temperature cycles used for samples and simulating the welding process. Essentially start from the real measured temperature cycles characteristic of the used welding

method and eventually move to temperature cycles predicted by means of experimentally verified numerical simulations.

- Design the shape and size of the test specimens so that the testing machine applying the temperature cycle to the sample is able to reach the desired T_{max}, heating, and cooling rates.

By following these recommendations, it is possible to summarize the results of the high-temperature HAZ achieved by the welding the S700MC steel as following:

1. The values of the temperature dependence of the proportionality constant K, which defines the grain growth kinetics and is obtained experimentally by the isothermal method, can be used with the sufficient precision to predict the grain size both for long-term soaking times at exposure temperatures and for dynamic welding temperature cycles that are characteristic of welding

2. While welding parts that are rigidly clamped or have higher rigidity, the thermal expansion of the part is compensated by plastic deformation. In this case, the yield strength R_e was reduced by 12%, and the KV_2 value was reduced by 32% compared to parts that can dilate freely.

3. The value of specific heat introduced into the weld should be limited. In this case, when the cooling time is double extended in the interval T_{max}-500 °C (i.e., also for temperature $t_{8/5}$), the yield strength R_e was reduced by 6%, and, above all, the KV_2 value decreased by more than 70% compared to the same samples clamped in the same manner.

Author Contributions: Conceptualization, J.M. and I.N.; methodology, J.M.; investigation, J.M., I.N., J.S. and H.N.; resources, J.M. and H.N.; data curation, J.M., I.N. and J.S.; writing—original draft preparation, J.M. and I.N.; visualization, J.M. and J.S.

Funding: This research was supported by the Ministry of Industry and Trade of the Czech Republic as grant project MPO FV 10709, as well as by the institutional funding of science and research at the Technical University of Liberec TUL 117/2200.

Acknowledgments: The authors would like to thank to Ing. Tomasz Kik and Ing. Martin Svec for technical support during evaluation of experiments.

Conflicts of Interest: The authors declare no conflict of interest.

References

1. Burke, J.E.; Turnbull, D. Recrystallization and grain growth. *Prog. Met. Phys.* **1952**, *3*, 220–292. [CrossRef]
2. Zöllner, D.; Streitenberger, P. Three-dimensional Normal Grain Growth: Monte Carlo Potts Model Simulation and Analytical Mean Field Theory. *Scr. Mater.* **2006**, *54*, 1697–1702. [CrossRef]
3. Zöllner, D.; Streitenberger, P. *Monte Carlo Potts Model Simulation and Mean-Field Theory of Normal Grain Growth*; Shaker Verlag GmbH: Herzogenrath, Germany, 2006; p. 144.
4. Abbaschian, R.; Abbaschian, L.; Reed-Hill, R.E. *Physical Metallurgy Principles*; Cengage Learning: Stamford, CT, USA, 2009.
5. Giumelli, A. Austenite Grain Growth Kinetics and the Grain Size Distribution. Ph.D. Thesis, University of British Columbia, Vancouver, BC, Canada, 1995.
6. Moravec, J. Metodické postupy využitelné kzískání vstupních veličin numerických simulací svařování a tepelného zpracování. Habilitation Thesis, Technical University of Liberec, Liberec, Czech Republic, 2017.
7. Priadi, D.; Mapitupulu, R.A.M.; Siradj, E.S. Austenite Grain Growth Calculation of 0.028% Nb steel. *J. Min. Metall.* **2011**, *47*, 199–209. [CrossRef]
8. Manohar, P.A.; Dunne, D.P.; Chandra, T.; Killmore, C.R. Grain Growth Predictions in Microalloyed Steels. *ISIJ Int.* **1995**, *36*, 194–200. [CrossRef]
9. Górka, J. Microstrukture and Properties of the High-Temperature (HAZ) of Thermo-Mechanically Treated S700MC High-Yield-Strength Steells. *Mater. Technol.* **2016**, *50*, 617–621.
10. Górka, J. Influence of the maximum temperature of the thermal cycle on the properties and structure of the HAZ of steel S700MC. *IOSR J. Eng.* **2013**, *3*, 22–28. [CrossRef]
11. Górka, J. An Influence of Welding Thermal Cycles on Properties and HAZ Structure of S700MC Steel Treated Using Thermomechanical Method. In Proceedings of the Scientific Proceedings IX. International Congress "Machines, Technologies, Materials", Varna, Bulgaria, 19–21 September 2012; Volume 3, pp. 41–44.

12. Górka, J.; Stano, S. Microstructure and Properties of Hybrid Laser Arc Welded Joints (Laser Beam-MAG) in Thermo-Mechanical Control Processed S700MC Steel. *Metals* **2018**, *8*, 132. [CrossRef]
13. Górka, J. Assessment of Steel Subjected to the Thermomechanical Control Process with Respect to Weldability. *Metals* **2018**, *8*, 169. [CrossRef]
14. Lahtinen, T.; Vilaça, P.; Peura, P.; Mehtonen, S. MAG Welding Tests of Modern High Strenght Steels with Minimum Yeild Stenght of 700 MPa. *Appl. Sci.* **2019**, *9*, 1031. [CrossRef]
15. Rahman, M.; Albu, M.; Enzinger, N. On the Modelling of Austenite Grain Growth in Micro-Alloyed Steel S700MC. In Proceedings of the 10th International Seminar on Numerical Analysis of Weldability, Mathematical Modelling of Weld Phenomena, Schloss Seggau, Germany, 24–26 September 2012; Volume 9, pp. 623–636.
16. Kik, T.; Górka, J. Numerical Simulations of Laser and Hybrid S700MC T-Joint Welding. *Materials* **2019**, *12*, 516. [CrossRef] [PubMed]
17. ESI Group. *Reference Manual Sysweld 2019*; ESI Group: Paris, France, 2019; p. 366.
18. Schmidová, E.; Bozkurt, F.; Culek, B.; Kumar, S.; Kucharíková, L.; Uhríčik, M. Influence of Welding on Dynamic Fracture Toughness of Strenx 700MC Steel. *Metals* **2019**, *9*, 494. [CrossRef]
19. Spanos, G.; Fonda, R.W.; Vandermeer, R.A.; Matuszeski, A. Microstructural changes in HSLA-100 steel thermally cycled to simulate the heat-affected zone during welding. *Metall. Mater. Trans. A* **1995**, *26*, 3277–3293. [CrossRef]
20. Welding of Strenx. Available online: https://ssabwebsitecdn.azureedge.net/-/media/files/en/strenx/ssab-strenx-welding-brochure-2018.pdf (accessed on 29 May 2019).

Article

Influence of Welding on Dynamic Fracture Toughness of Strenx 700MC Steel

Eva Schmidová [1,*], Fatih Bozkurt [1], Bohumil Culek [1], Sunil Kumar [1], Lenka Kucharíková [2] and Milan Uhríčik [2]

[1] Faculty of Transport Engineering, Department of Mechanic and Materials, University of Pardubice, Studentská 95, 53210 Pardubice, Czech Republic; st43852@student.upce.cz (F.B.); bohumil.culek@upce.cz (B.C.); sunilmr21@gmail.com (S.K.)

[2] Faculty of Mechanical Engineering, Department of Materials Engineering, University of Žilina, Univerzitná 8215/1, 01026 Žilina, Slovakia; lenka.kucharikova@fstroj.uniza.sk (L.K.); milan.uhricik@fstroj.uniza.sk (M.U.)

* Correspondence: eva.schmidova@upce.cz; Tel.: +420-466-036-444

Received: 7 February 2019; Accepted: 26 April 2019; Published: 28 April 2019

Abstract: Thermomechanically processed high-strength steels feature specific fracture behavior. One of the decisive criteria for their application is their stability against internal defects during impact loads, especially in connection with the welding. The work is focused on experimental analyses of the influence of welding on static and dynamic fracture toughness of Strenx 700MC steel. The fracture toughness was determined using the circumferentially notched round bar specimens during static loads and two dynamic load levels. To achieve a homogeneous zone for the requirements of fracture toughness tests, simulation of the welding influence was performed. Fractographic and metallographic analyses described a specific fracture behavior controlled by the internal structural heterogeneity. A limiting degradation process due to welding was identified by the microstructural analysis.

Keywords: dynamic fracture toughness; high strength steels; fracture behavior; welding influence; heat-affected zone; simulation of welding; heterogeneous carbide precipitation

1. Introduction

The use of high-strength steel in the construction of means of transport has grown over the years. The driving force is both technical and environmental, where passive safety requirements have brought about the development of ultra-high-strength steel based on specific processes of dynamic strengthening during the crash-strain rate [1,2]. Particularly for the application in means of transport, a variety of steels after thermomechanical treatment meet the increasing requirements for lightweight construction and therefore for an increased efficient loading capacity and lower fuel consumption [3–5].

One of the most effective processes to increase material resilience to a critical sudden fracture is to influence the natural crack propagation through microstructural heterogeneity. The thermomechanical treatment of steels with limited semi-product thickness leads to microstructural and mechanical heterogeneity, which can positively affect the fracture behavior of steel. The presented study is focused on the mentioned process using one of the prospective steels for lightweight constructions—Strenx 700MC. Strenx 700MC steel is characterized by the strain hardening and strain-rate hardening; the prospective restriction in the particular application can be a thermal softening effect after welding. The strength decrease caused by thermal impacts of specific technology is the subject of numerous studies [6–9]. The worst impact properties are commonly obtained in the coarse-grained zone near the fusion line [10,11].

A frequent limitation for the wider use of this type of steel, such as in the railway vehicle chassis design, is the lack of data on induced material sensitivity changes due to internal defects during impact loads. During heating, fine carbides dissolve and re-form with the intensity depending on the cooling rate. The resulting fracture behavior in the weld joint critical zone is a question of the interaction of the change in the grain structure and specified phase changes, which can counteract the stability against the development of defects.

The fracture mode of the tested steel is driven by local heterogeneity. The extension of the plastic zone at the tip of a crack strictly depends on numerous variables, such as the yield stress, the crack length, the strain rate, and the thickness of the cracked component. In the case of significant toughness anisotropy, crack growth tends to be in the direction of a weaker zone and may not be in the direction of the initial crack even for mode-I loading [12,13]. Determination of fracture toughness is a way to quantify the crack sensitivity, especially in connection with the thermal effect after welding. The crack depth, section thickness, specimen size, crack geometry and loading configuration have a strong effect on fracture toughness measurements. In order to obtain conservative, constraint-independent fracture toughness measurements, all fracture test standards prescribe strict specimen geometry requirements [14–17]. Despite a tendency to reduce the thickness of construction components, their real toughness is driven by the overall design, including welding joints.

The main objective of the experimental analyses was to obtain information about the fracture behavior of this steel at precisely defined load parameters and also at the defined initial structural steel state. Various loading rates and the impact of structural degradation by welding were included in the analyses.

Presented work has also verified the possibility of using the non-standardized circumferential pre-cracked round bar test method (CCRB) [18–24] to assess fracture toughness under static and dynamic loading. Advantageously, the positive influence of the peripheral initiated fatigue crack on the fracture propagation was applied. This method made it possible to perform the comparative tests to quantify the two main examined effects—dynamic load and structural degradation in the critical layer of welded Strenx 700MC steel.

2. Materials and Methods

2.1. Experimental Material

The Strenx 700MC steel used in this study belongs to the category of microalloyed high-strength steels widely used for trailers, containers, truck frame rails, dump truck cabs, etc. The mechanical properties of Strenx 700MC meet or exceed the requirements in EN 10149-2. Strenx 700MC is a hot-rolled structural steel with a minimum yield strength of 700 MPa, made for cold forming and intended for stronger and lighter structures.

This steel type is typical by a low carbon content, an increased content of manganese, a reduced sulphur content, and the addition of micro-alloyed elements (see Table 1). The sum of Nb, V, and Ti is maximal 0.22% of steel content, and together with thermomechanical processing results in grain refinement and increases the strength and toughness. As for the influential technical, manufacturing, economic, and environmental aspects, this steel has good prerequisites for use in railway transport to reduce weight and increase payload.

Table 1. Chemical composition (wt. %) of the used steels.

Used Steels/Standards		C	Si	Mn	P	S	Al_{tot}	Nb	V	Ti
Strenx	EN 10149-2 (max.)	0.12	0.21	2.1	0.020	0.010	0.015	0.09	0.20	0.15
700MC	Sample	0.047	0.028	1.75	0.009	0.0026	0.041	0.061	0.016	0.091
S355NJ	Standard (max.)	0.23	0.05	1.60	0.05	0.05	-	-	-	-
	Sample	0.15	0.18	1.32	0.017	0.0056	0.035	-	0.0016	0.001

Today's railway frames are primarily made of material with a yield strength of 235–355 MPa. For that reason, the S355NJ steel (see Table 1 for chemical composition) was used as a referential steel for the performed experimental analyses. The real mechanical properties may significantly vary depending on the rolling direction. The mechanical parameters of the analyzed steels in a longitudinal direction, i.e., the same as the direction of samples used for all performed analyses, are in Table 2.

Table 2. Static mechanical parameters of the used steels.

Material	Yield Strength (MPa)	Tensile Strength (MPa)
S355NJ	516	550
Strenx 700MC	623	683

2.2. Methodology of Fracture Testing

The Strenx 700MC steel is typical by significant structural heterogeneity, where the fracture response is controlled by local differences of plasticity. The structural state of material affects the usability of possible assessment methods, especially for dynamic fracture toughness assessments. One of the ways to evaluate the dynamic fracture toughness for steel plates of limited width is a measurement of fracture toughness at impact loading rates using pre-cracked Charpy-type test pieces. Testing single-edge bend specimens (SENB), fatigue pre-cracked and loaded in three-point bending, is introduced in the ISO 26843 [25] and ASTM E 1820 [26], standards, including the recommended shape, specimen dimensions, and fatigue pre-cracking requirements.

The pre-cracked Charpy-type test was employed to verify this methodology for the used steel with defined thickness. For the dynamic test, 10 mm × 5 mm × 55 mm samples were used with a chevron notch (according to EN ISO 12737), as shown in Figure 1. As shown below, the use of a 1-sided notched specimen has led to an unacceptable deflection of the crack face and false plasticity indication.

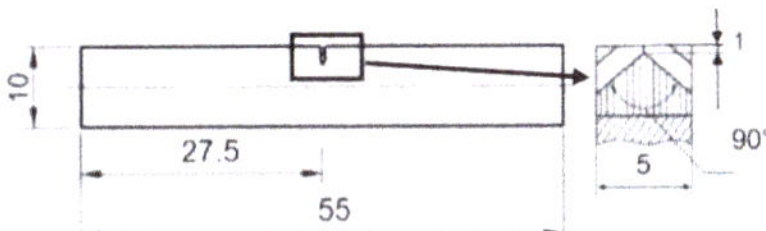

Figure 1. Chevron notched sample for dynamic fracture toughness testing (unit: mm).

Therefore, we used the variant method to assess the material sensitivity against the development of cracks using the round specimens. This approach enables radial pre-cracking and uniaxial static and also dynamic loading, e.g., research of strain rate sensitivity up to the standard crash rate loading [13–16]. The circumferential pre-cracked round bar test method (CCRB) is based on round specimens fatigue pre-cracked at a defined notch and loaded in a uniaxial tensile test (at different loading rates) until failure. The specimens with V-type notches with notch angles of 60° and 1 mm radial depth, mean notch radius 0.225, and with the dimensions displayed in Figure 2 were used in the performed analyses. The specimen was prepared from steel plates parallel to rolling.

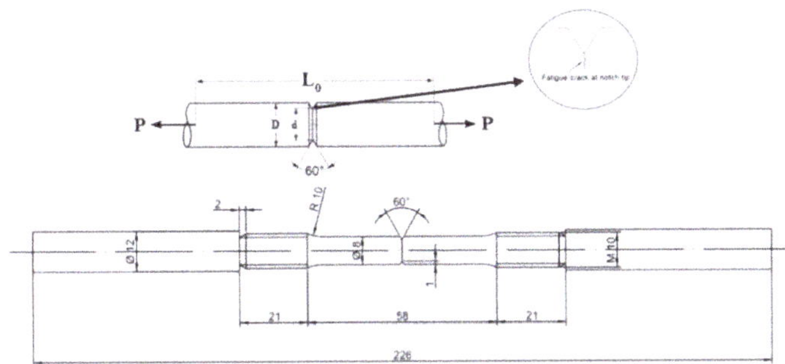

Figure 2. Round specimen for dynamic fracture toughness testing (unit: mm).

An R.R. Moore four-point rotating bending fatigue-loading machine (MEZIM, Moscow, Russian) was used for specimen pre-cracking as shown in Figure 3 and the crack propagated radially towards the center of the specimens. The specimens were subjected to cyclic tensile-compressive loads (R_{stress} = −1). The chosen bending loading (M) had to respond to the maximum stress intensity factor K_{max}, which should not exceed 60% of the minimum expected K_{IC} fracture toughness of the tested material.

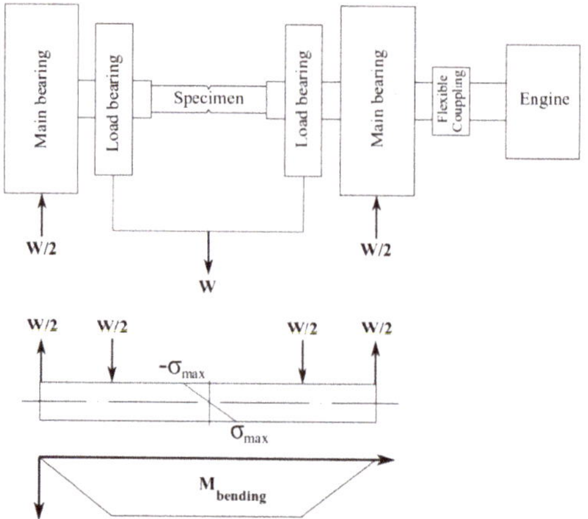

Figure 3. R. R. Moore four-point fatigue testing machine.

The pre-cracking process was controlled by a JK–1 Crack Depth Meter device (JaKM, Prague, Czech Republic), based on the measurement of the electrical resistance at the notched area. Particular resistance levels, according to the surface and initial notch geometry conditions of each sample, were measured during individual calibration processes for all samples. When a fatigue crack was detected, the specimen was ready for the static or dynamic tensile test.

For mode-I loading conditions, the pre-cracked specimen was loaded under both static and dynamic tension conditions. A crosshead displacement rate of 0.5 mm/min until failure was used for the evaluation of the static fracture response. In order to investigate the effect of the loading rate on the dynamic fracture toughness of the materials, the impact tensile tests were conducted at two different impact velocities, namely 3.48 and 5.23 m/s. In the present study, a Zwick/Roell RKP 450 connected to a PC with the testXpert testing software (TestXpert II Master, Zwick&Co.KG, Ulm, Germany, 2016) was chosen to conduct the instrumented impact tensile tests. The force and crosshead displacement were also recorded during the static and dynamic tensile tests. All tests were performed at room temperature. A minimum of three specimens of the material was tested under the same conditions for fracture toughness calculation. The maximum loads reached by each specimen were specified.

After completion of the uniaxial tests, the fracture surface of the specimen was investigated, and specified dimensions were measured with a stereomicroscope as shown in Figure 4. To calculate fracture toughness, defined dimensions were machined notch depth (a_m—region A) and length of fatigue pre-crack (a_f—region B). Stable vs. unstable fracture modes were distinguished inside the final fracture range (region C) in the case of different fracture modes action.

For determination of the fracture toughness of the material, the effective diameter (d_{eff}) was stated:

$$d_{eff} = D - 2(a_m + a_f) \tag{1}$$

Depending on the effective diameter (d_{eff}), unnotched section diameter (D) and maximum reached force during the impact tensile test, static fracture toughness (K_{IC}) and dynamic fracture toughness (K_{Id}) were calculated as follows [24]:

$$K_{IC} = \frac{P_f}{D^{3/2}}\left[1.72\frac{D}{d_{eff}} - 1.27\right] \tag{2}$$

$$K_{Id} = \frac{P_{dmax}}{D^{3/2}}\left[1.72\frac{D}{d_{eff}} - 1.27\right] \tag{3}$$

where P_f and P_{dmax} were the maximum static and dynamic fracture load, respectively. The valid range of Equations (2) and (3) is $0.46 < (d_{eff}/D) < 0.86$ [24].

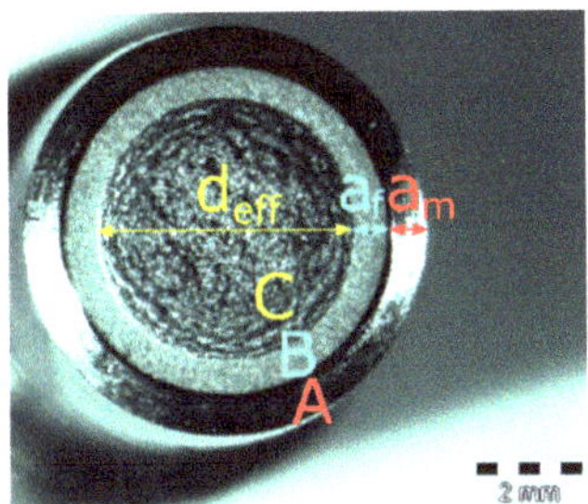

Figure 4. Defined fracture regions for fracture toughness determination.

2.3. Experimental Welding and Simulation of Impact of Welding on Samples for Fracture Tests

The above-mentioned approach, using round-bar specimens, was also used to provide a more detailed description and understanding of the degradation caused by welding. The welding effect evaluation was performed with the support of experimental welding and subsequent simulation of the maximal degradation effect in the area of fracture propagation.

Steel plates with dimensions of 120 mm × 300 mm × 10 mm and V-type grooves were prepared for the resistance welding process, as shown in Figure 5. The Kempact 253R welding equipment was used for MAG welding (CO_2 and Ar shielding gases), with three welding passes using Böhler UNION NiMoCr Ø1 mm filler wire (Table 3). The heat input for each welding pass was between 1.1 and 1.5 KJ/mm; no heat treatment process such as stress relief annealing was carried out after or before the welding process.

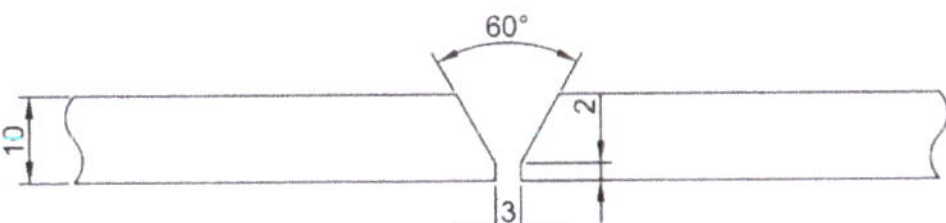

Figure 5. Details of V type configuration and dimensions of Strenx 700MC plate (unit: mm).

Table 3. Chemical composition (wt. %) of filler wire Böhler UNION NiMoCr Ø1.

C	Si	Mn	Cr	Mo	Ni
0.08	0.60	1.70	0.20	0.50	1.50

The real weld joint evaluation was used to identify the heat affected zone (HAZ) sublayer critical to the initial strength of the steel. The hardness measurement and microstructural evaluation served as parameters for identifying the critical degradation effects of the used experimental welding technology. The minimum hardness value of the heat affected zone showed a critical temperature for the next simulation of a critical degradation process.

The simulation of welding heat effect on the Strenx 700MC steel was carried out for two reasons. The first reason was the potential presence of microscopic defects that cannot be totally excluded and that affect the results of any weld joint tests. The purpose of the performed research was not to verify the technology itself, but to examine the degradation effect on the material in question, its substance and its influence on dynamic fracture behavior. Secondly, the welding effect simulation allows the creation of a geometrically optimized HAZ for fracture toughness analysis. As an accurate orientation of the developing crack to the zone with the maximum degradation impact of the welding is necessary for these tests, this area must be structurally homogeneous throughout the tested cross section. The simulation, by the localized critical heating, allows the controlled preparation of much wider critical HAZ sublayers with typical heat levels and corresponding microstructural effects. The used approach was chosen to suppress the natural structural gradient in the HAZ, which usually results in scattering results. The critical thermal effect was simulated to create the critical HAZ sublayer located in the crack propagation plane and, thereby, to accurately evaluate the fracture toughness after welding.

A welding simulation was performed using the WTU 315-3 welding equipment without the filler material. The specimens made for the fracture analyses were heated by an electric arc over copper rings. The critical temperature and thus the limit degradation effect were induced in the notched part of the samples. During the welding simulation process, the temperature was recorded by an Omega HH309A four-channel data logger thermometer over time. Time and temperature records were used to reproduce the heating regime used. The structural analysis and hardness measurement were performed to validate the simulation process by means of comparison with the real critical HAZ sublayer. The specimens before and after the welding simulation process and typical induced temperature gradient are shown in Figure 6a,b, respectively.

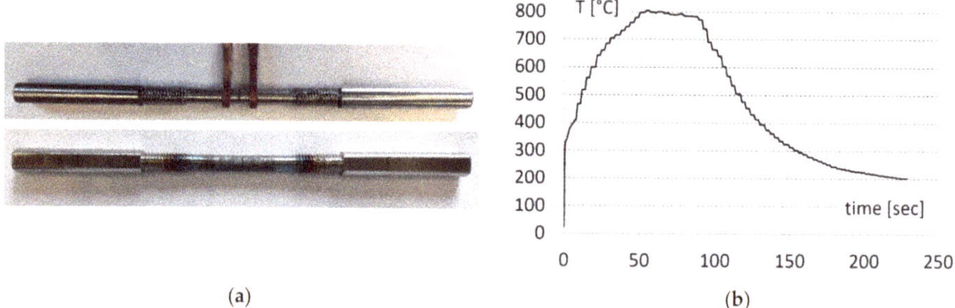

Figure 6. Simulation of the critical influence of welding: (**a**) Samples before/after heating; (**b**) induced temperature gradient.

3. Results

3.1. Verification Analysis of Welding Simulation

As a first step, the complex material analyses of experimentally welded Strenx 700MC steel were performed to identify the critical sublayer of HAZ in terms of structural degradation and to describe the involved degradation process. The local differences in hardness have shown the weakest area of the welding joint and the most intensive degradation effect of welding in the used high-strength steel. Identification of the most softened sublayer provided information about the critical temperature level for the design of the weld thermal influence simulation.

The Vickers hardness measurement of the welded specimen (according to EN ISO 6507-1) was conducted perpendicularly to the weld joint axis, on the HV1 scale, with a 0.2 mm interval—see Figure 7. The hardness measurement indicated the lowest values in the sublayer of the outer part of the grain refinement zone (236 HV1).

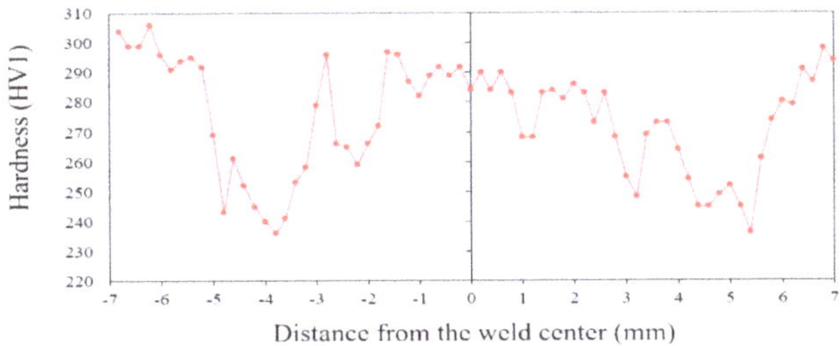

Figure 7. HV1 hardness vs. distance from weld center for welded Strenx 700MC specimen.

A simulation of the degradation process was designed based on the analyses of experimental welding process influences. During the thermal cycle simulation, the specimen, with geometry for static or dynamic uniaxial tests, was heated to a maximum temperature of 800 °C and held at this temperature for 5 s; the temperature was recorded as a function of time. For validation of the used simulation, the thermal cycle influence was evaluated by the metallography analyses and hardness measurement. The samples were extracted from the longitudinal axis of the circumferentially round bar specimen of the Strenx 700MC steel and prepared by a standard metallographic preparation process. Stable hardness values at the notch tip presented very important results in terms of material homogeneity in the fracture plane, i.e., they prove the suitable conditions for the fracture resistance evaluation. The hardness values varied from 216 to 233 HV1, and the mean value was approximately 225 HV1. As mentioned above, the main aim was to approach the lowest hardness level (236 HV1), which was induced in the experimental welding sample of Strenx 700MC. In this way, the suitability of the simulation used to assess the degradation process due to welding was verified.

3.2. Metallography Evaluation of Related Structural Effects

The microstructure of uninfluenced Strenx700MC steel consists of a fine-grained polygonal ferrite with a limited amount of pearlite. Outstanding strength and impact toughness are a result of the fine-grained microstructure in combination with thermomechanical rolling in a hot strip mill. The overall view of the experimental HAZ weld is presented in Figure 8. Typical microstructural effects of the applied welding technology, compared to the simulation effect, are documented in Figure 9.

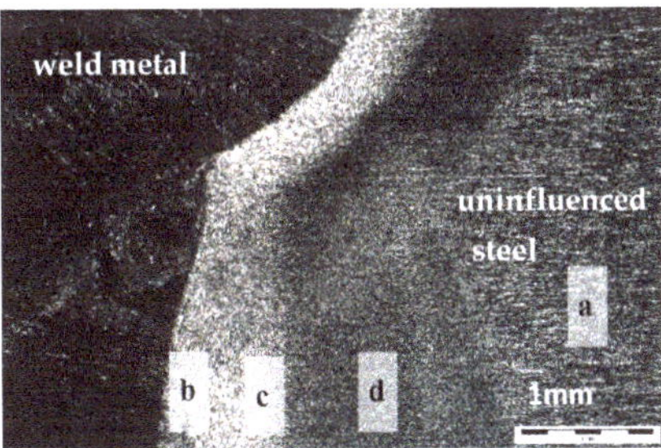

Figure 8. Macroscopic view of the HAZ; the positions of detailed views in Figure 9.

The ferritic structure of the uninfluenced test steel has shown a partial acicular morphology along with the distribution of very fine carbides (Figure 9a), which contributes to increased strength. Re-austenitisation caused a typical grain coarsening in the so-called superheated welding area adjacent to the fusion zone (Figure 9b). Grain refinement was observed in a substantial part of the HAZ; in addition to the influence on the grain size, the heat input led to the carbide re-precipitation. Significant distribution of relatively coarsened carbides (up to 1 μm thick) was found in the band immediately above the austenitisation temperature (Figure 9c), contrary to the partially austenitization zone (Figure 9d). Much finer and less globularised carbides were formed in the grain-coarsened zone.

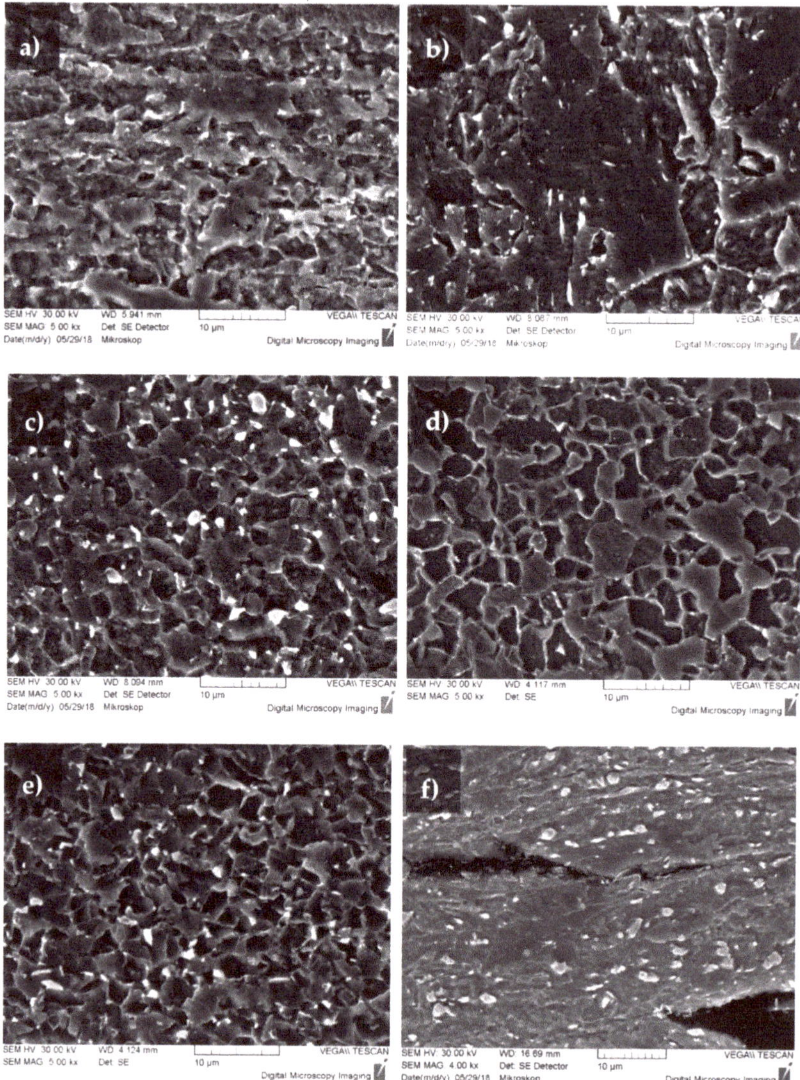

Figure 9. Strenx 700MC structural changes after used welding technology and simulation: (**a**) Uninfluenced steel; (**b**) grain coarsened zone along the fusion zone; (**c**) carbides coarsening in the grain refined zone; (**d**) partially austenitization zone; (**e**) microstructure obtained by simulation; (**f**) microstructure obtained by simulation near the fracture.

After completion of the welding influence simulation, a sample for the metallographic examination was prepared from the circumferential notched round bar specimen. Comparative structural analyses were conducted directly in the area of the notch tip, i.e., in the area of crack propagation within the fracture toughness examination. The carbide reprecipitation process was found to be the most effective degradation process by the metallography evaluation of the real welding influence. Due to that, the comparison of dissolution and reprecipitation of carbides was also an important validation effect of the performed simulation process. As shown in Figure 9e, coarsening of the primary carbides was induced by the simulation process. This observation confirms accordance with the critical structural influence of real welding processes. The initiation of microcracks along the coarsened carbides during mechanical testing was also visible (Figure 9f).

3.3. Fracture Response Evaluation

Chevron-notched sample testing led to unacceptable fracture responses in all tested positions. The plane of the fracture deflected immediately as the tip of the crack reached the carbide row, so the real distribution of carbides has de facto driven the following fracture plane and thus influenced all the measured results, namely the maximal force and total energy consumption. Figure 10a presents the influence of the position of crack deflection on the fracture resistance of the Strenx 700MC steel. Furthermore, the tendency to ductile fracture is evident from the detail in Figure 10b, i.e., plane deformation condition, was not achieved.

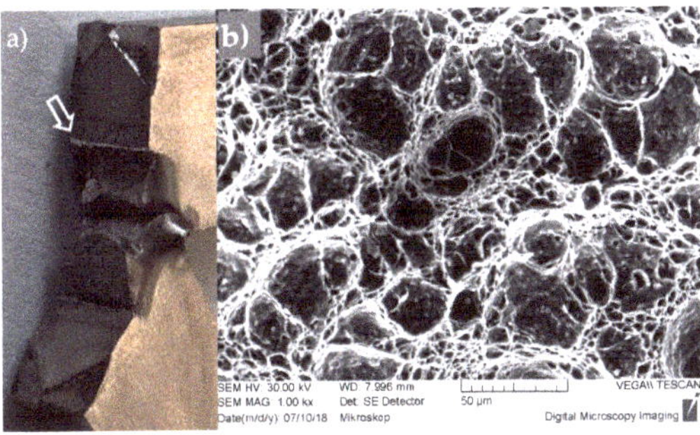

Figure 10. Fractured chevron notched samples: (**a**) Crack deflection due to carbide precipitation; (**b**) ductile fracture mode.

The results of static versus dynamic fracture toughness of the parent material, determined using the CCRB method, are listed in Table 4 compared to the S355NJ steel. Dynamic values obtained at the maximal loading rate are significantly higher compared to the static values, so a strengthening process has been involved. It points to the presence of an effective plastic zone on the tip of the moving fracture, which enables the dislocation hardening process. An increased scattering of dynamic strength and fracture toughness values was obtained at the maximal loading rate. The heat input of simulated welding caused significant change of static and also dynamic fracture responses; the fracture toughness values are listed in Table 5.

An uneven circumference fatigue crack caused by a stripe of carbide precipitation was typical for all loading conditions. The lamellar final fracture morphology (Figure 11) was observed and evaluated in both the macro and microscale to assess an increased loading rate influence on fracture behaviour. Figure 11a compared to Figure 11b presents the macroscopic appearance of the fracture surface after its static compared to dynamic loads. In both cases, the influence of the lamellar precipitation of very fine carbides is dominant, with the higher occurrence of secondary cracks typical of dynamic loading, which partly led to the fracture relief refinement. The simulation of the critical welding influence has led to the fundamental change in the morphology of the fracture surface. The lamellar character of the fracture was completely suppressed (Figure 11c), which corresponds to the observed structural effect, i.e., carbide reprecipitation. It also proves the complete removal of dislocation reinforcement after thermomechanical treatment.

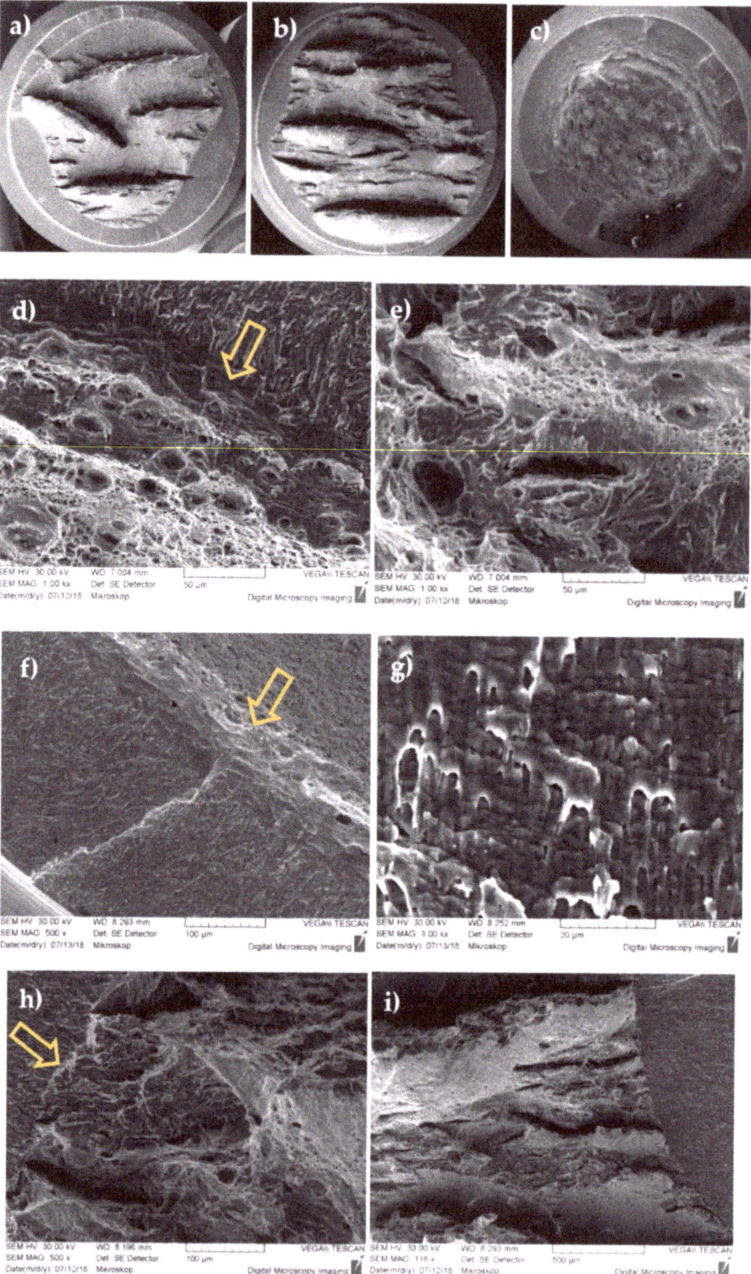

Figure 11. Fracture morphology of tested samples, (**a**) static—uninfluenced steel; (**b**) dynamic—uninfluenced steel; (**c**) dynamic—HAZ; (**d,e**) static loaded uninfluenced steel; (**f,g**) uninfluenced steel loaded by impact velocity 3.48 m/s; (**h,i**) uninfluenced steel loaded by impact velocity 5.23 m/s.

Table 4. Fracture response results of the Strenx 700MC at different loading rate—parent material.

Loading Rate	Sample No.	F_{max} (N)	D (mm)	a_m (mm)	a_f (mm)	d_{eff} (mm)	d_{eff}/D	K_I (MPa·m$^{1/2}$)
Static 0.02 m/s	S1	28,744	7.7	0.7	0.64	5.02	0.65	58.2
	S2	31,953	7.65	0.695	0.47	5.32	0.70	57.5
	S3	32,243	7.61	0.69	0.38	5.47	0.72	54.5
	Average value							57
	Average value—S355NJ							45
Dynamic 3.48 m/s	D-01	43,676	7.62	0.69	0.2	5.84	0.77	64
	D-02	41,164	8.11	0.75	0.42	5.77	0.71	64.7
	D-03	44,534	8.11	0.73	0.25	6.15	0.76	60.9
	Average value							63
	Average value—S355NJ							67
Dynamic 5.23 m/s	D-1	41,484	7.59	0.695	0.3216	5.56	0.73	67.7
	D-2	40,534	7.78	0.66	0.38	57	0.73	63.7
	D-3	43,833	6.95	0.535	0.211	5.46	0.79	69.6
	D-4	38,764	7.84	0.725	0.4865	5.42	0.69	68.1
	D-5	43,510	7.8	0.72	0.49	5.38	0.69	77.3
	Average value							69
	Average value—S355NJ							69

Table 5. Strenx fracture response at different loading rate—simulated heat affected zone (HAZ).

Loading Rate	Sample No.	F_{max} (N)	D (mm)	a_m (mm)	a_f (mm)	d_{eff} (mm)	K_I (MPa.m$^{1/2}$)	d_{eff}/D	UTS (MPa)
Static 0.02 m/s	WS-01	34,183	7.67	0.735	0.21	5.78	51.5	0.75	1303
	WS-02	24,784	6.98	0.74	0.35	4.8	52.3	0.69	1370
	WS-03	24,972	7.26	0.73	0.43	4.94	50.8	0.68	1303
	Average value						51.5	0.70	1325
Dynamic 5.23 m/s	DWS-01	36,057	7.22	0.83	0.45	4.66	82	0.65	2114
	DWS-02	22,769	7.74	0.805	1.68	2.77	118.2	0.36	3778
	DWS-03	32,839	7.2	0.575	0.67	4.71	73.1	0.65	1885
	Average value						91.1	0.55	2592

Our microscopic images document the effect of various load rates, i.e., deformation, on the mechanism of crack development in the thermally unaffected steel. The fracture mode in the initial phase of destruction, i.e., at the interphase of the fatigue fracture, was decisive for the further propagation of fractures and, thereby, for the total energy consumption. In the case of static loads (Figure 11d,e), we documented the onset of the ductile fracture, as determined over the entire fatigue crack circumference. When increasing the load speed to 3.48 m/s, a change in the orientation of microscopic shear bridges was observed. This effect may be attributed to the reduced opening of the V-notch prior to the crack development due to the increased load rate. Again, a ductile fracture mechanism was identified around the entire fatigue crack circumference. We have observed the quasi-cleavage fracture mode (Figure 11h,i) at the maximum tested load rate (5.23 m/s) on the fatigue crack face. The area of a ductile fracture mode, corresponding to the stable crack propagation, reached a decreasing area with increasing loading rate. The transition to a final unstable fracture was more affected by the carbide phase precipitation.

4. Discussion

Experimental welding revealed the substantial microstructural changes due to the applied metal active gas welding method. The hardness measurement indicated the lowest value in the sublayer at the rim of the grain refinement zone towards a lowered temperature (236 HV1). This result seemed to contradict the generally known grain refining effect. Based on the detailed microstructure evaluation, it could be concluded that it was a result of a partial austenitisation process, together with the loss of both primary-strengthening processes—dislocation hardening and precipitation hardening. The latter process was more effective, hence the reprecipitation of carbides leading to a substantial decrease in hardness. The minimal hardness level was an important parameter for the performed temperature cycle simulation (for the chosen maximal temperature). The region of grain coarsening was very narrow.

Microadditives of titanium, aluminium, and vanadium in this type of microalloyed steel strongly influenced the grain growth, recrystallization of austenite, and phase transformation as well as morphology of the transformed products. The most common effect of grain coarsening, i.e., the decreasing of hardness, was overcome by the partial transformation to acicular ferrite and bainite. The presence of nitrogen carbides, revealed in the HAZ by chemical microanalysis, indicated that there are sufficient levels of titanium needed to bind to free nitrogen, which is related to reduced aging processes, in the investigated steel.

4.1. Fracture Behaviour under Welding Influence, Specific Conditions for Methodology of Fracture Toughness Evaluation

The standardized pre-cracked Charpy-type test according to ISO 26843-2015 was used in the first step of testing. The distinctive row of carbide precipitation in the Strenx 700MC steel led to crack deflection, and de facto thereby caused the evaluation of the tests to be impossible. So the first step of experimental analyses underlined the need for non-standard tests. Uniaxial tests using round bar pre-cracked samples allowed the comparative static and dynamic fracture behaviour evaluation. Two main points—the effect of carbide reprecipitation and the sensitivity of steel to the strain rate—were found to be crucial parameters for the prospective application. Fracture toughness values, determined at the dynamic loading of up to 5.23 m/s, differ from values measured at quasi-static loading rates. Principally, it applies (and it is reported, e.g., in ISO 26843), that an increase in loading rate causes a decrease in fracture toughness when tests are performed in brittle or ductile-to-brittle fracture regimes; an increase in fracture toughness is observed in the fully ductile regime.

This generally acknowledged relation has been confirmed. The strengthening of the parent Strenx 700MC steel and the simulated critical sublayer of HAZ ratios under different loading rates is displayed in Figure 12. A substantial increase in sensitivity to the rate of loading was observed in HAZ. This can be explained by the reprecipitation of carbides; microstructure homogenization led to a more intensive dislocation hardening effect.

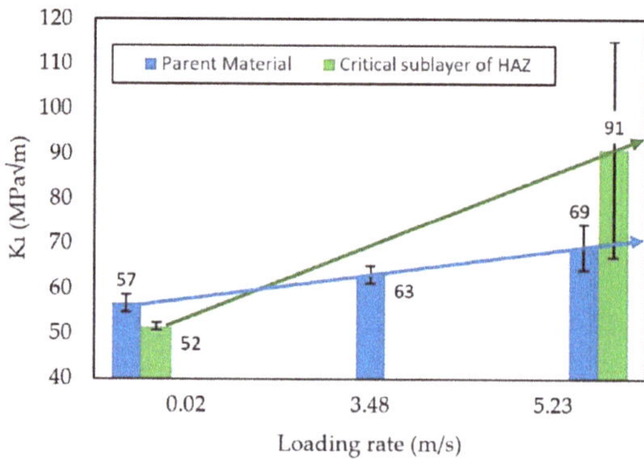

Figure 12. Fracture toughness of Strenx 700MC at different loading rate.

4.2. Effect of Structural Heterogenity

The observed fracture response revealed significant influence on the primary structural heterogeneity of the Strenx 700MC steel. Lamellar-like tearing was observed in the uninfluenced area; hence the obtained values of energy were over-valuated due to the branching crack propagation. Complex fracture patterns appeared in all tested samples. The simulation of the welding degradation effect caused substantial changes in the fracture response. A slight decrease in fracture toughness at the static loading was observed together with more intensive dynamic strengthening. This can be explained by the induced contradictional structural effects in the HAZ—carbide reprecipitation as a crucial process towards strength impairment versus microstructural homogenization (by suppressing the row of carbide distribution) leading to increased susceptibility to deformation hardening. The performed experimental analysis of the static and impact crack behavior of simulated welded joints showed the real level of degradation caused by welding. The homogenization effect in the HAZ is capable of suppressing the primary strip-like carbide precipitation.

The fracture behavior of the Strenx 700MC steel in the particular fracture stages is reflected in the force-displacement diagrams. Representative records at different loading rates of the uninfluenced steel are displayed in Figure 13. A substantially different displacement of the final fracture presents the above-discussed effect of crack deflection due to the lamellar carbide precipitation. The generally acknowledged approach to fracture toughness evaluation using round-notched samples does not reflect the significant energy consumption differences in the case of similar maximal force to fracture. In this specific case, however, there is no continuous development of macroplastic deformation. The decisive factor is that, as opposed to the fracture toughness test using the Charpy-type samples, the evaluation regime used here leads to a magistrate crack in the final plane and thus de facto allows for evaluation.

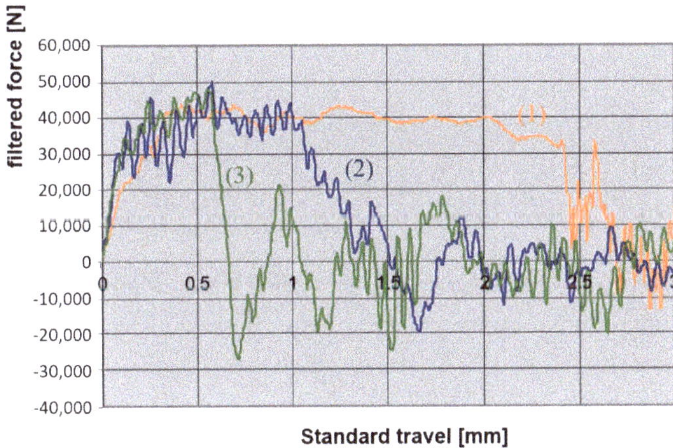

Figure 13. Typical fracture response at different loading rate ((**1**): 0.02 m/s, (**2**): 3.48 m/s, (**3**): 5.28 m/s).

5. Conclusions

The comparative experimental analyses revealed the specifics of fracture behavior in the Strenx 700MC steel. The fracture toughness, determined by CCRB samples under the static loading conditions, proved the higher values for the Strenx 700MC steel as compared to the S355NJ steel. The opposite ratio was observed when increasing the loading rate. The surprisingly lower dynamic fracture toughness of the Strenx 700MC steel compared to the S355NJ steel was observed despite the superiority of a brittle fracture mode in the S355NJ steel.

Increased fracture toughness due to an increased loading rate for both tested steels confirmed the decisive role of a plastic zone on top of the propagated crack. Intensive dynamic strengthening took place in the Strenx 700MC steel in the case of primary microstructure conditions as well as after the simulated degradation due to welding. The final dynamic fracture resistance of the temper-influenced zone even overcame the fracture resistance of the parent steel.

The critical sublayer of HAZ was defined as a maximal softening zone; the induced phase transformation and change of the structural components' morphology were evaluated in direct connection with the fracture response in this zone. Contrary to the common limited effect of the grain-coarsening zone adjacent to the fusion line, intensive reprecipitation was induced near the partial austenitisation zone.

The following advantages of the CCRB methodology have been confirmed:

- A uniaxial loading system as the optimal mode for monitoring all the circumstances of crack development;
- The ability to apply different loading rates, e.g., using the standard Charpy hammer system up to 5.23 m/s;
- Suppression of the plane stress state, typically influencing the crack propagation near the surface of the simple notched Charpy-type samples;
- Support for the plane orientation of the fracture due to the circumferential initiation of fatigue cracks.

The methodology used in this testing is especially effective for materials with heterogeneous microstructures and thereby with heterogeneous local mechanical parameters. Circumferential fatigue pre-cracking suppresses the fracture deflection towards weaker microvolumes and thus supports planar crack propagation. The CCRB methodology supports the plane strain condition; a transition of fracture to the main shear stress plane is suppressed significantly compared to Charpy-type samples.

Author Contributions: Conceptualization, E.S. and B.C.; Methodology, E.S., F.B. and B.C.; Investigation, E.S., F.B., S.K., M.U. and L.K.; Resources, E.S., and F.B.; Data curation, E.S., F.B. and S.K.; Writing—original draft preparation, E.S. and F.B.; Writing—review and editing, E.S. and F.B.; Visualization, E.S. and F.B.

Funding: This research was funded by the Railway Vehicle Competence Centre, project No.TE01020038.

Conflicts of Interest: The authors declare no conflict of interest.

References

1. Ulewicz, R.; Mazur, M.; Bokůvka, O. Structure and mechanical properties of fine-grained steels. *Periodica Polytech. Transp. Eng.* **2013**, *41*, 111–115. [CrossRef]
2. Mazur, M. Fatigue properties of fine-grained steels applied in components of semitrailers. *Czasopismo Techniczne* **2016**, *4-M*, 9–14.
3. Sperle, J.; Hallberg, L.; Larsson, J.; Groth, H. *The Environmental Value of High Strength Steel Structures, Environmental Research Programme for the Swedish Steel Industry, The Steel Eco-Cycle*; Scientific Report; Jernkontoret: Stockholm, Sweden, 2013; pp. 151–171.
4. Ulewicz, R.; Szataniak, P. Fatigue Cracks of Strenx Steel. *Mater. Today Proc.* **2016**, *3*, 1195–1198. [CrossRef]
5. Eckerlid1, J.; Åsell1, M.; Ohlsson1, A. *Use of Vanadium High-Strength Low-Alloy Steels in Trailers, in a Case Study*; The Steel Company Tunnplåt AB: Lulea, Sweden, 2009.
6. Laitila, J.; Larkiola, J.; Porter, D. Effect of forced cooling on the tensile properties and impact toughness of the coarse-grained heat-affected zone of a high-strength structural steel. *Weld. World* **2018**, *62*, 79–85. [CrossRef]
7. Kim, S.; Kang, D.; Kim, T.W.; Lee, J.; Lee, C. Fatigue crack growth behavior of the simulated HAZ of 800 MPa grade high-performance steel. *Mater. Sci. Eng. A* **2011**, *528*, 2331–2338. [CrossRef]
8. Mohyla, P.; Hlavatý, I.; Tomčík, P. Cause of Secondary Hardening in Cr-Mo-V Weld during Long-Term Heat Exposure. *Avtomaticheskaya Svarka* **2011**, *694*, 24–26.
9. Górka, J. Influence of the maximum temperature of the thermal cycle on the properties and structure of the HAZ of steel S700MC. *IOSR J. Eng.* **2013**, *3*, 22–28. [CrossRef]
10. Górka, J.; Stano, S. Microstructure and Properties of Hybrid Laser Arc Welded Joints (Laser Beam-MAG) in Thermo-Mechanical Control Processed S700MC Steel. *Metals* **2018**, *8*, 132. [CrossRef]

11. Shi, Y.; Han, Z. Effect of weld thermal cycle on microstructure and fracture toughness of simulated heat-affected zone for a 800 MPa grade high strength low alloy steel. *J. Mater. Proc. Technol.* **2008**, *207*, 30–39. [CrossRef]

12. Zhu, X.K.; Joyce, J.A. Review of fracture toughness (G, K, J, CTOD, CTOA) testing and standardization. *Eng. Fract. Mech.* **2012**, *85*, 1–46. [CrossRef]

13. Li, S.; He, J.; Gu, B.; Zeng, D.; Xia, Z.C.; Zhao, Y.; Lin, Z. Anisotropic fracture of advanced high strength steel sheets: Experiment and theory. *Int. J. Plast.* **2018**, *103*, 95–118. [CrossRef]

14. Leskov, V. Multi-Functional Kıc-Test Specimen for the Assessment of Different Tool- and High-Speed-Steel Properties. *Mater. Technol.* **2013**, *47*, 273–283.

15. Grant, T.J.; Weber, L.; Mortensen, A. Plasticity in Chevron-notch fracture toughness testing. *Eng. Fract. Mech.* **2000**, *67*, 263–276. [CrossRef]

16. Antolovich, S.D.; Saxena, A.; Gerberich, W.W. Fracture mechanics–An interpretive technical history. *Mech. Res. Commun.* **2018**, *91*, 46–86. [CrossRef]

17. ASTM-E-399 Standard Test Method for Plane-Strain Fracture Toughness of Metallic Materials. E 399-90. 2003. Available online: https://www.astm.org/Standards/E399 (accessed on 27 April 2019).

18. Bayram, A.; Uguz, A.; Durmus, A. Rapid determination of the fracture toughness of metallic materials using circumferentially notched bars. *J. Mater. Eng. Perform.* **2002**, *11*, 571–576. [CrossRef]

19. Londe, N.V.; Jayaraju, T.; Rao, P.R.S. Determination of Plane-Strain Fracture Toughness of AL 2014-T6 Alloy Using Circumferentially Cracked Round Bar Specimen. *Eng. E-Trans.* **2006**, *6*, 26–31.

20. Wilson, C.D.; Landes, J.D. Fracture toughness testing with notched round bars. In *Fatigue Fracture Mechanics: 30th Volume*; ASTM International: West Conshohocken, PA, USA, 2000; pp. 69–82.

21. Londe, N.V.; Jayaraju, T.; Naik, P.; Kumar, D.; Rajashekar, C.R. Determination of fracture toughness and fatigue crack growth rate using circumferentially cracked round bar specimens of Al2014T651. *Aerosp. Sci. Technol.* **2015**, *47*, 92–97.

22. Li, D.M.; Bakker, A. Fracture toughness evaluation using circumferentially-cracked cylindrical bar specimens. *Eng. Fract. Mech.* **1997**, *57*, 1–11. [CrossRef]

23. Kuruppu, M.D.; Chong, K.P. Fracture toughness testing of brittle materials using semi-circular bend (SCB) specimen. *Eng. Fract. Mech.* **2012**, *91*, 133–150. [CrossRef]

24. Barsom, J.M.; Rolfe, S.T. *Fracture and Fatigue Control in Structure: Applications of Fracture Mechanics*, 2nd ed.; Prentice Hall: Englewood Clifts, NJ, USA, 1987.

25. ISO 26843:2015(E), Metallic Materials—Measurement of Fracture Toughness at Impact Loading Rates Using Precracked Charpy-Type Test Pieces. Available online: https://www.iso.org/standard/65516.html (accessed on 27 April 2019).

26. ASTM Designation E1820 − 17a, Standard Test Method for Measurement of Fracture Toughness. 2018. Available online: https://www.astm.org/Standards/E1820 (accessed on 27 April 2019).

 *metals*

Article

Failure Mechanisms of Mechanically and Thermally Produced Holes in High-Strength Low-Alloy Steel Plates Subjected to Fatigue Loading

Carlos Jiménez-Peña [1,]*, Constantinos Goulas [2,3], Johannes Preußner [4] and Dimitri Debruyne [1]

1   Department of Materials Engineering, KU Leuven, Kasteelpark Arenberg 44, 3001 Leuven, Belgium; dimitri.debruyne@kuleuven.be

2   Department of Materials Science and Engineering, Delft University of Technology, Mekelweg 2, 2628CD Delft, The Netherlands; K.Goulas@tudelft.nl

3   Rotterdam Fieldlab Additive Manufacturing (RAMLAB), Scheepsbouwweg 8, 3089JW Rotterdam, The Netherlands

4   Fraunhofer Institute for Mechanics of Materials IWM, Wöhlerstr. 11, 79108 Freiburg, Germany; johannes.preussner@iwm.fraunhofer.de

*   Correspondence: carlos.jimenezpena@kuleuven.be; Tel.: +32-9265-86-10

Received: 11 February 2020; Accepted: 25 February 2020; Published: 28 February 2020

Abstract: High-strength low-alloy steels (HSLA) are gaining popularity in structural applications in which weight reduction is of interest, such as heavy duty machinery, bridges, and offshore structures. Since the fatigue behavior of welds appears to be almost independent of the base material and displays little improvement when more resistant steel grades are employed, the use of bolted joints is an alternative joining technique which can lead to an increased fatigue performance of HSLA connections. Manufacturing a hole to allocate the fastener elements is an unavoidable step in bolted elements and it might induce defects and tensile residual stresses that could affect its fatigue behavior. This paper studies and compares several mechanical (punching, drilling, and waterjet-cut) and thermal (plasma and laser-cut) hole-making procedures in HSLA structural plates. A series of 63 uniaxial fatigue tests was completed, covering three HSLA grades produced by thermomechanically controlled process (TMCP) with yield strength ranging from 500 to 960 MPa. Samples were tested at single load level, which was considered representative in HSLA typical applications, according to the input received from end users. The manufactured holes were examined by means of optical and electron microscopy, 3D point measurement, micro hardness tests, X-ray diffraction, and electron backscatter diffraction (EBSD). The results give insight on cutting processes in HSLA and indicate how the fatigue failure is dominated by macro defects rather than by the steel grade. It was shown that the higher yield strength of the HSLA grades did not lead to a higher fatigue life. Best fatigue results were achieved with laser-cut specimens while punched samples withstood the lowest amount of cycles.

Keywords: high-strength low-alloy; hole manufacturing; fatigue; drilling; punching; waterjet-cutting; plasma-cutting; laser-cutting

1. Introduction

The field of application for high-strength low-alloy (HSLA) steels ranges from heavy duty equipment to offshore and civil engineering applications [1–6]. The greater yield strength of higher strength structural steels allows slender structural designs and lower self-weight loads, which result in economic and environmental benefits. However, structural steel grades with extremely high yield strength (up to 1100 MPa) are generally associated with a large concentration of alloying elements and they are traditionally produced by quenching and tempering (QT) [7]. These HSLA grades have a high hardenability, which may lead to potential brittle fracture and, when used in welds, hydrogen-induced fracture [8]. Alternatively, HSLA grades produced by thermomechanically controlled process (TMCP) are considered a good compromise between mechanical properties and fatigue resistance. TMCP allows us to produce a wide range of HSLA grades with diverse microstructures and material properties which are often adopted in civil and mechanical structures. The applied rolling scheme is individually designed depending on the chemical composition, the thickness, and further parameters, which result in multiple TMCP HSLA microstructures types [9]. The leaner steel composition results in improved weldability compared to QT HSLA grades. The fatigue performance of welded HSLA, however, seems to be practically independent from the base material and it displays little improvement when more resistant steel grades are employed compared with milder steel grades [10]. In order to tackle this issue, special post-weld treatments are to be carried out [11].

The use of bolted joints over welded joints in HSLA structures is contemplated as an option to avoid the problems associated with welding. While microstructural changes or temperature-induced stresses are not an issue in bolted connections, specific parameters such as the bolt preload [12] or the hole-making process [13] play a critical role in the fatigue performance of HSLA structures and might influence additional failure modes such as fretting wear or fatigue associated with the relative displacement between bolted plates. Additional damaging phenomena such as wear or fretting fatigue due to the small relative displacement between two bolted elements might pose. An essential requirement in bolted connections is the presence of a hole to allocate the fastener elements and, therefore, the necessity of a cutting process to produce this hole. Cutting procedures are part of almost every steel product manufacturing process, with numerous cutting techniques being available, depending on application requirements such as the cut tolerance, edge quality, economical aspects, and so on. The cut-edges often form an integral part of the structure and sustain high stresses. In applications under cyclic loading, and due to the fact that fatigue is strongly influenced by the surface, the characteristics of the cut-edge play a crucial role in the life of the mechanical component. Mechanical cutting processes can induce cold-work hardening into the material [14] or surface defects due to cracking or material removal [13] while thermal cutting processes also induce a heat-affected zone (HAZ) surrounding the cut-edge [15]. The study of the quality and performance of HSLA as-cut edges has gained increased interest in recent industrial research. Two European-funded researches have studied the fatigue performance of thermally and mechanically cut edges in moderately thick steels with yield strength ranging from 355 to 890 MPa. In the Coldfossproject [16], the fatigue performance and hole quality of punched and drilled holes was investigated. It was reported that although mechanical cutting processes could lead to reduction of performance in some cases, this was not a concern in most conventional design situations. In the Hipercut project [17], the influence of laser beam and plasma-arc cutting processes on the edge quality and the process parameters was optimized for best in-service performance. The edge quality was found to be strongly dependent on the microstructure and the thickness of the material, and the process parameters had to be modified accordingly. A previous research by the authors [13] covered the effect of the most common hole-making processes (thermal and mechanical) on the fatigue behavior on the HSLA grade S500MC. The results revealed how different thermal and mechanical cutting processes affected the material, the fatigue performance, and the geometrical accuracy of the tested samples. For instance, drilling was found to produce the most geometrically accurate hole and the smoothest surface finish, while laser and waterjet cutting displayed the best fatigue performance. It was found that aspects like the hole morphology and geometrical distortions were found not to affect the fatigue performance as much as the presence of large individual defects. However, it was not clear how other HSLA grades, with increased resistance and decreased ductility, might behave.

Hole-making procedures are covered by many constructional standards. However, several standards refer only to holes made by punching or drilling [18,19] or include local hardness restrictions that can be excessively limiting for thermal cutting processes [18,20]. These constraints, together with the restrictions imposed already on HSLA with a yield strength exceeding 700 MPa [21], might limit even further the use of HSLA in structural applications. Additional research on hole-making procedures in HSLA grades and their performance under cyclic loading might extend the use of HSLA to further structural applications and reduce manufacturing costs.

This research aimed to extend previous knowledge on the HSLA grade S500MC [13] to higher strength grades, S700MC and S960MC, which are characterized by a delicate multiphase microstructure, containing a significant fraction of metastable phases, like martensite and bainite. These phases typically exhibit high strength, but they can also be brittle under high-strain rates. Thus, the presence of these phases is expected to affect the response of the material during the mechanical hole-making processes. Furthermore, the grades S700MC and S960MC contain an increased concentration of alloying elements for achieving the required hardenability. The complex microstructure of S700MC and S960MC, being metastable, is prone to softening or transformation under the influence of heat. A series of fatigue tests were performed to S700MC and S960MC samples with holes produced by the most commonly used mechanical and thermal hole-making procedures. The nature of the manufactured holes will be evaluated by means of optical and electron microscopy, topographic measurements, and hardness tests. Additional electron backscatter diffraction (EBSD) and residual stress measurements were performed on the S500MC and S960MC grades. The fatigue behavior and hole morphology of these steel grades will be compared with those obtained by the authors on the HSLA grade S500MC [13].

2. Materials and Methods

2.1. Material

Two widely used commercial TMCP HSLA grades were considered in this investigation, S700MC and S960MC, supplied in rolled plates of 1500 mm × 1000 mm × 5 mm. The steel grades are thermomechanically rolled (M) structural steels (S) with a specified minimum yield strength at ambient temperature of 700 and 960 MPa, respectively. These HSLA grades were compared with the HSLA grade S500MC which was evaluated in [13]. Digital image correlation was applied to extract the averaged strain field in the central region of dog-bone samples tested under uniaxial loading. The employed procedure is partly described in [22]. The chemical composition of the steels was provided by the manufacturers [23,24] and it is summarized in Table 1 together with the measured mechanical properties.

Table 1. Chemical composition and mechanical properties of HSLA grades.

Chemical Composition (wt.%)										
Grade	C	Si	Mn	P	S	Al	Nb	V	Ti	Mo	B
S500MC	≤0.120	≤0.500	≤1.700	≤0.025	≤0.015	0.015	≤0.090	≤0.200	≤0.150	-	-
S700MC	≤0.120	≤0.600	≤2.100	≤0.025	≤0.015	≥0.015	≤0.090	≤0.200	≤0.220	≤0.5	≤0.0050
S960MC	≤0.120	≤0.250	≤1.300	≤0.020	≤0.010	≥0.015	≤0.050	≤0.050	≤0.070	-	-

Mechanical properties				
Grade	Young modulus (GPa)	Yield stress (MPa)	UTS (MPa)	Elongation (%)	Hardness (HV0.2)
S500MC	210	562 ± 6	658 ± 4	13.7 ± 1.2	207 ± 7
S700MC	210	731 ± 3	801 ± 4	11.8 ± 0.6	266 ± 9
S960MC	210	977 ± 7	1061 ± 8	3.2 ± 0.2	344 ± 6

2.2. Sample Design

The sample design, shown in Figure 1, is similar to the sample design previously employed by the authors in [13]. The sample consists of a dog-bone shaped tensile specimen without fastener elements in order to isolate the effect of the hole-making process from the other parameters governing bolted connections, such as the bolt preload and the washer geometry. The distance from the hole edges was 1.5 times the hole diameter, as specified in the Eurocode standard EC3-1-8 for bolted connections under cyclic loading [25]. The hole diameter produced by every hole-making process was 18 mm, a value which is currently used by industrial partners in their bolted joint applications.

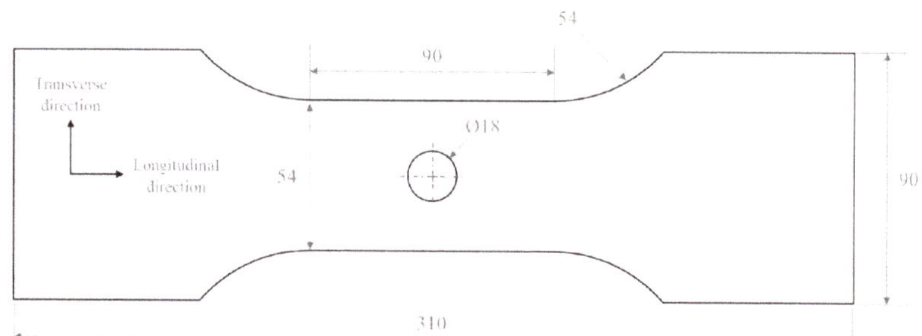

Figure 1. Sample design. Units are specified in mm.

2.3. Microstructural Characterization

Light optical microscopy images were taken with a VHX 5000 Keyence digital microscope (Keyence Corporation, Osaka, Japan) equipped with image analysis software. The characterization with scanning electron microscopy (SEM) was performed using a JEOL JSM-6500F (JEOL USA, Inc., Peabody, MA, USA) operated at 15 kV. A standard metallographic preparation procedure was followed and etching was performed with Nital 2% (98% Ethanol and 2% HNO3). The fracture surfaces were immersed with acetone in an ultrasonic bath to remove artifacts. The microstructure was further observed with a field emission gun scanning electron microscope (SEM) Zeiss Supra 40VP (Carl Zeiss SMT GmbH, Oberkochen, Germany) with 20 kV accelerating voltage and a secondary electron (SE) detector or the electron backscatter diffraction (EBSD) technique. For the acquisition of EBSD patterns, an aperture of 30 µm was used. The working distance needed to be adapted at each scan according to the sample size and the region of interest. The samples were tilted by 70° towards the EBSD detector, a DigiView 3 camera (AMETEK GmbH, Weiterstadt, Germany), which was operated with the EDAX-TSL-OIM-Data Collection software (EDAX Inc., MahWah, NJ, USA) version 6, was used.

2.4. Hole-Making Procedures

The hole-making procedures employed in this investigation corresponded with the methods employed in the S500MC study [13]: Punching (three conditions), drilling, waterjet cutting, laser cutting, and plasma cutting. Two punch geometries were investigated: A flat punch and a chamfered punch (shown in Figure 2). The orientation of the chamfered punch with respect to the axial loading was also investigated. The geometry of the punch is shown in Figure 2. Some of the hole-making techniques were modified from [13] to adapt to the new material properties and thickness. The optimal process parameters were provided by the industrial partners and are included in Table 2. All cutting processes were evaluated for all three HSLA steel grades.

Punching was executed with an EDEL Stanzomat 407-20 punching press (Edel stanztec gmbh & Co.K, Bietigheim-Bissingen, Germany). Drilling was performed in a Kunzmann WF 7/3-320 CNC machine (Robert Kunzmann GmbH & Co. KG, Tullastraße, Germany) for improved accuracy. The drilling process parameters were given by the drill manufacturer. The drill bit type used was a VDS201F18000 VariDrill (Kennametal Widia Produktions GmbH & Co. KG, Essen, Germany) solid carbide drill [26]. Waterjet-cut holes were produced with an OMAX 55100 Jet Machining Center (Omax Corporation, Kent, WA, USA). Laser cutting was performed with a Trumpf TruLaser 3040 laser cutting machine (Trumpf GmbH + Co. KG, Ditzingen, Germany). Plasma-cut samples were produced with a

ESAB Combirex DX 3500 plasma cutting machine (Elektriska Svetsnings-Aktiebolaget, Gothenburg, Sweden).

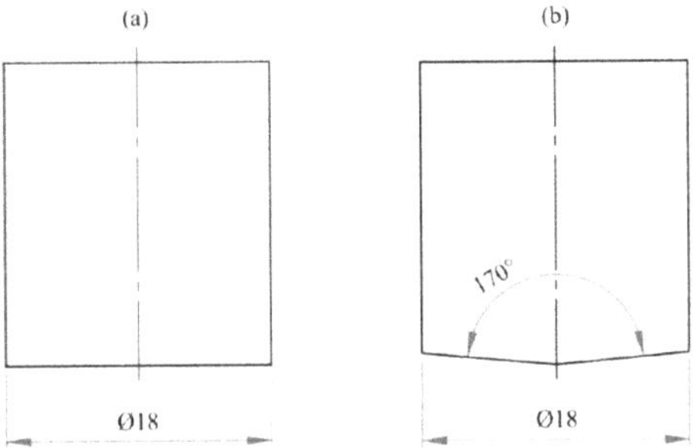

Figure 2. (**a**) Flat punch and (**b**) chamfered punch end geometry (Unit: mm).

Table 2. Cutting process parameters for S700MC and S960MC samples.

Punching			
Max. punching speed		20 m/min	
Cutting clearance		0.8 mm	
Punch type	Flat	Chamfered	Chamfered
Alignment	-	0°	90°
Drilling			
Cutting speed		80 m/min	
Feed		0.3 mm/rev	
Coolant		Flood	
Material Removal rate		108.02 cm^3/min	
Torque at tool		31.92 Nm	
Waterjet			
Normal offset		1.27 mm	
Radial offset		0.32 mm	
High pressure setting		379.2 MPa (55,000 psi)	
Low Pressure setting		137.9 MPa (20,000 psi)	
Mixing tube diameter		0.762 mm	
Jewel diameter		0.3048 mm	
Abrasive flow rate		0.3073 Kg/min	
Laser			
Beam power		4000 W	
Cutting speed		6.5 m/min	
Nozzle diameter		0.8 mm	
Nozzle distance		0.7 mm	
Focus diameter		−1.8 mm	
Plasma			
Current		100 A	
Nozzle		4.1/1.4 (mm)	
Speed		2159 mm/min	
Torche standoff		4 mm	
Height control		Not Active	

2.5. Hole Geometry Evaluation

The overall shape of the hole was investigated by probing the hole surface at different depths with a coordinate measurement apparatus. A Mitutoyo BX303 Manual Coordinate Measuring Machine (Mitutoyo Corporation, Kawasaki, Kanagawa, Japan) was used to measure the diameter of the hole along the thickness of the sample. The 12 point measurements were obtained at every millimeter of the hole depth. The hole diameter with respect to the hole depth was then calculated using a Gaussian fit.

2.6. Surface Profiling

Surface profiles of the different hole surfaces were obtained by means of laser scanning microscopy. The large measured area, compared to other tactile measurements techniques, provides a general overview of the topography of each hole type. A Keyence VK-9700 Laser Confocal microscope (Keyence Corporation, Osaka, Japan) was used to perform the measurements. A section of 2.5 mm by 0.5 mm was examined per hole, with an average amount of 2.5 million data points. The point clouds were filtered to eliminate outliers and measurement errors.

2.7. Micro Hardness Mapping

Hardness measurements were performed alongside the hole cross-section in order to evaluate the hardening induced by each hole-making process, either due to cold-work hardening or due to microstructural changes. The measurements covered a region of 5 mm by 5 mm (thickness of the hole) and they were distributed in steps of 500 μm for both directions. The spacing between micro-indentations complied with the Vickers hardness test standard for metallic materials (ISO 6507-1: 2005) [27]. The hardness measurements were performed with a Struers DuraScan G5 automated hardness tester (Struers ApS, Ballerup, Denmark).

2.8. Residual Stresses

Previous research by the authors [13] indicated the absence of hardening in the waterjet-cut and drilled samples. The abrasive cutting nature of these processes results in a very localized surface damage next to the produced hole. This was not the case for punching or laser-cut, where considerable hardening was measured near the hole. Since hardness measurements can be also affected by the presence of residual stresses induced during the hole-making process, a more detailed evaluation of the residual stress analysis was conducted with X-ray diffraction (XRD). By measuring the strain in the crystal lattice and assuming a linear elastic distortion, the residual stresses producing that distortion was calculated [28]. The residual stress analysis was conducted at laser cut and punched specimens at the S500MC and S960MC grades. To account for stress gradients, the residual stresses on the surface of the four samples were measured in a straight line perpendicular to the hole rim with increasing distance to the hole (1.0 mm, 2.0 mm, 3.0 mm, 4.0 mm, 5.0 mm). Because the punched samples had deformed zones directly at the hole rim, the center of the first measurement point was at 1.0 mm distance to the hole rims to achieve an equivalent measurement for both production processes. To account for singularities and to investigate the reproducibility in the measurements, both sides of the holes (left and right side to the hole) were measured. Both sides of the plates (top and bottom side) were measured to see influences from the entry and exit of the punching/laser cutting process. On some samples a few microns of the surface layer had to be removed using chemical etching to remove oxides and impurities that might affect the results. Measurements were performed on a Stresstech X3000-G3 XRD diffractometer (Stresstech GmbH, Rennerod, Germany) and in both the longitudinal and transverse direction (Figure 1). The residual stresses were determined with respect to the 211 peak of α-Fe ($2\theta_0 = 156.4°$) using Cr-Kα-radiation. The measurement was determined by the $\sin^2\psi$ method, collecting 15 ψ-angles that were equally distributed over a tilt between $-45°$ and $+45°$. The diameter of the X-ray spot was 1 mm at incident beam with an estimated penetration depth of 12 μm. The residual stresses were calculated using a Young's modulus of 211,000 MPa, a Poisson's ration of 0.3, and an absorption coefficient of 89.7 mm^{-1}.

2.9. Fatigue Testing

Fatigue tests were performed with a Zwick HA100 servo-hydraulic load frame (Zwick Roell Group, Ulm, Germany) with a maximal axial load of 100 kN. As in [13], the test stress ratio was set to R = 0.1 and the test frequency to f = 25 Hz. The stress range was selected after an initial experimental study with the objective of achieving an average number of 1 million cycles between all holing conditions and both grades, which was considered representative in typical heavy duty applications. This decision was motivated by the requirements of the industrial end users involved in the present research project. The loading level employed in [13] was insufficient to provoke fracture under 5 million cycles due to the lower fatigue resistance of the base material with respect to the more resistant HSLA grades. The chosen applied nominal stress range was 300 MPa. The run-out number of cycles was set to 5×10^6 fatigue cycles. Three repetitions were performed for each test condition, resulting in a total test number of 42 fatigue tests. In addition, the 21 fatigue samples made of S500MC and tested in [13] by the authors were also more extensively evaluated in this research, as previously mentioned in Sections 3.6 and 3.7, resulting in a total number of 63 specimens covering three HSLA yield strength ranging from 500 to 960 MPa.

3. Results and Discussion

3.1. Microstructural Characterization

The microstructure of the S700MC is shown in Figure 3a and it was formed by ferrite and bainite, together with carbides. The microstructure of the S960MC on the other hand (Figure 3b) was composed of a martensitic matrix, cementite, or M_7C_3 carbides and retained austenite.

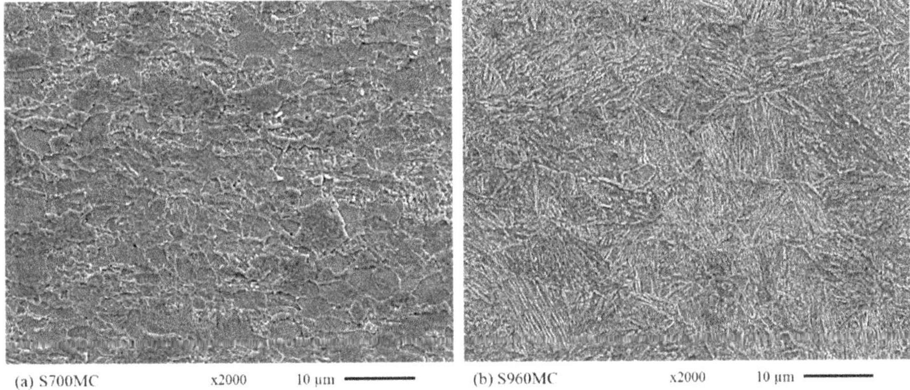

Figure 3. Microstructure micrograph for (**a**) 700MC and (**b**) S960MC.

The microstructures evaluated in this research differed significantly from the ferritic microstructure of the HSLA grade S500MC studied in [13]. The harder and finer martensitic phases of the S700MC and S960MC microstructure, together with the resulting lower formability, affected the hole quality and the fatigue properties of the cut edges.

3.2. Hole Surface Evaluation

The optical analysis of the hole surfaces revealed significant differences between hole-making conditions. In Figure 4, the surface of the punched holes (flat punch) is shown. The shear-cutting section was formed by dimples elongated in the direction of the punch travel. A large amount of shear dimples was observed in the S700MC and S960MC steel grades, in which large sections of the hole surface were removed during punching. This phenomenon is displayed in Figure 4a for the S700MC and it was observed in all punching conditions. The dimples were much less present in the S500MC punched samples [13], where most of the surface was covered by smooth and highly deformed material. During the initial penetration of the punch in hole making process, the S500 formed coarse dimples when the hole began to form. However, the dimples were smeared away during the penetration of the punch. On the cutting surface of S700MC, Figure 4b, and S960MC, Figure 4c, more dimples could be observed. This is because the microstructure in these grades was mostly of martensitic nature, therefore more difficult to smear away due to the increased hardness. The dimples in these cases were smaller, because the original microstructure was finer and contained carbides, which act as nucleation sites for the dimples. This fact makes the dimples more uniformly distributed in the higher strength grades than in the S500MC.

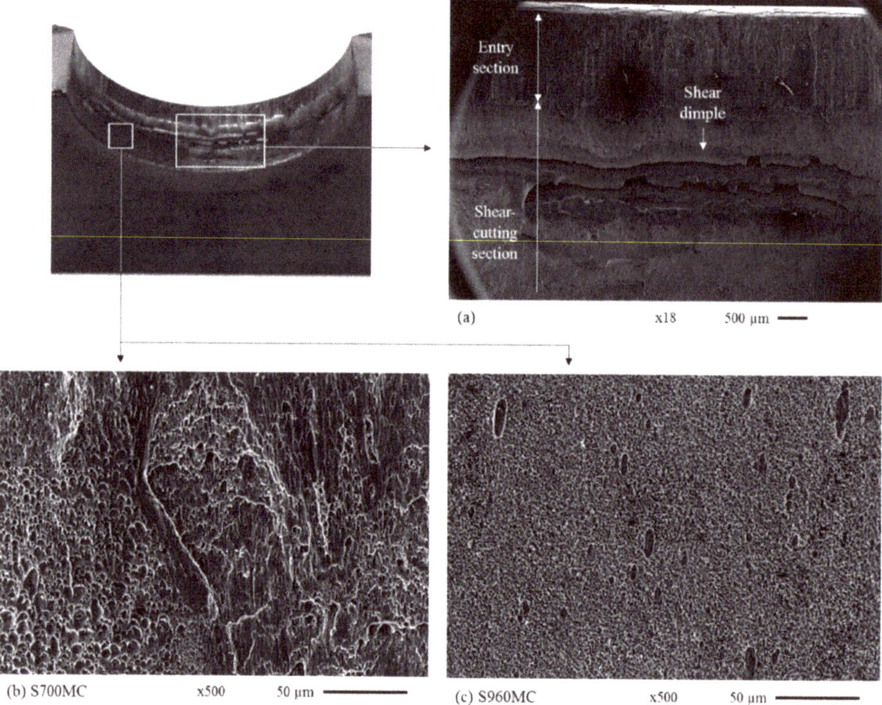

Figure 4. (**a**) SEM overview of entry and shear-cutting sections in punched holes. Below, a detailed view of the exit side is shown, where differences between (**b**) S700MC and (**c**) S960MC are evident.

The remaining hole-making techniques produced hole surfaces with similar features, independently of the steel grade. Drilling always resulted in a smooth, feature-less surface, with no visible characteristics apart from the drill-traces in the direction of the drill rotation, noted in Figure 5a. Drilling also produced a characteristic drill-chip, which is normally removed manually with a deburring tool. The chip was left intact, as in the previous research by the authors [13]. Waterjet-cutting produced highly abraded surfaces, with multiple dents at the locations in which the material was removed by the abrasive jet. A characteristic waterjet-cut surface is shown in Figure 5b. Both laser-cut (Figure 5c) and plasma-cut (Figure 5d) surfaces were covered by a regular crack pattern. Cracks were very superficial and likely to be formed on a thin surface layer covering the hole surface. Plasma-cut surfaces additionally displayed traces in the direction of the hole, which indicate that the material at that location was extensively melted and resolidified. The location in which the plasma first entered the material was also characterized by the presence of a large blob of solidified material. In order to avoid any effect caused by this distinct surface condition, the plasma-cutting process was adjusted so the entry point of the plasma jet was not located at the expected fracture location (midsection of the hole) but at an angle of 45 degrees with respect of the axial loading. Similar observations for the drilled, waterjet-cut, laser-cut, and plasma-cut holes were made for the HSLA grade S500MC in [13], hinting at the independence of the hole surface from the microstructure of the steel. In drilling and waterjet cutting, the surface was highly abraded and no distinct features were distinguishable between steel grades. In thermal cutting processes the surfaces were melt and resolidified into a new martensitic microstructure, which was also similar in all grades.

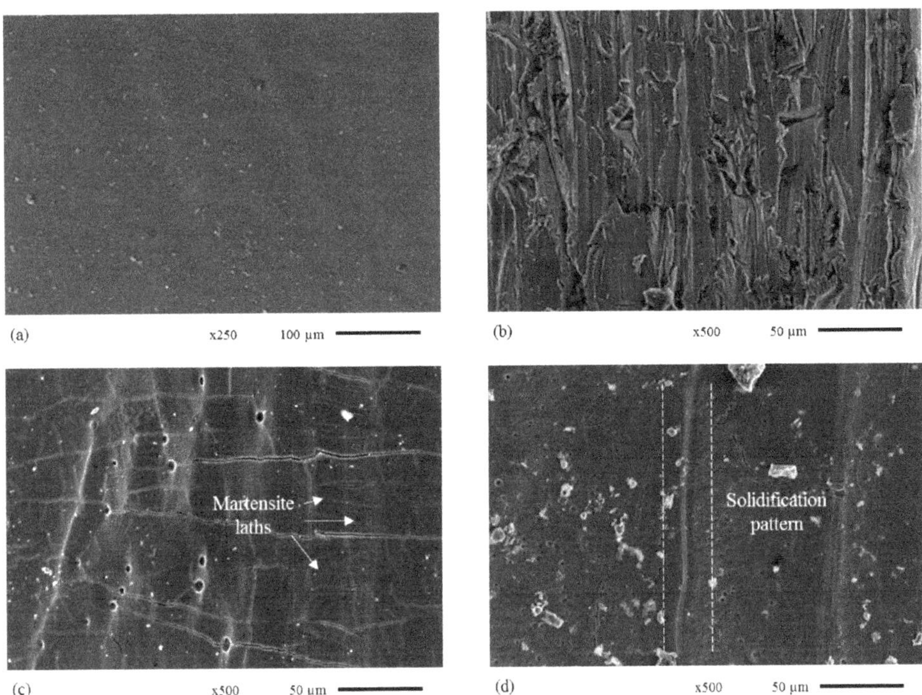

Figure 5. Hole surfaces produced by (**a**) drilling, (**b**) waterjet cutting, (**c**) laser cutting, and (**d**) plasma cutting.

Similarly to [13], protruding acicular structures were observed in the laser cut of all HSLA grades, indicating the presence of martensite formation. These structures were also found in the plasma-cut surfaces but not as clearly and in the quantity observed in the laser-cut samples. The cross-section of the thermally cut specimens was further studied to evaluate the layer of transformed material. Cross-sections of the laser- (above) and plasma- (below) cut holes are displayed in Figure 6. The analysis of the heat-affected zone (HAZ) revealed that there is a microstructurally affected layer surrounding the hole. This layer was characterized by martensitic microstructure. The dominant presence of martensite was a result of the large heat input from the cutting process and the subsequent fast cooling. The extension of the HAZ in the plasma-cut specimens was approximately three times larger than in the laser-cut specimens, in similarity with the S500MC grade [13].

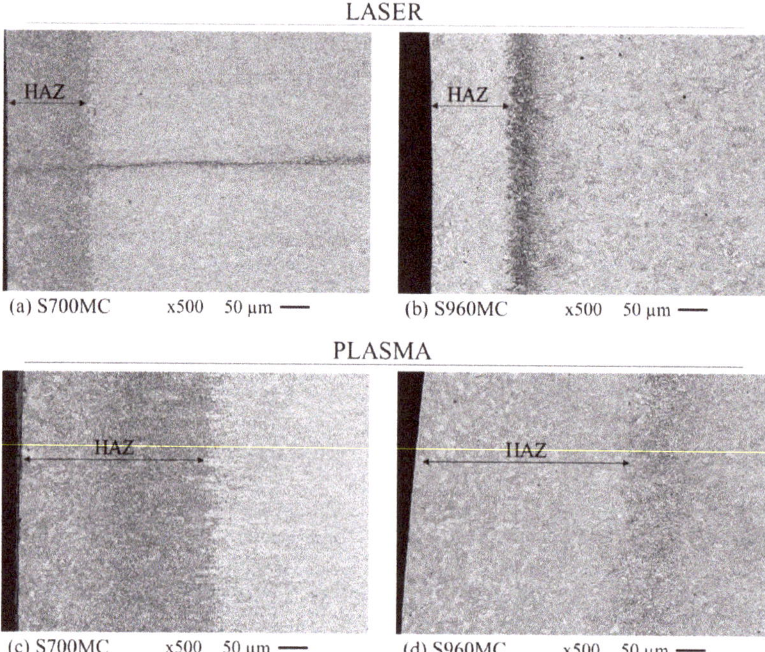

Figure 6. Cross-section of thermal-cut holes: (**a,b**) Laser and (**c,d**) plasma.

The average width of the HAZ was measured and it is displayed for all three HSLA grades in Figure 7. It can be noted that the difference between plasma- and laser-cut HAZ widths was bigger for S700MC and the S960MC and that the HAZ in both thermal processes tended to be wider with increasing material strength.

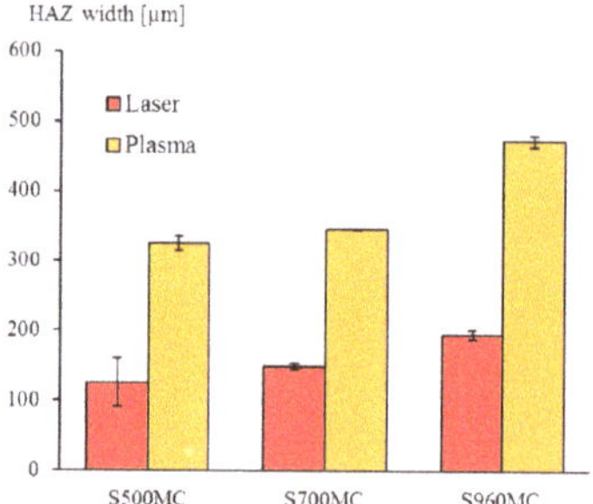

Figure 7. Average HAZ width in laser- and plasma-cut holes for S500MC, S700MC, and S960MC.

Additionally, the cross-sectional analysis revealed the presence of centerline segregation indicated in Figure 6a, in the all three HSLA grades. This is related to the continuous casting production process. The centerline segregation can lead to local embrittlement [29].

3.3. Surface Topology Evaluation

The surface profiles obtained by laser microscopy allowed to observe the predominant features that some of the hole-making techniques had in certain directions. A wavy pattern can be observed in laser-cut holes. This striation pattern is produced by the laser pulsations during the cutting process.

This waviness was not observed in the plasma-cut surface, which appeared flat, but was also covered by striations or ridges. These ridges were observed to be long and straight in the S500MC sample, see Figure 8b, and short and interconnected in the S960MC sample. They related with the solidification patterns previously discussed in Section 3.2. The analysis of the punched samples indicated that roughness differed locally in sections with a large amount of removed material (shear dimples). A distinct step was found in the punched S960MC sample (Figure 8c), which indicated that the morphology of the hole is largely dependent on the presence of shear dimples. Finally, the waterjet-cut samples were found to have a granular surface topology caused by the abrasive cutting process, as shown in Figure 8d.

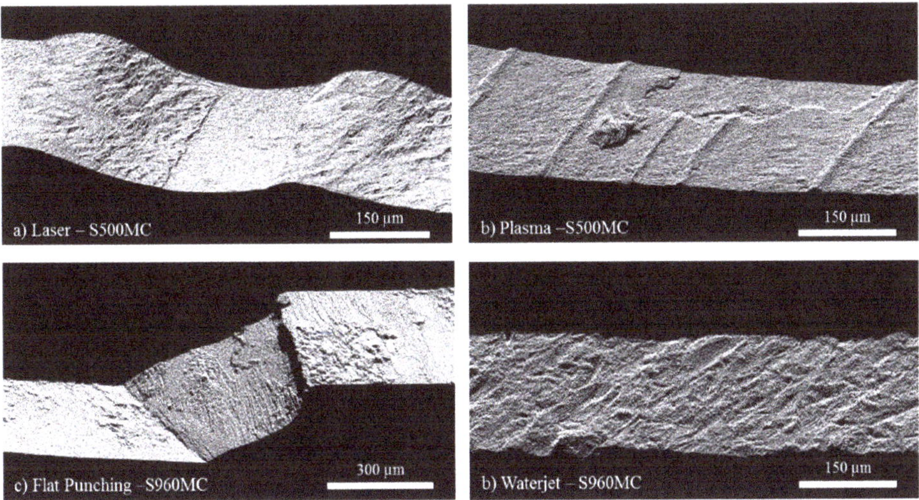

Figure 8. Surface profile of (**a**) laser-cut hole (S500MC), (**b**) plasma-cut hole (S500MC), (**c**) flat punching (S960MC), and (**d**) waterjet-cut hole.

3.4. Hole Geometry Evaluation

The averaged hole diameter for the S700MC and the S960MC steel grades is displayed in Figure 9. It is noted that the hole geometry was rather independent of the steel grade since both materials exhibited similar hole geometries for the same hole-making procedures. Drilling produced the most accurate hole geometry, with a constant hole diameter along the hole depth. Following, laser-cut resulted in a more deviated and undulating hole, with a less constant hole diameter. Waterjet-cutting displayed a tendency to produce an oversize hole diameter at the entry, which gradually shifted to a smaller hole diameter at the exit side of the hole. The waterjet-cut hole deviation did not exceed 0.2 mm in any case. Plasma-cut and punching resulted both in large deviations from the hole diameter. In the case of punched samples, an accurate hole diameter was produced at the hole start. However, at a depth of approximately 2 mm, the hole started to widen due to the shear cutting of the material. The hole produced by plasma-cutting had the most distortions, with a large oversize at the entry and an undersized hole at the exit.

The variation of the obtained results between repetitions was generally low (below 0.1 mm for most cutting procedures) and it has not been included in the graph to improve its clarity. Punched S960MC samples, however, displayed a substantial maximum scatter (0.203 mm) which was approximately eight times larger than punched S700MC and S500MC [13] samples. The higher hardness of this steel might make the forming process more difficult and, hence, produce less constant results when using a punching process. The rest of cutting techniques produced geometrically similar holes in all grades.

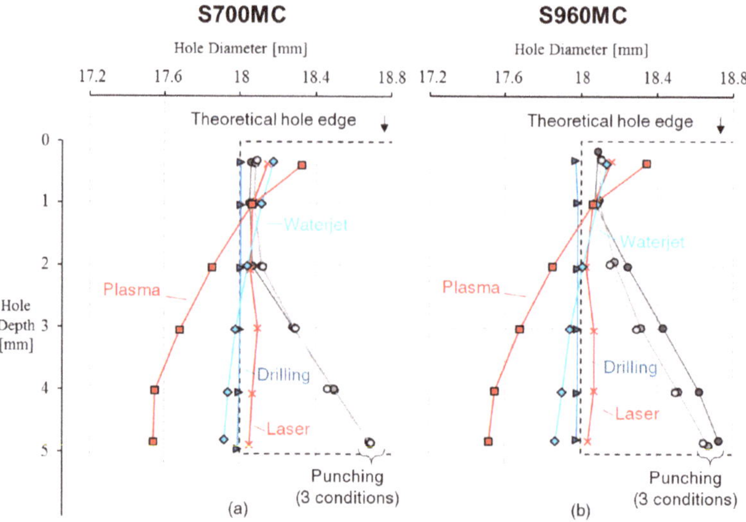

Figure 9. Hole diameter along sample thickness for waterjet cutting, laser cutting, drilling, and punching in (**a**) S700MC and (**b**) S960MC.

3.5. Micro-Hardness Evaluation

The produced hardening maps for the S700MC (Figure 10) and S960MC (Figure 11) show a clear increase in the material hardness near the edge for punching and the thermal cutting processes, similarly to the S500MC in [13].

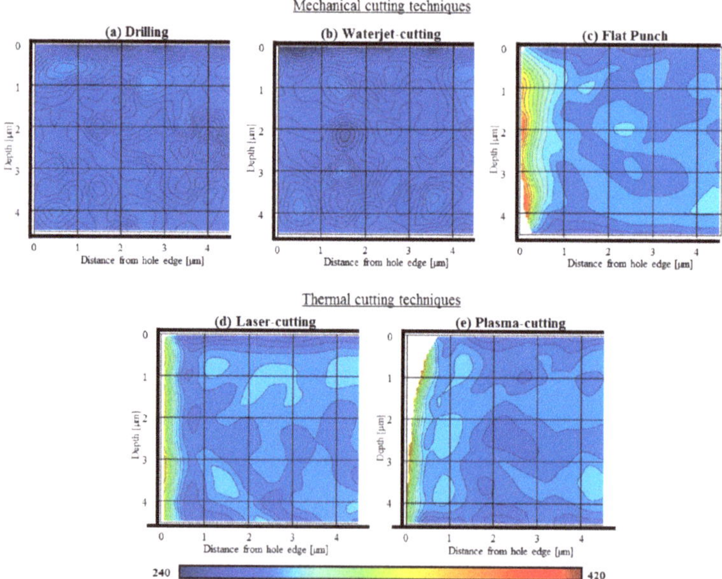

Figure 10. Hardness distribution near the hole edge in HSLA grade S700MC for various hole-making procedures: (**a**) drilling, (**b**) waterjet-cutting, (**c**) flat punching, (**d**) laser-cutting and (**e**) plasma-cutting. The hole edge is located on the left side of the images.

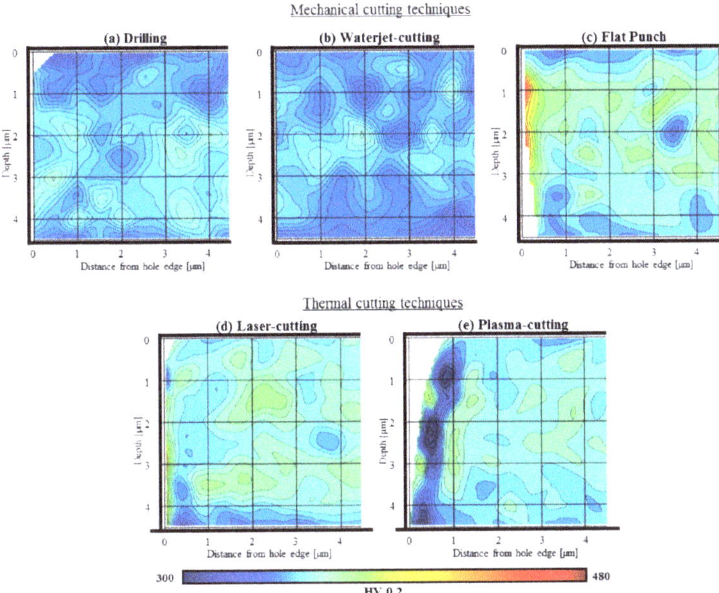

Figure 11. Hardness distribution near the hole edge in HSLA grade S960MC for various hole-making procedures: (**a**) drilling, (**b**) waterjet-cutting, (**c**) flat punching, (**d**) laser-cutting and (**e**) plasma-cutting. The hole edge is located on the left side of the images.

In the case of the punched samples, there was a clear hardness increase towards the hole edge. This phenomena has been already observed in punched specimens by Valtinat et al. [14] and can be attributed to the effect of cold-work hardening. The increase of hardness in punched samples (flat punch) with respect to the base material hardness was found to be 62%, 50%, and 28% for the S500MC, S700MC, and S960MC HSLA grades, respectively. The hardness increase was in accordance with the differences in ductility of the HSLA grades. The punch shape and orientation also played a role in the final hardening distribution near the hole. As displayed in Figure 12, when the chamfered punch was employed and its sharp edges were not facing the measured surface (0°), the resulting hardening values were lower than when the other punching conditions were adopted. The chamfered punch at 90° and the flat punch yielded, however, similar hardening distributions. This effect was mostly appreciable in the HSLA S960MC, while in the other two HSLA grades the punching conditions yielded similar results. This was related to the lower ductility and formability of the HSLA 960MC, which resulted in a less uniform strain accommodation along the entire hole.

A notable hardness increase was also found in the laser-cut samples, but mainly in the S700MC HSLA grade. The phase transformation triggered during the thermal cutting processes resulted in a thin layer of martensitic microstructure surrounding the hole edge. The hardness of this layer was significantly higher than the mixed bainitic-tempered martensitic of S700MC base material and similar to the martensitic matrix of the S960MC microstructure. The martensite formation could explain the notable hardness increase in the S700MC grades while in the S960MC grade this increase was barely significant. A similar hardening was found in the S700MC plasma-cut samples. However, the hardening effect for the S960MC grade differed from the rest of HSLA grades as regions of softer material were found just after the initial hardened HAZ layer. This is because the martensitic matrix of the S960MC was tempered extensively by the heat provided by the plasma. A more detailed hardness analysis was performed near the hole edge, with less distance between indentations, shown in Figure 13. It was observed that although there was a notable hardening increase at the hole edge, there was a decrease in hardness near the end of the HAZ layer. The measured Vickers hardness values at this region was 300 HV 0.2, which represented an approximate decrease in hardness of 15% with respect to the base material hardness. The lower hardness of the layer adjacent to the plasma accounted for the softer regions observed in Figure 11.

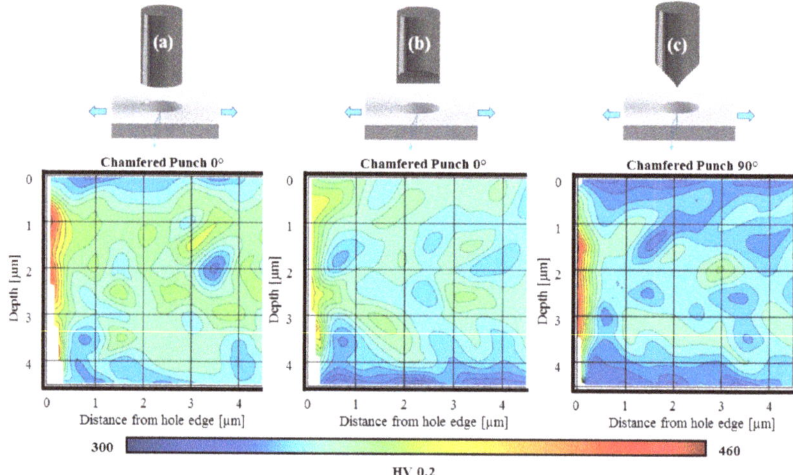

Figure 12. Hardness distribution produced in the S960MC by (**a**) flat punch, (**b**) chamfered punch at 0°, and (**c**) chamfered punch at 90°.

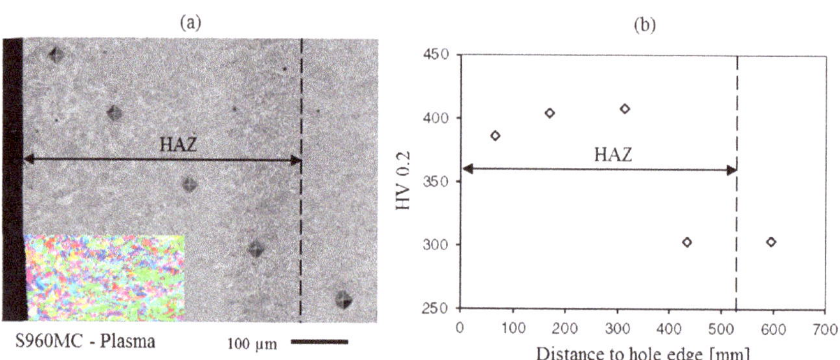

Figure 13. Hardness analysis near S960MC plasma layer: (**a**) micrograph displaying hardness measurements, (**b**) hardness values vs. distance to hole edge for each measurement.

3.6. EBSD

The heat-affected zone of laser-cut samples was analyzed with electron backscatter diffraction (EBSD). The hole edges of S500MC and S960MC are shown in Figure 14, measured with a step size of 0.15 μm. As can be seen, the hole edges were covered with a thin oxide layer followed by a thin, fine-grained zone.

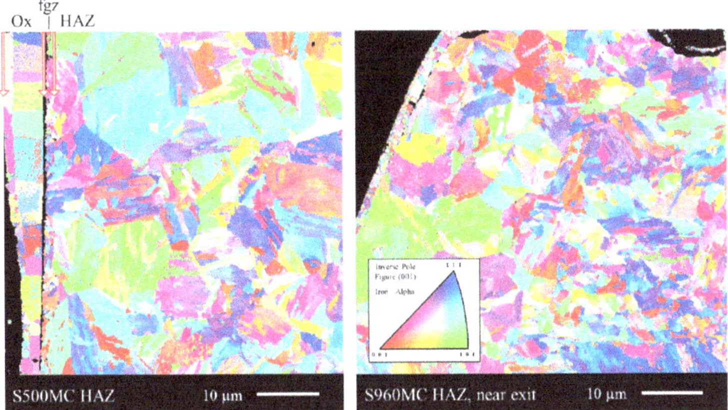

Figure 14. Hole edges of S500MC and S960MC measured with electron backscatter diffraction (EBSD). The EBSD inverse pole figure orientation map is given. The transition of different surface layers are marked with arrows in the S500MC grade: Ox, oxidation layer; fgz, fine-grained zone; HAZ, heat-affected zone.

In Figure 15 an overview of the hole edges is shown. The EBSD maps are measured with a step size of 0.50 μm for the laser-cut samples and 0.25 μm for the punched S960MC sample. For the S500MC grade, the transition to the ferritic microstructure is visible. To image highly deformed materials the spatial resolution needs to be smaller than the size of the dislocation cells [30]. The acquired EBSD patterns became weaker with high deformation, resulting in low confidence index values at the hole edge of the punched samples, but the transition from the base material to the deformed material is visible.

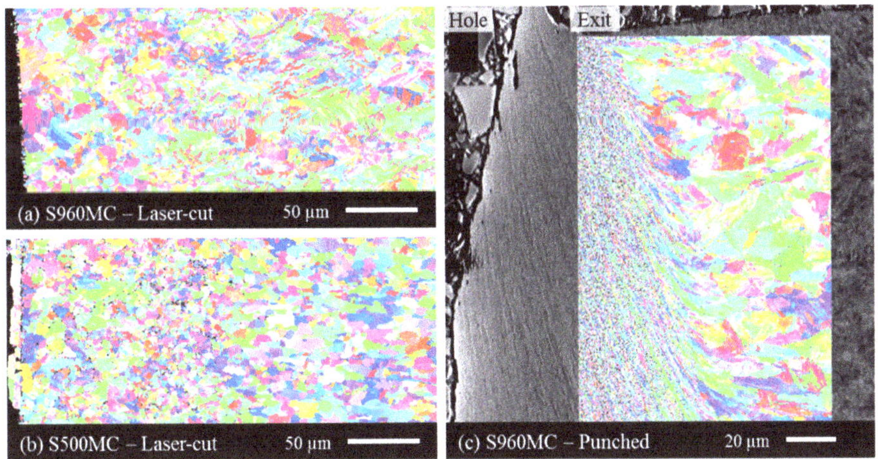

Figure 15. EBSD inverse pole figure orientation map of the two grades (**a**) S960MC and (**b**) S500MC. The laser-cut surfaces are on the left-hand side. (**c**) SEM image of a punched S960MC grade overlaid with an EBSD inverse pole figure orientation map. The exit side is shown on the top side.

3.7. Residual Stress

The results of the residual stress measurements of the punched sample in the longitudinal and the transversal direction are displayed in Figure 16a,b. Both the left and the right sides of the hole were measured. The measurements did not differ much and showed the same tendency. Here, only the XY side is shown. The residual stresses in longitudinal direction are, in comparison to the transversal direction, more shifted towards compressive stresses. The formation of the residual stresses during the punching process were dependent on various parameters, e.g., local plastic flow, evolution of damage, and clamping parameters, which can be reproduced and explained via simulations [31]. The highest stresses can be observed in the S960MC punched sample due to the higher flow stress of the material. The residual stresses of the laser-cut specimen in the longitudinal and transversal direction are shown in Figure 16c,d. The magnitude of the residual stresses was lower compared to punched samples and there was not such a large difference between the top and the bottom face. The residual stresses resulting from the laser cutting were, on average, larger for the S500MC. It is also important to note that, due to the difference in thickness (5 mm for the S960MC and 6 mm for the S500MC), the cutting parameters differed between both samples. In particular, the laser-cutting speed was 2.4 times faster for the 5-mm samples, which may result in a lower heat input and, hence, lower residual stresses around the hole.

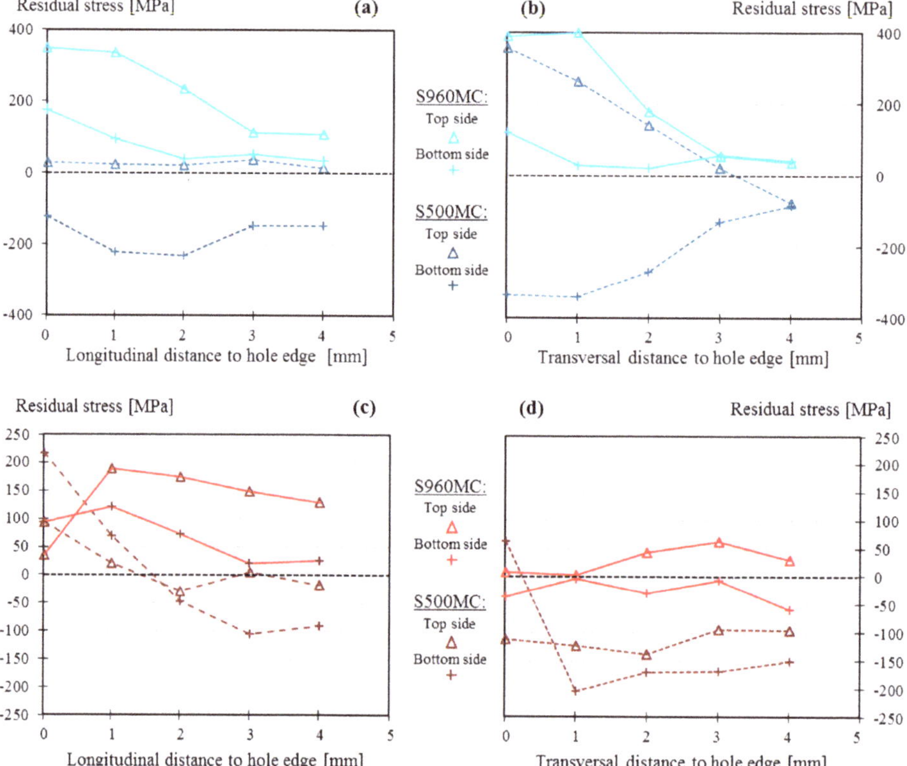

Figure 16. Residual stresses of punched S500MC and S960MC specimens in the (**a**) longitudinal and (**b**) transverse direction, and of laser-cut S500MC and S960MC specimens in the (**c**) longitudinal and (**d**) transverse direction. Here, error bars are smaller than the markers and are, therefore, not shown. Connecting lines between measurement positions are for guidance of the eye only.

3.8. Fatigue Results

The fatigue results for the holed specimens, tested at a nominal stress range of 300 MPa, are presented in Figure 17. The S700MC and S960MC fatigue results followed the same trend as the S500MC in [13]. Punched samples displayed the lowest fatigue performance, whereas the chamfered punch aligned with loading (with the sharp edge not facing the failure location) performed best among the punched samples, followed by both plasma-cut and drilled specimens, which showed a similar fatigue life. A significantly better performance was observed for waterjet-cut holes. The best fatigue performance was achieved with the laser-cut specimens, where run-out tests were found.

The difference in fatigue life for the different punch geometries related with the hardness distribution results previously presented in Section 3.5, in which the chamfered punch aligned with the load yielded lower localized cold-work hardening near the hole edge.

As in [13], failure in plasma-cut samples took place predominantly at plasma-blob locations, which did not seem to affect the number of cycles to failure. It was also found for the drilled-cut sample that the drill chip had a strong influence in the failure mechanisms and its removal resulted in run-out fatigue tests.

When the fatigue results of both HSLA grades were compared it can be noted that, although both material followed the same trend in results, S960MC had a slightly better fatigue performance than S700MC, although scarce to be statistically significant without additional testing.

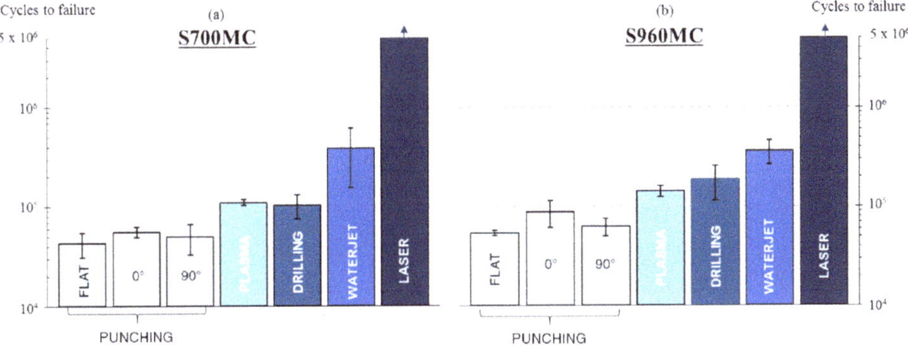

Figure 17. Fatigue results for (**a**) S700MC and (**b**) S960MC.

The presented fatigue results share many similarities with two researches performed in the framework of the Research Fund for Coal and Steel (RFCS): Coldfoss [16] and Hipercut [17]. Both research projects studied the influence of hole-making processes on the fatigue performance of steel grades with yield strength ranging from 350 to 900 MPa. Constant amplitude tests with a stress ratio of 0.1 were performed and the sample geometry also consisted of a plate with a hole located at its center. The fatigue behavior of punched specimens in the present study was compared with the punched and punched-plus-shot peened specimens in [16]. The results indicated that the fatigue performance of punched samples did not differ much at the present nominal stress range between different HSLA grades, which might indicate that the fatigue behavior was dominated by the presence of large macro defects. However, it is shown [16] that the shot-peening process treatment dramatically improved fatigue life in punched specimens. The laser-cut and plasma-cut specimens were compared with the laser-cut, plasma-cut, and oxy-fuel-cut specimens in [17]. Unlike in the present investigation, in which laser-cut specimen displayed the highest fatigue performance, the fatigue performance of laser-cut samples in [17] was significantly lower than oxy-fuel and plasma-cut samples. The authors referred to the large amount of draglines in the laser-cut edges as a reason for this lower performance.

3.9. Fracture Analysis

Fatigue failure typically occurred at the middle of the sample, starting from the hole edge. Fatigue cracks propagated perpendicular to the axial load direction. After a certain distance, when the cross-section was significantly reduced, ductile failure happened at the remaining connected material.

The characteristic cup and cone shape produced by the shear deformation at 45° with respect to the tensile axis was observed at the ductile failure region for most specimens.

Both the entry section and the shear cutting section of the punched holes were evaluated by SEM. Although all fractured punched specimens presented cracks along the shear cutting section, there was a significant difference between the milder grade S500MC studied in [13] and the other two grades (S700MC and S960MC). In Figure 18, the characteristic crack formation location is shown for the S700MC and the S960MC grades punched with a flat punch. The results in [13] indicated that cracks were likely to initiate from the lower end of the shear-cutting section for the S500MC grade. However, the presence of larger cracks propagating from the middle of the shear-cutting section was noted for the S700MC and the S960MC grades. As previously commented in Section 3.2, shear dimples were more frequent in the S700MC and S960MC grades, which may suggest larger damage and cracks produced during the punching process. These cracks, just as the shear dimples, would be located at the middle of the shear-cutting section and would further propagate due to the action of the induced cyclic loading. This middle area contained segregation of alloying elements, such as Mn. The alloying element segregation resulted in reduced ductility and toughness in this area, making it prone to cracking. The punched S500MC samples showed fewer dimples, which might explain why cracks tended to propagate from the lower region of the shear-cutting section.

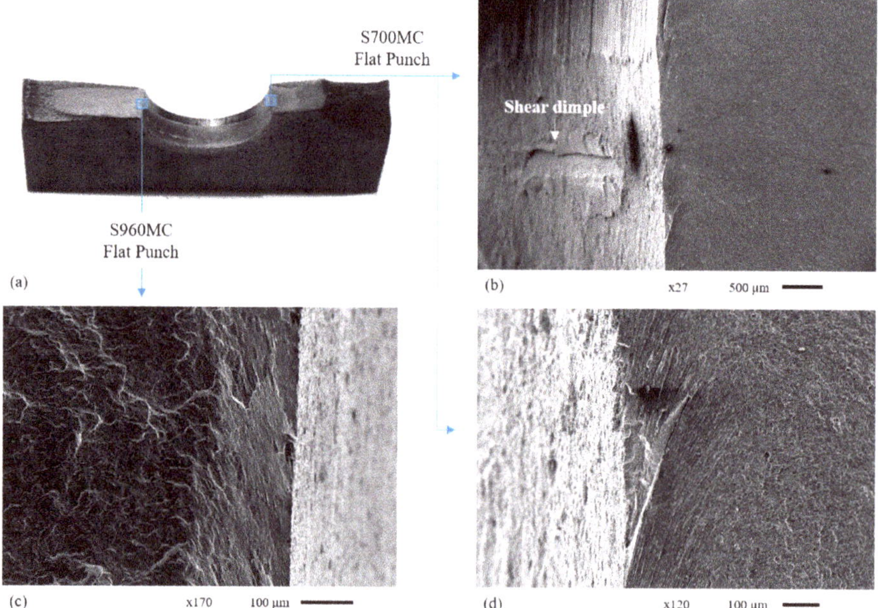

Figure 18. (a) Overview of fractured punched specimen. Fracture surfaces of flat-punched specimens: (b)(d) S700MC and (c) S960MC.

The SEM evaluation of drilled samples revealed that fracture in drilled specimens originated at the edges (top or bottom) of the hole. Both edges presented multiple sharp edges that acted as a notch during the fatigue cycles for cracks to originate and propagate. This can be observed in Figure 19a, where a crack propagated from the drill entry site to the rest of the cross-section in a S700MC specimen. The sharp edges at the entry and exit of the drill were the origin of cracks in all tested specimen and HSLA grades. Due to the high notch sensitivity of the HSLA grades, the deburring of the drilled edges was a necessary step to make the most of the smooth surface and the absence of hardening produced during drilling.

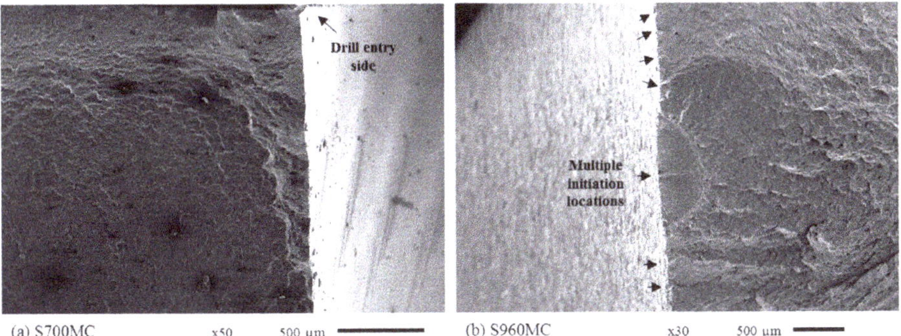

Figure 19. Fracture details of (**a**) entry drill chip (S760MC) and (**b**) waterjet-cut mid-hole surface (S960MC).

All observed waterjet-cut fractured samples were similar for all studied grades. A characteristic fractured section is shown in Figure 19b (S960MC). The waterjet-cut fractured surfaces presented multiple crack initiation sites along the hole depth, which originated from the geometrical features produced by the rough waterjet-cut surface finish, previously discussed in Section 3.2. A significant difference between waterjet-cut and punched specimens is that in the case of the first, cracks initiated from the geometrical features produced by the abrasive material while for punched specimens, cracks that were already introduced during the punching process propagated during the fatigue cycles.

The plasma blob was found to be responsible for the initiation of many of the fractured samples. As discussed in Section 3.2, the plasma blob was located at 45 degrees with respect of the hole section in which failure was expected to happen. However, due to the multiple notches induced by the plasma blob, failure occurred in this region in many of the tested specimens. As shown in Figure 20a, the plasma blob extended into the material and the pores present in the blob acted as notches for crack initiation. The surface defects produced during the plasma-cutting process were responsible for the remaining crack initiation locations. These defects were distributed along the entirety of the hole depth and their size was relatively large if compared with the defects induced during waterjet cutting, which could explain the lower performance of plasma-cut samples. An example of cracks originating from a plasma surface defect is shown in Figure 20b.

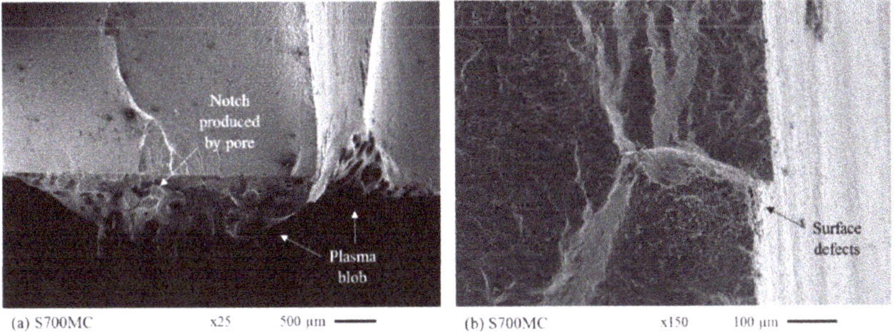

Figure 20. Fracture surfaces at (**a**) plasma blob and (**b**) mid-hole surface (S700MC).

4. Conclusions

The results of this research provided more insight on how different hole-making procedures affect the failure mechanisms of different HSLA steel plates. It was observed that the microstructural differences between HSLA grades produced variations in the hole morphology, such as different hardening distributions, residual stresses, HAZ extension, and presence of defects. However, the higher yield strength of the HSLA grades did not lead to a higher fatigue life. The results of the present research on the HSLA grades S700MC and S960MC followed a similar trend as previous research performed by the authors on S500MC samples [13], with punching samples exhibiting the lowest fatigue values. This can be attributed to the high residual tensile stresses and the large amount of shear dimples and defects along the punched-hole surface. The presence of shear dimples, however, was considerably higher in the S700MC and the S960MC compared to the S500MC. Plasma samples provided similar fatigue lives to punched specimens, followed by drilled samples. Finally, waterjet cutting and laser cutting resulted in the best fatigue performing samples.

The better performance of waterjet-cut samples and laser-cut samples offers an explanation of the failure mode of the tested samples. The abrasive nature of waterjet-cut samples produced a fairly constant amount of small defects and an absence of hardening near the hole, which were not large enough to create a large stress concentration. Laser-cut samples, on the other hand, did present a significant amount of hardening and tensile residual stresses near the hole edge. However, they also exhibited an even and defect-free surface. From these observations, it seems that the fatigue performance of the holed samples was dominated by large defects inherent to each hole-making procedure, which finally resulted in a similar number of fatigue cycles for the studied HSLA grades.

Author Contributions: Conceptualization, C.J.-P. and D.D.; methodology, C.J.-P. and C.G.; validation, C.J.-P. and C.G.; formal analysis, C.J.-P., C.G. and J.P.; investigation, C.J.-P., C.G., and J.P.; resources, C.G, J.P., and D.D.; data curation, C.J.-P., and J.P.; writing—original draft preparation, C.J.-P. and C.G.; writing—review and editing, C.J.-P., C.G., J.P., and D.D.; visualization, C.J.-P.; supervision, D.D.; project administration, D.D.; funding acquisition, D.D. All authors have read and agreed to the published version of the manuscript.

Funding: This work was supported by the European Research Project DURAMECH: "Towards Best Practice for Bolted Connections in High Strength Steels" (project number 709962 [2016]) and made possible through funding support of the KU Leuven Fund for Fair Open Access.

Conflicts of Interest: The authors declare no conflict of interest.

References

1. Gogou, E. Use of High Strength Steel Grades for Economical Bridge Design. Ph.D. Thesis, TU Delft, Delft University of Technology, Delft, The Netherlands, 2012.

2. Schröter, F.; Schütz, W. State of art in the production and use of high-strength heavy plates for hydropower applications. In Proceedings of the High Strength Steel for Hydropower Plants, Graz, Austria, 5–6 July 2005.

3. Jensen, L.; Bloomstine, M.L. Application of high strength steel in super long span modern suspension bridge design. In Proceedings of the Nordic Steel Construction Conference, Stockholm, Sweden, 2–4 September 2009.

4. Miki, C.; Homma, K.; Tominaga, T. High strength and high performance steels and their use in bridge structures. *J. Constr. Steel Res.* **2002**, *58*, 3–20. [CrossRef]

5. Willms, R. High strength steel for steel constructions. In Proceedings of the Nordic Steel Construction Conference-NSCC, Malmö, Sweden, 2–4 September 2009; pp. 597–604.

6. Gresnigt, A.; Steenhuis, C. High strength steels. *Prog. Struct. Eng. Mater.* **1997**, *1*, 31–41. [CrossRef]

7. Billingham, J.; Sharp, J.; Spurrier, J.; Kilgallon, P. *Review of the Performance of High Strength Steels Used Offshore*; Health and Safety Executive: Bootle, UK, 2003; p. 111.

8. Bailey, N.; Coe, F.R.; Gooch, T.; Hart, P.; Jenkins, N.; Pargeter, R. *Welding Steels without Hydrogen Cracking*; Woodhead Publishing: Sawston, UK, 1993.

9. Tamura, I.; Sekine, H.; Tanaka, T. *Thermomechanical Processing of High-Strength Low-Alloy Steels*; Butterworth-Heinemann: Oxford, UK, 2013.

10. Braconi, A.; Osta, A.; Cama, P.; Blasi, N.; Mordini, A.; Wenzel, H.; Chellini, G.; Lippi, F.; Salvatore, W.; Rauert, T.; et al. *Design for Optimal Performance of High-Speed Railway Bridges by Enhanced Monitoring Systems (Details)*; EU Publications: Bruxelles, Belgium, 2013. [CrossRef]

11. European Commission. *Improving the Fatigue Life of High Strength Steel Welded Structures by Post Weld Treatments and Specific Filler Material (FATWELDHSS)*; Directorate-General for Research and Innovation (European Commission): Brussels, Belgium, 2010.

12. Jiménez-Peña, C.; Talemi, R.H.; Rossi, B.; Debruyne, D. Investigations on the fretting fatigue failure mechanism of bolted joints in high strength steel subjected to different levels of pre-tension. *Tribol. Int.* **2017**, *108*, 128–140. [CrossRef]

13. Jiménez-Peña, C.; Goulas, C.; Rossi, B.; Debruyne, D. Influence of hole-making procedures on fatigue behaviour of high strength steel plates. *J. Constr. Steel Res.* **2019**, *158*, 1–14. [CrossRef]

14. Valtinat, G.; Huhn, H. Bolted connections with hot dip galvanized steel members with punched holes. In Proceedings of the Connections in Steel Structures V, Amsterdam, The Netherlands, 3–5 July 2004; pp. 297–310.

15. Goldberg, F. Influence of thermal cutting and its quality on the fatigue strength of steel. *Weld. J.* **1973**, *52*, 392–404.

16. Bannister, A.; Skalidakis, M.; Pariser, A.; Langenberg, P.; Gutierrez-Solana Salcedo, F.; Sánchez, L.; Pesquera, D.; Azpiazu, W. *Performance Criteria for Cold Formed Structural Steels*; EU Publications: Brussels, Belgium, 2006; pp. 1–241.

17. European Commission. *High Performance Cut Edges in Structural Steel Plates for Demanding Applications (HIPERCUT)*; Directorate-General for Research and Innovation (European Commission): Brussels, Belgium, 2016.

18. AASHTO. *AASHTO LRFD Bridge Design Specifications*; Fourth Edition with 2008 Interim Revisions; American Association of State Highway and Transportation Officials: Washington, DC, USA, 2012.

19. British Standards Institution. *Code of Practice for Fatigue Design and Assessment of Steel Structures*; British Standards Institution: London, UK, 1993.

20. Eurocode—Basis of Structural. *EN 1990: 2002—Basis of Structural Design*; British Standard Institution: London, UK, 2002.

21. CEN. *EN 1993-1-12. Eurocode 3: Design of Steel Structures—Part 1–12: Additional Rules for the Extension of EN 1993 Up to Steel Grades S 700*; European Committee for Standardization: Brussels, Belgium, 2007.

22. Denys, K.; Coppieters, S.; Seefeldt, M.; Debruyne, D. Multi-DIC setup for the identification of a 3D anisotropic yield surface of thick high strength steel using a double perforated specimen. *Mech. Mater.* **2016**, *100*, 96–108. [CrossRef]

23. SSAB. Strenx® 960 MC. Available online: https://www.ssab.com/products/brands/strenx/products/strenx-960-mc?accordion=downloads (accessed on 1 December 2019).

24. Arcelormittal. Amstrong—High Strength Steels. 2018. Available online: http://industry.arcelormittal.com/catalogue/A20/EN (accessed on 1 December 2019).

25. EN, B. *1-9 (2005): Design of Steel Structures*; Part 1.9: Fatigue; British Standards Institution: London, UK, 1993.

26. VariDrill™ Catalogue. Available online: https://www.widia.com/en/products/24019771/24064245/46993284/100006851.html (accessed on 21 February 2019).

27. ISO, E. *6507-1: 2005-Metallic Materials-Vickers Hardness Test*; European Committee for Standardization: Brussels, Belgium, 2006.

28. Prevey, P.S. X-ray diffraction residual stress techniques. *Asm Int. Asm Handb.* **1986**, *10*, 380–392.

29. Guo, F.; Wang, X.; Liu, W.; Shang, C.; Misra, R.; Wang, H.; Zhao, T.; Peng, C. The Influence of Centerline Segregation on the Mechanical Performance and Microstructure of X70 Pipeline Steel. *Steel Res. Int.* **2018**, *89*, 1800407. [CrossRef]

30. Schwartz, A.J.; Kumar, M.; Adams, B.L.; Field, D.P. *Electron. Backscatter Diffraction in Materials Science*; Springer: New York, NY, USA, 2000.

31. Janarthanam, H.; Sommer, S.; Carl, E.-R.; Preußner, J.; Huberth, F. Numerical prediction of damage in punching process using shear modified Gurson model. In Proceedings of the 4th European Steel Technology and Application Days, Düsseldorf, Germany, 24–28 June 2019.

Article

Controlling Variability in Mechanical Properties of Plates by Reducing Centerline Segregation to Meet Strain-Based Design of Pipeline Steel

Fujian Guo [1], **Wenle Liu** [1], **Xuelin Wang** [1], **R.D.K. Misra** [2] **and Chengjia Shang** [1,3,*]

[1] Collaborative Innovation Center of Steel Technology, University of Science and Technology Beijing, Beijing 100083, China

[2] Department of Metallurgical and Materials Engineering, University of Texas El Paso, El Paso, TX 799968, USA

[3] State Key Laboratory of Metal Materials for Marine Equipment and Applications, Anshan 114021, China

* Correspondence: cjshang@ustb.edu.cn; Tel.: +86-10-6233-2428

Received: 14 June 2019; Accepted: 2 July 2019; Published: 4 July 2019

Abstract: Low variability in mechanical properties is required for pipeline project designs to meet a strain-based design, which is used in regions of large ground movements. The objective of this study is to elucidate the influence of centerline segregation in continuously cast slab on variability in the mechanical property of pipeline steel, and controlling centerline segregation can meet the requirements of a strain-based design. Mannesmann rating method was used to evaluate the degree of segregation of two slabs and its effect on variability in mechanical properties of corresponding plates. Microstructural characterization indicated that bainite/martensite was formed in a segregated area where the content of C and Mn enriched. The mechanical property results indicated that controlling the degree of centerline segregation can reduce tensile strength variability and improve ductile-brittle transition temperature (DBTT).

Keywords: centerline segregation; segregation rating; pipeline steel; mechanical property variability; strain-based design

---

1. Introduction

With the rapid development of the oil and gas pipeline industry, pipeline projects need to traverse harsh environments [1,2], including discontinuous permafrost, seismic activity, iceberg scour, and landslides which require pipelines that can sustain plastic deformation as a result of large ground movements. To ensure pipeline integrity in environmentally sensitive areas and overall cost effectiveness, a strain-based design approach needs to be considered. Low fluctuations in strength, low yield ratios, and high uniform elongation are often used as new requirements for pipeline designs to meet a strain-based design [3].

The high performance of pipeline steel is determined by fiber structure. Microstructure uniformity is a necessary method for achieving the above mentioned high performance. However, for multi-phase microstructure uniformity control, continuous casting slab uniformity (macro internal quality) is an important factor affecting the fiber uniformity of hot rolled steel plates. Centerline segregation is the major internal defect of continuously cast slab. It is difficult to eliminate during heating and rolling [4,5]. Many studies [6–16] suggested that centerline segregation decreases the mechanical performance of final products, such as: Low temperature toughness, weldability, and resistance to sour gas worsen; delamination fracture and band structure increase. References [17–20] suggested that control of centerline segregation in continuously cast slab can effectively improve low temperature toughness and elongation of pipeline steel. Because it is impossible to make the continuous casting slab absolutely uniform, it is necessary to recognize the quantitative relationship between degree of segregation of slabs and the performance stability of plates. This paper presents a method to reduce fluctuation of mechanical properties through controlling centerline segregation, which can be applied in the strain-based design of pipeline steels.

2. Materials and Methods

The material used in this study is X70 pipeline steel, which chemical composition in weight percent (wt.%) is 0.045C-1.7Mn-0.1Si-0.41(Mo + Cr + Ni)-0.073(Nb + Ti)-0.012P-0.003S [20]. Two slabs and corresponding plates were studied in this paper and were named slab A and slab B, respectively. The steel plates (13 mm) were rolled using same rolling condition from the two continuously cast slabs with a thickness of 230 mm and a width of 1910 mm, which were produced by using the same continuous casting machine. Slab samples for the evaluation of centerline segregation were taken from the head part of the slab. Each cross section of slab and plate is divided into 5 smaller parts and marked with number 1# (left), 2# (left quarter), 3# (center), 4# (right quarter), and 5# (right), as shown in Figure 1a,b. Centerline segregation in slabs was etched with ~50 pct HCl at ~80 °C for ~30 min, and macro-etched images were taken by a digital camera. The plate samples were taken from the head of plates which were rolled from the corresponding slabs, used for determination of mechanical properties and microstructure. Impact toughness was determined using the Charpy V-notch specimen (55 mm × 10 mm × 10 mm) from each position of two plates, machined according to ISO 148-1 standard from the mid-thickness of plates in the cross-section (Figure 1c), conducting at temperatures of −40, −60, −80, and −100 °C. Vickers micro-hardness (by VMHT 30M, Leica, Wetzlar, Gemany) of the segregation band was measured using a load of 50 g and a dwell time of 15 s. The load was selected to ensure a relatively small indent size.

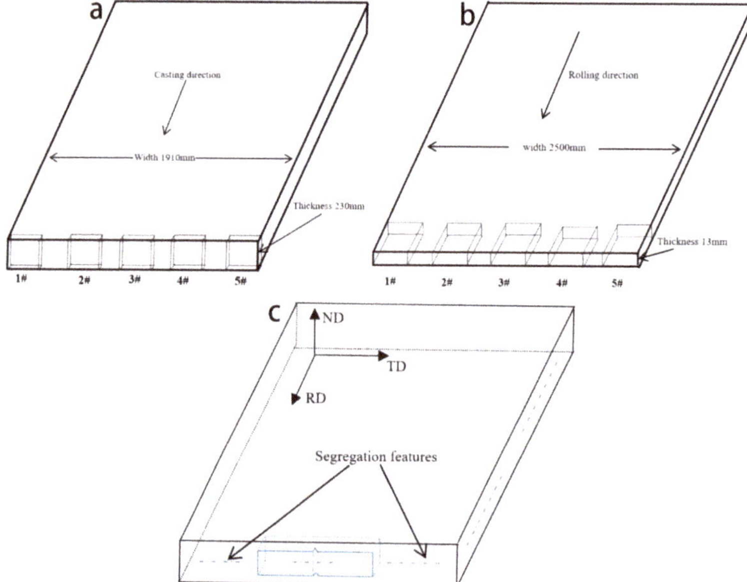

Figure 1. Sampling schematic of continuously cast slab and rolled plate: (**a**) Continuous casting slab; (**b**) rolled plate; (**c**) Charpy V-notch specimen.

Samples for microstructure observation were cut from the middle area of the thickness of the plates and then mounted and mechanically polished using standard metallographic procedure.

The microstructure of specimens was characterized via a combination of an optical microscope (OM), a scanning electron microscope (ULTRA-55field SEM, ZEISS, Jena, Germany), and a transmission electron microscope (Tecnai G2 F20, FEI, Hillsboro, OR, USA). TEM was carried out using 3 mm disks ground to a thickness of 50 μm and were then twin-jet electropolished with an electrolyte of 10% perchloric acid and 90% ethanol. The diagram of time temperature transformation was calculated by the software JMatPro 7.0 according to the alloy content measured by an electron probe micro-analyzer (EPMA).

3. Results

3.1. Segregation Evaluation of Slabs

Shown in Figure 2 are the macro-etched slab images with the distribution of the centerline segregation of the slab uneven. In slab A (Figure 2a), the degree of centerline segregation at position 2# and 3# was more severe and rated at Mannesmann class 3.5 [21], while position 1# was the best one in this slab rated at class 2.5. In slab B (Figure 2b), the degree of segregation for the majority of the positions was rated at class 2 except for position 5# which was rated at class 3. The degree of segregation of these two slabs is summarized in Table 1, the segregated severity of slab B was better than slab A, indicating that the distribution of centerline segregation was uneven across the transverse section.

Figure 2. Images of the macro-etched slab: (**a**) Slab A; (**b**) slab B.

Table 1. Centerline segregation degree rated by the Mannesmann rating method (class).

Slab	1#	2#	3#	4#	5#
A	2.5	3.5	3.5	3	3
B	2	2	2	2	3

3.2. Low Temperature Toughness

A total of two Charpy tests were performed for each position of two plates which were rolled from each Mannesmann sample location. The results of the Charpy impact test are shown in Figure 3. These ten samples with different segregation degrees and positions from two plates all exhibited good impact toughness. However, the low temperature toughness of these samples was significantly different. Sample A-2# (red line) and A-3# (blue line) with a high severity of centerline segregation (Mannesmann class 3.5) exhibited the lowest impact energy and the ductile-brittle transition temperature (DBTT) of these two samples was ~−65 °C, while sample B-1#, B-2#, B-3#, and B-4# with the lowest degree of segregation (class 2) had the highest impact energy and DBTT of these four samples were ~−90 °C. The DBTT of the remaining samples was between ~−65 °C and ~−90 °C. Therefore, impact energy decreases and DBTT increases as the degree of centerline segregation increases.

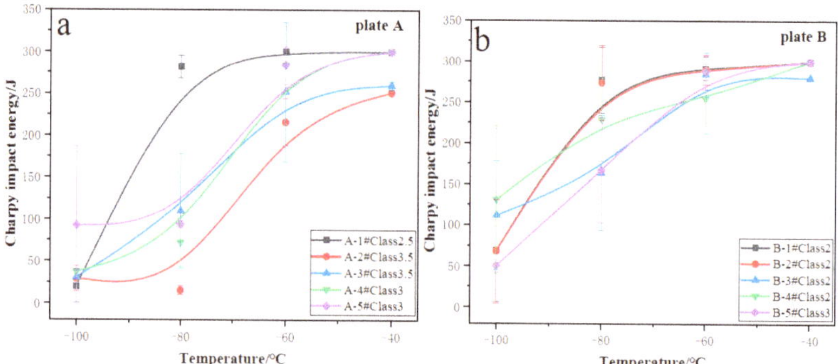

Figure 3. Charpy impact energy of strips from slabs with different segregation degrees: (**a**) Plate A; (**b**) plate B.

In plate A, sample A-2# (class 3.5) exhibited the lowest impact energy, while sample A-1# (class 2.5) had the highest impact energy in the same transverse section of the plate. In plate B, sample B-5# (class 3) exhibited the lowest impact energy, while the remaining three samples had similar impact energy, significantly higher than sample B-5#. At −60 °C, the variability range in toughness of plate A (~75 J) was higher than plate B (~40 J); at −80 °C, the variability range in toughness of plate A (~260 J) was higher than plate B (~75 J). Therefore, the variability in toughness in the same plate was caused by centerline segregation of the slab. This result is consistent with Su et al. [22,23] who suggested that centerline segregation in continuously cast slabs increased variability in mechanical properties of steel plates.

3.3. Tensile Properties

A similar trend was observed in tensile strength and yield strength (Figure 4a), which increase with the increase of the degree of segregation. In plate A, the degree of segregation was from Mannesmann class 2.5 to class 3.5, the variability range of tensile strength and yield strength were ~110 MPa and ~30 MPa, respectively, while in plate B, the degree of segregation was from Mannesmann class 2 to class 3, the variability range of tensile strength and yield strength were ~13 MPa and ~15 MPa, respectively. Therefore, the severity in centerline segregation not only increases the tensile and yield strength, but also increase the variability in tensile strength. Meanwhile, an identical trend was observed for total and uniform elongation, which decreased as the degree of segregation was increased. The total and uniform (Figure 4b) elongation of plate A were significantly lower than plate B because of severity in centerline segregation.

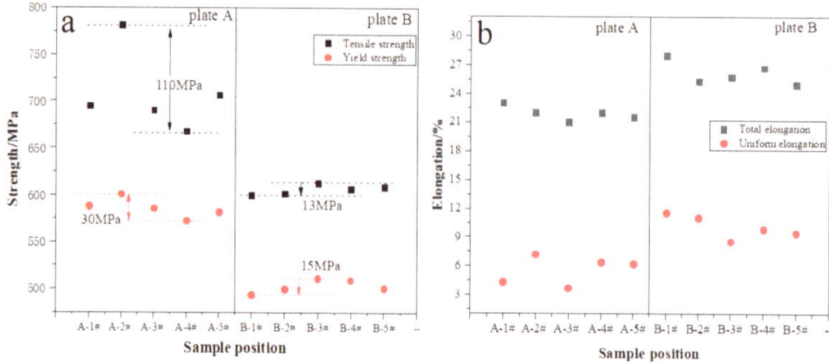

Figure 4. Tensile test result of plates from slabs with different segregation degree: (**a**) Tensile and yield strength, (**b**) total and uniform elongation.

3.4. Microstructure in Segregated Region

Three samples with different degrees of segregation were selected to investigate the impact of segregation on the microstructure. Samples B-4# (Mannesmann class 2), A-4# (class 3), and A-2# (class 3.5) were selected. The microstructure of the segregated area of plates as characterized by OM and SEM is shown in Figure 5. A large amount of lath-like martensite was present in sample A-2# (Figure 5c,f). Lath-like bainite can also be observed in sample A-4# (Figure 5b,e), but its proportion was less than sample A-2#. The microstructure was dominated by polygonal ferrite and a small amount of martensite/austenite constituent (MA) in sample B-4#.

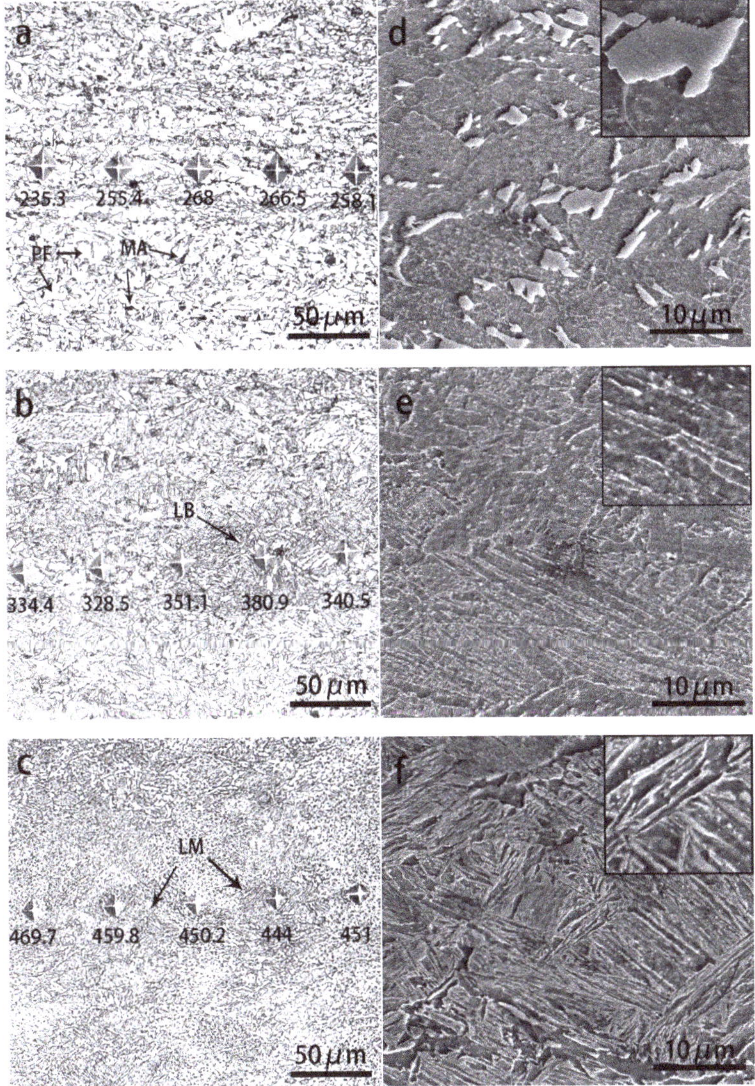

Figure 5. Microstructure in the segregated region as seen through an optical microscope and a scanning electron microscope: (**a**) And (**d**) for sample B-4#; (**b**) and (**e**) for sample A-4#; (**c**) and (**f**) for sample A-2#.

The detailed microstructure of the segregated areas was further characterized by TEM. The microstructure of sample B-4# was polygonal ferrite and MA (Figure 6d), and lath bainite and martensite with different width were observed in samples A-4# and A-2# (Figure 6e,f), respectively. In addition, the size of lath martensite/bainite was measured, with the average width of lath in samples A-4# and A-2# was ~1.24 μm and ~0.87 μm, respectively. Microhardness data confirmed that the microstructures in the segregated region were polygonal ferrite (~256.6 HV), lath bainite (~347.1 HV), and lath martensite (~454.9 HV), respectively.

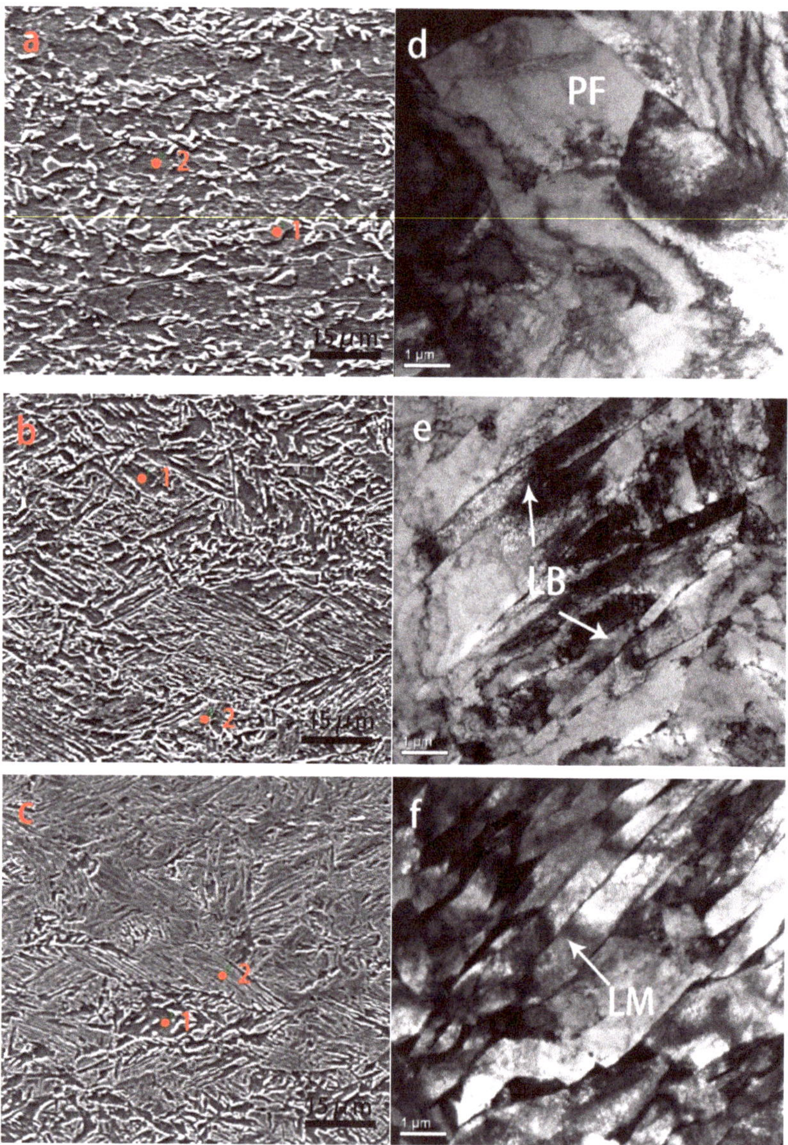

Figure 6. Measurements of segregated spots in the segregated area by an electron probe micro-analyzer and a transmission electron microscope: (**a**) And (**d**) for sample B-4#, (**b**) and (**e**) for sample A-4#, (**c**) and (**f**) for sample A-2#.

3.5. Enrichment of Elements in Segregated Region

Figure 6a–c shows that the concentration of alloying elements located in the segregated region was measured by EPMA, and the measured spots are the black dots with numbers. Table 2 shows the results of chemical composition (without S and P) in the segregated region of samples B-4#, A-4#, A-2#, S, P composition, which were difficult to be examined by EPMA. The results indicated that all the alloying elements were not segregated except for C and Mn. The content of C in the centerline of plates increased with the increase of the degree of segregation of the corresponding slabs, i.e., for sample B-4# (class 2), sample A-4# (class 3), and sample A-2# (class 3.5), the maximum carbon was 0.47 wt.%, 0.527 wt.%, and 0.732 wt.%, respectively. It should be pointed out that the carbon content is not accurate, because it is a light element, but the carbon content has a meaning. The content of Mn in the centerline of plates had the same trend as the carbon content, i.e., for sample B-4#, sample A-4#, and sample A-2#, the maximum Mn content was 1.813 wt.%, 2.382 wt.%, and 2.385 wt.%, respectively.

Table 2. Alloy component in the segregated region measured by electron probe micro-analyzer (wt.%).

Position		C	Mn	Al	Si	Ti	Cr	Ni	Nb	Mo
B-4#	point1	0.414	1.796	0.021	0.201	0.033	0.007	0.008	0.083	-
	point2	0.471	1.813	0.018	0.163	0.010	0.039	-	0.054	0.094
A-4#	point1	0.387	1.701	0.025	0.116	0.068	-	-	0.069	-
	point2	0.527	2.382	0.027	0.265	-	0.049	0.013	0.115	0.102
A-2#	point1	0.410	2.031	0.019	0.201	0.033	0.004	-	-	0.006
	point2	0.732	2.385	0.017	0.195	0.02	0.005	-	0.094	0.067

A previous study [24] reported that the content of Mn was at 1.82 wt.% or 1.76 wt.%, the corresponding microstructure was ferrite, while the content of Mn was up to 2.9 wt.%, its microstructure was bainite/martensite. Literature [25] also suggested that segregation is caused by an enrichment of Mn in the centerline region of plates.

4. Discussion

4.1. Influence of Degree of Segregation on the Microstructure

Since a direct correlation between the degree of segregation and mechanical properties was obtained, the following evaluation scale is proposed. The variability range of degree of segregation was from Mannesmann class 2.5 to class 3.5 in slab A, while from class 2 to class 3 in slab B. A total of three plate samples (B-4#, A-4#, and A-2#) with different degrees of segregation (class 2, class 3, and class 3.5, respectively) were selected for further study. The EPMA results (Table 2) show that the content of enriched elements (C and Mn) in the plate centerline region was higher with an increased degree of segregation of corresponding slabs. The transformation temperature can be varied because of the content of enrichment of alloying elements. Previous studies [26–28] suggested that Mn in solid solution has an important influence on supercooled austenite transformation. On the one hand, it changes the free energy difference between the phases and varies the temperature of phase transformation point, causing the pearlite transformation curve and the bainite transformation curve to move downward (Figure 7). On the other hand, C and Mn are stabilized austenite elements, while Mn increases the diffusion activation energy of C in austenite, affecting the nucleation and growth process of pearlite transformation and bainite transformation and causing the pearlite transformation curve and the bainite curve to move to the right (Figure 7). At the same cooling rate, the microstructure with different content of alloying elements could be different [22]. Therefore, the microstructure in the centerline region of plates can vary because of segregation.

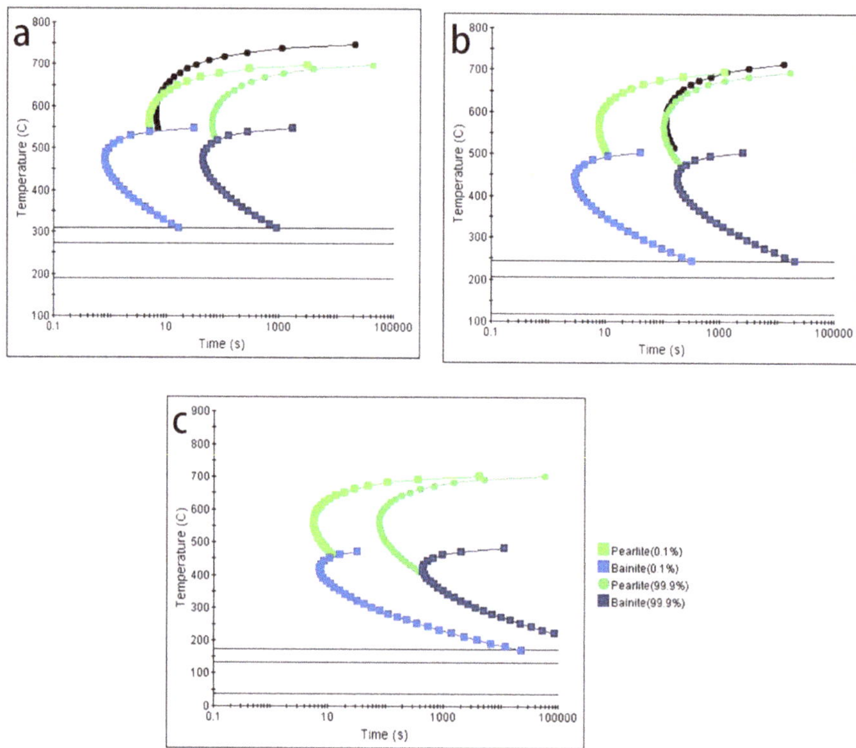

Figure 7. Time temperature transformation diagram of centerline region of strips: (**a**) Sample B-4#, (**b**) sample A-4#, (**c**) sample A-2#.

4.2. Relationship between Microstructure and Variability on Mechanical Properties

The variability in mechanical properties is often due to the difference in the microstructure, which was caused by the degree of segregation of corresponding slabs. Although samples in plate A all show good low temperature toughness, the variability in toughness was significant. The range of degree of segregation is from class 2.5 to class 3.5, the DBTT increased from −90 °C to −65 °C. However, the low temperature toughness of plate B was better than that of plate A, because the degree of segregation of samples in slab B were all class 2, except for B-5# with class 3, so the toughness variability in plate B was smaller than plate A. At the same time, a similar trend appeared in the tensile test. Comparing with plate B, variability in tensile strength was observed, the strength of plate A was significantly higher than the strength of plate B, and the uniform elongation and total elongation of plate A were lower than in plate B, but variability in uniform was not observed. The reason for variability in strength and toughness is caused by the difference in microstructure in the segregated region and explained by a different degree of segregation. The microstructure in the segregation area of sample B-4# was dominated by polygonal ferrite, while that of sample A-4# and A-2# were lath bainite and lath martensite, respectively. On the other hand, the elongation and toughness of ferrite was better than bainite and martensite. This also explains why the variability in toughness and strength of plate B was small, and the elongation was higher. Through the characterization of microstructure, it is also clearly explained that the variability in the performance of plates in industrial production is mainly caused by centerline segregation heredity of continuous casting slabs.

Based on the strain-based design, the anti-large deformation pipeline steel can withstand large deformations and has some characteristics in terms of structure and performance. Such as the requirements of DNV2000 [29] for pipeline steel, the measured yield strength cannot exceed 100 MPa than the standard specified value, the measured yield ratio is ≤0.85, and the total elongation is ≥25%. Obviously, sample B-4# of which the degree of segregation is at or lower than Mannesmann class 2 can meet the requirements, while the elongation of A-2# and A-4# cannot meet the requirements. Especially the yield strength of plate A-2# exceeds 126 MPa over the standard yield strength. Therefore, controlling the centerline segregation of the continuous casting slab can effectively improve the service performance of the pipeline steel for adapting harsh environments.

5. Conclusions

(1) The relationship between the degree of centerline segregation and mechanical properties indicates that controlling the centerline segregation of continuous casting slab can achieve the requirements of a strain-based design, such as small fluctuations in strength, low yield ratio, high elongation, and better toughness at low temperatures;

(2) As the degree of centerline segregation increases, impact energy decreases gradually and DBTT increases. Meanwhile, the variability in impact toughness becomes larger, especially when the degree of segregation exceeds Mannesmann class 2; i.e., at −60 °C, the variability range in toughness (~75 J) of plate A (segregation degree higher than class 2) was higher than plate B (~40 J, segregation degree almost at class 2); at −80 °C, the variability range in toughness of plate A (~260 J) is higher than plate B (~75 J);

(3) With an increase in the degree of segregation, the tensile and yield strength increased; elongation decreased, the variability in strength increased. When the degree of centerline segregation is higher than the Mannesmann class 2, the variability range of tensile strength ranges from ~13 MPa to ~110 MPa;

(4) When the degree of centerline segregation in the continuous casting slab exceeds the Mannesmann class 2, the microstructure in the centerline of the corresponding rolling plate is transformed from ferrite and pearlite into bainite/martensite.

Author Contributions: C.S., R.D.K.M., and X.W. designed and supervised the research; W.L. performed the experiments; F.G. analyzed the data and wrote the paper.

Funding: This research was funded by National Key Research and Development Plan project of China, grant number 2017YFB0304700.

Acknowledgments: This research was financially supported by Anshan Iron and Steel Group Company and National Key Research and Development Plan project of China (2017YFB0304700).

Conflicts of Interest: The authors declare no conflict of interest.

References

1. Macia, M.L.; Kibey, S.A.; Arslan, H. Approaches to Qualify Strain-Based Design Pipelines. In Proceedings of the 8th International Pipeline Conference, American Society of Mechanical Engineers, Calgary, AB, Canada, 27 September–1 October 2010; pp. 365–374.

2. Zhou, J.; Horsley, D.; Rothwell, B. Application of strain-based design for pipelines in permafrost areas. In Proceedings of the 2006 International Pipeline Conference. American Society of Mechanical Engineers, Calgary, AB, Canada, 25–29 September 2006; pp. 899–907.

3. Liu, B.; Liu, X.J.; Zhang, H.J. Strain-based design criteria of pipelines. *J. Loss Prev. Proc.* **2009**, *22*, 884–888. [CrossRef]

4. Beckermann, C. *Macrosegregation ASM Handbook*; American Society of Metals: Geauga County, OH, USA, 2008; Volume 15, pp. 4733–4738.

5. Flemings, M.C. Our understanding of macrosegregation: Past and present. *ISIJ Int.* **2000**, *40*, 833–841. [CrossRef]

6. Fujda, M. Centerline segregation of continuously cast slabs influence on microstructure and fracture morphology. *J. Metal Mater. Miner.* **2005**, *15*, 45–51.

7. Ueyama, S.; Niizuma, M.; Yonezawa, K. Development of High Quality Heavy Plates on Steelmaking Process at Kimitsu Works. *Nippon Steel Tech. Rep.* **2013**, *104*, 102–108.

8. Bor, A.S. Effect of pearlite banding on mechanical properties of hot-rolled steel plates. *ISIJ Int.* **1991**, *31*, 1445–1446. [CrossRef]

9. Osamu, H.; Hidenari, K.; Yasuhiro, H.; Setsuo, K.; Hajime, B.; Shoji, S. Macro-and semi-macroscopic features of the centerline segregation in CC slabs and their effect on product quality. *ISIJ Int.* **1984**, *24*, 891–898.

10. Mendoza, R.; Alanis, M.; Perez, R.; Alvarez, O.; Gonzalez, C.; Juarez-Islas, J.A. On the processing of Fe–C–Mn–Nb steels to produce plates for pipelines with sour gas resistance. *Mater. Sci. Eng. A* **2002**, *337*, 115–120. [CrossRef]

11. Lin, B.Y.; Chen, E.T.; Lei, T.S. The effect of segregation on the austemper transformation and toughness of ductile irons. *J. Mater. Eng. Perform.* **1998**, *7*, 407–419. [CrossRef]

12. Tushal, K.; Shant, J.R.; Rajesh, K.G.; Kathayat, T.S. Understanding the Delamination and Its Effect on Charpy Impact Energy in Thick Wall Linepipe Steel. *J. Mater. Metall. Eng.* **2014**, *4*, 31–39.

13. Bhattacharya, D.; Roy, T.K.; Mahashabde, V.V. A study to establish correlation between intercolumnar cracks in slabs and off-center defects in hot-rolled products. *J. Fail. Anal. Prev.* **2016**, *16*, 95–103. [CrossRef]

14. Loucif, A.; Fredj, E.B.; Harris, N.; Shahriari, D.; Jahazi, M.; Lapierre-Boire, L.-P. Evolution of A-type macrosegregation in large size steel ingot after multistep forging and heat treatment. *Metall. Mater. Trans. B* **2018**, *49*, 1046–1055. [CrossRef]

15. Tao, J.; Hu, S.; Yan, F.; Zhang, Y.; Langley, M. A study of the mechanism of delamination fracture in bainitic magnetic yoke steel. *Mater. Des.* **2016**, *108*, 429–439. [CrossRef]

16. Zhang, G.; Yang, X.; He, X.; Li, J.; Hu, H. Enhancement of mechanical properties and failure mechanism of electron beam welded 300M ultrahigh strength steel joints. *Mater. Des.* **2013**, *45*, 56–66. [CrossRef]

17. Ji, Y.; Lan, P.; Geng, H.; He, Q.; Shang, C.; Zhang, J. Behavior of Spot Segregation in Continuously Cast Blooms and the Resulting Segregated Band in Oil Pipe Steels. *Steel Res. Int.* **2018**, *89*, 1700331. [CrossRef]

18. Su, L.; Li, H. The assessment of centreline segregation in continuously cast line pipe steel slabs using image analysis. In Proceedings of the 6th International Pipeline Technology Conference, American Society of Mechanical Engineers Ostend, Ostend, Belgium, 7–9 October 2013; Volume 1.

19. Collins, L.E. Effects of Segregation on the Mechanical Performance of X70 Line Pipe. In Proceedings of the 11th International Pipeline Conference, American Society of Mechanical Engineers, Calgary, AB, Canada, 26–30 September 2016; Volume 1.

20. Guo, F.; Wang, X.; Liu, W.; Shang, C.; Misra, R.D.K.; Wang, H.; Zhao, T.; Peng, C. The Influence of Centerline Segregation on the Mechanical Performance and Microstructure of X70 Pipeline Steel. *Steel Res. Int.* **2018**, *89*, 1800407. [CrossRef]

21. No. SN 960: 2009. *Classification of Defects in Materials-Standard Charts and Sample Guide*; SMS Demag AG Mannesmann: Düsseldorf, Germany, 2009.

22. Nayak, S.S.; Misra, R.D.K.; Hartmann, J.; Siciliano, F.; Gray, J.M. Microstructure and properties of low manganese and niobium containing HIC pipeline steel. *Mater. Sci. Eng. A* **2008**, *494*, 456–463. [CrossRef]

23. Su, L.; Li, H.; Lu, C.; Li, J.; Fletcher, L.; Simpson, I.; Barbaro, F.J.; Zheng, L.; Bai, M.; Shen, J.; et al. Transverse and z-direction CVN impact tests of X65 line pipe steels of two centerline segregation ratings. *Metall. Mater. Tran. A* **2016**, *47*, 3919–3932. [CrossRef]

24. Bhattacharya, D.; Poddar, G.P.; Misra, S. Centreline defects in strips produced through thin slab casting and rolling. *Mater. Sci. Tech.* **2016**, *32*, 1354–1365. [CrossRef]

25. Lee, H.; Jo, M.C.; Sohn, S.S.; Zargaran, A.; Ryu, J.H.; Kim, N.J.; Lee, S. Novel medium-Mn (austenite + martensite) duplex hot-rolled steel achieving 1.6 GPa strength with 20% ductility by Mn-segregation-induced TRIP mechanism. *Acta Mater.* **2018**, *147*, 247–260. [CrossRef]

26. Mukherjee, M.; Chintha, A.R.; Kundu, S.; Misra, S.; Singh, J.; Bhanu, C.; Venugopalan, T. Development of stretch flangeable ferrite–bainite grades through thin slab casting and rolling. *Mater. Sci. Tech.* **2016**, *32*, 348–355. [CrossRef]

27. Zhang, H.; Cheng, X.; Bai, B.; Fang, H. The tempering behavior of granular structure in a Mn-series low carbon steel. *Mater. Sci. Eng. A* **2011**, *528*, 920–924. [CrossRef]

28. Feng, R. Research on Internal Quality and Its Influencing Factors of Ferrite/Pearlite Steel Plate. Ph.D. Thesis, Shandong University, Shandong, China, 30 October 2013.

Metals **2019**, *9*, 749

29. DNV-OS-F101: 2000. *Offshore Standard Submarine Pipeline Design*; Det Norske, Veritas: Oslo, Norway, 2000.

 *metals*

Article

Fatigue Examination of HSLA Steel with Yield Strength of 960 MPa and Its Welded Joints under Strain Mode

Tomasz Ślęzak 

Faculty of Mechanical Engineering, Military University of Technology, ul. gen. S. Kaliskiego 2, 00-908 Warsaw, Poland; tomasz.slezak@wat.edu.pl; Tel.: +48-261-837-685

Received: 30 December 2019; Accepted: 4 February 2020; Published: 7 February 2020

Abstract: The full benefits of application the high strength low alloyed steels HSLA can be achieved if the structures will be able to carry the alternate loads and fatigue cracks will not be formed, even in the vicinity of welded joints. For this reason the purpose of this study is to find and to explain the influence of different factors on fatigue crack initiation and the nature of crack propagation in HSLA steel and its welded joints. The S960QL steel and two types of welded joints were subjected to low cycle fatigue (LCF) tests at a strain mode and the received surfaces of fractures were analyzed using SEM microscope. Additionally, the microhardness measurements and the residual stress analyze in a cross-section of the joint were conducted. The maximum hardness was determined on the fusion line and more favorable hardness distribution was in the square joints than in single-V. Compiled maps of residual stresses have shown that the local orientation and values of the principal stress vector near the fusion line can influence negative the fatigue life. Finally, the square joints tested in the low cycle fatigue regime have shown a slightly higher fatigue life in comparison with single-V.

Keywords: high strength steel; welded joints; fatigue strength; cracking; residual stresses

---

1. Introduction

The main purpose of modern companies is focused on the gaining economic efficiency of the production processes and the costs of the final product maintenance. In order to meet the increasing customer demands the metallurgical industry is still developing modern advanced structural steels with increasingly better performance characteristics. Although the advanced methods for improving the performance of structural materials present exist or are being developed [1–3], the reduction of total weight as the main factor affecting the production costs is achieved through the use of steels with a reduced cross-section or smaller thickness, compensated by higher strength. This approach leads to lower production costs and improved operating characteristics, such as capacity, strength or energy efficiency.

One of the most popular steel grades used to fulfill the aforementioned requirements are high performance steels (HPS), which can be divided into high strength low alloyed steels (HSLA) and high strength steels (HSS). These steels are produced in processes of thermo-mechanical processing (TMCP) and liquid-quenching and tempering (Q&T). They have a low content of micro-alloying elements (Ti, V or Al) and carbon resulting in a low carbon equivalent value (CE) and good weldability [4,5]. Moreover, all this makes the structure ultra fine-grained, consisting of martensite and bainite [4,6]. For this reason, HSS steels are widely used in the design solutions for heavily loaded elements and entire structures, in particular those transferring tensile and bending loads, but the grades of S960 steel or above have presently little applicability in construction. They can be found in bridges, pipelines, pressure vessels, platforms or mobile cranes [7–9] and S960/S1100 grades were used for example in Swedish military bridge Fast Bridge 48 [10] or the MS-20 scissors type Polish mobile assault bridge [11].

HSS steels have higher yield-to-tensile strength ratios, which reduces their ability to carry loads that cause plastic deformation. The value of this ratio reaches the value of 0.90–0.99 and unintentional, local loads exceeding the yield point, may lead to rapid destruction of the structure [12–16]. Moreover, the martensite/bainite structures increase the sensitivity of the steels to the presence of notches. These properties of HSS steel are characterized by a greater crack propagation rate *da/dN* during the crack growth stage together with increased resistance to fatigue crack initiation characterized by the higher value of fatigue crack threshold ΔK_{th} [17–19]. For this reason it is essential to discover the mechanism of crack initiation in these steels, especially taking under consideration the presence of welded joints. It is commonly known that the presence of welds significantly reduces the carrying capacity under the variable loads, characteristic of engineering structures and significantly deteriorates fatigue life. While this phenomenon is very well recognized in the case of carbon steels, for HSS steels despite some research, i.e., [20–25], further deepening of this knowledge is needed.

The perspective of the wide application of HSLA steels in engineering constructions has led to extended research on the welded joints prepared with these grades of steel [26–30]. High-strength low-alloy steels, due to their specific microstructure and alloying design, are sensitive to welding and the full benefit of applications of those steels can only be obtained by suitable welding [28]. Additionally, the most problematic domain in terms of strength/hardness of welded joints is the recrystallized zone, moreover crack initiation is commonly observed at the fusion line at the face side, but in some cases the crack initiation is observed at the root side too [30]. The main aim of the presented research was to investigate the behavior of HSS and two types of their welded joints subjected to high loads controlled at strain mode. Additionally, the study was extended by the measurements of microhardness, a residual stress analysis and by the SEM observation of received fractures. Indicated research program has allowed to perform a comprehensive analysis of investigated material in the area of fatigue failure influenced by different factors.

2. Materials and Methods

The research has been conducted on the high strength steel S960QL (1.8933). The verification of the chemical composition and mechanical properties was executed. Obtained results, along with the data from the delivery certificate, are presented in Tables 1 and 2.

Table 1. Chemical composition of used steel S960QL; wt [%].

	C	Si	Mn	Cr	Mo	Ni	Al	V	Cu	Ti	Nb	B
Measurement	0.18	0.36	1.19	0.23	0.66	0.052	0.11	0.032	0.19	0.012	0.002	0.0028
Certificate	0.18	0.28	1.13	0.22	0.67	0.077	0.08	0.027	0.18	0.004	n.d.	n.d.

n.d.—not detected.

Table 2. Mechanical properties of used steel S960QL.

	E [MPa]	σ_Y [MPa]	σ_U [MPa]	EL [1] [%]	RA [2] [%]
Measurement	2.2×10^5	974	1070	14.2	45.6
Certificate	-	997	1069	13.0	-

[1] EL—elongation; [2] RA—reduction of area.

Chemical composition was determined by the energy-dispersive X-ray spectroscopy (EDS) method using a scanning electron microscope JSM-6610 (Jeol Ltd., Tokyo, Japan) equipped with an EDS spectrometer X-Max 50 (Oxford Instruments NanoAnalysis, High Wycombe, UK) whereas the strength properties were established during tensile tests carried out according to the standard [31] and the post-test made measurements. The microstructure of the investigated S960QL steel is shown in Figure 1.

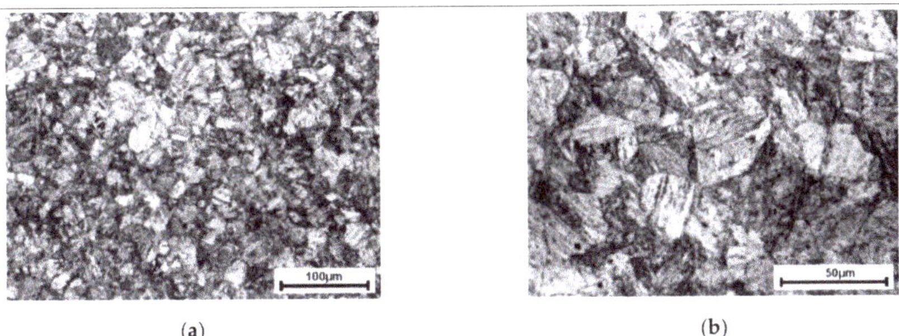

(a)   (b)

Figure 1. The microstructure of S960QL steel shown at two different magnifications: (**a**) view of fine grained structure; (**b**) selected grains with bainitic structure.

The photos of the structure were obtained on metallographic microsections after grinding (finished at grade 2000) and polishing (1 μm Al_2O_3) which were then subjected to etching for approx. 15 s with nital (4% solution of nitric acid in ethanol). This steel features a fine-grained martensitic-bainitic structure with equivalent grain diameter of 10–25 μm (the cross-section of etched material was observed using SEM and the surface areas of selected grains were measured; the equivalent grain diameters were calculated as the diameter of circle with the same measured surface area).

The main investigations were performed on prepared two types of butt welds, namely a square joint and a single-V joint. The welding parameters were selected in order to reduce the amount of input heat which, in accordance with the welding requirements for the tested material should not exceed 1 kJ/mm [32]. The welds were made using MAG welding with shielding gas containing 82% CO_2 and 18% Ar (EN ISO 14175-M21-ArC-18). Wire UNION X96 (EN ISO 16834-A-G Mn4Ni25CrMo) with a diameter of 1.2 mm was used to make the welds. Maximum heat did not exceed 0.75 kJ/mm. The second crucial difference in the execution of the welds was the type of welding process. The square joints were made using an automatic process, contrary to the single-V joints where a root was made manually, followed by automatic welding of a face. Detailed descriptions of the welding conditions, geometry of the welded joints before welding and macroscopic evaluations of the welds can be found in [33]. The quality assessment was made on the basis of X-ray tests and confirmed the high quality of the welds. The negligible welding defects were detected in the square joints, but in the single-V joints their concentration was significantly greater.

The assessment of fatigue properties was performed in terms of the low cycle fatigue (LCF) and carried out for the paternal material (PM) and welded joints (WJ). Fatigue tests were conducted on flat samples prepared on the basis of the ASTM E606-4 standard [34]. Test samples were cut from a slab with a nominal thickness of 6 mm. The length of reduced section in the samples has got 28 mm for PM (55 mm for WJ) and the width of 6 mm (24 mm for WJ). The technical drawings of the samples were shown in Figure 2.

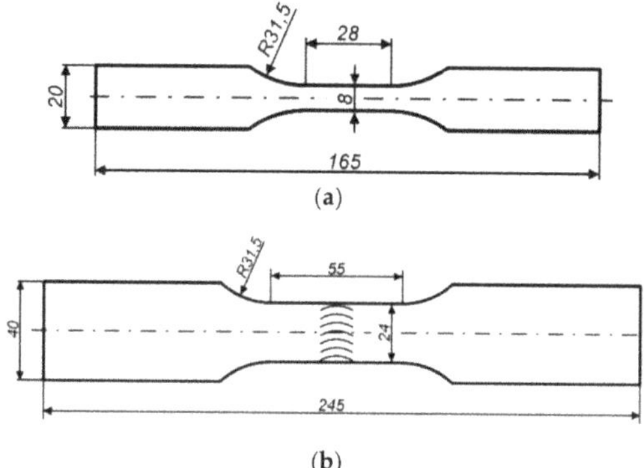

Figure 2. The dimensions of samples used during LCF tests of: (**a**) paternal material; (**b**) welded joints.

The fatigue tests were performed on the material as supplied to reflect the causes of fatigue crack initiation in real conditions. Fatigue tests were carried out using an Instron 8808 pulsator (Instron, Norwood, MA, US) equipped with a dynamic extensometer with a 25 mm and 50 mm gauge length, respectively, for the fatigue tests of PM and WJ. The tests were conducted in strain mode controlled with the value of the total strain amplitude ε_{ac} with a sinusoidal waveform. The values of ε_{ac} fit in the range from 0.30% to 1.5% for PM and from 0.15% to 0.40% for WJ. The value of the strain ratio R of 0.1 and an average strain rate of $\dot{\varepsilon} = 10^{-2}$ s^{-1} was adopted. Used value of $\dot{\varepsilon}$ has prevented the samples against the growth of temperature in at the higher strains, and has allowed for shorting the duration of testing. The adopted failure criterion was a 25% decrease of maximum load. During the research a minimum of ten reliable tests were performed for each case (PM and two types of WJ).

The main investigations were expanded by the measurements of microhardness, the measurements of residual stresses in the cross-section of the weldments and the fractographic analysis of the fatigue fracture surfaces in order to more accurate description the influence of welding on fatigue behavior of high strength steel.

The measurement of Vickers HV0.1 micro-hardness on cross-sections of the welds of examined HSS steel was carried out using a semi-automatic micro-hardness meter (Shimadzu, Kyoto, Japan).

Residual stresses have been measured in the cross-section of WJ-S joint in two directions: vertically (perpendicularly to the sample surfaces) and horizontally (parallel to the sample surfaces). The surface of metallographic micro section was prepared by grinding under final gradation not lower than 600 and electrochemical polishing. The aim of preparation was to remove surface layer disrupted by machining. Measurements of residual stresses were conducted by X-ray diffraction technique using the sin2ψ method. The angular position of the selected Bragg diffraction *hkl* was the measured value, whose change reflects the elastic deformation of the crystal lattice of the tested material. The measurement was performed using a Rigaku MSF-3M system (Rigaku, Tokyo, Japan) equipped with an X-ray tube with a cobalt anode (filtered radiation CoKα, λ = 1.79026 Å). The fractographic observations were conducted using a JSM-6610 scanning electron microscope equipped with two detectors: secondary electrons (SE) and backscattered electrons (BSE).

3. Results

3.1. LCF Tests for S960QL Steel and Its Welded Joints

The results of the performed low cycle fatigue tests are presented in the form of recorded exemplary courses of fatigue parameters: a maximum stress σ_{max}, a stress amplitude σ_a, and an amplitude of plastic strain component ε_{ap}, both for PM and two types of WJ. Obtained data are shown in Figures 3 and 4, nevertheless in the case of WJ the results of the amplitude of plastic strain component ε_{ap} and the maximum stress σ_{max} are presented.

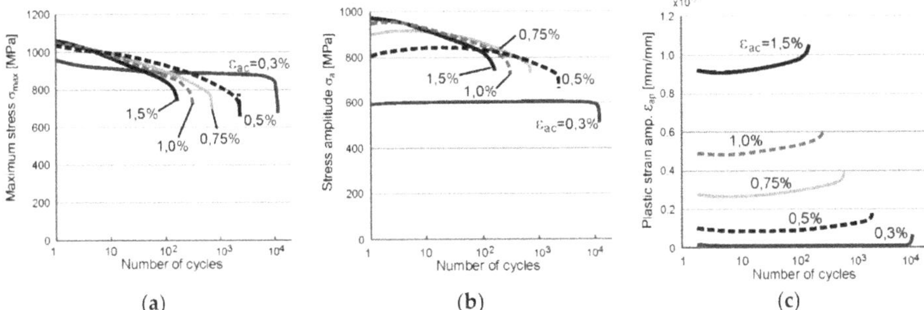

Figure 3. The results of LCF tests recorded for PM as the courses of: (**a**) maximum stress σ_{max}; (**b**) stress amplitude σ_a; (**c**) amplitude of plastic strain component ε_{ap}.

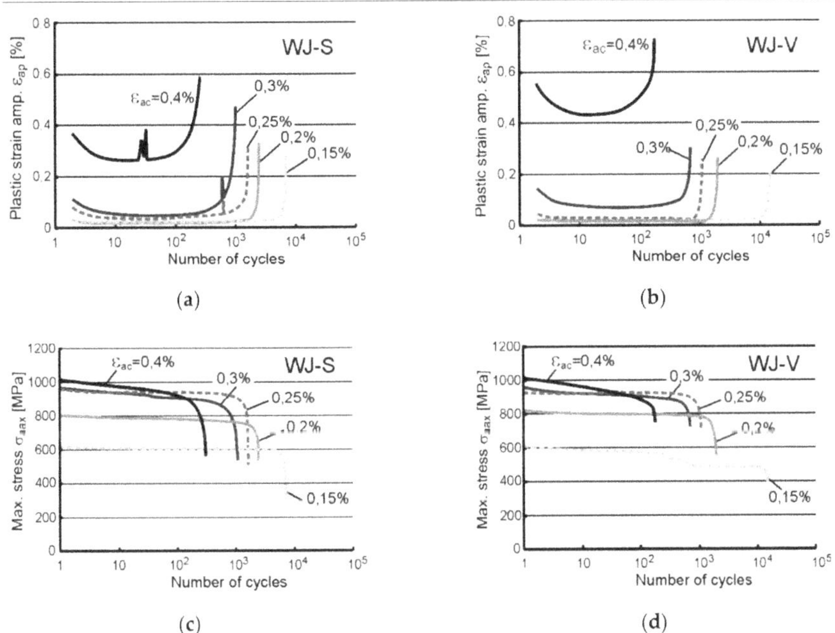

Figure 4. The results of LCF tests recorded for WJ as the courses of: (**a**) ε_{ap} recorded for WJ-S; (**b**) ε_{ap} recorded for WJ-V; (**c**) σ_{max} recorded for WJ-S; (**d**) σ_{max} recorded for WJ-V.

On the basis of the graphs presented in Figure 3 it can be stated that the steel S960QL is subjected to cyclic weakening at load amplitudes ε_{ac} higher than 0.5% (Figure 3a) while this phenomenon has not been observed at the amplitude of 0.3%. This indicates a low propensity of the examined steel to strengthen. A change in the behavior occurs after 5–10 cycles what can be seen in Figure 3a,c. It involves the stabilization of the parameters (at $\varepsilon_{ac} = 0.3\%$) or a change of their course (at other values of ε_{ac}) from increasing/decreasing to the opposite (both in case of σ_a and ε_{ap}).

Significant differences in the courses of the plastic strain amplitude ε_{ap} were revealed during the tests of both types of welds WJ-S and WJ-V (Figure 4a,b). At loads ε_{ac} equal to 0.3% and 0.4% the amplitude ε_{ap} is much higher in WJ-V than in WJ-S joints. At lower loads the recorded values of ε_{ap} are comparable, although it can be noted that in the case of WJ-S the values are slightly higher. Courses of σ_{max} value (Figure 4c,d) have not shown significant differences.

The further study of the cyclical behavior of examined PM and its welds, both the WJ-S and WJ-V, was performed on the basis of data obtained from each LCF fatigue test and the values of parameters determined from recorded stabilized hysteresis loops. This data have been used to conduct the fatigue analysis based on the Manson-Coffin-Basquin relationship (1):

$$\varepsilon_{ac} = \varepsilon_{ae} + \varepsilon_{ap} = \sigma'_f \cdot E^{-1} \cdot (2N_f)^b + \varepsilon'_f \cdot (2N_f)^c \tag{1}$$

where:

ε_{ac}—a total strain amplitude;

ε_{ae}—an elastic component of total strain;

ε_{ap}—a plastic component of total strain;

σ'_f—a fatigue strength coefficient;

ε'_f—a fatigue ductility coefficient;

N_f—a number of cycles to failure;

$2N_f$—a number of reversals to failure;

b—a fatigue strength exponent;

c—a fatigue ductility exponent.

In order to determine the values of fatigue coefficients and exponents, two curves are considered separately, namely elastic curve and plastic. These curves have courses similar to rectilinear in log-log system of strain amplitude (ε_{ae} or ε_{ap}) and number of reversals ($2N_f$). That means that they can be determined easily and than add up, according to relationship (1).

In Figure 5 the strain-life curves obtained for S960QL steel and two types of their WJ are presented. There are shown both the experimental data (points) and the curves described by relationship (1).

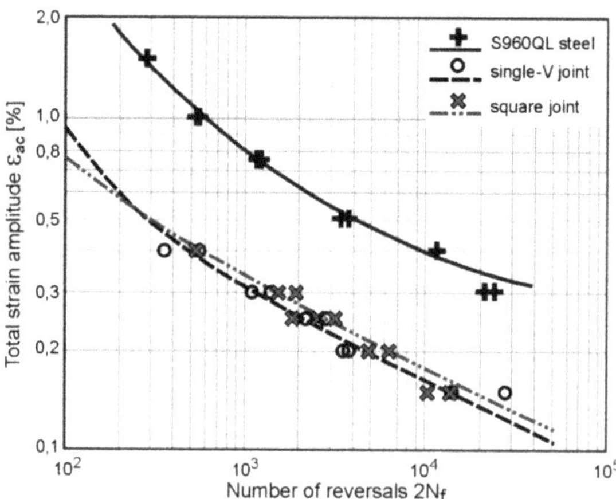

Figure 5. The strain-life curves obtained for PM and their welded joints: WJ-V and WJ-S.

Determined values of individual coefficients and the corresponding exponents, both for PM and welded joints WJ-S and WJ-V, are summarised in Table 3. There are also the values of the correlation coefficient R^2 corresponding to particular groups of parameters placed.

Table 3. Fatigue properties of examined HSS steel and two types of welded joints.

	σ'_f [MPa]	b [-]	ε'_f [-]	c [-]
PM	1976	−0.107	0.542	−0.782
	$R^2 = 0.998$		$R^2 = 0.991$	
WJ-V	4326	−0.263	0.264	−1.073
	$R^2 = 0.979$		$R^2 = 0.992$	
WJ-S	4150	−0.267	4.408	−1.523
	$R^2 = 0.950$		$R^2 = 0.999$	

3.2. Measurement of Microhardness

The measurements were carried out on the cross-sections of samples in two rows which were in a distance of approximately 1 mm from the surface. The distribution of micro-hardness obtained in the tested WJ-S and WJ-V joints are shown in Figures 6 and 7.

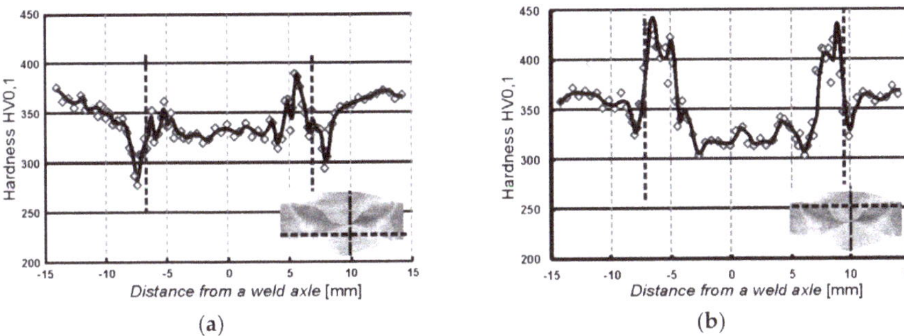

(a)   (b)

Figure 6. Distribution of micro-hardness measured in the WJ-S joint on: (**a**) the side of first welding run; (**b**) the side of second welding run.

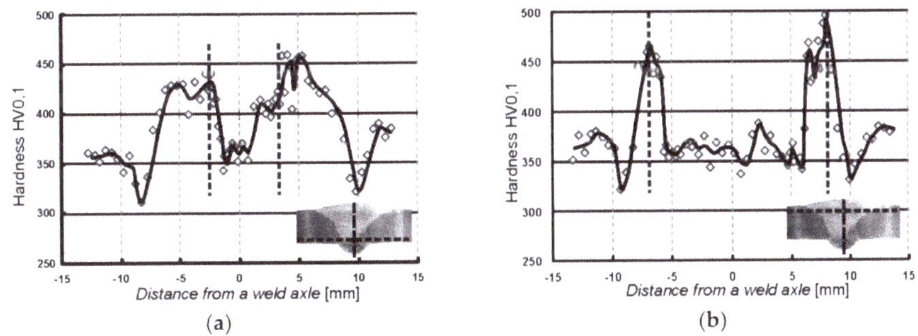

(a)   (b)

Figure 7. Distribution of micro-hardness measured in the WJ-V joint on: (**a**) root side; (**b**) face side.

The distributions of micro-hardness in measurement rows involving welding runs made first are shown in Figures 6a and 7a, contrary to Figures 6b and 7b, where the distributions in measurement rows covering the second runs are presented. On the weld thumbnails horizontal lines indicate the track along which the measurements were made and the vertical lines indicate the weld axis from with the distances were measured. Vertical dashed lines placed in the plots represent the position of the fusion line of individual welds.

On the basis of the presented results it can be stated that the greatest micro-hardness of almost 500 HV0.1 is located into welded area in the immediate vicinity of the fusion line. In addition, tempering zones (incomplete normalization) are visible, manifested by the maximum hardness decrease out of the weld. Moreover, the course of hardness in the first run is smoothed as the beneficial effect of the second welding run (Figures 6a and 7a).

3.3. Residual Stresses Analysis

A measurement grid was established on the cross section of the weld WJ-S with the distance between the measuring points of 1 mm. Almost one hundred measurement points were defined and the received results has allowed for preparing maps of projection of the principal stress vector, closer to individual direction, in the vertical and horizontal direction. Obtained maps of residual stresses are shown in Figure 8.

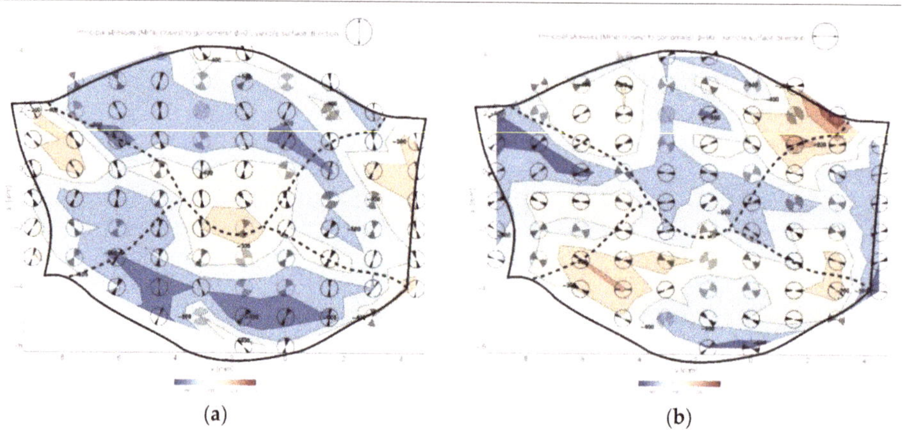

(a)   (b)

Figure 8. Residual stresses in WJ-S joint cross section in the form of projections of the principal stress vector in two directions: (**a**) vertical; (**b**) horizontal. The fusion lines are located by dashed lines.

The direction of the principal stress vector on the residual stresses maps is identified by pointing the directional marker located in individual circular areas. The value of the absolute uncertainty of this direction is specified by the opening angle of the marker, moreover its tone shows the uncertainty of determining the value (the brighter the greater uncertainty of measurement).

The obtained results clearly indicate that compressive stresses are present in the whole area of the weldment. However in the vicinity of the fusion line (near the surface) a reduction of compressive stresses in the horizontal direction can be noticed (Figure 8b). The stress reduction is also apparent in the vertical direction (Figure 8a) in the middle of the weld, which is directly related to the direction of heat flow resulting in the formation of temperature stresses. Furthermore, the directions of the principal stresses are characteristically oriented with respect to the fusion line, namely they are approximately perpendicular/parallel to this line.

3.4. Fractographic Analysis of Fatigue Fractures

The results of the observation of fatigue fracture surfaces of the tested samples, both PM and WJ, are presented. Representative photos of PM fractures were taken on the samples tested at the load ε_{ac} of 0.4% and 0.75% (Figures 9 and 10). The fractographic analysis of fatigue fractures of the welded joints WJ-S and WJ-V was carried out on the samples loaded with an amplitude $\varepsilon_{ac} = 0.3\%$ (shown in Figures 11 and 12).

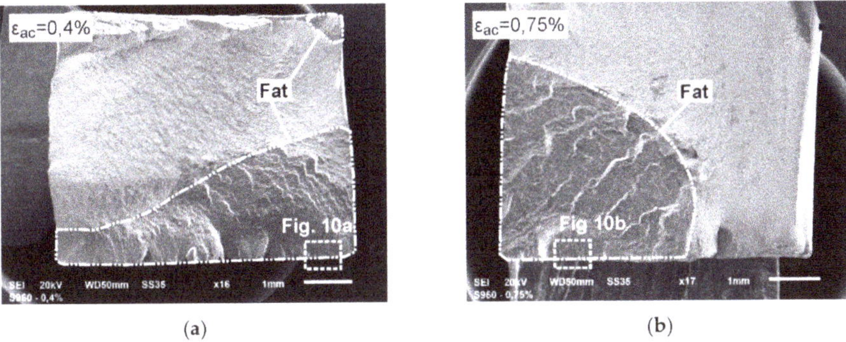

Figure 9. The fatigue fractures of PM tested at load ε_{ac}: (**a**) 0.4%; (**b**) 0.75%.

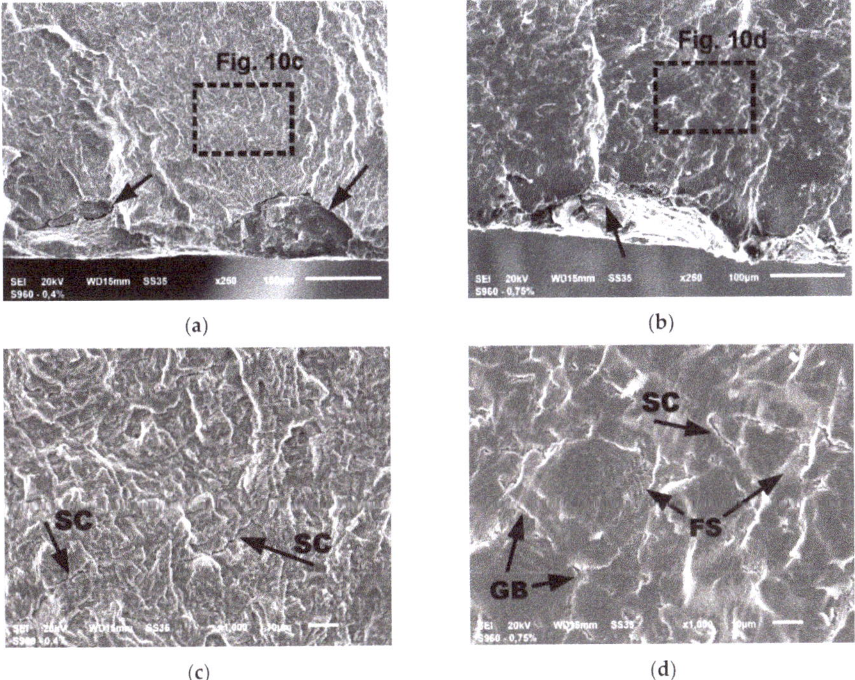

Figure 10. The details of fatigue fractures: (**a**) the origins in sample tested at ε_{ac} = 0.4%; (**b**) the origin in sample tested at ε_{ac} = 0.75%; (**c**) the stage of stable crack growth in sample tested at ε_{ac} = 0.4%; (**d**) the stage of stable crack growth in sample tested at ε_{ac} = 0.75%.

Envelopes marked in Figure 9 and named *FAT* indicate the areas of fatigue cracking. Both surfaces presented in Figure 9 are characterized by a very diverse morphology and system of numerous origins of fatigue cracking. A greater number of origins have occurred in the samples tested at lower value of ε_{ac} (one origin was also created at the opposite surface of the sample—Figure 9a). This is associated with inhibition of subsequent crack initiation due to the rapid growth of cracks that had appeared first. The structure of the presented fractures indicates the high crack propagation rate and is characterized by numerous offsets. In the case of lower loads development of cracking is more gentle (Figure 9a).

The photographs in Figure 10a,b show the site of fatigue crack initiation, while in Figure 10c,d the areas of the stable crack growth in a distance of about 0.1 mm from the initiators.

The initiation of fatigue cracks proceeded in local concentration of stress within the hard and brittle rolled-in scale particles on the surface of the material, what is indicated by arrows in Figure 10a,b. The ductile character of cracking is preserved in the initial stage of cracking as well as during stable growth. The most visible differences in the images of the fracture surface are the squashed areas in the sample tested at 0.75% resulting from the presence of large surface strains in the compression phase of the load. In both presented cases, cracking is accompanied by numerous secondary cracks (SC)—Figure 10c,d. Moreover, at high strains SC cracks run mainly along the grain boundaries (GB) as marked in Figure 10d. Frequently occurring fatigue striations (FS) were also discovered, nevertheless in many cases they have been crushed (Figure 10d).

The fractographic analysis of the fatigue fractures tested welded joints WJ-S and WJ-V was begun from the evaluation of panoramic photos (Figure 11). The details of analyzed surfaces are shown in Figure 12.

Fractures shown in Figure 11 display a varied course of cracking. A more rapid course of cracking with complex morphology was observed in the case of the single-V joint (Figure 11b) contrary to the square joint where the crack propagation is more gentle (Figure 11a). There were no dominating origins of fatigue cracks. Initiation in both cases occurred over the entire length of the fusion line (in WJ-V only from the side of the root), hence generating a uniform cracking front already at an early stage. There were also minor, local offsets connecting the developing local cracks.

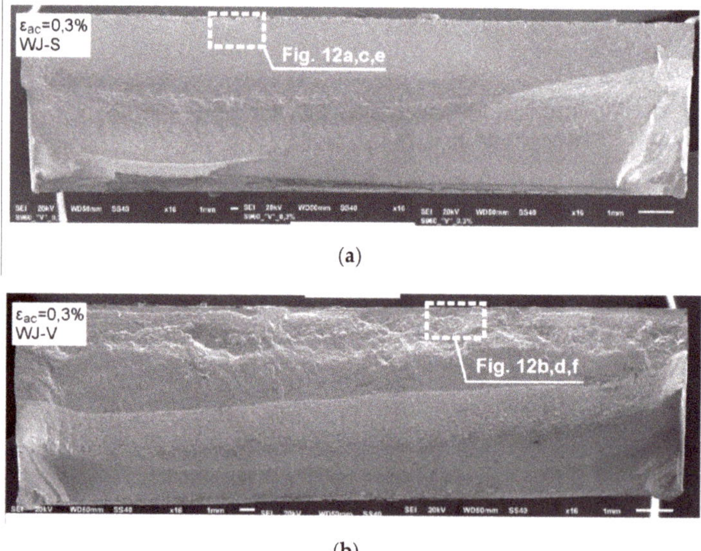

Figure 11. The panoramic photographs of fatigue fractures of the welded samples tested at ε_{ac} = 0.3%: (**a**) WJ-S joint; (**b**) WJ-V joint.

The photographs of the fracture areas shown in Figure 12 indicate also the regions of fatigue crack initiation (Figure 12a,b), the areas of the initial crack growth in a distance of about 0.15 mm from the sample surface (Figure 12c,d) and the areas of rapid crack growth in a distance of approximately 0.4 mm from the surface (Figure 12e,f) for samples WJ-S and WJ-V, respectively.

The mechanism of crack initiation in both cases is a little different. In the WJ-S the direct initiators are very small particles of mill scale (Figure 12a), while in WJ-V initiation is related to the discontinuity of the fusion line in the weld root, which was made manually. Furthermore, the occurrence of gas bubbles and numerous micropores (P) was found in the single-V joint (Figure 12b,d) affecting the crack propagation rate. Both at the initial stage of development of fatigue cracks (Figure 12c,d) as well as a further stage (Figure 12e,f) the ductile nature of cracking is preserved, what is evidenced by the structure of fractures presented with recognizable fatigue striations. The course of cracking in WJ-V at later stage is more rapid than in the WJ-S probably because it runs along the border of the weld and the HAZ. This is evidenced by the microstructure of the surface shown in Figure 12f with a varied topography and frequently occurring areas of quasi-static cracks (QS).

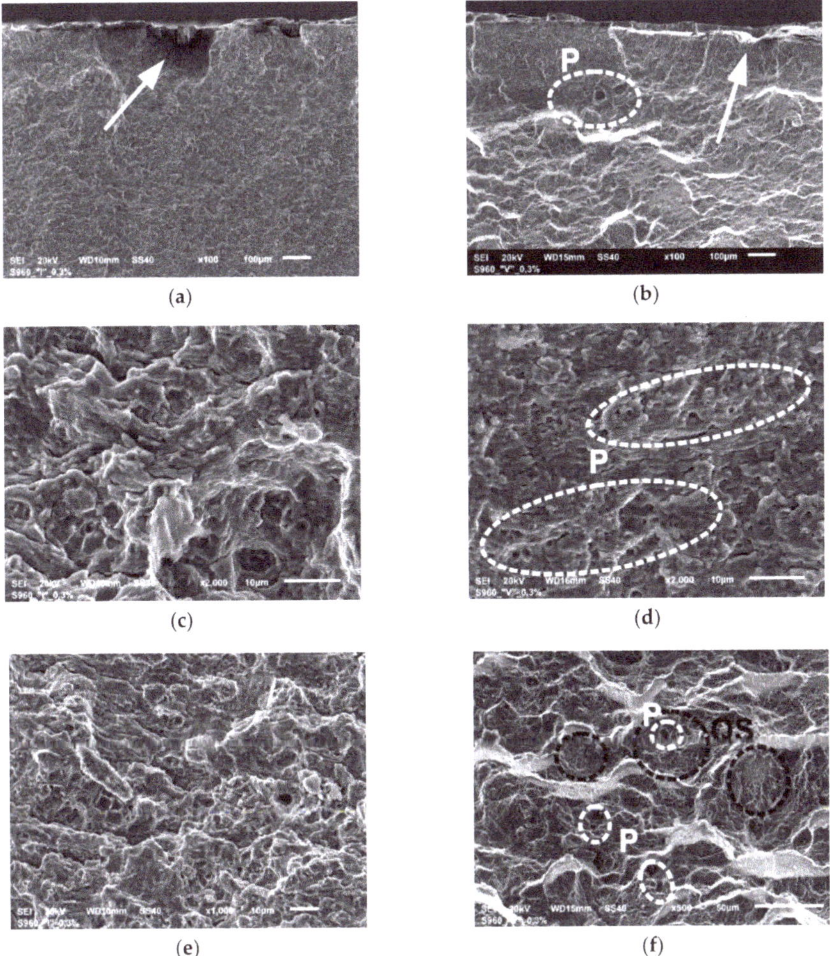

Figure 12. The details of fatigue fractures of WJ tested at ε_{ac} = 0.3%: (**a**) the region of initiation in sample WJ-S; (**b**) the region of initiation in sample WJ-V; (**c**) the stage of stable crack growth in sample WJ-S; (**d**) the stage of stable crack growth in sample WJ-V; (**e**) the stage of rapid crack growth in sample WJ-S; (**f**) the stage of rapid crack growth in sample WJ-V.

4. Discussion

The courses of microhardness values (Figures 6 and 7) verify that zones of significantly increased hardness on the fusion line are present in both types of welded joints. It is worth stressing the favorable distribution of hardness in the case of WJ-S joints. It is characterized by a smaller gradient, a lower maximum value of micro-hardness in the area of the fusion line (approx. 445 HV0.1) and lower hardness of the weld itself. This is probably due to the fact that a larger quantity of heat was introduced in the first run, which led to a higher interpass temperature and consequently contributed to a reduction in the cooling rate. The greater gradient and higher values of hardness were measured in WJ-V weld which comparing with generally known fact indicate an increase in material strength, while local decreasing ductility and cracking resistance.

Residual stresses which arise in welded joint and their vicinity as a consequence of thermal strains caused during heating/cooling cycles, also effected the fatigue behavior of welded structures. The results of conducted measurements (Figure 8) show their not entirely positive influence, with the highest magnitude (red region) occurring in the vicinity of the fusion line. The local orientation of the principal stress vector in this region has negative influence on the fatigue life, because one of the component coincides with the direction of the stress resulted from external applied load. It results directly from the sum of post-welded residual stresses and stresses induced by external load.

LCF fatigue tests conducted in strain mode have allowed for the assessment of fatigue behavior of examined HSS S960QL steel and its welded joints. Moreover, it was possible to observe the influence on fatigue life due to weld type and automation of welding process while maintaining the same other welding conditions. The courses of plastic strains curves (Figure 3c, Figure 4a,b) state that this component of strain is substantially higher in WJ compared to PM. Additionally the plastic strains are slightly higher in WJ-V joint in comparison with WJ-S. The obtained results indicate slightly better fatigue properties of welded joints made of S960QL steel using WJ-S joints. It was observed that their fatigue life is on average higher by 20–50% than the fatigue life of WJ-V. Fatigue properties describing the elastic component (σ'_f and b) are very similar in both types of welded joints, while those describing the plastic component (ε'_f and c) how significant differences. This demonstrates similar fatigue characteristics in conditions of stress component dominance, i.e., at lower loads, and at the same time a higher fatigue life. A different character of fatigue occurs under significant influence of plastic component.

The fractographic investigation of fatigue fracture has revealed the mechanism of crack initiation, localization of the origins and the differences between concerned cases. In paternal material (PM) the initiators of fatigue cracks have occurred in the immediate vicinity of the brittle particles of mill scale rolled-in on the surface of material (Figure 10). They became notches inducing stress concentration caused by local permanent plastic strains, already at the production phase. Fractographic investigation of the fractures of welded joints has shown the absence of clear origins of fatigue cracking. Initiation occurred over the entire length of the fusion line, whereas in WJ-V initiation followed always on the side of the root (Figure 11b). It was caused by the presence of welding imperfections induced during manual welding of first pass from the root side. The structure of fracture surface exhibits the characteristics of ductile cracking, and faster development of cracks in the WJ-V resulted from the presence of gas bubbles and micro-pores. In WJ-S crack propagation is more stable which proves that mechanized welding allows to achieve much better quality of welds and improve fatigue behavior already at stage of execution of welding.

Summarizing the above discussion, we can state that the fatigue life of high strength steels can be extended by reducing the presence of crack initiators introduced during production process. It was disclosed that fully mechanized welding in the case of considered steel grade produces better results than partly mechanized thus it should point toward to introduce mechanized welding as widely as possible. Additionally, square welded joints show better properties than single-V joints what influence the fatigue properties.

5. Conclusions

The paper presents the results of fatigue tests of S960QL steel and its welded joints supplemented by the measurement of residual stresses in the cross-section of the weld, hardness measurements and fractographic examination of fatigue fractures. The analysis of the results leads to the following conclusions:

- Increased resistance to fatigue cracks initiation of structural HSS steel is mishandled due to the presence of induced crack initiators in a form of rolled-in mill scale. Therefore fatigue behavior of S960QL steel in delivery state can be improved by refinement the production processes influencing the quality of surface.

- Concerning the microhardness distribution in examined welded joints, the square joints show more favorable courses than single-V. The maximum value of HV0.1 in HAZ is clearly lower what influence the ductility of this zone and their fatigue properties, especially crack resistance.

- In the low cycle fatigue regime, square joints show a higher fatigue life in comparison with single-V. The superior fatigue life corresponds to a more favorable microhardness distribution and less number of welding imperfections. Conducted fractographic investigation has revealed that the origins of crack initiation were the surface imperfections from the root side and the presence of gas pores additionally increase a cracking rate.

- Higher quality of square joints was caused primarily by fully-mechanized execution of welds in contrast to the second type of considered joint. Reduction of imperfections amount and its magnitude is especially important in high strength steels sensitive to notches.

Funding: This research was funded by Faculty of Mechanical Engineering, Military University of Technology, Warsaw, internal grant number RMN 08-878 aimed to support scientific development of young scientists. The APC was funded by Faculty of Mechanical Engineering, Military University of Technology, Warsaw.

Acknowledgments: The author is grateful to Lucjan Śnieżek for advice and guidance during planning and realization of final research program and also to Janusz Torzewski, who have supported the technical realization of LCF tests.

Conflicts of Interest: The author declares no conflict of interest.

References

1. Kocanda, D.; Hutsaylyuk, V.; Slezak, T.; Torzewski, J.; Nykyforchyn, H.; Kyryliv, V. Fatigue crack growth rates of S235 and S355 steels after friction stir processing. *Mater. Sci. Forum* **2012**, *726*, 203–210. [CrossRef]

2. Weglowski, M.S.; Kopyscianski, M.; Dymek, S. Friction stir processing multi-run modification of cast aluminum alloy. *Key Eng. Mat.* **2014**, *611–612*, 1595–1600. [CrossRef]

3. Zasimchuk, E.; Markashova, L.; Baskova, O.; Turchak, T.; Chausov, N.; Hutsaylyuk, V.; Berezin, V. Influence of Combined Loading on Microstructure and Properties of Aluminum Alloy 2024-T3. *J. Mater. Eng. Perform.* **2013**, *22*, 3421–3429. [CrossRef]

4. Ray, P.K.; Ganguly, R.I.; Panda, A.K. Optimization of mechanical properties of an HSLA-100 steel through control of heat treatment variables. *Mat. Sci. Eng. A* **2003**, *346*, 122–131. [CrossRef]

5. Show, B.K.; Veerababu, R.; Balamuralikrishnan, R.; Malakondaiah, G. Effect of vanadium and titanium modification on the microstructure and mechanical properties of a microalloyed HSLA steel. *Mat. Sci. Eng. A* **2010**, *527*, 1595–1604. [CrossRef]

6. Hildebrand, J.; Werner, F. Change of structural condition of welded joints between high-strength fine-grained steels and structural steels. *J. Civ. Eng. Manag.* **2004**, *X*, 87–95. [CrossRef]

7. Günther, H.P. *Use and Application of High-Performance Steels for Steel Structures (Structural Engineering Documents 8)*; IABSE: Zurich, Switzerland, 2005.

8. Muller, T.; Straetmans, B. High strength seamless tubes and steel hollow sections for cranes and machine building applications—Production and properties. *Stahlbau* **2015**, *84*, 650–654. [CrossRef]

9. High-Strength and Ultra-High-Strength Heavy Plates. Available online: https://www.voestalpine.com/alform/en/content/download/4494/file/voestalpine_heavy_plate_TTD_ALFORM_620-1100_E_0319.pdf (accessed on 12 December 2019).

10. Gogou, E. Use of High Strength Steel Grades for Economical Bridge Design. Master's Thesis, Delft University of Technology, Delft, The Netherlands, April 2012. Available online: https://www.scia.net/sites/default/files/thesis/reportgogou.pdf (accessed on 12 December 2019).

11. Mosty. Available online: http://www.obrum.gliwice.pl/mosty (accessed on 12 December 2019).

12. Lachowicz, M.; Nosko, W. Welding of Structural Steel Weldox 700. *Prz. Spaw.* **2010**, *1*, 13–18. (In Polish)

13. Bjorhovde, R. Performance and Design Issues for High Strength Steel in Structures. *Adv. Struct. Eng.* **2010**, *13*, 403–411. [CrossRef]

14. Shi, G.; Hu, F.; Shi, Y. Recent research advances of high strength steel structures and codification of design specification in China. *Int. J. Steel Struct.* **2014**, *14*, 873–887. [CrossRef]

15. Björklund, O.; Larsson, R.; Nilsson, L. Failure of high strength steel sheets: Experiments and modelling. *J. Mater. Process. Technol.* **2013**, *213*, 1103–1117. [CrossRef]

16. Branco, R.; Berto, F. Mechanical Behavior of High-Strength, Low-Alloy Steels. *Metals* **2018**, *8*, 610. [CrossRef]

17. Wei, D.Y.; Gu, J.L.; Fang, H.S.; Bai, B.Z.; Yang, Z.G. Fatigue behavior of 1500 MPa bainite/martensite duplex-phase high strength steel. *Int. J. Fatigue* **2004**, *26*, 437–442. [CrossRef]

18. Neimitz, A. Ductile Fracture Mechanisms in the High-strength Steel Hardox-400. Microscopic Observations and Numerical Stress-strain Analysis. *Procedia Mater. Sci.* **2014**, *3*, 270–275. [CrossRef]

19. Simunek, D.; Leitner, M.; Grün, F. In-situ crack propagation measurement of high-strength steels including overload effects. *Procedia Eng.* **2018**, *213*, 335–345. [CrossRef]

20. Dunder, M.; Vuherer, T.; Samardzic, I. Weldability of microalloyed high strength steels TStE 420 and S960QL. *Metalurgija* **2014**, *53*, 335–338.

21. Gałkiewicz, J. Simulation of Tensile Test of The 1/2Y Welded Joint Made of Ultra-High Strength Steel. *Mater. Sci. Forum* **2012**, *726*, 110–117. [CrossRef]

22. Slezak, T.; Sniezek, L. Fatigue Properties and Cracking of High Strength Steel S1100QL Welded Joints. *Key Eng. Mat.* **2014**, *598*, 237–242. [CrossRef]

23. Sołtysiak, R. Effect of Laser Welding Parameters of DUPLEX 2205 Steel Welds on Fatigue Life. *Solid State Phenom.* **2014**, *223*, 11–18. [CrossRef]

24. Ślęzak, T.; Śnieżek, L. A Comparative LCF Study of S960QL High Strength Steel and S355J2 Mild Steel. *Procedia Eng.* **2015**, *114*, 78–85. [CrossRef]

25. Ślęzak, T. Prediction of Fatigue Life of Welded Joints Made of Fine-Grained Martensite-Bainitic S960QL Steel and Determination of Crack Origins. *Adv. Mater. Sci. Eng.* **2019**, *201*, 9520801. [CrossRef]

26. Lan, X.; Chan, T.-M. Recent research advances of high strength steel welded hollow section joints. *Structures* **2019**, *17*, 58–65. [CrossRef]

27. Berdnikova, O.; Pozniakov, V.; Bernatskyi, A.; Alekseienko, T.; Sydorets, V. Effect of the structure on the mechanical properties and cracking resistance of welded joints of low-alloyed high-strength steels. *Procedia Struct. Integr.* **2019**, *16*, 89–96. [CrossRef]

28. Ilić, A.; Ivanović, L.; Lazić, V.; Josifović, D. Welding method as influential factor of mechanical properties at high-strength low-alloyed steels. *IOP Conf. Ser. Mater. Sci. Eng.* **2019**, *659*, 012036. [CrossRef]

29. Lahtinen, T.; Vilaça, P.; Peura, P.; Mehtonen, S. MAG Welding Tests of Modern High Strength Steels with Minimum Yield Strength of 700 MPa. *Appl. Sci.* **2019**, *9*, 1031. [CrossRef]

30. Vimalraj, C.; Kah, P.; Layus, P.; Belinga, E.M.; Parshin, S. High-strength steel S960QC welded with rare earth nanoparticle coated filler wire. *Int. J. Adv. Manuf. Technol.* **2019**, *102*, 105–119. [CrossRef]

31. ISO 6892-1. *Metallic Materials—Tensile Testing—Part 1: Method of Test at Room Temperature*; ISO: Geneva, Switzerland, 2019.

32. SuperElso®960. A High Yield Strength Steel for Welded and Weight-Saving Structures. *ArcelorMittal—Industeel.* Available online: http://www.solid.cl (accessed on 3 July 2012).

33. Ślęzak, T. Characteristics of MAG Welded Joints Made in Fine-Grained High-Strength Steel S960QL. *Inst. Weld. Bull.* **2018**, *62*, 45–53. [CrossRef]

34. ASTM E606-04. *Standard Practice for Strain-Controlled Fatigue Testing*; ASTM: Philadelphia, PA, USA, 2005.

Article

Low Cycle Fatigue Behavior of Elbows with Local Wall Thinning

Muhammad Faiz Harun [1], Roslina Mohammad [1,*] and Andrei Kotousov [2] 

[1] Razak Faculty of Technology and Informatics, Universiti Technologi Malaysia, Jalan Sultan Yahya Petra, Kuala Lumpur 54100, Malaysia; mfaizharun@yahoo.com.my

[2] School of Mechanical Engineering, The University of Adelaide, Adelaide SA 5005, Australia; andrei.kotousov@adelaide.edu.au

* Correspondence: mroslina.kl@utm.my; Tel.: +60-176-571-769

Received: 24 December 2019; Accepted: 11 February 2020; Published: 17 February 2020

Abstract: There have been a number of studies concerning the integrity of high-strength carbon steel pipe elbows weakened by local pipe wall thinning, the latter can be typically caused by flow accelerated erosion/corrosion. In particular, the focus of several recent studies was on low cycle fatigue behavior of damaged elbows, mainly, in relation to strength and integrity of piping systems of nuclear power plants subjected to extreme loading conditions, such as earthquake or shutdown. The current paper largely adopts the existing methodology, which was previously developed, and extends it to copper-nickel elbows, which are widely utilized in civil infrastructure in seismically active regions. FE (finite element) studies along with a full-scale testing program were conducted and the outcomes are summarized in this article. The overall conclusion is that the tested elbows with various severity of local wall thinning, which were artificially introduced at different locations, demonstrate a strong resistance against low cycle fatigue loading. In addition, elbows with wall thinning defects possess a significant safety margin against seismic loading. These research outcomes will contribute to the development of strength evaluation procedures and will help to develop more effective maintenance procedures for piping equipment utilized in civil infrastructure.

Keywords: elbow; low cycle fatigue; wall thinning; steel; copper-nickel alloy; safety margin; seismic load

---

1. Introduction

Elbows are common components in piping systems, primarily used to change the direction of flow and, therefore, are frequently subjected to flow-induced corrosion and erosion. These structural elements often represent the weakest link in the piping systems due to higher stress concentration and more severe loading conditions in comparison with straight pipe segments. One significant and challenging technical issue in the evaluation of integrity and strength of elbows with local wall thinning is their behavior in the case of extreme events, such as earthquake or shutdown. As elbows are very flexible and, when subjected to extreme loading condition, these structural elements can demonstrate cross-sectional ovalization and local buckling as well as failure in the form of cross-thickness cracking. It was also reported that multi-axial stress state due to bending and internal pressure can lead to reduction of fatigue life [1].

During normal operation, elbows as well as the rest of piping equipment can be subjected to internal pressure, bending loads induced by deadweight or misalignment and thermal expansion, and, in many instances, to temperature gradients. Early research efforts were largely focused on steel elbows under monotonic loading. Numerous experimental studies for different elbow geometries and applied loading conditions, e.g., [2–4] supported by extensive finite element studies e.g., [5], were performed with an aim of developing a unified methodology for evaluating integrity and strength of elbows.

Local wall-thinning due to flow-accelerated corrosion/erosion has been recently realized as one of the main degradation mechanisms. It was stated that the effect of the wall thinning on strength and integrity represents a significant issue, specifically for nuclear power plants, since wall-thinning has already led to the failures of carbon steel piping systems in several nuclear power plants in the past [6]. A serious concern is also associated with the operation, behavior and integrity of piping undergoing progressive wall thinning under extreme loading conditions [7], such as the 2011 massive earthquake in Japan. The analysis of damage due to this earthquake has demonstrated the importance of maintaining structural integrity and low cycle fatigue resistance of piping equipment weakened by various defects and accumulated mechanics damage.

There were several recent studies both experimental and numerical addressing the problem of structural integrity and life of carbon steel elbows under severe loading conditions [6–16]. Fatigue life was mainly considered around 50–300 cycles, as this range is representative of a typical earthquake loading. This cycle range can be regarded as occasional loading according to the provisions of both ASME B31.3 [17] and EN 13480-3 [18] standards for process piping. Both standards adopt a similar approach to low-cycle fatigue design, which is based on the allowable stress concept and linear-elastic stress analysis.

Despite extensive previous studies, more experimental and theoretical investigations are still required in order to develop adequate predictive models and incorporate these models into the standards and design codes [1]. The difficulties in prediction methodologies are largely attributed to numerous factors, which influence the strength and integrity of elbows with wall thinning defects [7]. Some of these factors are currently disregarded in design and evaluation procedures or incorporated through the use of large safety factors [6–8]. In addition, the current standards, which are oriented for the design of nuclear piping components [17,18], are mainly focused on carbon steel materials providing design fatigue curves for steel elbows in high and low-cycle fatigue regimes. Lack of design guidelines and experimental data for other structural materials is the main motivation behind the current work.

The focus of the present study are elbows made of copper-nickel alloys, which are widely used along with steel elbows in civil infrastructure, in particular, in fuel gas transportation in Malaysia. In this paper, the effect of the localized wall thinning on low cycle fatigue resistance of elbows was investigated by conducting a full-scale test program for elbows with artificially introduced defects. The experimental approach in this work was largely adopted from the similar studies conducted for carbon steel elbows in relation to the nuclear power industry. The fabricated elbow specimens with local wall thinning were subjected to displacement-controlled loading conditions (± 20 mm) until failure. Similar loading conditions have also been utilized in previous studies providing the fatigue life of elbows between 50 and 500 cycles [1]. The present experimental studies were supported by high-fidelity finite element (FE) simulations, which identified main deformation and failure mechanisms. This paper summarizes main outcomes of the experimental program and numerical studies with a focus on the influence of the dimensions and location of thinning defects on low cycle fatigue life. Further, the safety margin against seismic load of elbows with wall thinning was evaluated in accordance with guidelines provided in ASME Boiler and pressure Vessel Code Section III. However, many experimental and numerical results are not presented in this paper due to space limitations and the specific focus of this article. These additional results will be presented and discussed in other publications.

2. Test Specimens and Procedures

2.1. Material and Specimens

The short and long bend elbows of OD = 108 mm and nominal thickness, t_n = 2.5 mm, as shown in Figure 1a were fabricated by die casting. The elbows were made of C70600 Copper Nickel 90/10 (90%

copper and 10% nickel) as specified in the Malaysia Standards MS930 and MS830. As mentioned in the Introduction, these elbows are widely used in fuel gas pipelines and other industrial applications. For mechanical testing, 90° elbows were welded using Tungsten Inner Gas (TIG) welding machine to straight pipe segments as shown in Figure 1a. Two types of elbows were tested: short radius (R = 102 mm) and long radius (R = 142 mm) elbows; the latter are usually utilized to transport high-viscosity liquids. Table 1 lists basic mechanical properties of the elbow material obtained using standard tensile specimens machined from the intrados and extrados of elbows.

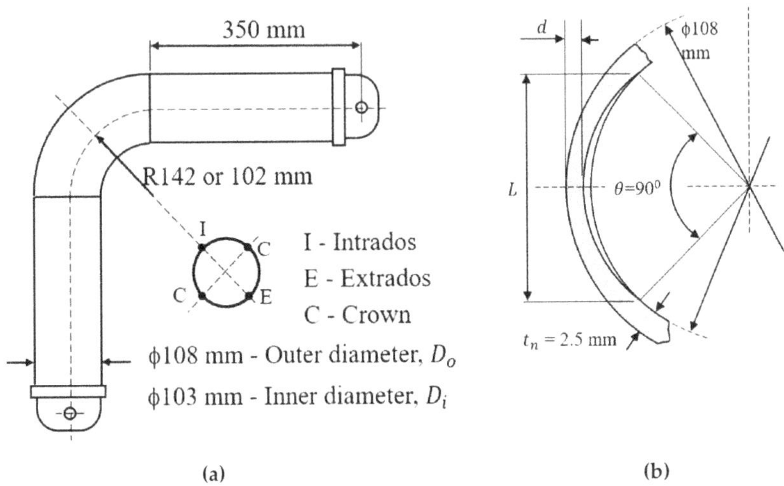

Figure 1. (a) Test specimen and main dimensions; (b) Wall thinning defect sizes.

Table 1. Material properties.

Mechanical Properties	
Young's modulus	125 GPa
Poisson's ratio	0.3
Ultimate strength	300 MPa
Yield strength	110 MPa
Hardness	80 HB

The wall-thinning defects were artificially introduced on the inner side of the elbow wall. These defects were machined at the intrados, crown or extrados center regions, as shown in Figure 1b (one defect for each elbow specimen). The circumferential and axial shapes of the defects had circular geometry. Then, the defect geometry can be specified by the following parameters: L is the equivalent thinning length, θ is the circumferential angle of the defect, which was kept at 90° in the present study and eroded ratio, d/t_n.

It is well-known from the classical elastic stress analysis of thin-walled donut shells (or membrane shell theory) that the hoop stress largely remains constant, and the magnitude of the hoop stress is similar to the hoop stress in circular pipes loaded by inner pressure. The maximum circumferential stress takes place at the intrados, the minimum stress appears at the extrados region of elbows, while the crown is subjected to an intermediate stress level. However, from our present elasto-plastic FE results, see Figure 2, as well as from previous studies [19], it follows that the crown region can be subjected to the highest stress under closing-mode in-plane bending (in the absence of the inner pressure). Therefore, these results as well as the previous studies justify the consideration of various locations of wall thinning defects as the inner pressure and in-plane bending loading can change the critical locations with the highest stress and strain.

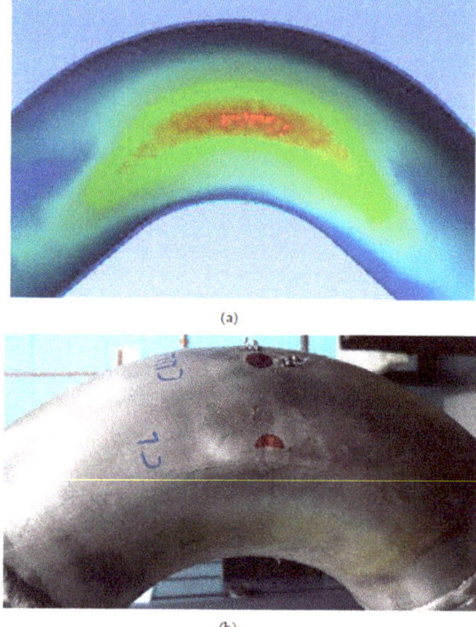

(a)

(b)

Figure 2. Distribution of effective stresses under in-plane bending (red area–at crown location shows the highest stress intensity), adapted from [19] with permission from wiley.com, 2019. (**a**) The FEA three-dimension failure illustration; (**b**) fatigue failure captured on the elbow after cyclic loading experiment.

2.2. Testing Apparatus

The testing apparatus used in this study included INSTRON 5982 universal testing machine, the pressurization system, and the data acquisition system. Elbow specimens were equipped with strain gauge rosettes at three critical locations on the outer surface to monitor deformations at intrados, extrados and crown regions during the displacement-controlled loading with and without inner pressure, see Figure 3.

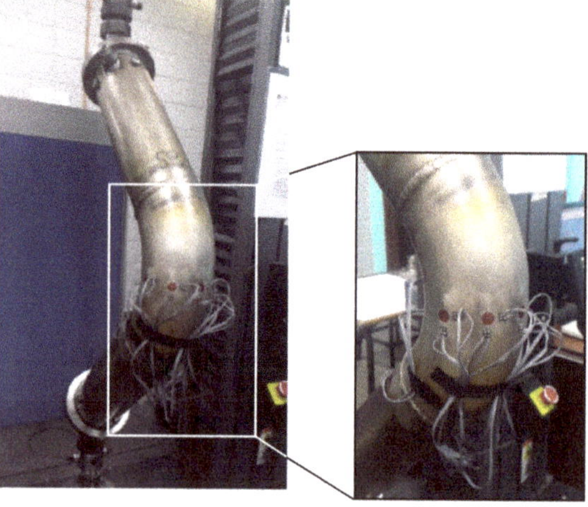

Figure 3. Test specimen equipped with strain gauge rosettes.

One end of the test specimen was supported by a hinge allowing free rotation, and displacements with the rate of 1 mm/second were applied to the other (top) end of the test specimen, which led to a closing- and opening-mode of in-plane bending. A hand pressure pump was used to pressurize elbow specimens with hydraulic oil. The data acquisition system consisted of a data logger and a PC to capture the experimental data (displacement-load diagram and strain gauges readings). In addition, a digital camera was utilized to monitor the defect formation and failure progression. The summary of the testing conditions of elbows with wall thinning defects is presented in Table 2.

Table 2. Summary of test conditions for short and long bend elbows.

Defect Location	d/t_n	L	θ	Pressure, MPa					δ
	0.25			0	0.5	1.0	1.5	2.0	
Extrados	0.5			0	0.5	1.0	1.5	-	
	0.75	80		0	0.5	1.0	1.5	-	
	0.25			0	0.5	1.0	1.5	2.0	
Crown	0.5		90°	0	0.5	1.0	1.5	-	± 20
	0.75			0	0.5	1.0	1.5	-	
	0.25			0	0.5	1.0	1.5	2.0	
Intrados	0.5	60		0	0.5	1.0	1.5	-	
	0.75			0	0.5	1.0	1.5	-	

3. Results and Discussion

3.1. Low Cycle Behavior of Sound Elbows

Table 3 shows the low cycle fatigue data (number of cycles to failure) obtained for sound elbows at different values of inner pressure. The maximum values of the internal pressure (4 MPa) were limited due to the use of the hand pressure pump, as at high pressures it was impossible to avoid large pressure fluctuations. Similar problems were reported in [6], where both manual and high-pressure hydraulic pumps were used to pressurize elbows for low and high pressure levels, respectively. Therefore, higher as well as burst pressures were not investigated in the current experimental program.

Table 3. Total fatigue life (number of cycles to failure) and location of failure for sound elbows.

Types of Elbow Radius	Long Radius Elbows					Short Radius Elbows				
Inner Pressure, MPa	0	1	2	3	4	0	1	2	3	4
Cycles to Failure	208 C	220 C	210 C	204 C	196 C	165 C	166 C	166 C	166 C	166 C

Notations: C—failure at crown region.

In all cases for sound elbows failure was observed at crown locations in the circumferential direction as shown in Figure 4. Cracks were not observed at intrados and extrados regions. The failure was identified when the reaction load was decreased or when a crack penetration was observed at the outer surface region of the crown, after that the fatigue test was terminated. These results were used as a benchmark for analysis of damaged elbows.

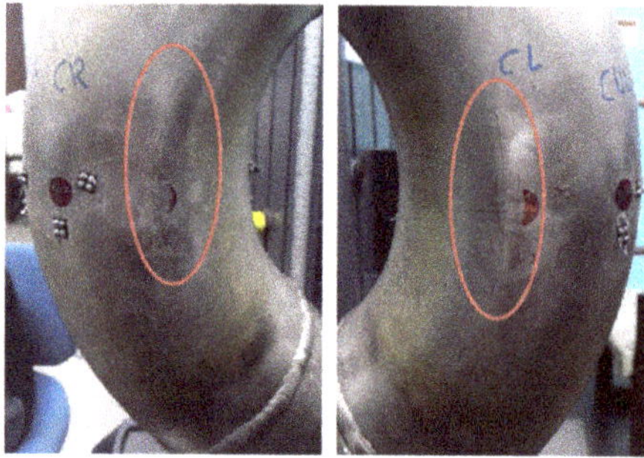

Figure 4. Fracture (crack penetration) at crown of sound elbow.

The experimental data indicates that the excessive inner pressure in the case of long radius elbows slightly reduces the fatigue life of elbows when the inner pressure is above 2 MPa. Based on the life pattern (Table 3), it can be noticed that the inner pressure below this critical value can increase the fatigue life. Indeed, the inner pressure reduces the possibility of local buckling by increasing the overall rigidity of the structure. In other words, the applied inner pressure can reduce the ovalization of elbows, thus lowering the local strain accumulation during cycling, which, in turn, may lead to some increase in fatigue life.

For short radius elbows, the fatigue life is largely unaffected by the inner pressure. This situation can be explained as follows: as the ratio of the short elbow radius to the diameter is quite small (1:1), this makes the elbow wall quite rigid when it was bent towards the in-plane direction. As a result, short elbows have a weaker tendency to local ovalization and buckling. Therefore, the inner pressure has almost no effect on the local strain accumulation during cycling and, subsequently, does not influence fatigue life of elbows.

3.2. Low Cycle Behavior of Elbows with Local Wall Thinning

Table 4 shows the low cycle fatigue data (number of cycles to failure) obtained for elbows with artificial defects at different values of the applied inner pressure.

Table 4. Total fatigue life (number of cycles to failure) and location of failure for elbows with artificial defects.

Location of Defects		Inner Pressure, MPa									
		Long Bend Radius Elbows					Short Bend Radius Elbows				
	d/t_n	0.0	0.5	1.0	1.5	2.0	0.0	0.5	1.0	1.5	2.0
Extrados	0.25	193 C	195 C	196 C	198 C	200 C	176 C	179 C	181 C	185 C	188 C
	0.5	187 C	188 C	188 C	189 C	-	171 C	171 C	172 C	173 C	-
	0.75	197 C	179 C	180 C	175 E/T	-	174 C	175 E/T	176 E/T	175 E/T	-
Crown	0.25	184 C	187 C	191 C	195 C	190 C	169 C	170 C	170 C	171 C	172 C
	0.5	172 C	173 C	174 C	175 C/T	-	166 C	167 C	168 C	168 C/T	-
	0.75	166 C/T	166 C	166 C/T	165 C/T	-	163 C	163 C	160 C	155 C/T	-
Intrados	0.25	194 C	195 C	196 C	198 C	200 C	176 C	177 C	178 C	178 C	179 C
	0.5	198 C	199 C	200 C	201 C	-	177 C	178 C	179 C	180 C	-
	0.75	201 C	196 C	192 I/T	189 I/T	-	170 C	170 C	169 I/T	166 I/T	-

Notations: C—failure at crown; I—failure at intrados; E—failure at extrados; T—failure in the region of the artificial defect (wall thinning).

In almost all elbow specimens with local wall thinning at intrados or extrados, fatigue crack initiated at crown region and normally propagated in the axial directions as it was observed for sound elbows, see Figure 4. It is noted that in many cases, the crack initiated at crown, even for quite severe local wall thinning. Table 4 demonstrates that wall thinning at crown regions has the largest impact on fatigue life, reducing it by approximately 30 percent. The failure initiation at different locations occurred only at a very severe level of wall thinning, when the eroded ratio was 0.75. Wall thinning at extrados and intrados had a very similar effect on low cycle fatigue life of elbows. As expected, a higher eroded ratio, d/t_n, led to a shorter fatigue life.

The most interesting observation is the increase of fatigue life with the increase of the applied inner pressure. This behavior has also been found in carbon steel elbows, e.g., [6,20,21], and it can be explained by the inner pressure shielding effect: as the elbow specimen is pressurized, the bending moment is reduced and the overall stiffness of the structure is increased even though the displacement is maintained at an initial set value. In other words, it means that the applied bending load is relaxed by the inner pressure. Another beneficial effect for fatigue life is the reduction of the ovalization and local stress accumulation as discussed previously for sound elbows. Similar to sound elbows, short bend radius elbows with defects exhibit a weaker tendency to local ovalization and buckling, which generally leads to a longer fatigue life as it can be noted from experimental results, Table 4.

It is also can be seen from comparison of Tables 3 and 4 that the life of the defected short radius elbows is generally longer than the sound elbows (160 to 188 cycles against 166 cycles for sound elbows). The counterintuitive increase in the fatigue life, which also happens for some values of the applied pressure, can be attributed to the decrease of stiffness of the damaged elbows. Under the displacement control loading, this leads to the lower applied stress and less accumulation of fatigue damage in the critical area (crown area).

4. Safety Margin Evaluation

The safety margin of elbows with wall thinning against seismic load can be evaluated with guidelines provided in ASME Boiler and pressure Vessel Code Section III. The current ASME Section III rules for seismic evaluation permit an alternative method for thin-wall piping ($D_o/t_n > 40$). In accordance with this method the fictitious stress amplitude for elbows:

$$S_a = \frac{B_1 P_D D_0}{2t_n} + B_2 \frac{D_o}{2I} M_E,\tag{1}$$

is compared with allowable stress, where P_D is inner pressure, D_0 is outside diameter of pipe and I is the moment of inertia of the elbow cross-sectional area. The purpose of the present calculations is to evaluate the fictitious stress amplitude and, in the absence of values of the allowable stress, compare this calculated fictitious stress with the yield stress of the material. In the calculations $B_1 = 0.5$ and:

$$B_2 = \frac{1.30}{h^{2/3}}; h = \frac{t_n R}{r^2},\tag{2}$$

were utilized [6], where R is the bend radius and r is the nominal mean radius of the elbow.

$M_E = P_E l$ is the moment amplitude corresponding to the maximum displacement of 20 mm, which can be obtained from the fictitious force amplitudes, P_E, corresponding to the first $\frac{1}{4}$ cycle as illustrated in Figure 5 and l is the moment arm, which can be evaluated from Figure 1 for short and long radius elbows.

The fictitious stress amplitudes, S_a, of the conducted tests were found between 700 and 800 MPa, and the differences of the value of S_a between tests are rather small. A comparison of these values of S_a with the material yield stress of 110 MPa indicates a sufficiently large safety margin against occasional loading as the elbow specimens with wall thinning did not collapse or did not reach the maximum loading capacity during the testing [7].

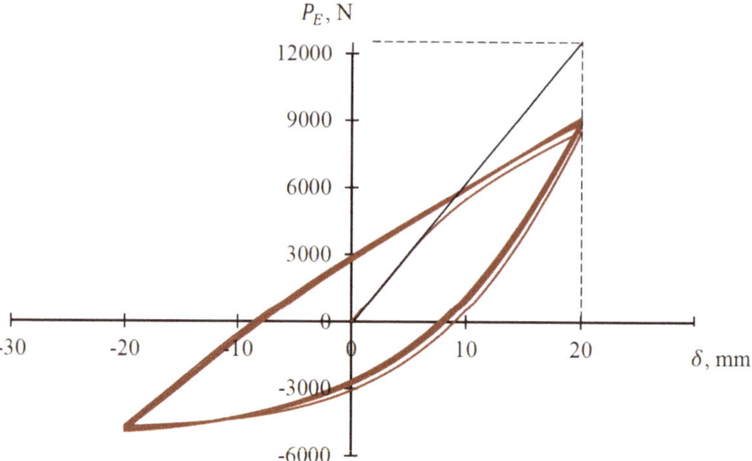

Figure 5. Typical load-displacement diagram.

5. Conclusions

Some general conclusions from the experimental results can be summarized as follows:

(1) Local wall thinning at extrados and intrados regions does not significantly affect low cycle fatigue life of damaged elbows. Fatigue cracks do not initiate at the weakened locations even at large eroded ratios;

(2) In most cases and regardless of the applied inner pressure, fatigue cracks initiate at the crown regions, and the presence of local wall thinning defect at this location only marginally reduces low cycle fatigue life;

(3) When the allowable stress is assumed as the material yield stress, and the applied stress level is calculated based on the ASME Boiler and Pressure Vessel Code, Section III, the elbow with wall thinning defects withstood 6–7 times of the allowable stress level, even when the eroded ratio has reached the extreme value of 0.75. This result indicates a sufficiently large safety margin against occasional loading.

More work, however, is still needed, specifically to clarify the explicit safety margin of elbows with defects, the influence of location and severity of wall thinning, eroded ratio, eroded angle and applied pressure. In addition, it has to be noted that the current test results have been obtained for relatively low pressure levels.

Author Contributions: M.F.H. designed, fabricated the elbow sample and performed the experiment. M.F.H. processed the experimental data, performed the analysis, drafted the manuscript and designed the figures together with A.K.; M.F.H. performed the analytic calculations, performed the numerical simulations and verified the analytical methods. R.M. devised the project, the main conceptual ideas and proof outline. R.M. encouraged M.F.H. to investigate the fatigue issues on pipe than supervised the project. M.F.H. and A.K. wrote the manuscript with support from R.M.; A.K. took the lead in writing the manuscript. All authors provided critical feedback and helped shape the research, analysis and manuscript. All authors have read and agreed to the published version of the manuscript.

Funding: The project was funded by the Ministry of Higher Education under Fundamental Research Grant Scheme, MyBrain15 MyPhD, the Ministry of Higher Education, Razak Faculty of Technology and Informatics and Universiti Teknologi Malaysia (UTM). FRGS UTM Vote No.: R.K130000.7840.4F852, UTM Q.K130000.2540.19H15 and Q.K130000.2656.16J42.

Acknowledgments: The authors would like to express their greatest appreciation and utmost gratitude to the Ministry of Higher Education for awarding the Fundamental Research Grant Scheme, MyBrain15 MyPhD, the Ministry of Higher Education, Razak Faculty of Technology and Informatics and Universiti Teknologi Malaysia (UTM) for all the support towards making this study a success.

Conflicts of Interest: The author(s) declared no potential conflicts of interest with respect to the research, authorship, and/or publication of this article.

References

1. Varelis, G.E.; Karamanos, S. Low-cycle fatigue of pressurized steel elbows under in-plane bending. *ASME J. Press. Vessel Technol.* **2015**, *137*. [CrossRef]
2. Sobel, L.H.; Newman, S.Z. Comparison of Experimental and Simplified Analytical Results for the In-Plane Plastic Bending and Buckling of an Elbow. *J. Press. Vessel. Technol.* **1980**, *102*, 400–409. [CrossRef]
3. Hilsenkopf, P.; Boneh, B.; Sollogoub, P. Experimental study of behavior and functional capability of ferritic steel elbows and austenitic stainless steel thin-walled elbows. *Int. J. Press. Vessel. Pip.* **1988**, *33*, 111–128. [CrossRef]
4. Gresnigt, A.M.; Van Foeken, R. Strength and deformation capacity of bends in pipelines. *Int. J. Offshore Polar Eng.* **1995**, *5*, 294–307.
5. Karamanos, S.A.; Giakoumatos, E.; Gresnigt, A.M. Nonlinear Response and Failure of Steel Elbows Under In-Plane Bending and Pressure. *J. Press. Vessel. Technol.* **2003**, *125*, 393–402. [CrossRef]
6. Kim, J.-W.; Na, Y.-S.; Lee, S.-H. Experimental Evaluation of the Bending Load Effect on the Failure Pressure of Wall-Thinned Elbows. *J. Press. Vessel. Technol.* **2009**, *131*. [CrossRef]
7. Takahashi, K.; Watanabe, S.; Ando, K.; Urabe, Y.; Hidaka, A.; Hisatsune, M.; Miyazaki, K. Low cycle fatigue behaviors of elbow pipe with local wall thinning. *Nucl. Eng. Des.* **2009**, *239*, 2719–2727. [CrossRef]
8. Nagamori, H.; Takahashi, K. The Revised Universal Slope Method to Predict the Low-Cycle Fatigue Lives of Elbow and Tee Pipes. *J. Press. Vessel. Technol.* **2017**, *139*. [CrossRef]
9. Foroutan, M.; Ahmadzadeh, G.; Varvani-Farahani, A. Axial and hoop ratcheting assessment in pressurized steel elbow pipes subjected to bending cycles. *Thin-Walled Struct.* **2018**, *123*, 317–323. [CrossRef]
10. Jeon, B.-G.; Kim, S.-W.; Choi, H.-S.; Park, N.-U.; Kim, N.-S. A Failure Estimation Method of Steel Pipe Elbows under In-plane Cyclic Loading. *Nucl. Eng. Technol.* **2017**, *49*, 245–253. [CrossRef]
11. Kim, S.-W.; Choi, H.-S.; Jeon, B.-G.; Hahm, D.-G. Low-cycle fatigue behaviors of the elbow in a nuclear power plant piping system using the moment and deformation angle. *Eng. Fail. Anal.* **2019**, *96*, 348–361. [CrossRef]
12. Kiran, A.R.; Reddy, G.; Agrawal, M. Experimental and numerical studies of inelastic behavior of thin walled elbow and tee joint under seismic load. *Thin-Walled Struct.* **2018**, *127*, 700–709. [CrossRef]
13. Liu, C.; Shi, S.; Cai, Y.; Chen, X. Ratcheting behavior of pressurized-bending elbow pipe after thermal aging. *Int. J. Press. Vessel. Pip.* **2019**, *169*, 160–169. [CrossRef]
14. Udagawa, M.; Li, Y.; Nishida, A.; Nakamura, I. Failure behavior analyses of piping system under dynamic seismic loading. *Int. J. Press. Vessel. Pip.* **2018**, *167*, 2–10. [CrossRef]
15. Kiran, A.R.; Reddy, G.; Agrawal, M. Seismic fragility analysis of pressurized piping systems considering ratcheting: A case study. *Int. J. Press. Vessel. Pip.* **2019**, *169*, 26–36. [CrossRef]
16. Firoozabad, E.S.; Jeon, B.-G.; Choi, H.-S.; Kim, N.-S. Failure criterion for steel pipe elbows under cyclic loading. *Eng. Fail. Anal.* **2016**, *66*, 515–525. [CrossRef]
17. American Society of Mechanical Engineers. Process Piping, B31.3. In *ASME Code for Pressure Piping*; American Society of Mechanical Engineers: New York, NY, USA, 2006.
18. Comité Européen de Normalisation. *Metallic Industrial Piping—Part 3: Design and Calculation, EN13480-3*; Comité Européen de Normalisation: Brussels, Belgium, 2002.
19. Urabe, Y.; Takahashi, K.; Ando, K. Low Cycle Fatigue Behavior and Seismic Assessment for Elbow Pipe Having Local Wall Thinning. *J. Press. Vessel. Technol.* **2012**, *134*, 041801. [CrossRef]
20. Takahashi, K.; Ando, K.; Matsuo, K.; Urabe, Y. Estimation of Low-Cycle Fatigue Life of Elbow Pipes Considering the Multi-Axial Stress Effect. *J. Press. Vessel. Technol.* **2014**, *136*, 041405. [CrossRef]
21. Harun, M.F.; Mohammad, R.; Kotousov, A. Effect of wall thinning on deformation and failure of copper-nickel alloy elbows subjected to low cycle fatigue. *Mater. Des. Process. Commun.* **2019**. [CrossRef]

 *metals*

Article

The Effect of Mean Load for S355J0 Steel with Increased Strength

Roland Pawliczek * and Dariusz Rozumek 

Department of Mechanics and Machine Design, Opole University of Technology, Mikolajczyka 5, 45-271 Opole, Poland; d.rozumek@po.edu.pl
* Correspondence: r.pawliczek@po.edu.pl; Tel.: +48774498404

Received: 12 December 2019; Accepted: 28 January 2020; Published: 1 February 2020

Abstract: The paper presents an algorithm for calculating the fatigue life of S355J0 steel specimens subjected to cyclic bending, cyclic torsion, and a combination of bending and torsion using mean stress values. The method of transforming cycle amplitudes with a non-zero mean value into fatigue equivalent cycles with increased amplitude and zero mean value was used. Commonly known and used transformation dependencies were used and a new model of the impact of the mean stress value on the fatigue life of the material was proposed. The life calculated based on the proposed algorithm was compared with the experimental life. It has been shown that the proposed model finds the widest application in the load cases studied, giving good compliance of the calculation results with the experimental results.

Keywords: fatigue life; steel; bending and torsion; stress ratio

1. Introduction

Algorithms for calculating fatigue life are built from a series of steps that determine how to perform calculations to take into account many factors describing the load parameters of an element. Numerous papers present the results of fatigue tests describing the behavior of the material and experimental verification of the proposed models [1–8]. In the case of fatigue tests during bending and torsion with the participation of the mean load value, such an algorithm is widely presented in [6,9–11] and allows determining the fatigue life of tested specimens under various load conditions. The results of such analyzes require accurate knowledge of the behavior of the tested material under various loading conditions. It is necessary to have a wide spectrum of results of fatigue tests conducted in different conditions, which take into account the influence of the average load value and different specimen geometry [9–11]. Such tests are designed to understand the properties of the material and allow verification of the proposed calculation algorithms in terms of compliance with the results of the experiment. Kluger and Pawliczek [6] shows and discusses on the results of a comparison involving mathematical models applied for fatigue life calculations where the mean load value is taken into account. The specimens were subjected to bending, torsion and a combination of bending with torsion with mean value of the load. Analysis of the calculation results show that the best fatigue life estimations are obtained by using models that are sensitive to the changes of material behavior under fatigue loading in relation to the specified number of cycles of the load. Branco et al. [9] described studies the effect of different loading orientations on fatigue behavior in severely notched round bars under pulsating in-phase combined bending-torsion loading. Fatigue life is predicted via the Coffin-Manson model. The paper [11] presents results on the fatigue crack in specimens made of the 2017A-T4 alloy under bending. The specimens had different notch geometry. The tests were performed by imposing a constant load amplitude value and different stress ratio $R = -1.0$ and $R = 0$. In [12] the failure analysis of a coupled shaft from a shredder is presented, which was subjected to bending and torsion. The shaft failure occurred due to fatigue, on a perpendicular plane to the rotation axis, in the place of the hole. The shaft fracture surface presents characteristics of fatigue due to torsion combined with bending high loads.

The purpose of the tests described in the work is to learn about the behavior of specimens made of typical structural steel under conditions of cyclic loads with different stress ratios. An approach consisting of the reduction of the multiaxial (complex) load state to a simple load state (uniaxial tension-compression) and prediction of fatigue life of elements using the linear Palmgren-Miner damage hypothesis is presented. The results of calculations based on fatigue characteristics at uniaxial tensile-compression are compared with the results of experimental tests of S355J0 steel specimens subjected to cyclic bending, cyclic torsion and a combination of bending and torsion using non-zero mean stress values.

2. Analyzed Fatigue Models

The models described below relate to stress amplitude change σ_a (τ_a) and mean stress value σ_m (τ_m). Figure 1 shows the algorithm for calculating fatigue life under bending and torsion for courses with different values of the stress ratio R. In the algorithm, the equivalent stress amplitude was calculated using the Huber–Misess–Hencky hypothesis.

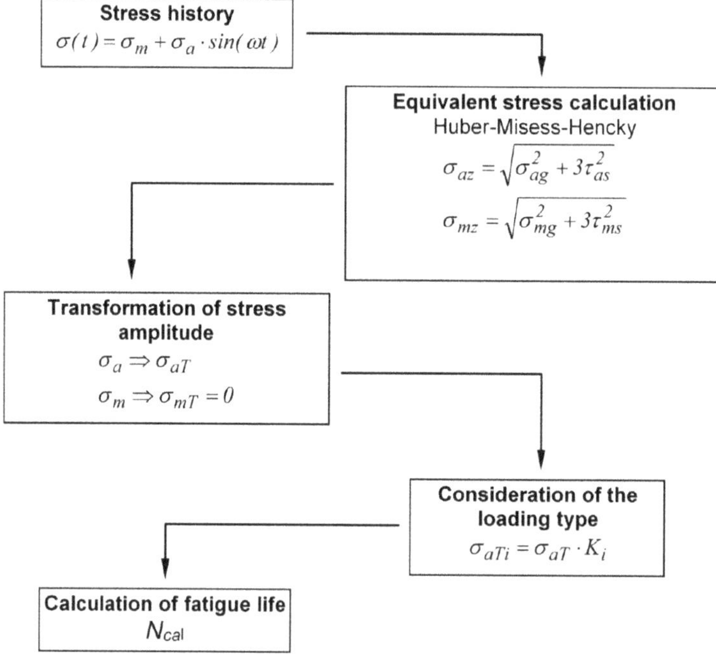

Figure 1. Algorithm for calculating fatigue life.

Where

$$\sigma_{ag} = \tfrac{M_a}{W_x} \cos \alpha; \quad \sigma_{mg} = \tfrac{M_m}{W_x} \cos \alpha,$$
$$\tau_{as} = \tfrac{M_a}{W_0} \sin \alpha; \quad \tau_{ms} = \tfrac{M_m}{W_0} \sin \alpha,$$

where: σ_{ag}, σ_{mg}—bending stress amplitude and stress mean value respectively, τ_{as}, τ_{ms}—torsional stress amplitude and stress mean value respectively, M_a, M_m—amplitude of the moment and mean value of the load moment, respectively, W_x—bending indicator, W_0—torsion indicator, α—angle between bending and torsion moments.

When transforming stress amplitudes, due to the non-zero mean stress values, commonly known transformation relationships were used. These are equations [13–15]

$$\text{Goodman}: \quad \sigma_{aT} = \frac{\sigma_a}{1 - \frac{\sigma_m}{R_m}}, \tag{1}$$

$$\text{Gerber}: \quad \sigma_{aT} = \frac{\sigma_a}{1 - \left(\frac{\sigma_m}{R_m}\right)^2}, \tag{2}$$

$$\text{Marin}: \quad \sigma_{aT} = \frac{\sigma_a}{\sqrt{1 - \left(\frac{\sigma_m}{R_m}\right)^2}}, \tag{3}$$

where R_m—ultimate strength, σ_m—mean stress, σ_a—stress cycle amplitude with a mean value $\sigma_m \neq 0$, σ_{aT}—fatigue equivalent stress cycle amplitude (about mean value $\sigma_{mT} = 0$).

In the next transformation dependencies used in the calculations, material constants characterizing the fatigue properties of the material are used. These are the relationships [16–18]

$$\text{Morrow}: \quad \sigma_{aT} = \frac{\sigma_a}{1 - \frac{\sigma_m}{\sigma'_f}}, \tag{4}$$

$$\text{Kliman}: \quad \sigma_{aT} = \frac{\sigma_a}{\left(1 - \frac{\sigma_m}{\sigma'_f}\right)^{\frac{c+1}{b+c+1}}}, \tag{5}$$

where σ'_f—fatigue strength coefficient, b, c—fatigue strength and ductility exponent, respectively.

Material sensitivity to the effect of average load can be described by the material sensitivity factor for the asymmetry of cycle [2,19]. As a result of developing and transforming dependencies based on the mean loading value on Haigh diagram $\sigma_a = f(\sigma_m)$ and $\tau_a = f(\tau_m)$, their limit for stress level corresponding to the unlimited life (fatigue limit) is indicated. It was indicated that material sensitivity on mean loading is not a material constant and depends on the number of the cycles corresponding to the failure of an element. The amplitude of the corresponding normal and shear stress components is calculated as follows:

$$\sigma_a = \sigma_{-1} - \psi_\sigma \sigma_m, \tag{6}$$

$$\tau_a = \tau_{-1} - \psi_\tau \tau_m \tag{7}$$

where σ_{-1}—fatigue strength and ductility exponent, respectively fatigue limit at the oscillatory cycle, ψ_σ, ψ_τ—material sensitivity factor for the asymmetry of cycle.

The material sensitivity factor for the asymmetry of cycle ψ_σ for bending is defined as

$$\psi_\sigma = \frac{2\sigma_{-1} - \sigma_0}{\sigma_0}, \tag{8}$$

and for torsion

$$\psi_\tau = \frac{2\tau_{-1} - \tau_0}{\tau_0}, \tag{9}$$

where σ_0, τ_0—fatigue limit at one-sided cycle for bending and torsion.

The value of the factor ψ_σ and ψ_τ determined by the relationships (8) and (9) is correct for the fatigue limit. Determining the function of changing the value of the material sensitivity factor for the asymmetry of cycle depending on the number of cycles N allows the assessment of the effect of the mean stress value depending on the material life. By making the factor ψ_σ and ψ_τ dependent on the number of cycles until failure N, in Formulas (8) and (9), stress amplitudes will appear depending on the number of cycles:

for bending

$$\psi_\sigma(N) = \frac{2\sigma_{-1}(N) - \sigma_0(N)}{\sigma_0(N)}, \tag{10}$$

and for torsion

$$\psi_\tau(N) = \frac{2\tau_{-1}(N) - \tau_0(N)}{\tau_0(N)}, \tag{11}$$

where $\sigma_{-1}(N)$, $\tau_{-1}(N)$—stress amplitude at oscillatory bending and torsion for a fixed number of N cycles, $\sigma_0(N)$, $\tau_0(N)$—one-sided cycle stress amplitude under bending and torsion for a fixed number of N cycles.

After transformation Formulas (6), (7) and taking into account Relationships (10) and (11), it is possible to determine the equivalent stress amplitude at a symmetrical cycle for given values of stress amplitude σ_a and τ_a and mean values of stress σ_m and τ_m with the assumed number of cycles N

$$\sigma_{-1}(N) = \sigma_a + \psi_\sigma(N)\sigma_m, \tag{12}$$

$$\tau_{-1}(N) = \tau_a + \psi_\tau(N)\tau_m, \tag{13}$$

Assuming the stress amplitude for the symmetrical cycle, determined in accordance with the relationship (12) and (13), as the transformed amplitude and transforming the relationship (12), (13) the model based on cycle asymmetry sensitivity factor (CASF model) may be written as:

$$\sigma_{aT} = \sigma_{-1}(N) = \sigma_a\left(1 + \psi_\sigma(N)\frac{1+R}{1-R}\right), \tag{14}$$

$$\tau_{aT} = \tau_{-1}(N) = \tau_a\left(1 + \psi_\tau(N)\frac{1+R}{1-R}\right), \tag{15}$$

The proposed criterion (13) and (14) allows to determine the equivalent stress amplitude in a symmetrical cycle ($R = -1$) for the given amplitude values and the average stress value of the asymmetrical cycle ($R \neq -1$) using the function of changing the material sensitivity factor for the asymmetry of cycle.

Relations (6) to (15) have been developed for uniaxial loads and cannot be directly used for multiaxial loads, when there is also uneven stress distribution in the tested specimens.

In this paper, it is assumed that the parameters describing the Wöhler curve in the area of limited fatigue life for uniaxial loads and for the analyzed complex load cases are similar to each other, i.e., the graphs are parallel. Therefore, a constant K_i coefficient is adopted, which allows to refer the case of multiaxial loads to uniaxial tension-compression

$$K_i = \frac{Z_{rc}}{Z_i}, \tag{16}$$

where Z_{rc}—fatigue limit under cyclic tension-compression, Z_i—fatigue limit expressed by the equivalent stress amplitude for bending, torsion or a combination of bending and torsion for zero mean load, determined using the Huber-Misess-Hencky hypothesis.

Equations (1)–(6) and (15) were modified to form

$$\sigma_{aTi} = \sigma_{aT}\,K_i, \tag{17}$$

Computational fatigue life was determined from the relationship

$$\log N_{cal} = B + A \log \sigma_{aTi}, \tag{18}$$

where A, B—regression coefficients of the fatigue characteristics of the material determined at symmetrical tensile-compression.

3. Material and Methods

The smooth specimens of S355J0 steel were subjected to fatigue testing (Figure 2). The starting material was a drawn rod. The chemical composition of the material (obtained from the certificate attached to the material) is presented in Table 1. Static and fatigue properties obtained under cyclic tensile-compression are given in Table 2.

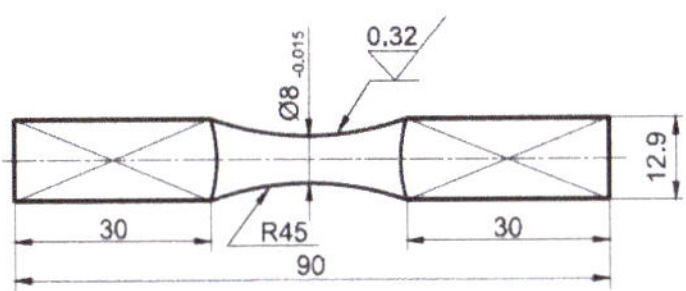

Figure 2. Shape and dimensions of specimens, dimensions in mm.

Table 1. Chemical composition of experimental material (%).

C	Mn	Si	P	S	Cr	Ni	Cu	Fe
0.21	1.46	0.42	0.019	0.046	0.09	0.04	0.17	Bal.

Table 2. Static and fatigue properties of the material.

R_e, MPa	R_m, MPa	A_{10}, %	Z, %	E, GPa	v	σ'_f, MPa	b	c	n'	K' MPa
357	535	21	50	210	0.30	782	−0.118	−0.410	0.287	869

The tests were carried out on the fatigue test stand MZGS-100 [20,21]. The stand used in experiment allows to perform cyclic of bending, torsion and both combined. The tests of cyclic bending and torsion were conducted within the range of low and high number of cycles with controlled force (in the considered case, the moment amplitude was controlled). The loads were of a sinusoidal nature with a frequency of about 25–29 Hz. The amplitudes and the mean value of the load were changed according to the test requirements. The nominal stress amplitude and nominal mean stress value were used in calculations.

The tests included three load states of the samples determined by the angle α determining the combination of bending and torsion (Figure 3). It is easy to find, that for $\alpha = 0$ we have $M_{g\alpha}(t) = M(t)$—the specimen is subjected to bending. When $\alpha = 90°$, we have $M_{g\alpha}(t) = M(t)$—the specimen is subjected to torsion. The combined bending with torsion is applied to specimen for every value of the angle in range $0 < \alpha < 90°$. For combined bending with torsion, torsion moment $M_{g\alpha}(t)$ is proportional to bending moment $M_{g\alpha}(t)$, where proportion is equal to tgα.

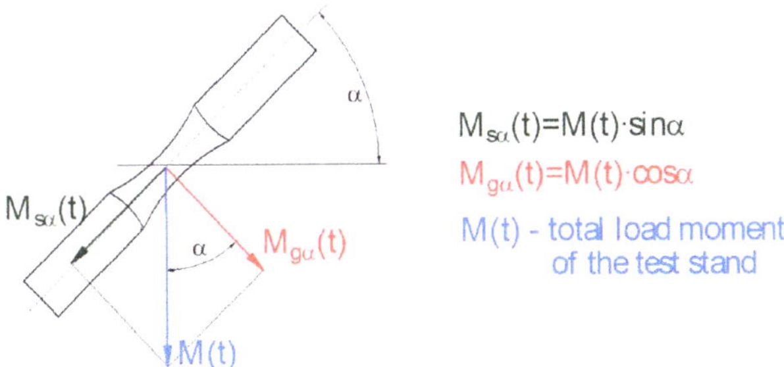

Figure 3. The load scheme of the fatigue test stand MZGS 100.

For test three values of the angle α were used:

$$\alpha = 0 \text{ (rad)}, (0°) \quad \text{(bending)},$$
$$\alpha = 1.107 \text{ (rad)}, (63.5°) \quad \text{(combined bending with torsion)},$$
$$M_{s\alpha} = 2M_{g\alpha}; \; \sigma_a(t) = \tau_a(t),$$
$$\alpha = \pi/2 \text{ (rad)}, \quad (90°) \quad \text{(torsion)}.$$

Experimental studies were conducted at a constant value of the stress ratio $R = -1, -0.5, 0$ by changing the moment amplitude value M_{ag} and the mean moment value M_{mg} for bending and M_{as} and M_{ms} respectively for torsion. The tests were performed for 4–5 levels of stress amplitudes, for each combination of load at least two or three specimens were used. As a result, for the given load levels, the corresponding fatigue life was obtained. Obtaining fracture of specimen was adopted as the failure criterion for the sample.

Standard fatigue characteristics for cyclic tension–compression are described by equation [22]

$$log\, N_{cal} = B + A\, log\, \sigma_a = 24.32 - 7.91 log \sigma_a, \tag{19}$$

where the limits of confidence intervals with a probability of 0.95 for individual parameters are as follows: $-6.16 \le A \le n - 9.67$; $20.04 \le B \le 28.60$. The fatigue limit obtained is $Z_{rc} = 204$ MPa and correspond her $N_0 = 1.12 \times 10^6$ cycles.

4. Results and Discussion

The results of bending fatigue tests were approximated by the regression equation in the form (in the case of torsion, the equation contains τ_a)

$$log\, N_{exp} = B + A log \sigma_a, \tag{20}$$

Figure 4 show the results of tests on bending, torsion and combination of bending with torsion as well as fatigue characteristics determined based on Equation (20).

Table 3 contains the values of the parameters of the regression equation and the correlation coefficient r values of logN and logσ_a (logτ_a), while Table 4 contains the fatigue limits for individual load combinations.

Table 3. Parameters of the regression equation.

R	Bending			Torsion			Bending and Torsion		
	B	A	r	B	A	r	B	A	r
−1	23.93	−7.19	−0.97	32.81	−11.82	−0.94	29.73	−10.62	−0.98
−0.5	23.71	−7.40	−0.96	27.60	−10.01	−0.93	36.73	−14.67	−0.99
0	31.40	−10.73	−0.96	59.76	−25.25	−0.95	31.62	−12.42	−0.94

Table 4. Fatigue limits.

Moment and Stress	Bending	Bending and Torsion	Torsion
	$\alpha = 0°$	$\alpha = 63.5°$	$\alpha = 90°$
M_a (N·m)	13.59	16.82	17.53
$\sigma_{-1}\,(\tau_{-1})$ (MPa)	$\sigma_{-1} = 271$	$\sigma_{-1} = \tau_{-1} = 152$	$\tau_{-1} = 175$

Analyzing the graphs in Figure 4 and the parameters contained in Table 3, it can be concluded that the values of the coefficient A for bending are similar in all cases except $R = 0$ ($A = -10.73$). During torsion for $R = -1$ and -0.5, the values of coefficient A are close to each other, while for $R = 0$ its value clearly differs from the others. In the case of a combination bending with torsion, the values of the coefficient A are similar for R from -1 to 0.

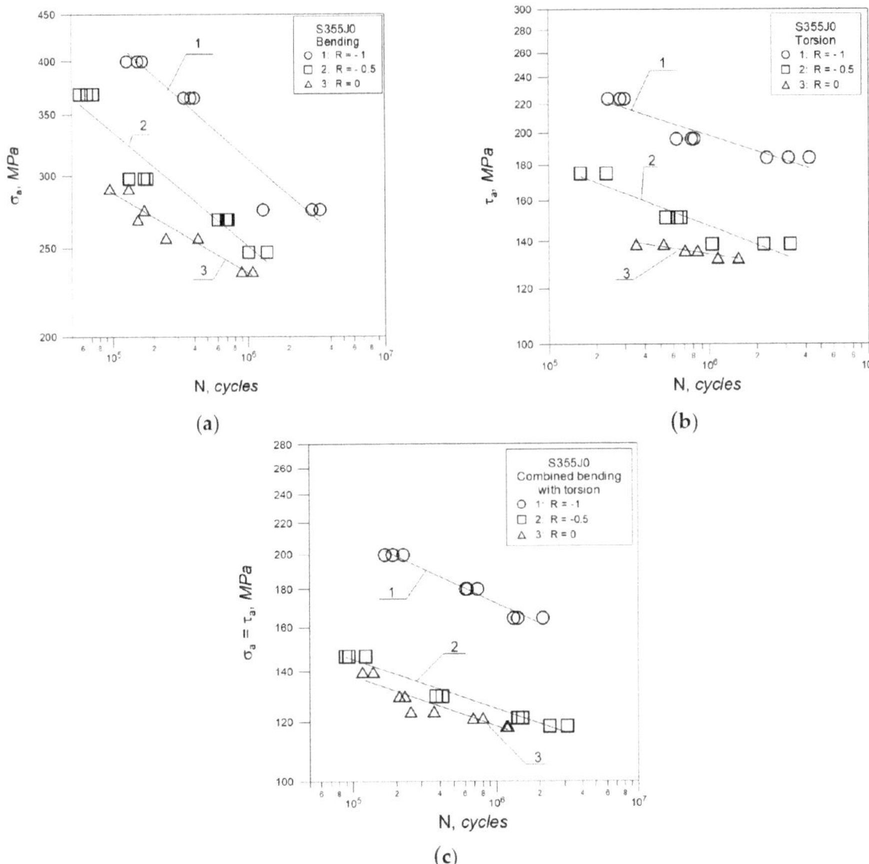

Figure 4. The results of fatigue tests on: (**a**) bending, (**b**) torsion, (**c**) bending with torsion.

Increasing the value of the stress ratio R from -1 to 0 causes a significant decrease in the permissible (acceptable) amplitudes. In the case of bending, R = -0.5 allowable amplitudes are about 20% lower than for $R = -1$ (at $\sigma_m = 0$) and for $R = 0$ they are 35% lower. During torsion for $R = -0.5$, the allowable stress amplitudes are about 19% smaller compared to the stress amplitudes for $R = -1$, while for $R = 0$ this decrease is from 40% for a small number of cycles and up to 27% for a large number of cycles. For the combination bending with torsion under $R = -0.5$ the stress amplitudes are reduced by 35% for a small number of cycles and to 27% for a large number of cycles, while for $R = 0$ the allowable stress amplitudes are reduced for a small and a large number of cycles respectively 37% and 32%.

Figure 5 show the impact of the mean value of bending stress σ_m and torsional stress τ_m on the change in allowable stress amplitudes σ_a and τ_a. Lines in Figure 5 present the interpolation of the data on the graph.

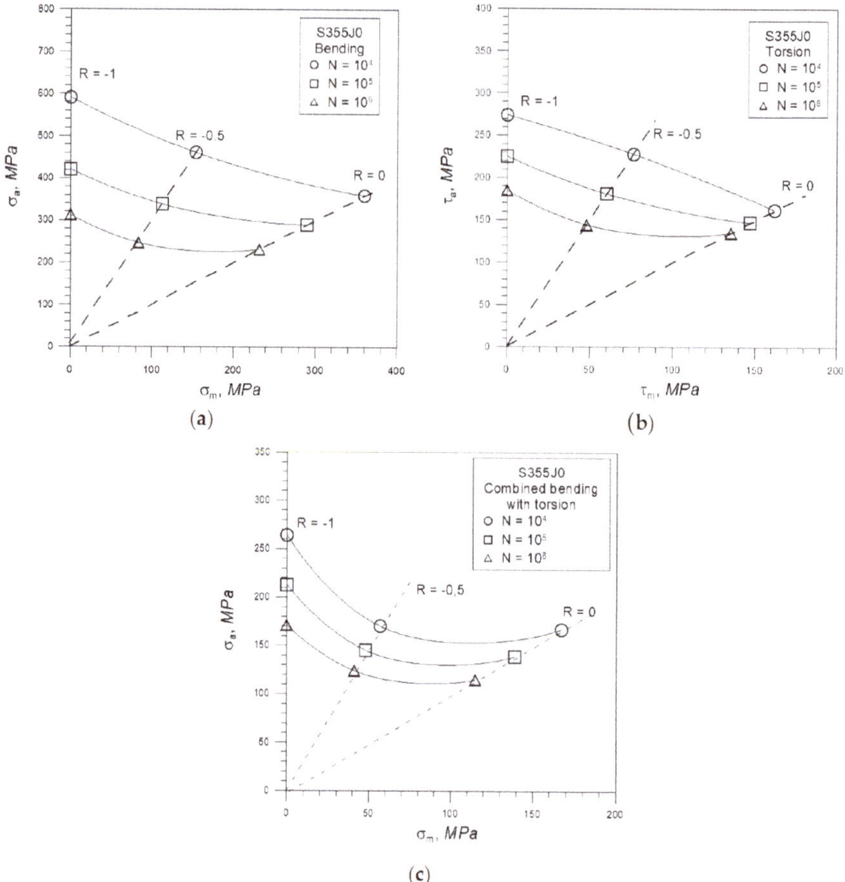

Figure 5. The dependence of the stress amplitude on the mean stress for: (**a**) bending, (**b**) torsion, (**c**) bending with torsion.

In the case of bending (Figure 5a) for a lifetime of 10^4 cycles, an increase in the mean stress value σ_m in relation to $\sigma_m = 0$ causes a decrease in the stress amplitude σ_a to varying degrees, e.g., for $R = 0.5$ by 22% and for $R = 0$ by 39%. For durability $N = 10^5$ cycles, the decreases in σ_a are 20% and 30% for $R = -0.5$ and 0, respectively. These differences for all R values are no longer as large as for $N = 10^4$ cycles. However, in the case of durability at the level of 10^6 cycles, there is a decrease in the value of stress amplitude σ_a by an average of 23% in relation to $\sigma_m = 0$, while for $R = -0.5$ and 0 the value of allowable stress amplitude practically does not depend on the value of the mean stress.

During torsion (Figure 5b), for life $N = 10^4$ cycles, a significant decrease in stress τ_a due to the increase in value τ_m is also visible, e.g., in relation to loads $\tau_m = 0$ the decrease is 17% and for $R = -0.5$ and about 40% for $R = 0$. At durability level $N = 10^5$ cycles these decreases are 20% and 34% respectively. For a large number of cycles, this difference virtually disappears and the decrease in the permissible amplitude values is very similar (23% and 27%).

For the combination of bending with torsion (Figure 5c) and durability $N = 10^4$ cycles, it can be seen that with an increase in average stress, there is a large decrease in the allowable stress amplitudes (about 35%) for both $R = -0.5$ and $R = 0$. Similarly, for $N = 10^5$ cycles- the decrease in the allowable stress amplitudes are 31% and 34% for $R = -0.5$ and $R = 0$, respectively. For $N = 10^6$ cycles, an increase in the mean stress value reduces the stress amplitude by 32% for $R = -0.5$ and 27% for $R = 0$ compared to the case $\sigma_m = \tau_m = 0$. The small influence of the mean stress value from the durability level $N = 10^5$ cycles is characteristic. A further increase in mean stress practically does not cause major changes in the values of allowable stress amplitudes.

Using the relationships (10) and (11) on the basis of fatigue tests, the value of the ψ coefficient was determined in terms of life $N = (5 \cdot \times 10^4 - 2.5 \times 10^6)$ cycles. The change of the value of the factor ψ depending on the number of cycles N is shown in Figure 6. For bending, this relationship was described by the function $\psi_\sigma(N) = 3.1621 N^{-0.164}$. There is a visible decrease in the value of the factor in terms of durability $N = (5 \cdot \times 10^4 - 7.5 \cdot \times 10^5)$ cycles. In the case of torsion, the function has the form $\psi_\tau(N) = 2.897 N^{-0.131}$, while the nature of the curve is similar to that for bending. In the case of a combination of bending and torsion, the function takes the form $\psi_\tau(N) = \psi_\sigma(N) = 0.854 N^{-0.044}$. From this function one can notice a milder course than for bending and torsion, with the largest decreases in the coefficient value also occurring in the durability range $N = (5 \cdot \times 10^4 - 7.5 \cdot \times 10^5)$ cycles.

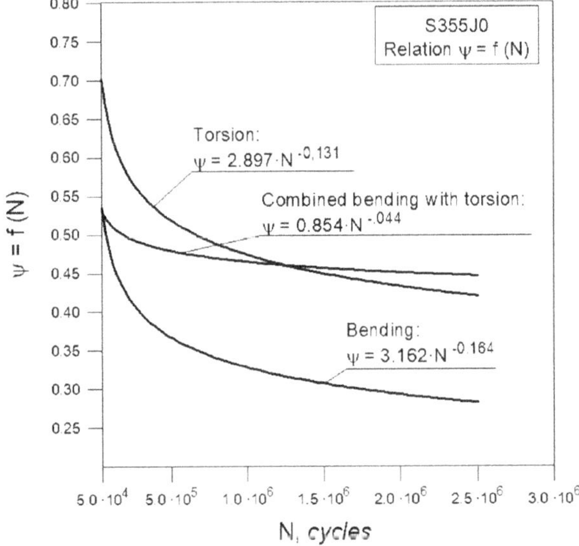

Figure 6. Change of the material sensitivity factor $\psi = f(N)$ for the asymmetry of cycle.

In the algorithm of calculating fatigue life for bending, torsion and combination of bending with torsion with the participation of mean stress values, the transformation relationships described by Equations (1)–(5) and (14), (15) modified to form (17), were used.

The values of K_i (16) coefficients were determined assuming the fatigue limits of individual types of loads at $\sigma_m = \tau_m = 0$ (Table 4) respectively:

for bending

$$K_g = \frac{Z_{rc}}{\sigma_{-1}}; \ \sigma_{-1} = 271 \ \text{MPa}, \tag{21}$$

for torsion

$$K_s = \frac{Z_{rc}}{\sqrt{3} \cdot \tau_{-1}}; \ \tau_{-1} = 152 \ \text{MPa}, \tag{22}$$

for bending and torsion

$$K_{gs} = \frac{Z_{rc}}{\sqrt{\sigma_{-1}^2 + 3 \cdot \tau_{-1}^2}}; \ \sigma_{-1} = \tau_{-1} = 175 \ \text{MPa}. \tag{23}$$

Stress amplitudes σ_{aTi} were determined from Equation (17), computational stability N_{cal} according to Equation (18), which is the fatigue characteristics of the tested steel under uniaxial tension-compression (where there is an even distribution of stress across the specimen cross-section).

Figure 7 shows the results of calculations for bending, torsion and a combination of bending and torsion with zero mean value of the load course. Figures 8–10 show a comparison of the design fatigue life with the experimentally obtained life for bending, torsion and combination of bending and torsion for non-zero mean stress values. The solid line means the perfect match between the computational stability N_{cal} and the experimental N_{exp}, while the dashed lines represent the scatter bands of results with the coefficients $N_{cal}/N_{exp} = 3$ (1/3).

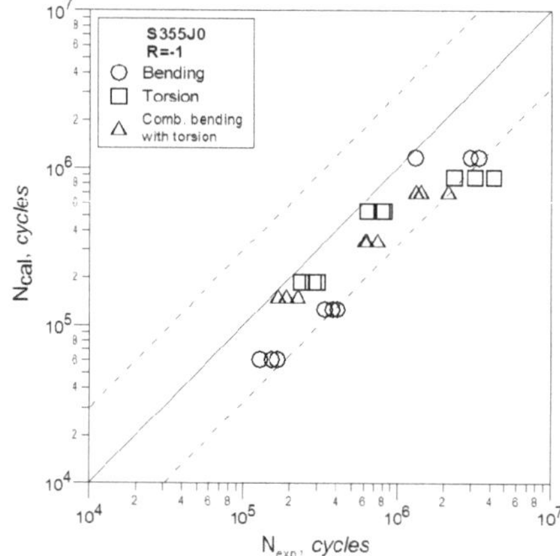

Figure 7. Comparison of the computational life N_{cal} with the experimental N_{exp} for $R = -1$.

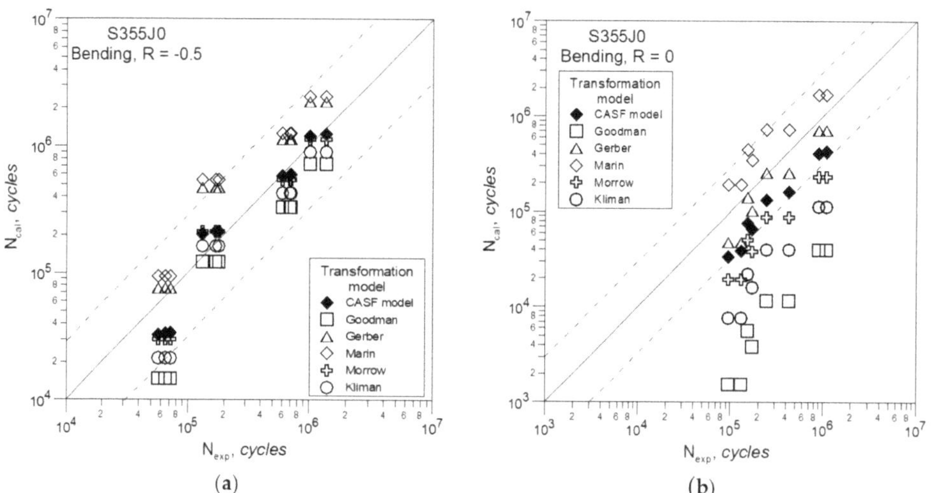

(a)   (b)

Figure 8. Comparison of the computational life N_{cal} with the experimental N_{exp} under bending for: (a) $R = -0.5$, (b) $R = -0$.

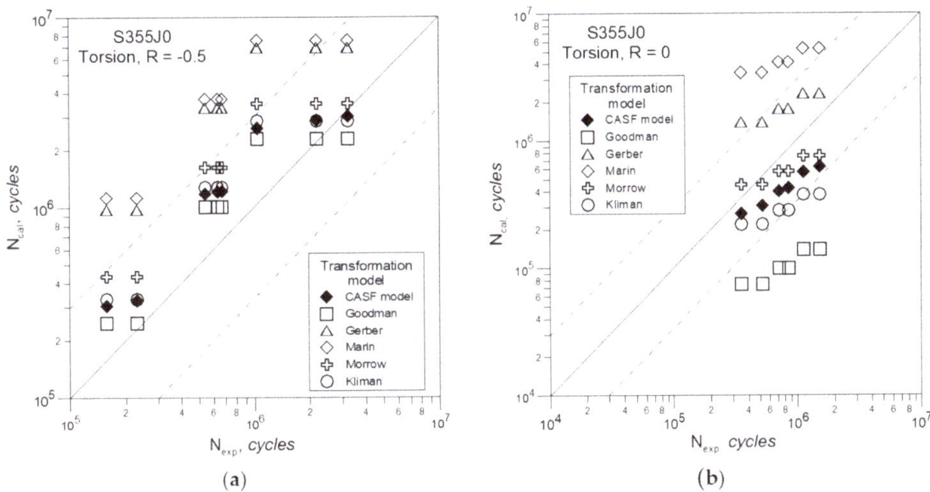

Figure 9. Comparison of the computational life N_{cal} with the experimental N_{exp} under torsion for: (**a**) $R = -0.5$, (**b**) $R = -0$.

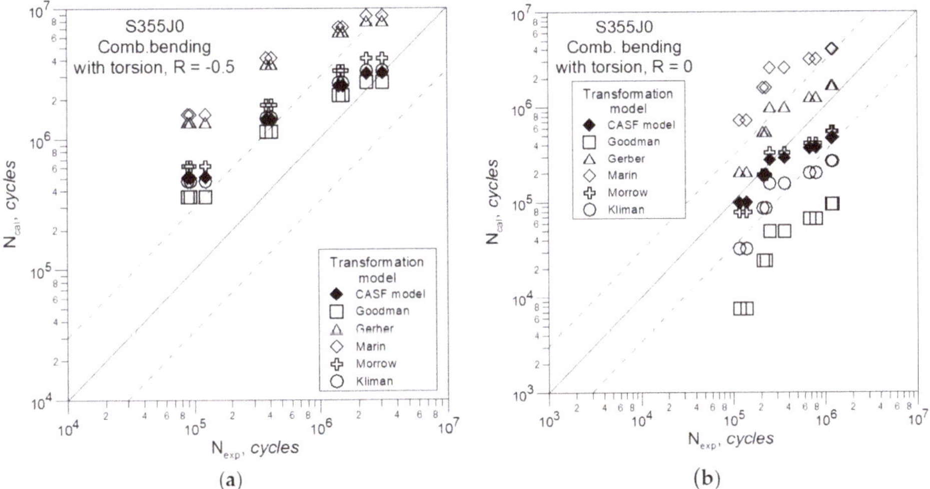

Figure 10. Comparison of the computational life N_{cal} with the experimental N_{exp} under bending and torsion for: (**a**) $R = -0.5$, (**b**) $R = -0$.

Figure 7 shows that the computational life are mostly in the result spread band regardless of the type of load and its value. The exception is torsion for a large number of cycles. This allows to state that the assumption adopted in Formula (16) gives satisfactory results of calculations for zero-mean loads and allows to believe that it will be used for other load cases.

The analysis of Figure 8a shows that in the case of bending for $R = -0.5$, the best agreement was obtained thanks to the Morrow transform dependence and the model described by Equation (14). Satisfactory results were also obtained for the remaining relationships. However, their dispersion is greater than for the previously indicated.

At bending loads for $R = 0$ (Figure 8b), correct results of fatigue life calculations were obtained using Gerber, Marin and CASF model (Equation (14)). The calculation results are in the scatter band with the $N_{cal}/N_{exp} = 3$ (1/3) coefficient for all three cases. It should be noted, however, that this model underestimates computational durability compared to experimental. The remaining transformation formulas clearly underestimate the calculated fatigue life; the calculation results are outside the scatter band $N_{cal}/N_{exp} = 3$ (1/3).

Figure 9 shows the results of calculations for torsion. In the case of $R = -0.5$ (Figure 9a), the best results are obtained by the relationship between Goodman and Kliman and the calculation of fatigue life using Equation (15), satisfactory results were obtained for the Morrow relationship. Gerber and Marina's transformation dependencies significantly overstate the results of calculations compared to the results obtained experimentally. For the stress ratio $R = 0$ cycle (Figure 9b), the best results of fatigue life calculations were obtained for Morrow's formula and calculated according to Equation (15).

In the case of a combination of bending and torsion for $R = 0.5$ (Figure 10), the best results of fatigue life assessment were obtained for Goodman and Kliman relationships. Calculations using Equation (14) overstate the computational life for a small number of cycles. For $R = 0$ when bending with torsion (Figure 10b), the best compatibility is given by the assessment of durability using the model defined by Equation (14) and the Morrow and Gerber relationships.

5. Conclusions

The following conclusions can be drawn from the tests performed on S355J0 steel specimens at proportional bending with torsion loads for different values of the R:

1. In the case when $R = -1$ satisfactory results of the fatigue life assessment for bending, torsion and a combination of bending and torsion were obtained using the K_i coefficient, which is the ratio of the fatigue limit at cyclic tension-compression to the fatigue limit, calculated in accordance with the Huber–Misess–Hencky hypothesis for analyzed loads.
2. For loads with a non-zero mean value, it was found that:

    - satisfactory results of the fatigue life assessment for bending were obtained using the Gerber transformation dependence and the CASF model described by Equation (14),
    - in the case of torsion, the best compatibility of computational fatigue life with experimental life is obtained by using Equation (15) and Morrow's formula,
    - for the combination of bending and torsion for both $R = -0.5$ and $R = 0$, the CASF model gives the best results of fatigue life assessment.

3. The CASF model described by Equations (14) and (15), including the effect of the mean stress value on fatigue life using the material sensitivity factor for the asymmetry of cycle, gives good results for predicting fatigue life for all tested load cases.

Author Contributions: Conceptualization, R.P.; formal analysis, R.P., D.R.; investigation, R.P.; methodology, R.P.; writing–original draft, R.P., D.R. All authors have read and agreed to the published version of the manuscript.

Funding: This research received no external funding.

References

1. Wehner, T.; Fatemi, A. Effect of mean stress on fatigue behavior of a hardened carbon steel. *Int. J. Fatigue* **1991**, *13*, 241–248. [CrossRef]
2. Gołoś, K.M.; Eshtewi, S.H. Multiaxial Fatigue and Mean Stress Effect of St5 Medium Carbon Steel. In Proceedings of the 5th International Conference on Biaxial/Multiaxial Fatigue and Fracture, Cracow, Poland, 8–12 September 1997; Volume 1, pp. 25–34.
3. Glinka, G.; Shen, G.; Plumtree, A. Mean stress effect in multiaxial fatigue. *Fatigue Fract. Eng. Mater. Struct.* **1995**, *18*, 755–764. [CrossRef]

4. Skibicki, D.; Sempruch, J.; Pejkowski, L. Model of non-proportional fatigue load in the form of block load spectrum. *Mater. Sci. Eng. Techonl.* **2014**, *45*, 68–78. [CrossRef]

5. Liu, Y.; Mahadevan, S. Multiaxial high-cycle fatigue criterion and life prediction for metals. *Int. J. Fatigue* **2005**, *27*, 790–800. [CrossRef]

6. Kluger, K.; Pawliczek, R. Assessment of validity of selected criteria of fatigue life prediction. *Materials* **2019**, *12*, 2310. [CrossRef] [PubMed]

7. Rozumek, D.; Marciniak, Z. Fatigue properties of notched specimens made of FeP04 steel. *Mater. Sci.* **2012**, *47*, 462–629. [CrossRef]

8. Prażmowski, M.; Rozumek, D.; Paul, H. Static and fatigue tests of bimetal Zr-steel made by explosive welding. *Eng. Fail. Anal.* **2017**, *75*, 71–81. [CrossRef]

9. Branco, R.; Costa, J.D.; Berto, F.; Antunes, F.V. Effect of loading orientation on fatigue behaviour in severely notched round bars under non-zero mean stress bending-torsion. *Theor. Appl. Fract. Mech.* **2017**, *92*, 185–197. [CrossRef]

10. Carpinteri, A.; Kurek, M.; Łagoda, T.; Vantadori, S. Estimation of fatigue life under multiaxial loading by varying the critical plane orientation. *Int. J. Fatigue* **2017**, *100*, 512–520. [CrossRef]

11. Rozumek, D.; Faszynka, S. Influence of the notch radius on fatigue crack propagation in beam specimens of 2017A-T4 alloy. *Mater Sci.* **2017**, *53*, 417–423. [CrossRef]

12. Vicente, C.M.S.; Sardinha, M.; Reis, L. Failure analysis of a coupled shaft from a shredder. *Eng. Fail. Anal.* **2019**, *103*, 384–391. [CrossRef]

13. Marin, J. Interpretation of fatigue strengths for combined stresses. In Proceedings of the International Conference on Fatigue of Metals, London, UK, 10–14 September 1956; pp. 184–194.

14. Goodman, J. *Mechanics Applied to Engineering*, 9th ed.; Longmans, Green and Co.: New York, NY, USA, 1954.

15. Gerber, W. Determination of permissible stresses in iron constructions. *Zeitschrift des Bayerischen Architekten-und Ingenieur-Vereins* **1974**, *6*, 101–110. (In German)

16. Pawliczek, R.; Kluger, K. Influence of irregularity coefficient of loading on calculated fatigue life. *J. Theor. Appl. Mech. Lett.* **2013**, *51*, 4.

17. Morrow, J. *Fatigue Design Handbook, Advances in Engineering*; Society of Automotive Engineers: Warrendale, PA, USA, 1968; Volume 4, p. 29.

18. Kliman, V. Prediction of Random Load Fatigue Life Distribution. In *Fatigue Design, ESIS 16*; Mechanical Engineering Publications: London, UK, 1993; pp. 241–255.

19. Kocańda, S.; Szala, J. *Basics of Fatigue Calculations*; PWN: Warsaw, Poland, 1997. (In Polish)

20. Achtelik, H.; Kurek, M.; Kurek, A.; Kluger, K.; Pawliczek, R.; Łagoda, T. Non-standard fatigue stands for material testing under bending and torsion loadings. *AIP Conf. Proc.* **2018**, *2029*, 020001.

21. Rozumek, D.; Marciniak, Z.; Lachowicz, C.T. The energy approach in the calculation of fatigue lives under non-proportional bending with torsion. *Int. J. Fatigue* **2010**, *32*, 1343–1350. [CrossRef]

22. ASTM E 739-80: Standard practice for statistical analysis of linearized stress-life (S-N) and strain-life (-N) fatigue data. In *Annual Book of ASTM Standards*; ASTM International: Philadelphia, PA, USA, 1989; Volume 03.01, pp. 667–673.

Article

Multiaxial Fatigue Life Prediction on S355 Structural and Offshore Steel Using the SKS Critical Plane Model

Alejandro S. Cruces [1], Pablo Lopez-Crespo [1,*] , Belen Moreno [1] and Fernando V. Antunes [2]

[1]   Department of Civil and Materials Engineering, University of Malaga, C/Dr Ortiz Ramos s/n, 29071 Málaga, Spain; ascruces@uma.es (A.S.C.); bmoreno@uma.es (B.M.)

[2]   Department of Mechanical Engineering, University of Coimbra, 3000-370 Coimbra, Portugal; fernando.ventura@dem.uc.pt

*   Correspondence: plopezcrespo@uma.es; Tel.: +34-951952308

Received: 13 November 2018; Accepted: 6 December 2018; Published: 13 December 2018

Abstract: This work analyses the prediction capabilities of a recently developed critical plane model, called the SKS method. The study uses multiaxial fatigue data for S355-J2G3 steel, with in-phase and 90° out-of-phase sinusoidal axial-torsional straining in both the low cycle fatigue and high cycle fatigue ranges. The SKS damage parameter includes the effect of hardening, mean shear stress and the interaction between shear and normal stress on the critical plane. The collapse and the prediction capabilities of the SKS critical plane damage parameter are compared to well-established critical plane models, namely Wang-Brown, Fatemi-Socie, Liu I and Liu II models. The differences between models are discussed in detail from the basis of the methodology and the life results. The collapse capacity of the SKS damage parameter presents the best results. The SKS model produced the second-best results for the different types of multiaxial loads studied.

Keywords: critical plane model; multiaxial fatigue; non-proportional; S355-J2G3

1. Introduction

Performing accurate fatigue design is crucial for making a difference between the efficient use of a material with a reliable security factor and a catastrophic failure that could endanger human lives. Since most mechanical systems are subjected to complex multiaxial loadings, it seems appropriate to apply specific theories, instead of uniaxial-based theories.

When a system is subjected to multiple loads, it is necessary to consider not only the stress state but also the response of the material, the interaction between the loads, and their effect on the nucleation and growth of cracks. Effects of mean stresses and strains on multiaxial fatigue were investigated by Socie and Kurath [1], observing the sensible detrimental effect over fatigue life of the mean stress. Studies about the material hardening for different loading paths in 304 stainless steel were carried out by Itoh et al. [2]. Interactions between sub-cycles of the loads were investigated by Erickson in order to develop a suitable damage parameter which includes such effects [3]. Fatigue life and crack growth planes predictions around notched specimens were also studied by Branco [4] under multiaxial conditions. Multiaxial fatigue theories try to encapsulate such information in a damage parameter that describes what happens throughout the load cycle. Instead of developing a single theory to cover all materials, load configurations and external factors, various theories have been postulated aimed at improving the predictions for a range of scenarios. Multiaxial fatigue theories can be classified into three main groups: (i) extended static yield criteria to fatigue [5–8]; (ii) cycle energy parameters [9–12]; and (iii) critical plane models [13–16]. Previous studies and observations have shown better results when applying critical plane models [13]. Critical plane models evaluate the damage in each cycle with a parameter, normally as combination of several stress-strain values, in the plane where the crack nucleates and grows. Critical plane models have shown positive results for various materials and industrial applications [17–19] and are already implemented in some recognised computer aided engineering tools widely used across structural, offshore and mechanical industries. For example, Fatemi-Socie and Wang-Brown models are included in MSC Fatigue and Comsol numerical software [20,21]. As in uniaxial fatigue, models with damage parameters based on stress data are more suited to high cycle fatigue and those methods based on strain data present better results for low cycle fatigue, as they account for plastic strain and other low cycle life effects [22]. Depending on the dominant cracking mode for the material, some models are more suitable for mode I [16] and other for mode II or III [13].

In this work, a recently introduced critical plane model is evaluated. It is a stress based model that was introduced previously by Sandip, Kallmayer and Smith [23]. We will refer to this recently introduced model as the SKS model. The SKS damage parameter is able to account for the material hardening/softening, the mean shear stress effect and the interaction between shear and normal stresses [23]. The SKS damage parameter includes two material parameters to be fitted with fatigue experimental data. Life is also estimated using other previously introduced critical plane models, namely Fatemi-Socie [13]; Wang-Brown [15]; Liu I and Liu II [14]. The fitness of the SKS critical plane model is assessed by comparing the results with the other well-established critical plane models on S355-J2G3 steel subjected to both in-phase and out-of-phase loadings.

2. Materials and Methods

The material utilised in this investigation is a low carbon steel S355-J2G3, commonly used in industry for both structural and offshore applications. This material exhibits good fatigue resistance and has environmental impacts for applications where no energy is consumed during the use phase of the component [24]. Its typical chemical composition is shown in Table 1.

Table 1. Chemical composition (%).

C	Mn	Si	P	S	Al	Cr	Ni	Mo
0.17	1.235	0.225	0.01	0.0006	0.032	0.072	0.058	0.016

The monotonic properties of the material are shown in Table 2 and were obtained from experimental testing, including tension, compression and torsion. Plane specimens were used to characterise tension properties and tubular specimens were used to characterise torsion/compression properties.

Table 2. Monotonic properties of S355-J2G3 steel.

Property	Value
Yield strength, σ_y	386 MPa
Ultimate strength, σ_u	639 MPa
Young's Modulus, E	206 GPa
Shear Modulus, G	78 GPa
Critical buckling stress, σ_{cr}	348 MPa

The cyclic uniaxial properties (Tables 3 and 4) were obtained following the ASTM recommendations [25]; for ε-N curves, fifteen samples were tested with five different strain levels (i.e., three samples for each strain level), and for γ-N curves, twelve samples were tested with four different strain levels (i.e., three samples for each strain level). In both series of tests, the criterion to stop the test was a 20% drop from the maximum load.

Table 3. Uniaxial properties of S355-J2G3 steel.

Property	Value
Cyclic strength coefficient, K'	630.6 MPa
Cyclic hardening exponent, n'	0.10850
Cyclic yield strength, σ'_y	321.3 MPa
Fatigue strength coefficient, σ'_f	564.4 MPa
Fatigue strength exponent, b	−0.0576
Fatigue ductility coefficient, ε'_f	0.1554
Fatigue ductility exponent, c	−0.4658

Table 4. Torsional properties of S355-J2G3 steel.

Property	Value
Cyclic torsional strength coefficient, K'_γ	593.8 MPa
Cyclic torsional hardening exponent, n'_γ	0.1553
Cyclic torsional yield strength, τ'_γ	594.2 MPa
Fatigue torsional strength coefficient, τ'_f	486.9 MPa
Fatigue torsional strength exponent, b_γ	−0.0668
Fatigue torsional ductility coefficient, γ'_f	0.0662
Fatigue torsional ductility exponent, c_γ	−0.3191

Multiaxial fatigue tests were conducted under strain control with total inversion with tubular specimens (Figure 1) [26]. The use of strain control rather than load control is more appropriate in order to define the stabilised stress-strain hysteresis loop, especially in the low cycle fatigue (LCF) regime when yield stress is exceeded and cross section reduction could be important [27]. A way to reduce the test time is to obtain the load stabilised values, and finish the test under load control with a higher frequency [28].

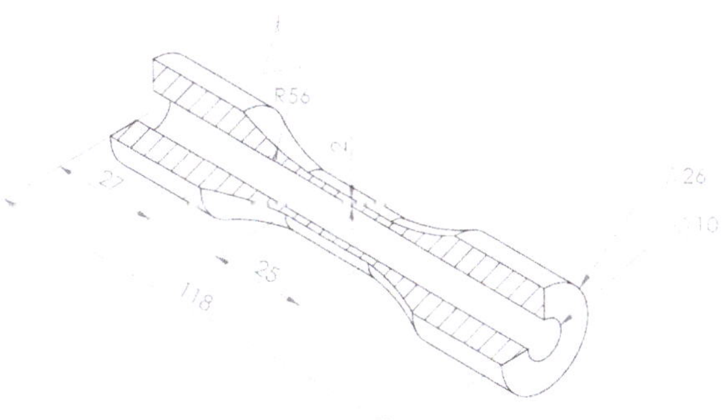

Figure 1. Geometry of the specimen. All dimensions are in mm.

3. Critical Plane Models

Critical plane models are based on observations of crack nucleation and growth during loading. Therefore, depending on the material and service conditions, fatigue life will be calculated using parameters at planes dominated by shear loading parameters, axial loading parameters or a combination of both types of parameters [29,30]. The general procedure for implementing a critical plane model consists of three steps: (i) finding the critical plane φ^*, along the cycle; (ii) quantifying the damage by the so-called damage parameter; and (iii) estimating the fatigue life with function depending on the damage parameter.

The critical plane is defined as the plane where a certain variable or the model damage parameter reaches its maximum value. Some implementations of the critical plane models assume that the critical plane is where one stress or strain component is at its maximum. Such a component can be the shear strain range [13,31], the shear stress range [32,33], the normal stress range [34] or the normal strain range [16]. Nevertheless, recent works tend to consider the plane where the maximum value of a damage combination is achieved [18,35,36]. The damage parameter often includes two of the above components. In this work, we have used the critical plane as it was defined by its authors.

3.1. Sandip-Kallmeyer-Smith Model (SKS)

The SKS critical plane damage parameter was initially proposed by Suman, Kallmeyer and Smith [23]. The SKS damage parameter is defined on the critical plane using the maximum shear strain range $\Delta\gamma$ as:

$$DP_{exp} = (G\Delta\gamma)^{w} \tau_{max}^{(1-w)} \left(1 + k\frac{(\sigma\tau)_{max}}{\sigma_o^2}\right) \tag{1}$$

where G is the shear modulus, $\Delta\gamma$ is the shear strain range, τ_{max} is the maximum shear stress, $(\sigma\tau)_{max}$ is the maximum shear and tensile stress product value, σ_o is a factor used to maintain unit consistency, w and k are material fitting parameters.

The strain hardening effect that takes place in the LCF regime is taken into account by $\Delta\gamma$ and τ_{max}. The mean shear stress effect in the high cycle fatigue (HCF) regime is also taken into account by the shear ratio τ_{min}/τ_{max}. At the HCF regime, plastic strain is reduced, thereby the damage parameter could be rearranged as is shown in Equation (2). The parameter w weights the hardening and mean shear stress effect. The product $(\sigma\tau)_{max}$ introduces the detrimental effect over fatigue life observed when sub-cycle load peaks are applied simultaneously. The parameter k gauges the interaction effect between the shear and the normal stresses.

$$DP_{exp} = \left(1 - \frac{\tau_{min}}{\tau_{max}}\right)^{w} \tau_{max} \left(1 + k\frac{(\sigma\tau)_{max}}{\sigma_o^2}\right) \tag{2}$$

Although some experiments have shown that mean shear stresses have a limited damaging effect for values of τ_{max} below the shear yield stress [7], it really depends on the material under study [37]. Nevertheless, its effect is lower than the effect of mean normal stress. For mild-steels such as the S355-J2G3 studied in the present work, Sines observed a low effect of the mean shear stress [7]. The SKS critical plane was developed for assessing the multiaxial fatigue crack behaviour. No information was found regarding the usefulness of the SKS model for notched geometries.

An effective damage parameter should be able to collapse into a single curve all the experimental data with minimum scatter. The following expression used a double exponential law to relate the damage parameter with fatigue life N_f [3,23]:

$$DP_{calc} = AN_f^{b} + CN_f^{d} \tag{3}$$

where A, b, C, and d are material dependent parameters and N_f is the fatigue life in cycles. In order to analyse the effectiveness of the SKS damage parameter, it is compared to other well-established models, relating them with the same type of curve (Equation (3)), obtaining all the material parameter for the best fitting. The parameters are evaluated with an optimisation process based on a least square error minimisation [3] (Equation (4)). This reduces the error between the experimental damage parameter (DP_{exp}), which defines the stress state on the critical plane and the calculated damage parameter (DP_{calc}) based on the mean life curve (Equation (3)). The initial values for the parameters are set to be close to the material properties defined on each model. For SKS, DP_{calc} exponent parameters are set to keep the negative exponential curve growth.

$$error = \left[\frac{(DP_{exp} - DP_{calc})}{DP_{exp}}\right]^2 \tag{4}$$

The SKS damage parameter is assessed based on the fatigue life prediction capacity, by fitting the 6 parameters required to fully define a model with Equations (1) and (3). The model works by predicting the fatigue life based on experimental fatigue data obtained under multiaxial conditions. If the experimental data employed for fitting the variables are not enough and/or have poor quality, the predictions are expected to be poor.

The material parameters are fitted using the information provided by different experiments conducted both in-phase and 90° out-of-phase (Table 5). Two different loading paths were used, and these are illustrated in Figure 2. The parameters of the SKS model were fitted using the tests shown in Table 5. Since the number of parameters required in the SKS model is six, a total of six multiaxial tests were used to obtain their values. Tests combine three different levels of normal strain for a constant shear strain and three levels of shear strain for a constant normal strain, between LCF and HCF regime [38]. The same number of tests with proportional and non-proportional loadings was used in order to avoid a polarised fit of the model. The parameters are fitted to minimise the fatigue life error rather than the error in the damage parameter.

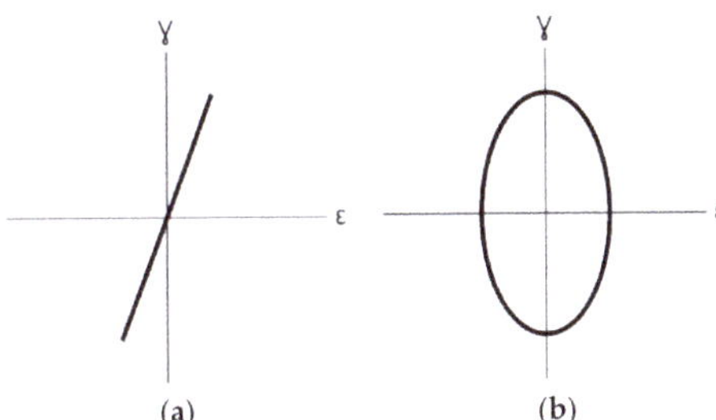

Figure 2. Schematic of the loading paths studied. In-phase (**a**) and out-of-phase load path (**b**).

The following values were obtained for the parameters: $w = -0.35$, $k = 3.25$, $\sigma_0 = 500$, $A = 193.1$, $b = -0.0687$, $c = 205.04$ and $d = -0.0682$. The negative value of w increases the weight of τ_{max} over $\Delta\gamma$. This is also in accordance with the material hardening behaviour, i.e., for a fixed shear stress amplitude, the shear strains decrease and for a fixed shear strain amplitude, the shear stresses will increase. Both cases will produce an increase in the damage. The positive value for k produces a negative effect over the fatigue life as this product increases and the magnitude is related to the value defined by σ_0.

Table 5. Summary of the six experimental data used to fit the six parameters included in the SKS model.

Path	ID	ε_a	γ_a	σ_a	τ_a	N_f
a	F-1	0.0015	0.0026	238	148	162119
	F-2	0.0011	0.0026	185	154	662706
	F-3	0.0009	0.0026	152	162	870886
b	F-4	0.0009	0.0032	189	203	65674
	F-5	0.0009	0.0028	190	192	158248
	F-6	0.0009	0.0026	189	182	248540

3.2. Wang-Brown Model (WB)

The Wang-Brown model (Equation (5)) is a strain-type model. The damage parameter is defined on the plane φ^* with the maximum shear strain range $\Delta\gamma$ [15]. Both shear and normal strains in the critical plane φ^* are taken into account in the damage parameter. Wang and Brown argued that the cyclic shear strain promotes crack nucleation and that the crack growth is a consequence of normal strain.

$$\frac{\Delta\gamma_{max}}{2} + S\Delta\varepsilon_n = \left[(1+\nu_e) + S(1-\nu_e)\right]\frac{(\sigma'_f - 2\sigma_{n,mean})}{E}(2N_f)^b + \left[(1+\nu_p) + S(1-\nu_p)\right]\varepsilon'_f(2N_f)^c \tag{5}$$

where $\Delta\gamma_{max}/2$ is the maximum shear strain amplitude, $\Delta\varepsilon_n$ is the range of normal strain, ν_e and ν_p are the Poisson's ratios in the elastic and plastic regimes, respectively, E is the Young modulus, σ'_f is the fatigue strength coefficient, b is the fatigue strength exponent, ε'_f is the fatigue ductility coefficient and c is the fatigue ductility exponent, $\sigma_{n,mean}$ is the mean normal stress at half fatigue life. S parameter defines the material sensitivity to the normal strain in the crack growth and can be estimated from the fatigue life, N_f, (Equation (6)).

$$S = \frac{\frac{\tau'_f}{G}(2N_f)^{b_\gamma} + \gamma'_f(2N_f)^{c_\gamma} - (1 + \nu_e)\frac{\sigma'_f}{E}(2N_f)^{b} - (1 + \nu_p)\varepsilon'_f(2N_f)^{c}}{(1 - \nu_e)\frac{\sigma'_f}{E}(2N_f)^{b} + (1 - \nu_p)\varepsilon'_f(2N_f)^{c}} \tag{6}$$

where G is the shear modulus and τ'_f is the shear fatigue strength coefficient, b_γ is the shear fatigue strength exponent, γ'_f is the shear fatigue ductility coefficient and c_γ is the shear fatigue ductility exponent.

3.3. Fatemi-Socie Model (FS)

The Fatemi-Socie model (Equation (7)) is a strain type model [13], based on the model proposed by Brown and Miller [39]. They suggested substituting the normal strain component by a normal stress component. The damage parameter is defined on the plane φ^* that maximises the shear strain range, $\Delta\gamma$.

$$\frac{\Delta\gamma_{max}}{2}\left(1 + K\frac{\sigma_{n,max}}{\sigma_y}\right) = \frac{\tau'_f}{G}(2N_f)^{b_\gamma} + \gamma'_f(2N_f)^{c_\gamma} \tag{7}$$

where $\Delta\gamma_{max}/2$ is the maximum shear strain amplitude, $\sigma_{n,max}$ is the maximum tensile stress at φ^*, σ_y is the yield stress, G is the shear modulus, τ'_f is the shear fatigue strength coefficient, b_γ is the shear fatigue strength exponent, γ'_f is the shear fatigue ductility coefficient and c_γ is the shear fatigue ductility exponent. K parameter can be estimated from the fatigue life, N_f, (Equation (8)).

$$K = \left[\frac{\frac{\tau'_f}{G}(2N_f)^{b_\gamma} + \gamma'_f(2N_f)^{c_\gamma}}{(1 + \nu_e)\frac{\sigma'_f}{E}(2N_f)^{b} + (1 + \nu_p)\varepsilon'_f(2N_f)^{c}} - 1\right]\frac{\sigma'_y}{\sigma'_f(2N_f)^{b}} \tag{8}$$

where ν_e and ν_p are the Poisson's ration in the elastic and plastic regimes, respectively, E is the Young modulus, σ'_f is the fatigue strength coefficient, b is the fatigue strength exponent, ε'_f is the fatigue ductility coefficient, c is the fatigue ductility exponent and σ'_y is the cyclic yield stress.

3.4. Liu I and Liu II Models

Liu I (Equations (9) and (10)) and Liu II (Equations (11) and (12)) are energy type models [14]. Depending on the failure mode, Liu presents two parameters, one for a normal tension failure, ΔW_I, and another one for shear tension failure, ΔW_{II}. In this way, Liu models can also take into account whether the failure is ductile or brittle. Shear stresses and strains will normally show a higher weight for ductile materiales and normal stresses and strains will normally show a higher weight for brittle materials [14,40]. For normal tension failure, the plane φ^* will be the one that maximises the axial work, ΔW_I (Equation (9)). For shear tension failure, the plane φ^* will be the one that maximises the shear work, ΔW_{II} (Equation (11)). Once φ^* is determined, the damage parameter is obtained as the sum of both the axial work and the shear work. The relation between the Liu damage parameters and the fatigue life, N_f, is given by Equations (10) and (12), respectively.

$$\Delta W_I = (\Delta\sigma_n\Delta\varepsilon_n)_{max} + (\Delta\tau\Delta\gamma) \tag{9}$$

where $\Delta\sigma_n$ is the normal stress range, $\Delta\varepsilon_n$ is the normal strain range, $\Delta\tau$ is the shear stress range and $\Delta\gamma$ is the shear strain range.

$$\Delta W_I = 4\sigma_f'\varepsilon_f'(2N_f)^{b+c} + \frac{4\sigma_f'^2}{E}(2N_f)^{2b} \tag{10}$$

where σ_f' is the fatigue strength coefficient, b is the fatigue strength exponent, ε_f' is the fatigue ductility coefficient, c is the fatigue ductility exponent and E is the Young modulus.

$$\Delta W_{II} = (\Delta\sigma_n\Delta\varepsilon_n) + (\Delta\tau\Delta\gamma)_{max} \tag{11}$$

$$\Delta W_{II} = 4\tau_f'\gamma_f'(2N_f)^{b\gamma+c\gamma} + \frac{4\tau_f'^2}{G}(2N_f)^{2b\gamma} \tag{12}$$

where τ_f' is the shear fatigue strength coefficient, b_γ is the shear fatigue strength exponent γ_f' is the shear fatigue ductility coefficient, c_γ is the shear fatigue ductility exponent and G is the shear modulus.

4. Results and Discussion

Seventeen multiaxial tests were used in order to assess the SKS model. The key parameters of these tests are summarised in Table 6, with increasing fatigue life. The tests are classified into sinusoidal normal-shear strain with total in-phase inversion (Figure 2a), and 90° out-of-phase (Figure 2b). Strain amplitudes ε_a and γ_a were chosen based on previous results to obtain a number of cycles to failure in the range 10^4 to 10^6, so that both LCF and HCF regimes are evaluated. All tests employed in the evaluation are described in Table 6. Table 6 includes the following information of each test: axial ε_a and shear γ_a amplitudes; stress amplitudes σ_a and τ_a calculated from the axial and torsion load measured at half-life; and the total life obtained, N_{exp}, in number of cycles. A clear hardening effect can be observed between the tests with the same amplitude strains when the loads are applied out-of-phase (tests 10 to 17).

Table 6. S355-J2G3 Fatigue experimental data.

Path	ID	ε_a	γ_a	σ_a	τ_a	N_{exp}
a	1	0.0011	0.0032	180	185	72011
	2	0.0015	0.0028	234	151	103138
	3	0.0015	0.0028	238	151	141938
	4	0.0011	0.0028	183	165	179446
	5	0.0011	0.0032	177	176	179628
	6	0.0009	0.0032	143	183	188219
	7	0.0009	0.0032	146	184	248009
	8	0.0011	0.0028	178	163	268051
	9	0.0009	0.0028	151	172	624521
b	10	0.0015	0.0032	291	200	9838
	11	0.0015	0.0032	305	213	19078
	12	0.0009	0.0032	195	204	38376
	13	0.0011	0.0028	229	196	44319
	14	0.0011	0.0028	229	193	44800
	15	0.0015	0.0026	298	192	46196
	16	0.0015	0.0026	298	193	47996
	17	0.0011	0.0026	226	182	249996

Figure 3 shows the correlation obtained for each damage parameter with the curve defined in Equation (3) for DP_{calc}. The experimental data obtained under proportional load appear as green triangles and the data obtained under non-proportional load appear in Figure 3 as blue asterisks. Each point is defined with the experimental fatigue life N_{exp} and the experimental damage parameter DP_{exp} obtain from the experimental stresses and strain for each model. The experimental data that fall above the calculated damage parameter curve are considered conservative. The collapse ability is studied with the sum of relative error, as defined in Equation (4). Table 7 summarises the relative error obtained for each model. Table 7 indicates that the best correlation is obtained by SKS and FS models, followed by Liu II. It is worth noting that the damage parameter defined in Equation (11) does not include any material dependent parameter. Figure 3b shows that one of the non-proportional tests shifts the WB curve upwards, thus increasing the total error. Finally, Liu I (Figure 3c) shows the worst result, with a larger deviation in the LFC regime. The better collapse capacity of the SKS damage parameter is probably due to the additional number of material parameters that improve the model's adaptability to different loading conditions.

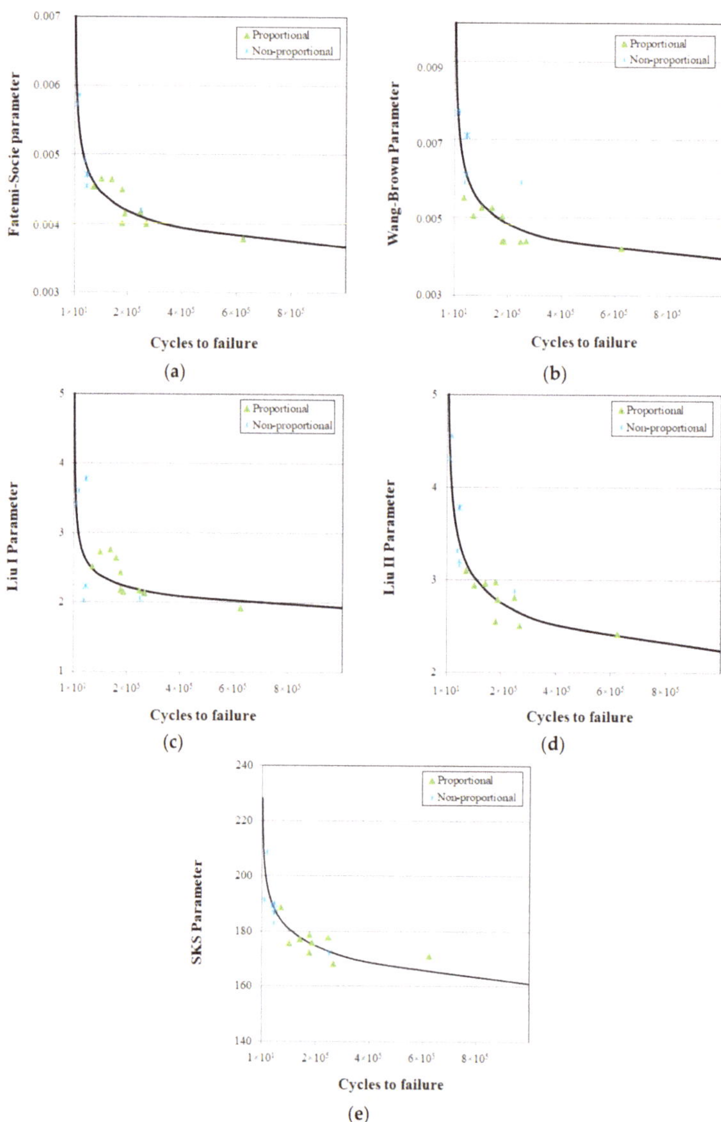

Figure 3. Correlation of S355-J2G3 fatigue data with (**a**) FS, (**b**) WB, (**c**) Liu I, (**d**) Liu II and (**e**) SKS damage parameters.

Table 7. Summary of the relative error between DP_{exp} and DP_{calc} for each model.

Model	FS	WB	Liu I	Liu II	SKS
In-phase test error	0.01469	0.05385	0.07246	0.02077	0.00475
Out-of-phase test error	0.01817	0.09774	0.39104	0.06065	0.00824
Total error	0.03287	0.15159	0.46351	0.08142	0.01300

Figure 4 shows the life estimation for each model under in-phase (a) and out-of-phase (b) loading. The points falling right on the solid line have a perfect coincidence between the experimental fatigue life (N_{exp}) and the calculated fatigue life by each model (N_{mod}). The dashed lines indicate the twice (100%) and half (-50%) deviation from the calculated life with respect to the experimental life. The estimation of SKS is shown with purple triangles, FS with green circles, WB with blue asterisks, Liu I with red squares, and Liu II with black crosses. Figure 5 shows the sum of the relative fatigue life deviation for each model, for in phase (a) and out-of-phase (b) loading.

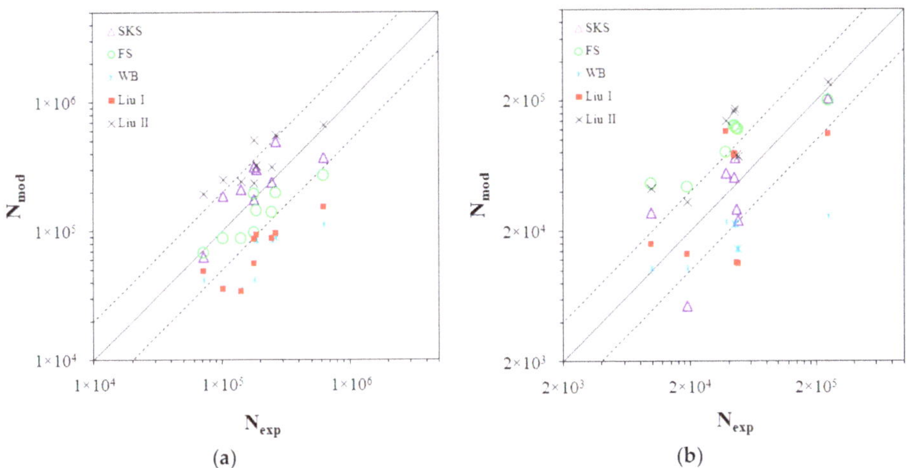

Figure 4. Fatigue life predicted by each model, N_{mod}, versus experimental fatigue life, N_{exp}, for in-phase loading (**a**) and out-of-phase loading (**b**).

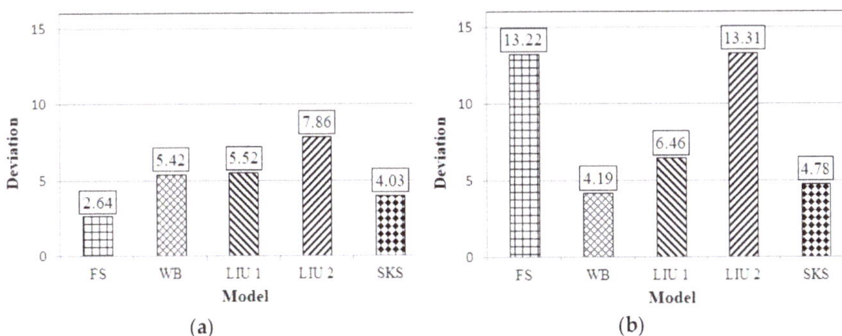

Figure 5. Fatigue life deviation for in-phase loading (**a**) and out-of-phase loading (**b**) for the five different critical plane models under study.

For in-phase loading, Figure 4a, SKS gives a good estimation, with values on the non-conservative side for most tests. This is probably caused by an underestimation of k value, because k parameter controls the shear and normal stress interaction. FS returns better predictions than SKS as can be seen on the sum of deviation shown on Figure 5a, with a conservative tendency with increasing fatigue life. Both WB and Liu I show similar results, mostly on the conservative side. Equations (5) and (9) indicate that WB and Liu I models increase the weight of $\Delta\varepsilon$ on the damage parameter. Equation (11) indicates the opposite behaviour in Liu II model, where the influence of the normal energy is smaller than for the Liu I model.

A comparison between Figure 4a,b indicate that out-of-phase loading shifts upwards the majority of the predictions given by all the models, thus making the predictions more on the non-conservative side. Figure 5 shows that the deviation results get worst for non-proportional loading (Figure 5b) as compared to proportional loading (Figure 5a) for all models with the exception of WB. A comparison between Figure 4a,b shows that the scatter increases for SKS model as a consequence of the non-proportionality of the loads. This is numerically confirmed by the 19% increment in the deviation, from Figure 5a to Figure 5b in the SKS values. Nevertheless, this increment is very similar to the increment exhibited by the Liu I model (17%), and much smaller than the increment displayed by the FS and Liu II models. It was found that the FS model yields better results by using a constant k value of 1, as recommended by Socie [22] instead of evaluating k according to Equation (8). WB results remain in the conservative side and predicted life present the lower sum of life deviation. Liu I shows a higher scatter with low fatigue life deviation. FS shows a better collapse capacity for some of the results compared to the SKS model (see Table 6, experiments 13 to 16). Table 6 shows that small variations are introduced in τ_a, unlike in γ_a. For the SKS fitted parameter, the compensation of $\Delta\gamma$ effect over the damage parameter is accounted for by $(\sigma\tau)_{max}$ factor only, and not $(\sigma\tau)_{max}$ or τ_{max}. Taking into account that the 90° shift between torsional and axial loads induces a very high level of hardening [41], if the shift between loads was smaller, the fatigue predictions would most likely improve. Hence, the results show that SKS model could be competitive both for proportional and non-proportional by fitting good quality experimental data. The results also indicate that the SKS model produces better estimations under in-phase loading than under out-of-phase loading, where a larger scattering of the data was observed.

5. Conclusions

The SKS critical plane damage parameter has been assessed on a S355 steel for in-phase and out-of-phase loadings. The SKS damage parameter has been compared to other established models, namely Fatemi-Socie, Wang-Brown, Liu I and Liu II. The results indicate that overall the collapse capacity of the SKS damage parameter appears to be better than that of the other models for the cases studied. The SKS damage parameter produces overall good predictions both under proportional and under non-proportional combinations of tension, compression and torsion loads. Among the different critical plane models evaluated, SKS produced the second-best results for both types of biaxial loads. The SKS model also showed good consistency in being able to adapt to different load scenarios. In a similar way to the other models, the quality of the data used for fitting the model parameters is crucial. One of the strengths of the SKS models is the simplicity of data fitting for evaluating the model parameters. The authors are currently working on testing the SKS model on other metals and different loading conditions.

Author Contributions: P.L.-C. and B.M. designed the experiments and reviewed the results; A.S.C. and F.V.A. analysed the data; B.M. performed the experiments; A.S.C. and P.L.-C. wrote the paper; P.L.-C., B.M. and F.V.A. reviewed the manuscript.

Funding: This research and the APC was funded by Ministerio de Economia y Competitividad (Spain) through grant reference MAT2016-76951-C2-2-P.

Acknowledgments: The authors would like to thank Jaime Dominguez from University of Seville (Spain) for interesting discussions and suggestions. Industrial support from Sandip Suman and UTC Aerospace Systems (CA, USA) is acknowledged.

Conflicts of Interest: The authors declare no conflict of interest.

References

1. Fatemi, A.; Kurath, P. Multiaxial fatigue life predictions under the influence of mean-stresses. *J. Eng. Mater. Technol.* **1988**, *110*, 380–388. [CrossRef]
2. Itoh, T.; Sakane, M.; Ohnami, M.; Socie, D.F. Nonproportional low-cycle fatigue criterion for type 304 stainless steel. *J. Eng. Mater. Technol. ASME* **1995**, *117*, 285–292. [CrossRef]
3. Erickson, M.; Kallmeyer, A.R.; Van Stone, R.H.; Kurath, P. Development of a multiaxial fatigue damage model for high strength alloys using a critical plane methodology. *J. Eng. Mater. Technol.* **2008**, *130*, 0410081–0410089. [CrossRef]
4. Branco, R.; Costa, J.; Antunes, F. Fatigue behaviour and life prediction of lateral notched round bars under bending-torsion loading. *Eng. Fract. Mech.* **2014**, *119*, 66–84. [CrossRef]
5. Gough, H.J. Engineering steels under combined cyclic and static stresses. *J. Appl. Mech.* **1950**, *50*, 113–125. [CrossRef]
6. Gough, H.J. The strength of metals under combined alternating stress. *Proc. Inst. Mech. Eng.* **1935**, *131*, 3–103. [CrossRef]
7. Sines, G. *Failure of Materials under Combined Repeated Stresses with Superimposed Static Stresses*; National Advisory Committee for Aeronautics: Washington, DC, USA, 1955.
8. Sines, G. *Behavior of Metals under Complex Static and Alternating Stresses*; McGraw-Hill: New York, NY, USA, 1959.
9. Garud, Y. A new approach to the evaluation of fatigue under multiaxial loadings. *J. Eng. Mater. Technol. ASME* **1981**, *103*, 118–125. [CrossRef]
10. Ellyin, F.; Kujawski, D. Plastic strain energy in fatigue failure. *J. Press. Vessel Technol.* **1984**, *106*, 342–347. [CrossRef]
11. Macha, E.; Sonsino, C.M. Energy criteria of multiaxial fatigue failure. *Fatigue Fract. Eng. Mater. Struct.* **1999**, *22*, 1053–1070. [CrossRef]
12. Park, J.; Nelson, D. Evaluation of an energy-based approche and a critical plane approche for predicting constant amplitude multiaxial fatigue life. *Int. J. Fatigue* **2000**, *22*, 23–39. [CrossRef]
13. Fatemi, A.; Socie, D.F. A critical plane approach to multiaxial fatigue damage including out-of-phase loading. *Fatigue Fract. Eng. Mater. Struct.* **1988**, *11*, 149–165. [CrossRef]
14. Liu, K.C.; Wang, J.A. An energy method for predicting fatigue life, crack orientation, and crack growth under multiaxial loading conditions. *Int. J. Fatigue* **2001**, *23*, 129–134. [CrossRef]
15. Wang, H.; Brown, W. A path-independent parameter for fatigue under proportional and non-proportional loading. *Fatigue Fract. Eng. Mater. Struct.* **1993**, *16*, 1285–1298. [CrossRef]
16. Smith, Kn.; Topper, T.H.; Watson, P. A stress strain function for the fatigue of metals(Stress strain function for metal fatigue including mean stress effect). *J. Mater.* **1970**, *5*, 767–778.
17. Chen, X.; Gao, Q.; Sun, X.F. Damage analysis of low-cycle fatigue under non-proportional loading. *Int. J. Fatigue* **1994**, *16*, 221–225. [CrossRef]
18. Chu, C. Multiaxial fatigue life prediction method in the ground vehicle industry. *Int. J. Fatigue* **1997**, *19*, 325–330. [CrossRef]
19. Kim, K.S.; Park, J.C. Shear strain based multiaxial fatigue parameters applied to variable amplitude loading. *Int. J. Fatigue* **2000**, *21*, 475–483. [CrossRef]
20. MSC Fatigue. Available online: http://www.mscsoftware.com/product/msc-fatigue (accessed on September 2018).
21. Comsol. Available online: https://www.comsol.com/fatigue-module (accessed on September 2018).
22. Socie, D.F.; Marquis, G.B. *Multiaxial Fatigue*; Society of Automotive Engineers, Inc.: Warrendale, PA, USA, 2000.
23. Suman, S.; Kallmeyer, A.; Smith, J. Development of a multiaxial fatigue damage parameter and life prediction methodology for non-proportional loading. *Frattura ed Integrità Strutturale* **2016**, *10*, 224–230.
24. Chaves, V. Ecological criteria for the selection of materials in fatigue. *Fatigue Fract. Eng. Mater. Struct.* **2014**, *37*, 1034–1042. [CrossRef]
25. ASTM. *Standard Test Method for Measurement of Fatigue Crack Growth Rates*; ASTM E739; ASTM International: West Conshohocken, PA, USA, 2003.

26. Lopez-Crespo, P.; Moreno, B.; Lopez-Moreno, A.; Zapatero, J. Study of crack orientation and fatigue life prediction in biaxial fatigue with critical plane models. *Eng. Fract. Mech.* **2015**, *136*, 115–130. [CrossRef]

27. Lopez-Crespo, P.; Garcia-Gonzalez, A.; Moreno, B.; Lopez-Moreno, A.; Zapatero, J. Some observations on short fatigue cracks under biaxial fatigue. *Theor. Appl. Fract. Mech.* **2015**, *80*, 96–103. [CrossRef]

28. Roessle, M.L.; Fatemi, A. Strain-controlled fatigue properties of steels and some simple approximations. *Int. J. Fatigue* **2000**, *22*, 495–511. [CrossRef]

29. Ellyin, F.; Golos, K.; Xia, Z. In-phase and out-of-phase multiaxial fatigue. *J. Eng. Mater. Technol.* **1991**, *113*, 112–118. [CrossRef]

30. Karolczuk, A.; Macha, E. A review of critical plane orientations in multiaxial fatigue failure criteria of metallic materials. *Int. J. Fract.* **2005**, *134*, 267–304. [CrossRef]

31. Kanazawa, K.; Miller, K.J.; Brown, M.W. Low-Cycle Fatigue under Out-of-Phase Loading Conditions. *J. Eng. Mater. Technol.* **1977**, *99*, 222–228. [CrossRef]

32. Stulen, F.; Cumming, H. A failure criterion for multi-axial fatigue stresses. *ASTM* **1954**, *52*, 822–835.

33. Findley, W.N.; Coleman, J.J.; Hanley, B.C. Theory for combined bending and torsion fatigue with data SAE 4340 steel. In Proceedings of the International Conference on Fatigue of Metals, London, UK, 10–14 September 1956; pp. 138–149.

34. Forsyth, P. A two-stage process of fatigue crack growth. In *Symposium on Crack Propagation*; The College of Aeronautics: Cranfield, UK, 1961; pp. 76–94.

35. Anes, V.; Reis, L.; Li, B.; Freitas, M. Crack path evaluation on HC and BCC microstructures under multiaxial cyclic loading. *Int. J. Fatigue* **2014**, *58*, 102–113. [CrossRef]

36. Shang, D.-G.; Wang, D. A new multiaxial fatigue damage model based on the critical plane approach. *Int. J. Fatigue* **1998**, *20*, 241–245.

37. Findley, W.N. *Combined Stress Fatigue Strength of 76S-T61 Aluminum Alloy with Superimposed Mean Stresses and Corrections for Yielding*; National Aeronautics and Space Administration: Washington, DC, USA, 1953.

38. Suresh, S. *Fatigue of Materials*; Cambridge University Press: Cambridge, UK, 2004.

39. Brown, M.W.; Miller, K.J. A theory for fatigue failure under multiaxial stress-strain conditions. *Proc. Inst. Mech. Eng.* **1973**, *187*, 745–755. [CrossRef]

40. Reis, L.; Freitas, M.J. Crack initiation and growth path under multiaxial fatigue loading in structural steels. *Int. J. Fatigue* **2009**, *31*, 1660–1668. [CrossRef]

41. Kanazawa, K.; Miller, K.J.; Brown, M.W. Cyclic deformation of 1% Cr-Mo-V steel under out-of-phase loads. *Fatigue Fract. Eng. Mater. Struct.* **1979**, *2*, 217–228. [CrossRef]

metals

Article

A New Electron Backscatter Diffraction-Based Method to Study the Role of Crystallographic Orientation in Ductile Damage Initiation [†]

Behnam Shakerifard [1,*], Jesus Galan Lopez [1,2] and Leo A. I. Kestens [1,3]

[1] Materials Science and Engineering Department, Faculty 3mE, Delft University of Technology, Mekelweg 2, 2628 CD Delft, The Netherlands; j.galanlopez@tudelft.nl (J.G.L.); Leo.Kestens@UGent.be (L.A.I.K.)
[2] Materials innovation institute M2i, Elektronicaweg 25, 2628 XG Delft, The Netherlands
[3] EEMMECS department, Metals Science and Technology Group, Faculty of Engineering and Architecture, Ghent University, Technologiepark Zwijnaarde 903, 9052 Ghent, Belgium
* Correspondence: b.shakerifard@tudelft.nl or behnam.shakerifard@gmail.com; Tel.: +31-15-27-822-02
† This paper is an extended version of our paper published in the 18th International Conference on Textures of Materials (ICOTOM 18), St George, UT, USA, 5–10 November, 2017.

Received: 19 October 2019; Accepted: 6 January 2020; Published: 12 January 2020

Abstract: The third generation of advanced high strength steels shows promising properties for automotive applications. The macroscopic mechanical response of this generation can be further improved by a better understanding of failure mechanisms on the microstructural level and micro-mechanical behavior under various loading conditions. In the current study, the microstructure of a multiphase low silicon bainitic steel is characterized with a scanning electron microscope (SEM) equipped with an electron backscatter diffraction detector. A uniaxial tensile test is carried out on the bainitic steel with martensite and carbides as second phase constituents. An extensive image processing on SEM micrographs is conducted in order to quantify the void evolution during plastic deformation. Later, a new post-mortem electron backscatter diffraction-based method is introduced to address the correlation between crystallographic orientation and damage initiation. In this multiphase steel, particular crystallographic orientation components were observed to be highly susceptible to micro-void formation. It is shown that stress concentration around voids is rather relaxed by void growth than local plasticity. Therefore, this post-mortem method can be used as a validation tool together with a crystal plasticity-based hardening model in order to predict the susceptible crystallographic orientations to damage nucleation.

Keywords: steels; bainite; crystallographic orientations; ductile failure; void initiation.

1. Introduction

Advanced high strength steels (AHSSs) and, in particular, the third generation of AHSSs with bainitic matrix have shown remarkable potential for automotive applications [1,2]. Complex multiphase bainitic steels with second phase constituents martensite and retained austenite are promising candidates to reach an optimum balance between strength, ductility, and formability [3–9]. Developing bainitic steels with enhanced properties would enable weight reduction of the car body, thus reducing fuel consumption and pollution and saving energy. Studying the micro-mechanical behavior and damage micro-mechanisms of such multiphase steels is essential for developing a bainitic microstructure with improved properties. In multiphase steels, it has been well reported that microstructural heterogeneities, such as different phases with various mechanical contrast, play a significant role in the partitioning of stress (and/or strain) during plastic deformation [10–13]. Hence, at the microstructure level, these local deformation incompatibilities may induce void formation at the critical moment in order to dissipate the local plastic work.

The role of the second phase (i.e., martensite) and its morphology (size, shape, and distribution) in ductile damage initiation in dual phase steels has been extensively studied by advanced experimental and crystal plasticity-based modelling techniques [14–18]. Moreover, in polycrystalline materials, the anisotropic response of each individual crystallographic orientation to stress (and/or strain) is another important factor that needs to be considered in damage initiation mechanisms. Jia et al. [19] have shown for a duplex stainless steel that, at high macroscopic stresses, when all grains in both constituent phases deform plastically, the role of the crystallographic orientation anisotropy on stress–strain partitioning is more pronounced than the phase contrast. The role of crystallographic orientation on damage initiation at relatively high levels of plastic deformations was also observed on metal–matrix [20], rigid fiber composites [21], as well as in an austenitic stainless steel under the cyclic loading condition [22]. Eventually, in multiphase polycrystalline materials, the simultaneous effect of phase contrast, with various morphologies, and crystallographic orientation on damage initiation is a frequent topic of investigation [15,22–24].

In the current study, firstly, a comprehensive and quantitative microstructural characterization is performed using a scanning electron microscope (SEM) equipped with an electron backscatter diffraction (EBSD) detector. Second, a quantitative analysis on void distribution and its correlation to the macroscopic strain is conducted. This enables to understand at which macroscopic strain level damage initiates in the microstructure. In this research work, particular attention is given to the crystallographic orientation aspect of damage initiation, while the mechanical phase contrast and its morphological aspect is reported elsewhere by Shakerifard et al. [18]. An EBSD analysis is employed to observe the orientations around the initiated voids. Afterwards, textures are calculated and compared in order to study the correlation between damage and crystallographic orientation.

2. Materials and Methods

The bainitic steel used in the current study has a low silicon content. The chemical composition is shown in Table 1. The 1 mm thick sheet of cold rolled ferritic–pearlitic steel is annealed above the Ac3 temperature in the Vatron Annealing Simulator® (Vatron, Linz, Austria), and quenched to the bainite holding temperature of 450 °C for the holding time of 120 s. Following bainite transformation, the material is quenched to room temperature. The annealing cycle of the developed bainitic steel is depicted in Figure 1.

Table 1. Chemical composition of the studied bainitic steel (wt.%).

Elements	C%	Mn%	Si%	Cr%	Al%	P%	S%	N%
Steel	0.215	1.96	0.10	0.61	0.020	0.007	0.0027	0.005

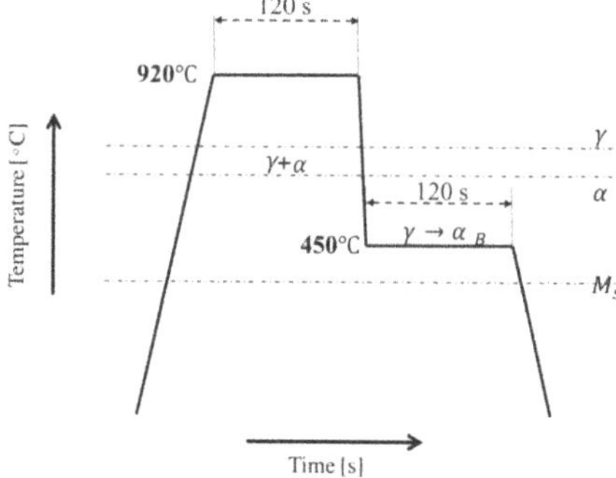

Figure 1. Schematic illustration of the annealing cycle used to develop the bainitic steel.

In order to analyze the microstructure of this bainitic steel, as reported by Shakerifard et al. [9], samples are prepared by the standard metallographic steps of grinding, diamond polishing, and polishing with colloidal SiO_2. A scanning electron microscope FEI® (SEM-Quanta FEG 450, FEI, Hillsboro, OR, USA) equipped with EBSD detector is employed to characterize the microstructure of the samples prior to and after tensile testing. EBSD maps are recorded with a step size of 0.05 μm and post processed with the EDAX-TSLOIM Analysis™ software.

The investigated steel consists of a bainitic matrix with two main second phase constituents of martensite and carbide. However, two types of martensites are observed: (i) auto-tempered martensite (ATM) and (ii) MA islands, which are composed of isolated martensite blocks (M) with a negligible fraction of retained austenite (A) between the martensite laths [18]. The EBSD technique is employed to quantify the grain size and fraction of bainite and martensite (ATM and MA) separately. For this purpose, the grain average image quality (GAIQ) parameter, directly correlated to the sharpness of the Kikuchi pattern, is used in order to distinguish these two constituents [14,15,25]. This is possible because the martensite phase demonstrates a lower IQ parameter compared with bainite, as a result of intrinsic distortion in the martensitic crystal structure, and higher dislocation density compared with bainite. Grains are defined with a minimum number of five pixels and a misorientation threshold of five degrees. The grain size is measured based on the linear intercept method and by averaging lines along both the rolling and normal directions (RD and ND) for each phase.

The fraction of MA islands is measured separately from the total martensite (ATM and MA) fraction by a specific etching method. Initially, the surface of the samples is activated by ethanol, which is quickly followed by Klemm etchant (50 mL saturated aqueous sodium thiosulfate, 1 g potassium metabisulfite) color etching. Later, ten backscatter electron (BSE) images, which provide an enhanced contrast between matrix and MA islands compared with secondary electron images, are captured from the sample and binarized into black and white images. In addition, the fraction of MA islands is measured by image processing.

Uniaxial tensile tests are performed at room temperature with strain rates of 2×10^{-5} and 8×10^{-3} s^{-1} in the elastic and plastic regions of the tensile curve, respectively, in order to examine the tensile properties of the bainitic steel. Tensile tests are carried out based on the DIN EN 6892-1 standard using a Zwick Z100® tensile machine (ZwickRoell AG, Fürstenfeld, Austria). Dog-bone specimens with a width and gauge length of 6.25 and 25 mm, respectively, are loaded along the rolling direction (RD). Two tests are performed, and elongations are measured by the extensometer. Additionally, two tensile tests are interrupted in the uniform and non-uniform regions, respectively, in order to investigate the evolution of damage within the microstructure at various strain levels.

Tasan et al. [16] have conducted an extensive study on various techniques for the quantification of voids evolution during plastic deformation. They have reported that microscopic techniques such as 2D SEM and 3D X-ray micro-tomography (XμT) provide significant information regarding the strain levels and the involved micro-mechanisms in damage initiation and evolution compared with mechanical-based techniques. The possible drawback of the 2D SEM technique is an underestimation in damage quantities, where voids may smear-out by mechanical polishing. Lai et al. [26] have demonstrated that the advantage of the SEM-based technique is its high resolution and its ability to detect smaller voids compared with XμT. However, as reported by Landron et al. [27,28], the XμT can provide 3D visual information with regard to growth characteristics of individual void.

In the current study, the quantification of voids is performed by a high resolution FEG-SEM Nova 600 (FEI, Hillsboro, OR, USA). The evolution of voids as a function of the plastic strain is studied on the RD–ND section in the middle width of the fractured tensile samples. Two variables are considered to quantify the evolution of voids: (i) the void number density (VD) and (ii) the void area fraction (VAF) in the 2D section of microscopic observation. The VD indicates the nucleation activity of voids during plastic deformation, while VAF includes void growth in addition to nucleation. This SEM-based technique provides the possibility to visualize the dispersion of the voids within the microstructure. True plastic strain $\left(\varepsilon_{\text{longitudinal}} \right)$ values are calculated in equally distanced intervals from the fracture surface based on the normal ($\varepsilon_{\text{thickness}}$) and transversal strains ($\varepsilon_{\text{transveral}}$) using optical microscopy images and the following equations:

$$\varepsilon_{\text{fracture}} = \ln \frac{A_0}{A_f},\tag{1}$$

$$\varepsilon_{\text{longitudinal}} = -(\varepsilon_{\text{thickness}} + \varepsilon_{\text{transveral}}) = \ln \frac{t_f}{t_0} + \ln \frac{w_f}{w_0},\tag{2}$$

where A_0 is the initial cross-section area of the tensile sample and A_f is the fracture surface area (measured by optical microscope image) in Equation (1). The t_0, t_f, w_0, and w_f are the thickness and width values of the tensile sample prior and after tensile test, respectively. A field of approximately 1 mm^2 is examined by running an automated image acquisition software to capture over 1000 SEM micrographs. The SEM micrographs are captured with a magnification of 4000×, having a width and height of 1024 × 884 pixels, respectively. These settings provide a pixel resolution of 31 nm and enable to detect the micro-voids within the microstructure. Matlab$^@$ (R2017(a), MathWorks, Natick, MA, USA) and ImageJ$^@$ software (1.5 1P, NIH, WA, USA) with specific image processing algorithm is used to stich and study all micrographs. A void size criterion is defined in which all the voids larger than 0.1 μm^2 (≈33 pixels) are counted.

3. Results and Discussion

3.1. Microstructure Characterization

The microstructures of the low Si bainitic steel are shown in Figure 2. The bainitic steel reveals a bainitic matrix in which small carbides are observed, whereby isolated MA islands are also observed within the bainitic matrix. Table 2 quantitatively illustrates the microstructural variations of the bainitic steel.

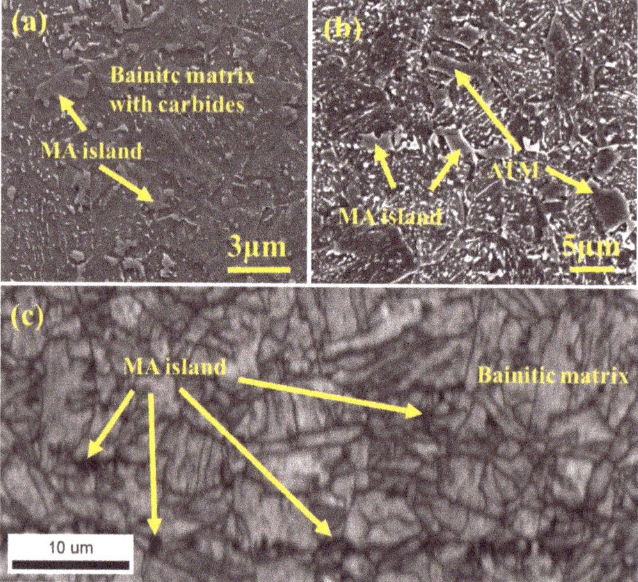

Figure 2. Scanning electron micrographs of the bainitic steel under consideration: (**a**) SEM micrograph of the microstructure etched by Klemm; (**b**) SEM micrograph of the microstructure etched by Nital; (**c**) image quality map and the fraction of retained austenite is negligible (<1%). The rolling direction is along the horizontal axis and the normal direction is along the vertical axis. ATM, auto-tempered martensite; SEM, scanning electron microscope.

Table 2. Microstructural variables of the bainitic steel.

Sample	Phases	Grain Size (µm)	Fraction (%)	Block Size (µm)	Cementite Size (µm)
Bainitic steel	Bainite	1.44 ± 0.01	Bal	-	-
	MA	-	3.4 ± 1.7	0.7 ± 0.35	-
	Cementite	-	3.6 ± 1.0	-	0.07 ± 0.04
	ATM[1]	0.5 ± 0.1	4.8 ± 1.8	-	-

[1] Auto-tempered martensite (ATM) phase was exposed by Nital etchant and quantified by the subtraction of whole martensitic region measured by electron backscatter diffraction (EBSD) from MA regions.

3.2. Tensile Test

The tensile curve and mechanical properties of the material are depicted in Figure 3. The engineering tensile curve of the material demonstrates the small non-uniform region, which implies that the steel could not accommodate considerable elongation beyond the uniform plastic domain. The localized necking behavior (cf. Figure 4) of the material also confirms the small post-necking region of the tensile curve.

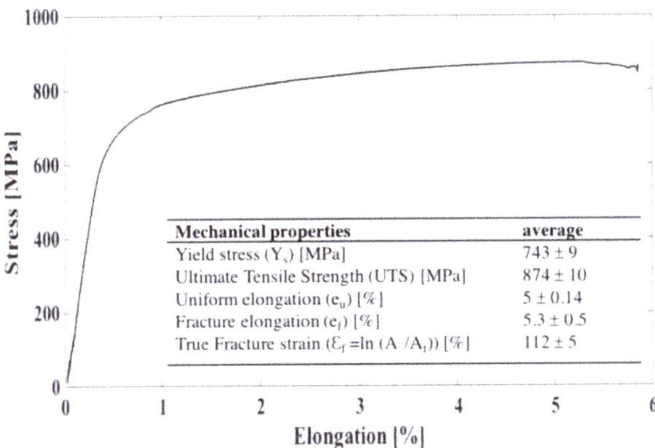

Mechanical properties	average
Yield stress (Y_s) [MPa]	743 ± 9
Ultimate Tensile Strength (UTS) [MPa]	874 ± 10
Uniform elongation (e_u) [%]	5 ± 0.14
Fracture elongation (e_f) [%]	5.3 ± 0.5
True Fracture strain ($\varepsilon_f = \ln(A/A_t)$) [%]	112 ± 5

Figure 3. Engineering stress and elongation curve with corresponding tensile properties.

3.3. Damage Analysis

3.3.1. Quantitative Analysis of Voids' Evolution

Figure 4 demonstrates the RD–ND cross section at half width of the fracture tensile sample before and after image processing. The size of the voids and their shape fully correspond to their real geometry. It is observed that voids are only present at the neck region of the fractured tensile sample. Voids are distributed heterogeneously and more densely present in the vicinity of the fracture surface. The microstructural observation of the interrupted tensile sample at the elongation of 5%, reported elsewhere by Shakerifard et al. [18], confirmed the absence of void formation during the uniform plastic deformation. In contrast to this bainitic martensitic steel, for ferritic and martensitic dual phase (DP) steels, it has been reported that voids can nucleate in the early stage of plastic deformation prior to necking [29].

It is observed in Figure 5 that VD and VAF parameters increase with the plastic deformation. The exponential trend of these two damage-related parameters demonstrates a sudden increase at the strain level of 74%. The localized necking behavior of the fracture tensile sample close to fracture surface (cf. Figure 4) resulted in higher stress triaxiality, and thus increases VAF significantly.

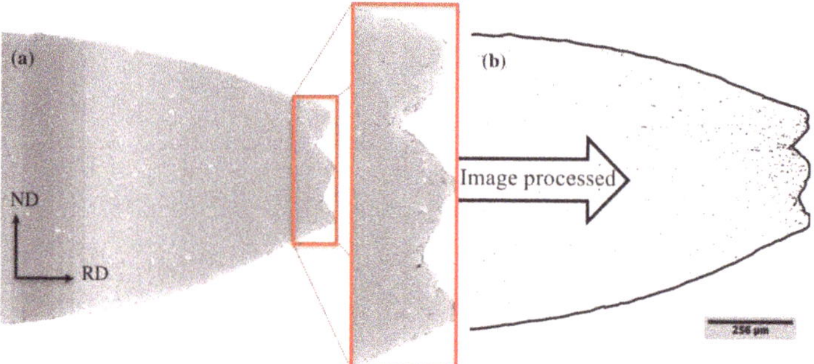

Figure 4. Roughly 1000 contiguous high-resolution SEM micrographs (each $\approx 27 \times 32$ µm^2); (**a**) after stitching all SEM micrographs and (**b**) image processed, whereby voids are magnified 50× with respect to matrix pixels for enhanced visualization.

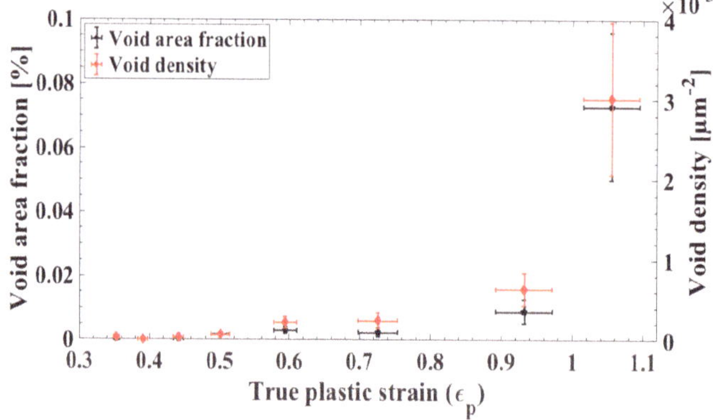

Figure 5. Quantitative analysis of voids' evolution as a function of true plastic strain.

3.3.2. Crystallographic Orientation Aspect of Void Initiation

The individual role of martensite as a hard phase and its topological impact on micro-mechanisms of void initiation in ductile failure in this bainitic multiphase steel has been reported in detail by Shakerifard et al. [18]. They have shown that in this steel grade voids nucleate at the interface of bainite and martensite (B/M). This shows the effect of mechanical phase contrast between B and M (soft and hard phase, respectively), which leads to stress and strain partitioning. At any moment when incompatibilities reach a critical level, voids nucleate in order to locally relax the sharp stress gradients among soft and hard phases. Although the effect of second phase constituents and their topology is significant, the role of crystallographic orientation in the evolution of stress (and/or) strain is inevitable. In this regard, in order to investigate the role of crystallographic orientation in void initiation, 33 voids were analyzed locally by the EBSD technique. The method of capturing orientations of grains is depicted schematically in Figure 6. Figure 6b particularly reveals the procedure of capturing the local surrounding orientations of grains around a void. This rectangular area has faces with approximately three times the short and long diameters of a typical void (approximating the void by an elliptical area with a major and minor axis). This method is used consistently for all 33 of the studied voids.

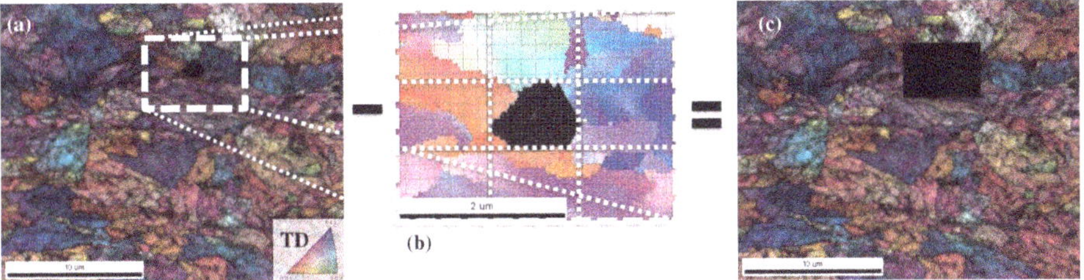

Figure 6. Schematic representation of orientation selection on an overlapped image quality and inverse pole figure map; (**a**) complete region of analysis, (**b**) region selected around the void with faces approximately equal to three times the void diameter; and (**c**) remaining region of analysis. The rolling direction is along the horizontal axis and the normal direction is along the vertical direction.

Owing to severe plastic deformation and distortion of the crystal lattices in the vicinity of the voids, a poor pattern quality was normally observed. Only orientations with a confidence index above 0.1 are used for further analysis. Later, the orientation distribution function (ODF) is calculated from approximately 131,000 void neighboring pixels (with an equivalent area of ~642 μm^2 surrounding 33 voids). The calculated 3D-ODF of 33 voids using MTEX software (5.1.1, Chemnitz, Germany) [30] is shown in Figure 7 and denominated as the void-ODF (VODF) using the Euler space.

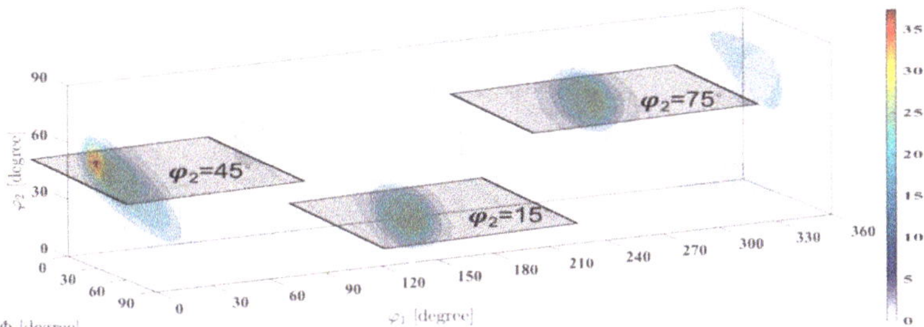

Figure 7. Three dimensional (3D) void orientation distribution function (ODF) (3D-VODF). Three φ_2 sections (φ_2 = 15°, 45°, and 75°) consist of the dominant components with regard to voids orientations.

Figure 7 shows three sections of φ_2 = 15°, 45°, and 75°, which exhibit the dominant components of VODF. In the neck region of the fracture tensile sample where voids are present, the orthorhombic sample symmetry no longer exists, and thus the Euler space is confined into $0 < \varphi_1 < 360°$ and $0 < \varphi_2$, $\Phi < 90°$ (triclinic symmetry condition). Figure 8c shows the texture of the regions with no voids, which will be denominated here as the no-void-ODF (NVODF). The NVODF reveals a smoother distribution, which is because of the fact that regions with no voids have more dispersed orientations compared with regions selected around voids (cf. Figure 6c).

In Section 3.3.1, it was illustrated that, prior to the necking point at elongation of 5% (e_u = 5%), no microstructural voids were initiated in the tensile sample. Therefore, the deformation texture of the material at the ultimate tensile strength (UTS) point, which is denominated here as the uniform ODF (UODF), cf. Figure 8a, can be considered as a reference texture from which voids will nucleate. Furthermore, the neck-ODF will be considered, cf. Figure 8d, which is calculated by merging the textures of VODF and NVODF, which means it includes all the orientations of void neighboring grains and void distant grains. It needs to be emphasized that this method is a post-mortem analysis, which implies that all orientations related to the VODF and NVODF are derived from the tensile sample after its fracture. Therefore, it is not obvious which interface was initially at the origin of the void nucleation. The rectangular region of interest around voids includes all the potentially susceptible orientations from which voids might have been initiated. Therefore, this rectangular geometry also includes the grains that were indirectly involved in the void initiation process by providing the local constraint to the void nucleation process.

The visual comparison of the VODF and NVODF with respect to the neck-ODF reveals that that the NVODF more closely resembles the texture observed in the necked region (neck-ODF). However, by the comparison between the VODF and NVODF with the UODF, it appears that the VODF better resembles the UODF. In order to further investigate the resemblance of the VODF and UODF textures, the contour maps of VODF are overlaid with the UODF at the three sections of $\varphi_2 = 15°$, 45°, and 75°, cf. Figure 9. The contour maps represent the VODF, while the gradient maps in the backgrounds represent the UODF, and the black arrows indicate the main components of the VODF. Figure 9 also reveals that, for the $\varphi_2 = 15°$ and 75° sections, the VODF closely resembles the UODF, as shown earlier for the $\varphi_2 = 45°$ section, cf. Figure 8a,b.

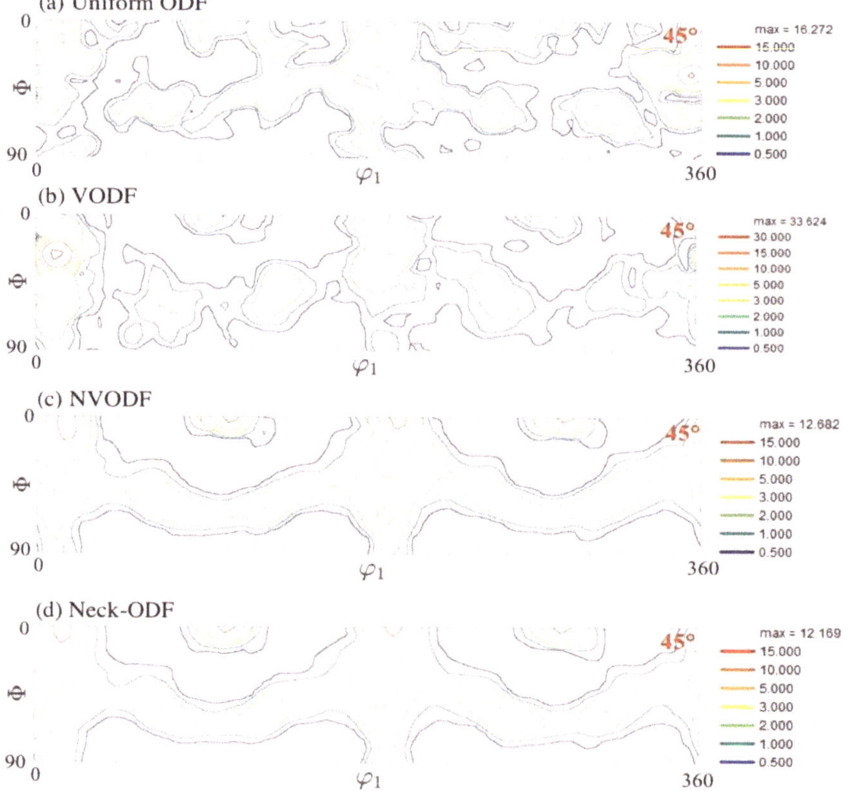

Figure 8. $\varphi_2 = 45°$ section of ODFs of the steel at two various deformation levels; (**a**) uniform deformation union ODF (UODF) (calculated from 796,000 data points), (**b**) post uniform VODF (131,000 data points), (**c**) post uniform no-void-ODF (NVODF) (460,000 data points), and (**d**) necking or post uniform ODF consist of the merged ODFs in **b** and **c**.

The texture similarity between the UODF and VODF can be understood by considering the stress relaxation processes in the microstructural level. According to Figure 5, voids are nucleated at a high rate at the strain level of approximately 74% and, knowing that voids act as stress concentrators, one may expect that the local stress peaks can be relaxed further either by plastic deformation of the grains surrounding a void and/or by further void growth. However, it is observed in Figure 9 that, upon void initiation, the local stress relaxation is rather controlled by the void growth mechanism than local plasticity inside the grains. This mechanism acts as a preferred local stress relaxation process as the VODF clearly resembles the UODF texture. However, it is also apparent that the components of the VODF and UODF do not correlate one to one, which can be explained either by further plasticity after void initiation at the neighboring grains around voids or including orientations that are not directly involved in void initiation. In addition, a void represents a free surface trapped in the bulk of the material that has appeared in order to relax the stresses at the microstructural level. However, once it has nucleated, the competitive role between the local hardening and free surface growth determines the major contribution of these two mechanisms on local stress relaxation, which is an energetically driven process. As void growth seems to be the main mechanism of local stress relaxation around voids, the dominant components of the VODF can be considered as susceptible orientations to void nucleation.

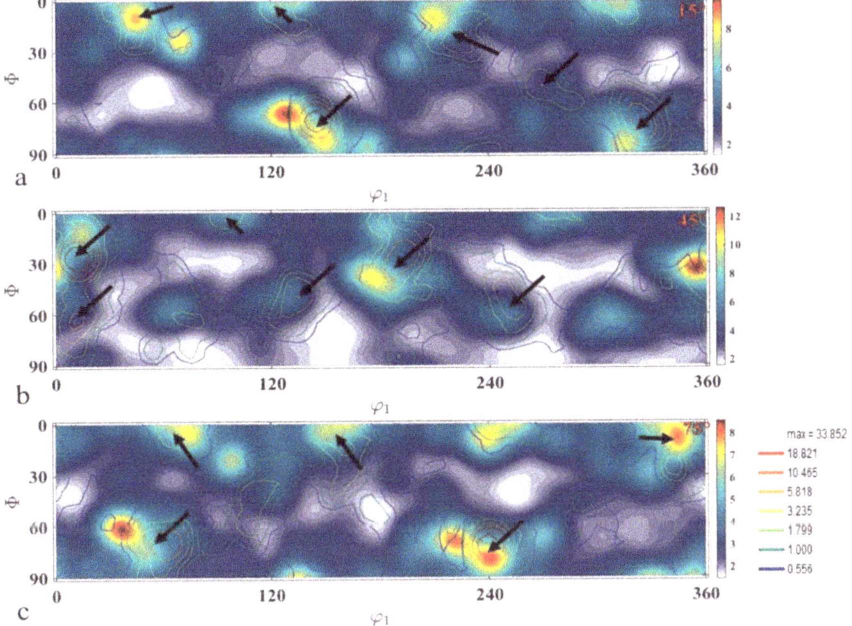

Figure 9. The overlapped VODF (contour map with arrows indicating dominant components) and UODF (color gradient map) of the material at two different deformation levels (UODF at 5% and VODF at ≥74%); (a) $\varphi_2 = 15°$, (b) $\varphi_2 = 45°$, (c) $\varphi_2 = 75°$.

4. Summary and Conclusions

In this research, a new EBSD-based method was used to study the role of crystallographic orientation in void nucleation. In this regard, the microstructure of the fractured tensile sample in the vicinity of the fracture surface, where voids were concentrated, was measured by EBSD. Later, ODFs were calculated based on their relative locations with respect to voids: (i) the crystallographic orientations of the grains surrounding 33 voids were selected and represented by the void-ODF, (ii) whereas the regions away from the voids represented the no-void-ODF (NVODF). It was observed that the VODF more closely resembles the UODF. The post-mortem crystallographic orientation analysis around voids revealed that, after void initiation, the local stress relaxation process is rather dissipated by void growth than by further local plastic deformation. In other words, it can be concluded, that upon void initiation, orientations around voids may not go through significant rotations. This can be explained by the hardening saturation process of grains at certain macroscopic strain level due to plasticity prior to damage initiation. Consequently, the increased macroscopic deformation until failure does not lead to considerable orientation change in the VODF components. Therefore, the major components of the uniform ODF at the onset of necking correspond to ODF observed in the vicinity of voids (the VODF).

Author Contributions: Methodology, B.S. and L.A.I.K.; Resources, B.S.; Supervision, J.G.L.; Visualization, B.S.; Writing—original draft, B.S.; Writing—review & editing, L.A.I.K. All authors have read and agreed to the published version of the manuscript.

Funding: This work has been conducted in the framework of the BaseForm project, which has received funding from the European Union's Research Fund for Coal and Steel (RFCS) research programme under grant agreement #RFCS-CT-2014-00017.

Acknowledgments: The authors would like to acknowledge Tata Steel, IJmuiden (The Netherlands), for providing the material in the current study. Special thanks to Frank Hisker for providing the tensile test results.

Conflicts of Interest: The authors declare no conflict of interests.

References

1. Matlock, D.; Speer, J.G.; de Moor, E.; Gibbs, P.J. Recent developments in Advanced High Strength Sheet Steels for automotive applications—An Overview. *JESTECH* **2012**, *15*, 1–12.
2. Bouaziz, O.; Zurob, H.; Huang, M. Driving force and logic of development of advanced high strength steels for automotive applications. *Steel Res. Int.* **2013**, *84*, 937–947. [CrossRef]
3. Caballero, F.G.; Allain, S.; Cornide, J.; Velásquez, J.D.P.; Garcia-Mateo, C.; Miller, M.K. Design of cold rolled and continuous annealed carbide-free bainitic steels for automotive application. *Mater. Des.* **2013**, *49*, 667–680. [CrossRef]
4. Caballero, F.G.; Bhadeshia, H.K.D.H. Design of novel high-strength bainitic steels. Thermec. *Mater. Sci. Forum* **2003**, *426–432*, 1337–1342. [CrossRef]
5. Caballero, F.G.; Bhadeshia, H.K.D.H.; Mawella, K.J.A.; Jones, D.G.; Brown, P. Design of novel high strength bainitic steels: Part 1. *Mater. Sci. Technol.-Lond.* **2001**, *17*, 512–516. [CrossRef]
6. Caballero, F.G.; Bhadeshia, H.K.D.H.; Mawella, K.J.A.; Jones, D.G.; Brown, P. Design of novel high strength bainitic steels: Part 2. *Mater. Sci. Technol.-Lond.* **2001**, *17*, 517–522. [CrossRef]
7. Caballero, F.G.; Garcia-Mateo, C.; Chao, J.; Santofimia, M.J.; Capdevila, C.; de Andres, C.G. Effects of morphology and stability of retained austenite on the ductility of TRIP-aided bainitic steels. *ISIJ Int.* **2008**, *48*, 1256–1262. [CrossRef]
8. Wang, Y.; Zhang, K.; Guo, Z.; Chen, N.; Rong, Y. A new effect of retained austenite on ductility enhancement in high strength bainitic steel. *Mater. Sci. Eng. A-Struct.* **2012**, *552*, 288–294. [CrossRef]
9. Shakerifard, B.; Lopez, J.G.; Legaza, M.C.T.; Verleysen, P.; Kestens, L.A.I. Strain rate dependent dynamic mechanical response of bainitic multiphase steels. *Mater. Sci. Eng. A-Struct.* **2019**, *745*, 279–290. [CrossRef]
10. Ahmad, E.; Manzoor, T.; Ali, K.L.; Akhter, J.I. Effect of microvoid formation on the tensile properties of dual-phase steel. *J. Mater. Eng. Perform.* **2000**, *9*, 306–310. [CrossRef]
11. Avramovic-Cingara, G.; Ososkov, Y.; Jain, M.K.; Wilkinson, D.S. Effect of martensite distribution on damage behaviour in DP600 dual phase steels. *Mater. Sci. Eng. A-Struct.* **2009**, *516*, 7–16. [CrossRef]
12. He, X.J.; Terao, N.; Berghezan, A. Influence of Martensite Morphology and Its Dispersion on Mechanical-Properties and Fracture Mechanisms of Fe-Mn-C Dual Phase Steels. *Meter. Sci.* **1984**, *18*, 367–373. [CrossRef]
13. Steinbrunner, D.L.; Matlock, D.K.; Krauss, G. Void Formation during Tensile Testing of Dual Phase Steels. *Metall. Trans. A* **1988**, *19*, 579–589. [CrossRef]
14. Tasan, C.C.; Diehl, M.; Yan, D.; Zambaldi, C.; Shanthraj, P.; Roters, F.; Raabe, D. Integrated experimental-simulation analysis of stress and strain partitioning in multiphase alloys. *Acta Mater.* **2014**, *81*, 386–400. [CrossRef]
15. Tasan, C.C.; Hoefnagels, J.P.M.; Diehl, M.; Yan, D.; Roters, F.; Raabe, D. Strain localization and damage in dual phase steels investigated by coupled in-situ deformation experiments and crystal plasticity simulations. *Int. J. Plast.* **2014**, *63*, 198–210. [CrossRef]

16. Tasan, C.C.; Hoefnagels, J.P.M.; Geers, M.G.D. Identification of the continuum damage parameter: An experimental challenge in modeling damage evolution. *Acta Mater.* **2012**, *60*, 3581–3589. [CrossRef]

17. Yan, D.S.; Tasan, C.C.; Raabe, D. High resolution in situ mapping of microstrain and microstructure evolution reveals damage resistance criteria in dual phase steels. *Acta Mater.* **2015**, *96*, 399–409. [CrossRef]

18. Shakerifard, B.; Lopez, J.G.; Hisker, F.; Kestens, L.A.I. Effect of banding on micro-mechanisms of damage initiation in bainitic/martensitic steels. *Mater. Sci. Eng. A-Struct.* **2018**, *735*, 324–335. [CrossRef]

19. Jia, N.; Peng, R.L.; Wang, Y.D.; Chai, G.C.; Johansson, S.; Wang, G.; Liaw, P.K. Interactions between the phase stress and the grain-orientation-dependent stress in duplex stainless steel during deformation. *Acta Mater.* **2006**, *54*, 3907–3916. [CrossRef]

20. Needleman, A.; Tvergaard, V. Comparison of Crystal Plasticity and Isotropic Hardening Predictions for Metal-Matrix Composites. *J. Appl. Mech. Trans. ASME* **1993**, *60*, 70–76. [CrossRef]

21. Nugent, E.E.; Calhoun, R.B.; Mortensen, A. Experimental investigation of stress and strain fields in a ductile matrix surrounding an elastic inclusion. *Acta Mater.* **2000**, *48*, 1451–1467. [CrossRef]

22. Li, R.G.; Xie, Q.G.; Wang, Y.D.; Liu, W.J.; Wang, M.G.; Wu, G.L.; Li, X.W.; Zhang, M.H.; Lu, Z.P.; Geng, C.; et al. Unraveling submicron-scale mechanical heterogeneity by three-dimensional X-ray microdiffraction. *Proc. Natl. Acad. Sci. USA* **2018**, *115*, 483–488. [CrossRef] [PubMed]

23. de Geus, T.W.J.; Maresca, F.; Peerlings, R.H.J.; Geers, M.G.D. Microscopic plasticity and damage in two-phase steels: On the competing role of crystallography and phase contrast. *Mech. Mater.* **2016**, *101*, 147–159. [CrossRef]

24. Shakerifard, B.; Lopez, J.G.; Hisker, F.; Kestens, L.A.I. Crystallographically resolved damage initiation in advanced high strength steel. In Proceedings of the 18th International Conference on Textures of Materials (Icotom-18), St. George, UT, USA, 6–10 November 2017; Volume 375.

25. Choi, S.H.; Kim, E.Y.; Woo, W.; Han, S.H.; Kwak, J.H. The effect of crystallographic orientation on the micromechanical deformation and failure behaviors of DP980 steel during uniaxial tension. *Int. J. Plast.* **2013**, *45*, 85–102. [CrossRef]

26. Lai, Q.; Bouaziz, O.; Goune, M.; Brassart, L.; Verdier, M.; Parry, G.; Perlade, A.; Brechet, Y.; Pardoen, T. Damage and fracture of dual-phase steels: Influence of martensite volume fraction. *Mater. Sci. Eng. A-Struct.* **2015**, *646*, 322–331. [CrossRef]

27. Landron, C.; Maire, E.; Adrien, J.; Bouaziz, O.; di Michiel, M.; Cloetens, P.; Suhonen, H. Resolution effect on the study of ductile damage using synchrotron X-ray tomography. *Nucl. Instrum. Meth. B* **2012**, *284*, 15–18. [CrossRef]

28. Landron, C.; Maire, E.; Bouaziz, O.; Adrien, J.; Lecarme, L.; Bareggi, A. Validation of void growth models using X-ray microtomography characterization of damage in dual phase steels. *Acta Mater.* **2011**, *59*, 7564–7573. [CrossRef]

29. Han, S.K.; Margolin, H.; Formation, V. Void Growth and Tensile Fracture of Plain Carbon-Steel and a Dual-Phase Steel. *Mater. Sci. Eng. A-Struct.* **1989**, *112*, 133–141. [CrossRef]

30. Bachmann, F.; Hielscher, R.; Schaeben, H. Texture Analysis with MTEX—Free and Open Source Software Toolbox. *Solid State Phenom.* **2010**, *160*, 63–68. [CrossRef]

Article

Effect of Sign-Alternating Cyclic Polarisation and Hydrogen Uptake on the Localised Corrosion of X70 Pipeline Steel in Near-Neutral Solutions

Alevtina Rybkina [1], **Natalia Gladkikh** [1], **Andrey Marshakov** [1], **Maxim Petrunin** [1] **and Andrei Nazarov** [2],*

[1]   Frumkin Institute of Physical Chemistry and Electrochemistry, Russian Academy of Sciences, 119071 Moscow, Russia; aa_rybkina@mail.ru (A.R.); fuchsia32@bk.ru (N.G.); mar@ipc.rssi.ru (A.M.); mmvp@bk.ru (M.P.)

[2]   French Corrosion Institute, 29200 Brest, France

*   Correspondence: andrej.nazarov@institut-corrosion.fr; Tel.: +33-(0)-2-98-05-15-52

Received: 12 December 2019; Accepted: 7 February 2020; Published: 12 February 2020

Abstract: The effect of sign-alternating cycling polarisation (SACP) on the localised corrosion of X70pipeline steel in solutions of various compositions was studied. Localised corrosion of steel at anodic potentials was accelerated with an increase in the duration of the cathodic half-cycle, in the presence of a promoter of hydrogen absorption in aqueous electrolyte, and with an increase in the concentrations of chloride and bicarbonate ions. It was pointed out that the corrosion rate is determined by the amount of hydrogen absorbed by the steel. A quantitative indicator to determine the intensity of localised corrosion under SACP was suggested.

Keywords: pipeline steel; sign-alternating polarisation; pitting corrosion; soil electrolyte; hydrogen uptake

1. Introduction

Corrosion of carbon and low-alloy steels under the effect of stray current is among the most hazardous types of corrosion damage of metal structures operating in soils or in natural waters. As a rule, the effect of stray current from a DC source can be different from the effect of current induced by an AC source (e.g., high-voltage power lines) [1]. On the other hand, stray current from DC sources vary with time. If cathodic protection of a structure is used, it results in oscillations in the polarisation potential (E), and the instantaneous value of E can be either more positive or more negative than the free corrosion potential of steel [2,3]. Hence, the effect of stray current from DC and AC sources leads to alternating-sign polarisation of a metal in a structure, which differs mainly in the frequency of potential oscillations.

It was initially supposed that the corrosion rate of steel is determined by dissolution during the anodic half-cycle of the alternating current. The effect of the cathodic on the anodic reaction was not taken into account, and corrosion was considered as a case of Faraday rectification [4–6]. In some cases, a qualitative relationship between the amplitude of the AC potential and the rate of AC corrosion was found. However, no quantitative agreement was observed for many corrosion systems [7–11]. In addition, the AC corrosion of steels is of a mainly localised nature, and corrosion in pits can occur at sufficiently negative potentials such as are applied for cathodic protection [4–16]. These features of AC corrosion of steels have been explained by a significant increase in the pH of the near-electrode solution [1,17]. In alkaline media, passivation of the metal occurs in the AC anodic half-period. In the negative half-wave period, the passive film is reduced to Fe (II) hydroxide/oxide species and a new passive film grows again during the next anodic cycle. When the newly formed passive film is reduced, the amount of Fe (II) compound increases. As a result, a certain amount of the metal is oxidised during each cycle, which results in a considerable mass loss of the steel structure due to corrosion.

A similar process scheme was suggested to explain the effect of rectangular potential pulses on steel corrosion, which are supposed to simulate the alternating-sign zone of stray currents from a DC source [18–20]. According to [20], at the cathodic protection potential (−0.65 V vs. saturated hydrogen electrode, SHE), steel is in the passive state due to the high pH. A shift of potential in the anodic direction gradually activates the metal surface due to a decrease in near-surface pH. A cycling of the potential only in the anodic direction from E = −0.65 V (SHE) accelerates the steel corrosion to a smaller extent comparing to a 'two-sided' potential shift both in the negative and positive directions around the cathodic protection potential [20].

Electrode activation-passivation caused by a variation in near-surface pH is not the only reason for localised steel corrosion under alternating polarisation [11]. It has been shown that under cathodic polarisation, 'pit-like' defects could form on the surface of pipeline steel [21,22]. This was explained as a result of hydrogen charging and the formation of high hydrogen pressure at metal/ non-metallic inclusion interfaces [22]. As the potential is shifted in the anodic direction, intense dissolution of the hydrogen-saturated metal layer inside the pits occurs. If the potential step is cycled, a corrosion location grows and is transformed to a micro-crack in mechanically stressed metal [23]. Dai et al. demonstrated that an increase in the frequency of the rectangular potential pulses in a simulated soil electrolyte increases the coverage of the surface of X100 pipeline steel with pits [24]. This was explained by a shift in the potential of the double electric layer in a positive direction. However, the effect of solution composition on the intensity of localised corrosion of steel was not considered. The effect of hydrogen adsorption in the cathodic cycle can also be important for the corrosion of steel at the rest potential or during anodic polarisation. Thus, the objective of this work was to study the effect of sign-alternating cycling polarisation (SACP) and the composition of a pH-neutral solution on the localised corrosion of pipeline steel.

2. Materials and Methods

The studies were carried out using samples of X70 pipeline steel machined from a pipe (manufactured by KhTZ Du Enterprise, Kharkov, Ukraine, pipe diameter 1420 mm and thickness 18.7 mm) along the centreline at a distance of 120 mm from the longitudinal weld. The chemical composition of the steel is presented in Table 1. Samples of discs with a working surface area of 1.93 cm^2 were used. Samples were polished using emery paper and finished with a diamond paste of grit size down to 0.5 μm. The samples were degreased for 25 min in an ultrasonic bath using ethanol/toluene 1:1.

Table 1. The chemical composition of X70 pipeline steel (wt.%).

C	Si	Mn	P	S	Cr	Ni	Cu	Al	Ti
0.115	0.34	1.63	0.021	0.003	0.04	0.02	0.007	0.030	0.07

The following working electrolytes were used: borate buffer (BB) with the composition: 0.4 M H_3BO_3 + 5.5 mM $Na_2B_4O_7$, pH 6.7; a mixture of BB and solution (NS4) with the composition 1.64 mM KCl + 5.75 mM $NaHCO_3$ + 1.23 mM $CaCl_2 \cdot 2H_2O$ + 0.74 mM $MgSO_4 \cdot 7H_2O$ [25]; and the solutions with addition of 0.01 M thiourea (TU), 0.1 M and 0.6 M NaCl, 0.1 M $NaHCO_3$, and IFKhAN-29 inhibitor (1 g/L). The compositions of the solutions are shown in Table 2. All solutions were prepared from reagents of chemical grade and de-ionised water. The experiments were carried out with free access to oxygen at room temperature of 22 ± 2 °C. This indicates electrochemical studies, the samples were subjected to cathodic pre-treatment for 15 min at a potential of −0.65 V (SHE).

Table 2. Specimen surface characteristics (total area S_p, density ρ, mean diameter d and maximum depth h of pits) after 72 cycles of SACP (-1 V $\leftrightarrow$ -0.3 V vs. SHE) with duration of anodic polarization τ_a 3 min and various durations of cathodic polarization τ_c in different solutions.

Electrolyte Composition	τ_c, min	$S_p \times 10^3$, mm^2/cm^2	ρ, pits/mm^2	d, μm	h, μm
BB	10	0.006	2	2	1
BB + NS4	0	0.73	22	6.5	3
BB + NS4	10	1.13	32	6.7	4
BB + NS4	60	1.66	30	8.4	4
BB + NS4 + TU	0	1.63	23	9.5	5
BB + NS4 + TU	10	4.72	68	9.4	7.5
BB + NS4 + TU	60	17.90	100	15.1	9
0.1 M NaCl	10	3.56	60	8.7	10
BB + 0.1 M NaCl	10	2.72	60	7.6	6.5
BB + 0.6 M NaCl	10	5.17	39	13	7.5
BB + 0.1 M NaHCO$_3$	10	0.84	17.5	7.8	2
BB + 0.1 M NaHCO$_3$ + TU	10	2.27	43	8.2	3
BB + NS4 + TU + IFKhAN-29	60	1.76	22	10.1	4

The potential was cycled from a cathodic value ($E_c = -1$ V) to an anodic value ($E_a = -0.3$ V) using an IPC Pro-MF potentiostat ("Volta", Saint Petersburg, Russian Federation). A standard three-electrode electrochemical cell described in [26] was used. The potentials are reported versus SHE. The duration of the cathodic SACP period (τ_c) was 10 or 60 min and the duration of the anodic period (τ_a) 3 min. The number of potential cycles in all the experiments was the same (72 cycles). At the end of an experiment, the sample was withdrawn from the electrochemical cell and the corrosion products removed with hydrochloric acid/methylamine (1:1). The sample was washed in distilled water, degreased with ethanol and the surface was photographed using a Biomed PR-3 optical microscope (manufactured by Biomed Service LLC, Moscow, Russia) and an AC-300 digital video camera recorder (Amoyca, Ho Chi Minh City, Vietnam) mounted on the ocular. The camera resolution was 2048 × 1536 pixels. Data from the camera were evaluated with Scope Photo 3.0 software ("Scope Tec", Munich, Germany) to determine the number of pits (N) and the surface area of the sample occupied by each pit (S_i). Initially, N and S_i were determined from surface images with low magnification and refined using images with higher magnification (Figure 1). The pit depth (h) was determined using a Neophot-2 metallographic microscope. The mean maximum value of h was determined from the values observed for the three deepest pits.

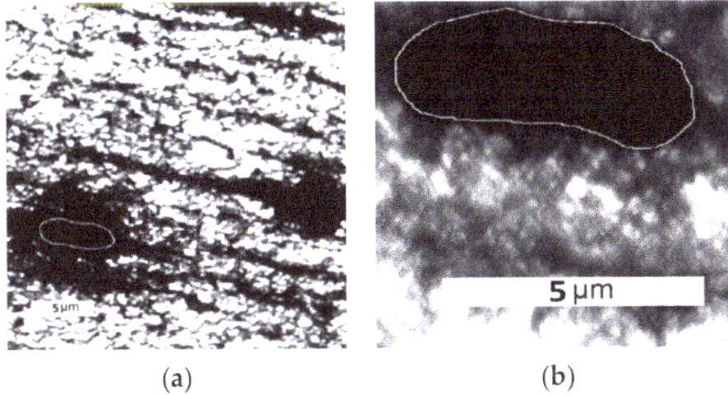

(a)   (b)

Figure 1. Images the sample surface after SACP at $\tau_c = 10$ min in 0.1 M NaCl solution. Magnification is (a) 5×; (b) 20×.

The total surface area of a sample covered by pits was calculated as:

$$S_P = \sum S_i / S \tag{1}$$

The coverage of the sample surface with defects was determined as:

$$p = N/S \qquad (2)$$

where S is the visible area of the sample (1.00 mm^2).

Assuming that pits have a hemispherical form, the mean defect diameter was calculated:

$$d = \sum d_i/N \qquad (3)$$

where $d_i = \sqrt{S_i/\pi}$.

The difference (tolerance) in the values of S_p, ρ, and d obtained on different parts (1 mm^2) of a working surface area did not exceed 15% of the average values. The rate of hydrogen permeation into the metal was measured using a steel hydrogen pickup sensor whose sensing principle was based on the Devanathan-Stachurski setup [27]. The sensor membrane was made of steel foil with a thickness of 100 μm, and the working area was 33.1 cm^2. A Pd film was deposited cathodically on the membrane exit side for 100 s at a constant current density of 25 mA/cm^2. The Pd plating solution contained 25 g/L PdCl$_2$ and 20 g/L NH$_4$Cl. The solution pH was adjusted to 8.5 by adding the required amount of NH$_4$OH. Prior to the tests, the Pd-plated membranes were degassed in a dry-air desiccator at room temperature for 2 days. The diffusing part of the cell was filled with 0.1 M NaOH and the permeation current was measured at a potential of 0.2 V (SHE). The electrochemical experiments for hydrogen sensing and optical microscopy were carried out using the same setup.

3. Results and Discussion

Based on images of the sample surfaces similar to those shown in Figure 1, the total area occupied by pits (S_p, mm^2), their density (ρ, mm^{-2}) and mean diameter (d, μm) were determined (Table 2). The same table shows the maximum depth of pits, (h, μm) and data obtained at a constant anodic potential $E = -0.3$ V ($\tau_c = 0$) and different durations of the cathodic half-cycle ($\tau_c = 10$ min and $\tau_c = 60$ min). In all setups, the total duration of anodic polarisation cycle was 216 min.

Figure 2 shows the variation in the total area (a), density (b), an average diameter (c) of pits versus time of cathodic part of the cycle (τ_s) in NS4 solution in borate buffer (BB + NS4) as the background and with the addition of 10^{-2} M thiourea (BB + NS4 + TU). The same setup of SACP was used either with or without addition of tiourea. The addition to the solution of a promoter of hydrogen absorption (thiourea) significantly increases the total area and density of pits (Figure 2a,b), and at $\tau_c = 60$ min, the diameter of pits (Figure 2c). The maximum pit depth also increases markedly with increasing τ_c in the solution containing thiourea (Table 2). As can be seen, the total area of pits increases upon SACP and the greatest effect is observed with the longest cathodic period ($\tau_c = 60$ min). Localised corrosion occurs more intensely as the duration τ_c increases. Figure 2 shows that the increase of the duration of cathodic polarization increases the corrosion of the steel with or without the addition of thiourea in borate electrolyte. It was determined that in all cases thiourea accelerates the corrosion of the steel comparing with reference electrolyte. This allows us to conclude that hydrogen charging of steel during the cathodic period leads to its accelerated localised dissolution during the anodic period [23,28].

It can be supposed that pits appear as a result of hydrogen charging of the pipeline steel and the formation of locations of high hydrogen pressure, predominantly at the metal/non-metallic inclusion interfaces [22,23]. Hydrogen absorbed by the metal accelerates the dissolution of steel in pH-neutral media at anodic potentials [29]. Due to the dissolution of the hydrogen-saturated metal layer inside the pits, the diameter and depth of these should increase.

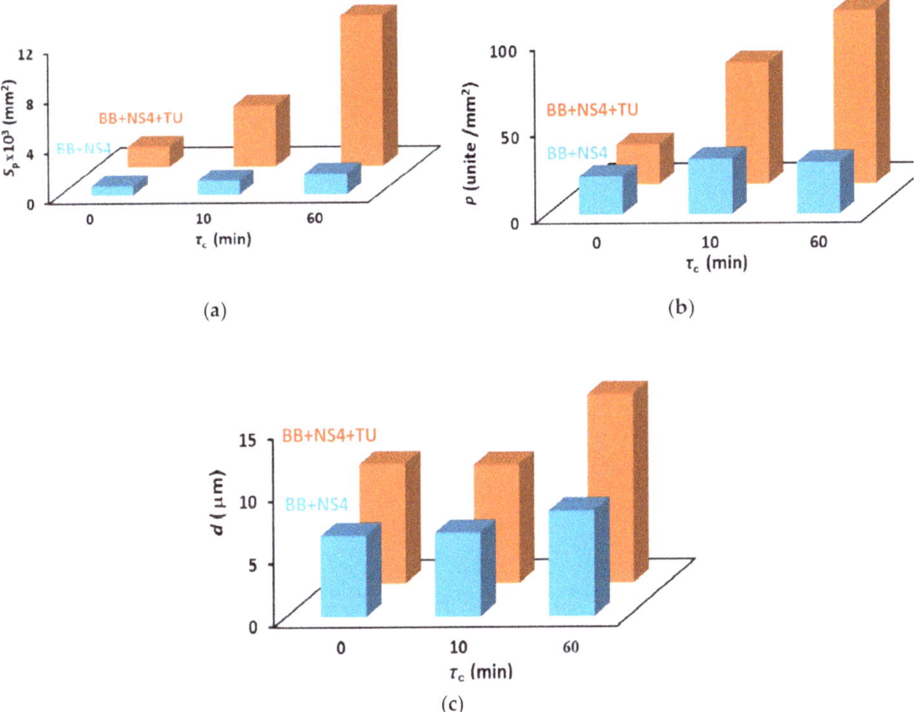

Figure 2. The effect of addition of 10^{-2} M of thiourea on the total pit area S_p (**a**); pit density on the sample surface (**b**) and their average diameter d (**c**) versus duration of the cathodic period of the cycle in SACP. The setup consists from 72 cycles, the duration of anodic period of one cycle 3 min.

As follows from Table 2, the intensity of localised corrosion increases with increasing concentration of chloride ions [Cl⁻] in the solution. Figure 3 (curves 1) shows the impact of chloride concentration in BB electrolyte on the total area, density, diameter, and depth of pits after cathodic polarisation for 10 min and the following anodic cycle. As can be seen, the values of S_p, ρ, d, and h increase abruptly on transition from the pure borate buffer to the BB + NS4 mixture, which can be a result of NS4 solution containing not only chloride but also bicarbonate and sulphate ions. However, with an increase in [Cl⁻] to 0.1 M, the area, density and depth of pits increase significantly compared to the values observed in the BB + NS4 solution (Figure 3a,b) containing a low concentration of aggressive ions. In BB + 0.6 M NaCl solution, the density of pits decreases (Figure 3b) but their diameters increase (Figure 3c). As a result, the total pit area increases (Figure 3a). Apparently, the density of pits decreases because they are merged together.

In non-buffered 0.1 M NaCl electrolyte, the values of S_p, ρ, d and h (Figure 3, points 2) are much greater than in borate buffered electrolyte or in the same solution with NS4. This confirms that chloride ions accelerate the localised corrosion of steel during cathodic-anodic cycling. Borate buffer does not affect the density of pits but slows down their growth (Figure 3b). As a result, the values of S_p, d and h in the BB + 0.1 M NaCl solution are smaller than in a pure chloride solution of the same concentration (Figure 3a,c,d).

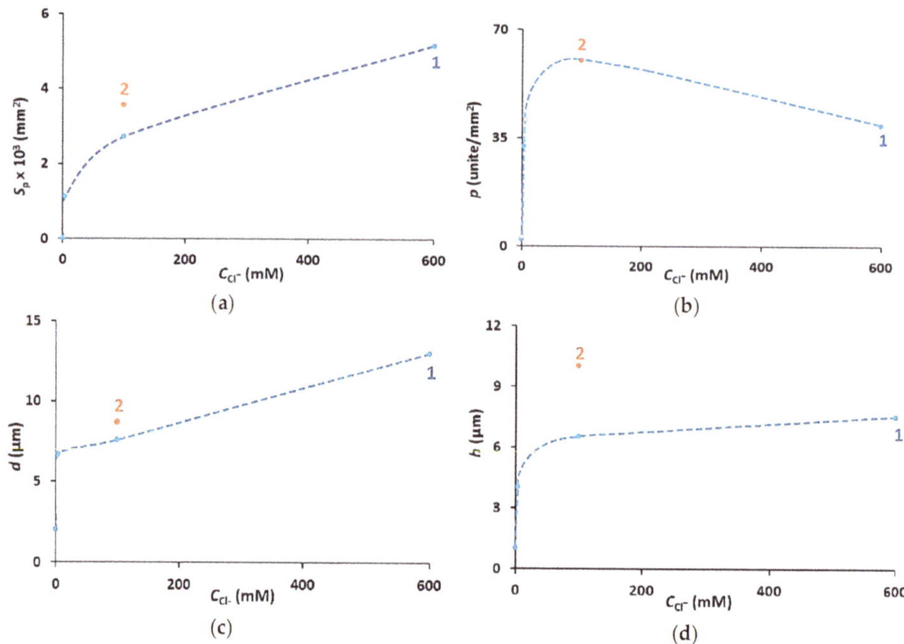

Figure 3. Variation in the total area S_p (**a**); density ρ (**b**); mean diameter d (**c**); and maximum pit depth h (**d**) as a function of the concentration of chloride ions (curves 1) in borate buffer electrolyte. Point 2—non-buffered 0.1 M NaCl aqueous solution. The setup consists from 72 cycles of SACP (the potential switched between −1 V vs. SHE (duration 10 min) and −0.3 V vs. SHE (duration 3 min).

Corrosive electrolytes in soils often contain carbonate and bicarbonate ions, and one of the objectives was to determine the effect of these ions on pitting corrosion. Table 2 shows that the addition of 0.1 M $NaHCO_3$ affects the localised corrosion of X70 steel. Addition of bicarbonate to borate electrolyte significantly increases both the total area S_p and the density of the pits ρ. The mean pit diameter d increases almost four-fold. However, bicarbonate affects the characteristics of localised corrosion to a smaller extent than does chloride. Thus, the ratios of S_p and ρ values in BB + 0.1 M NaCl and BB + 0.1M $NaHCO_3$ solutions are 3.3 and 3.4, respectively (Table 2). The pit diameter in these solutions is approximately the same (Table 2); i.e., chloride favours the initiation of more numerous pits.

Cathodic polarisation can lead to damage of steel constructions due to hydrogen-assisted steel cracking. The addition of a hydrogen uptake promoter such as thiourea to BB electrolytes containing 0.1 M $NaHCO_3$ significantly activates the localised corrosion, increasing the density of pits, but variation in their diameter is not observed. On the other hand, localised corrosion occurs more intensely in the BB + NS4 + TU solution that in BB + 0.1 M $NaHCO_3$ + TU (Table 2). This also confirms that chloride ions have a stronger effect on pit nucleation and growth relatively bicarbonate ions. The pit depth in BB + 0.1 M $NaHCO_3$ and BB + 0.1 M $NaHCO_3$ + TU solutions is insignificant: 2 and 3 μm respectively (Table 2). Hence, a sufficiently high concentration of bicarbonate inhibits the growth in the depth. This is in line with the results for the corrosion of pipeline steel in simulated soil electrolytes [30].

The addition of IFKhAN-29 corrosion inhibitor to the BB + NS4 + TU solution inhibits local steel dissolution during cathodic-anodic cycling (τ_c = 60 min, Table 2). In this case, the density of pits decreases significantly (by a factor of almost 5). Hence, the inhibitor primarily affects the pit nucleation process. The growth of pits is also inhibited; the diameters and depths decrease by factors of 1.5 and 2.2, respectively (Table 2). Thus, it is possible to point out that the intensity of the localised corrosion, which is characterised by the S_p, ρ, d and h values, increases with an increase in the duration of the cathodic period of the cycle and in the presence of a promoter of hydrogen absorption. It depends also on the ionic composition of the electrolyte, such as by the addition of chloride or bicarbonate anions.

The results show that the hydrogen charging of steel promotes both the nucleation of pits during the cathodic period and their growth during the anodic period of the cycle, which is in line with the literature [22,23,28]. It can be expected that the rate of localised corrosion is determined by the amount of hydrogen absorbed by the steel at the cathodic potential. It is important to find a quantitative integral parameter that would determine the occurrence of localised corrosion of pipeline steel due to sign-alternating polarisation. To determine the hydrogen flux into the steel during cathodic polarisation, an electrochemical technique to measure the hydrogen penetration current (i_p) across a membrane was applied. Table 3 shows i_p as function of the solution composition during steel cathodic polarisation.

Table 3. The stationary rate of hydrogen penetration (i_p) into steel at $E = -1$ V (SHE), the stationary density of anodic current ($i_{a,st}$) at $E = -0.3$ V (SHE) and the anodic current density in 0.09 s after switching the potential (i_0) under SACP with $\tau_c = 10$ min in solutions of various compositions.

Electrolyte Composition	i_p, A/cm^2	$i_{a,st}$, mA/cm^2	i_0, mA/cm^2
BB	6	0.020	0.17
BB + NS4	10	0.020	0.20
BB + NS4 + TU	90	0.116	0.57
0.1 M NaCl	14.5	0.041	1.85
BB + 0.1 M NaCl	18	0.100	1.27
BB +0.6 M NaCl	38	0.082	3.8
BB + 0.1 M NaHCO$_3$	18	0.067	0.44
BB + 0.1 M NaHCO$_3$ + TU	27.6	0.113	1.05

Figure 4 correlates the total area of the corroded surface during SACP (time of cathodic polarisation 10 min) versus the hydrogen permeation current as a function of electrolyte composition. Figure 4 brings together the data obtained in BB, BB + NS4, BB + NS4 + TU solutions (curve 1), BB, BB + 0.1 M NaCl, BB + 0.6 M NaCl solutions (curve 2), and BB, BB + 0.1 M NaHCO$_3$, BB + 0.1 M NaHCO$_3$ + TU solutions (curve 3). It is obvious that hydrogen entry to the steel during cathodic polarisation accelerates the pitting dissolution that is taking place during the anodic period of the cycle.

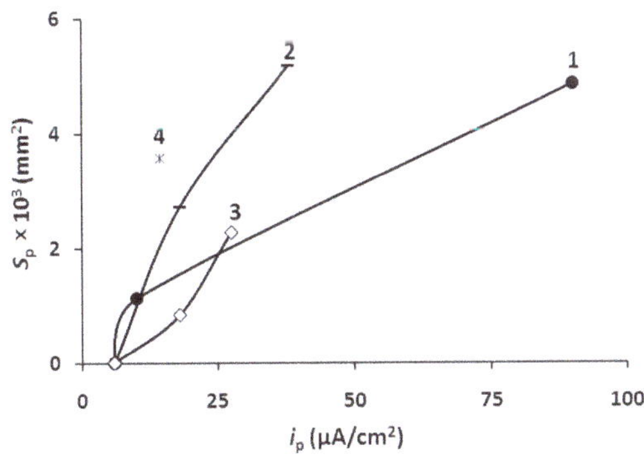

Figure 4. Correlation between the total pits area S_p (duration of cathodic part of SACP 10 min and anodic part 3 min, total 72 cycles) and the hydrogen penetration current i_p in solutions: 1 (·) BB, BB + NS4, BB + NS4 + TU; 2 (-) BB, BB + 0.1 M NaCl, BB + 0.6 M NaCl; 3 (◊) BB, BB + 0.1 M NaHCO$_3$, BB + 0.1 M NaHCO$_3$ + TU; point 4 (∗) is non-buffered 0.1 M NaCl aqueous electrolyte.

As can be seen in Figure 4, an increase in the concentrations of chloride and bicarbonate results in an increase in the rate of hydrogen incorporation into the steel. At the similar hydrogen permeation current (i_p), an increased area of pitting corrosion (S_p) was observed in chloride solutions, while a lower corroded area was observed in bicarbonate-containing solutions. Thus, the corrosion rate is not influenced only by the hydrogen uptake but also the kind of electrolyte species.

Current transients obtained after switching cathodic and anodic potentials can give additional information about the mechanism of corrosion of the pipeline steel at SACP. Figure 5 shows the transients of the anodic (solid lines) and cathodic (dashed lines) currents in logarithmic coordinates obtained during potential cycling for some electrolytes and shows significantly different results. One can see that at the initial period (up to about 0.1 s) after switching of the potential, the cathodic current (i_c) is slightly higher than the anodic current (i_a), but the difference increases with time. The slope of the log i_c-log τ curves decreases with time and the cathodic current tends to reach a constant value. The slope of the log i_a-log τ curves remains nearly constant for some time (the duration of this section on the curves depends on the solution composition). Subsequently, the anodic current stabilises or increases with time. A increase in the slope of log i_a-log τ curves could be due to the formation of a primary passive film on the metal. Similar anodic current transients were observed on iron in BB solution [31]. The formation of a layer of iron oxides on the electrode surface was shown in solution by method of quartz resonator [32]. However, at $E = -0.3$ V, a continuous barrier layer of iron oxide/hydroxide compounds is not formed [32]. Hence, the current at the end of the anodic SACP half-cycle ($i_{a,st}$) relates to the uniform dissolution of the metal and localised dissolution in the pits formed at the cathodic potential.

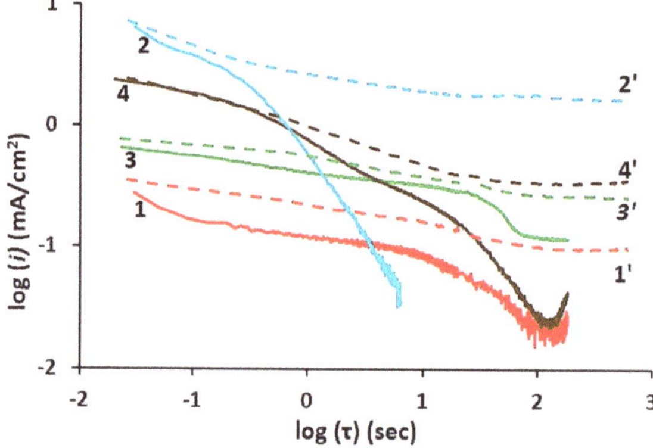

Figure 5. Transients of the anodic currents at −0.3 V vs. SHE (1–4) and cathodic currents at −1 V vs. SHE (1′–4′) in solutions: (1,1′) BB; (2,2′) BB + 0.6 M NaCl; (3,3′) BB + NS4 + TU; (4,4′) 0.1 M NaCl.

It is important to discuss the anodic current transients at the beginning of the passivation process in more details. Anodic current transients (Figure 5) were presented in i_a-$\tau^{-0.5}$ coordinates in Figure 6a. One can see that, in a certain time range, the current transients obtained in BB + 0.6 M NaCl (Figure 6a, curve 2) and 0.1 M NaCl solutions (Figure 6a, curve 4) are linearized in Cottrell coordinates, and the extrapolated linear section passes through the origin of coordinates at $\tau \to \infty$. This is an indication that nonstationary diffusion in semi-infinite approximation is the limiting step of the anodic process. The anodic process of hydrogen-charged steel can involve the ionisation of atoms of iron, alloying elements and the hydrogen atoms absorbed by the alloy. It has been reliably established that the active dissolution of iron under stationary conditions is limited by the kinetic of electron transfer [33]; therefore, the nonstationary anodic process cannot be limited by the diffusion of Fe^{2+} ions in the liquid phase. In principle, the dissolution of alloying components of steel might be limited by their solid-phase diffusion in iron. However, due to small values of the diffusion coefficients of alloying components in the metal matrix, this possibility seems unlikely, since the linear portions of the i_a-$\tau^{-0.5}$ curves can be rather long, for example, up to 20 s in 0.1 M NaCl solution (Figure 6b, anodic curve).

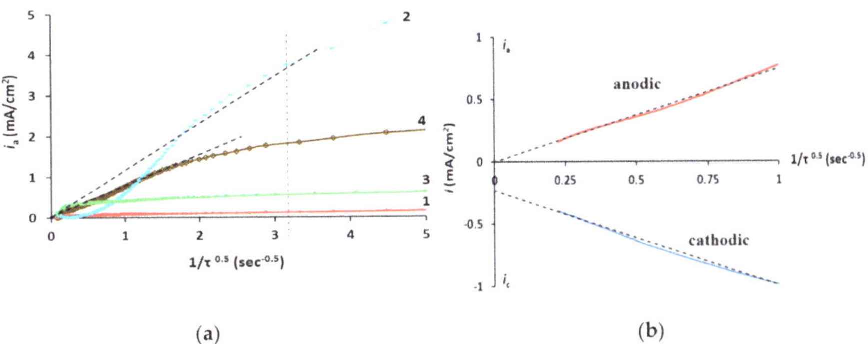

Figure 6. (**a**) Transients of anodic current at −0.3 V (SHE) in solutions: (1) BB; (2) BB + 0.6 M NaCl; (3) BB + NS4 + TU; (4) 0.1 M NaCl. (**b**) Initial parts of anodic (−0.3 V vs. SHE) and cathodic (−1 V vs. SHE) current transients in 0.1 M NaCl aqueous electrolyte.

The cathodic current transients obtained in BB + 0.6 M NaCl and 0.1 M NaCl solutions are also linearized in the i_c-$\tau^{-0.5}$ coordinates in a certain time range. For example, Figure 6b shows the linear section of the cathodic current transient obtained in 0.1 M NaCl solution. The linear sections of the i_a-$\tau^{-0.5}$ and i_c-$\tau^{-0.5}$ plots (Figure 6b) are observed in the same time range; their slopes show opposite signs and are nearly equal. This indicates that the anodic and cathodic current transients are related to the same process, such as nonstationary solid-phase diffusion of hydrogen atoms in the metal [33]. However, the cathodic evolution and entry of hydrogen on iron and steels consist of a number of reactions. The reactions of the evolution of hydrogen gas proceed in parallel with the incorporation of hydrogen into the metal. Therefore, the extrapolated linear section of the i_c-$\tau^{-0.5}$ dependence (Figure 6b, cathodic curve) does not pass through the origin but shifts down on the Y-ordinate axis. This shift in cathodic current should be close to the rate of hydrogen gas evolution. In fact, it is nearly equal to the stationary cathodic current and to the rate of cathodic reaction because the hydrogen penetration current into the metal is relatively small (if the electrode thickness is sufficient). In addition to hydrogen oxidation, the anodic process in hydrogen-charged steel should include a parallel reaction of metal ionisation. However, extrapolation of the linear section of the i_a-$\tau^{-0.5}$ curve to the origin (Figure 6b, anodic curve) shows that in the present experimental setup, the ionisation rate of the metal in 0.1 M NaCl solution is small compared to the rate of oxidation of absorbed hydrogen.

The initial section of the i_a-$\tau^{-0.5}$ curves deviates from a linear plot at lower currents (Figure 6a). As a rule, this effect results from the kinetic limitations of the reaction rate. For example, hydrogen extraction from Pd and its alloys over a short time is limited by the process of hydrogen desorption from the metal phase [34]. It can be assumed that the weak dependence of the current on $\tau^{-0.5}$ observed in our experiments in the initial period (Figure 6a) is also due to the kinetic limitations of the interface reaction of hydrogen desorption. This is in line with the data on anodic current transients obtained on an iron electrode in BB that is described by the equation of diffusion-phase-boundary kinetics of hydrogen extraction from the metal [35,36]. In fact, for different reasons, a quantitative interpretation of the current transients obtained on steel is difficult. The hydrogen charging of steel under SACP should occur locally in the corrosion defects. Additionally, the gradient of hydrogen concentration in the steel is nonlinear because hydrogen is mainly absorbed by the surface layer of a metal [37,38]. A primary passivating film is formed during SACP, which should inhibit both iron dissolution and hydrogen effusion from the metal (the diffusion coefficient of hydrogen in the iron oxide layer is lower by several orders of magnitude than in the metal [37,38]). Nevertheless, it can be pointed out that in the initial period the anodic current is primarily determined by the rate of hydrogen extraction (oxidation) from the steel.

It can be expected that the hydrogen extraction from local corrosion events, such as the bottom and the 'banks' of pits would also accelerate dissolution of the metal due to defectiveness of the subsurface layer of the alloy. A similar effect is observed for selective dissolution of a number of alloys; the dissolution of the electropositive alloy component is activated due to dissolution of the electronegative component [39,40]. Thus, it can be supposed that a relationship should exist between the intensity of localised steel corrosion under SACP and the current value at the beginning of the anodic half-cycle (Figure 5). Figure 7a shows the correlation between the total area of pits and the anodic current taken at $\tau = 0.09$s. The value of $\tau = 0.09$s was taken arbitrarily, but the curve (Figure 7a) does not change significantly for the currents measured at a lower initial time. The S_p-i_0 plot (Figure 7a) is nonlinear with a coefficient of regression R^2 of 0.63. The point TU obtained in the BB + NS4 + TU solution containing the promoter of hydrogen absorption is far from the S_p-i_0 curve. The reason for the intensive localised corrosion of steel in this solution is a result of a high rate of hydrogen penetration into the metal (i_p, see Table 3). Therefore, for an understanding of the mechanism of pitting corrosion of X70 steel in SACP cycles, the current of hydrogen penetration (i_p) has also been taken into account.

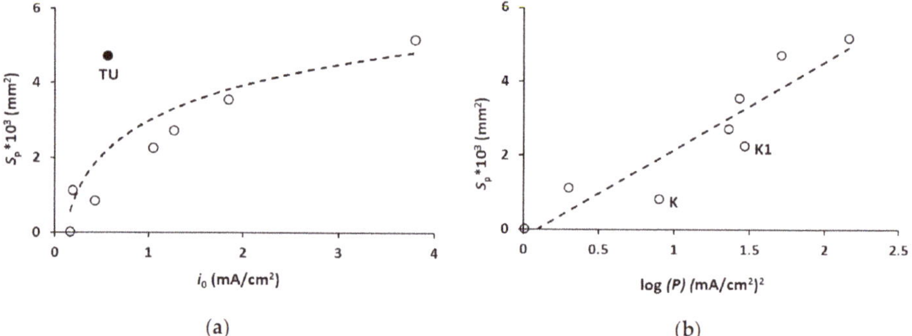

Figure 7. (**a**) Correlations between total area of the pits S_p and anodic current (i_0) at the beginning of anodic cycle; (**b**) total area of the pits S_p vs. parameter P (Equation (4)) related to the intensity of steel corrosion during SACP cycles.

Figure 7b shows the relationship between surface area containing pits (S_p) and the product (P) of anodic current (i_0) at the beginning of an anodic cycle and the current (i_p) of hydrogen penetration across the membrane (Equation (4)). P is the empirical parameter determining the rate (surface area) of pitting corrosion that shows the most promising correlation. Considering the nonlinear nature of S_p vs. P, the plot is presented in semi-logarithmic coordinates:

$$i_0 \times i_p = P \tag{4}$$

$$S_p = 2.37 \log P - 0.22 \tag{5}$$

The resulting regression (Equation (5)) relates to all points presented in Figure 7b with a coefficient of determination R^2 of 0.85. The points K and K1 (Figure 7b), corresponding to solutions BB + 0.1 M $NaHCO_3$ and BB + 0.1 M $NaHCO_3$ + TU, are located below the linear regression line. Apparently, in solutions with a high content of bicarbonate ions, insoluble $FeCO_3$ can precipitate during steel dissolution and inhibit the growth of corrosion defects at the bottom of pits [30]. Thus, it can be supposed that the intensity of localised corrosion of pipeline steel under SACP (Figure 7b) depends both on the amount of hydrogen absorbed during the cathodic half-period and on the rate of hydrogen extraction during the anodic half-period.

An integral quantitative indicator P (Equation (4)), determining the intensity of localised corrosion of steel during SACP, contains the stationary hydrogen penetration current through a steel membrane (at the potential of the cathodic period) and the anodic current measured at the beginning of the anodic period of SACP. However, this indicator does not take into account less-controlling factors occurring during cyclic alternating polarisation, such as the formation of a layer of insoluble corrosion products.

4. Conclusions

(1) The intensity of localized corrosion of pipeline steel X70 increases with an increase of duration of the cathodic period of sign-alternating cycling polarization. This effect was found either with or without the addition of a promoter of hydrogen absorption (thiourea) to background electrolyte. The thiourea increases the total area and density of pits significantly. Thus, effective hydrogen charging of steel during the cathodic period accelerates localized dissolution during the anodic period of SACP.

(2) The corrosion of pipeline steel during anodic period of SACP increases with an increase of concentration of chloride and bicarbonate ions in solution. Bicarbonate ions influence on the rate of localized corrosion to a smaller extent than chloride ions. Chloride ions accelerate the nucleation of pits and increase their maximum depth.

(3) The IFKhAN-29 corrosion inhibitor affects the nucleation of the pits and hinders the localised corrosion of pipeline steel in the anodic period of an SACP cycle. The inhibitor reduces the surface density of pits by a factor of nearly five.

(4) The nature of current transients in SACP as a result of switching of the potential was investigated. It was shown that in the initial period, cathodic and anodic currents are determined by the rates of hydrogen absorption and extraction from the metal.

(5) An integral quantitative indicator of the intensity of localised corrosion of X70 steel in different solutions during SACP was proposed with a satisfactory coefficient of determination ($R^2 = 0.85$). This is the product of two currents: the current of hydrogen penetration through a steel membrane in the cathodic period and the anodic current of hydrogen extraction measured at the beginning of the anodic period of the SACP.

Author Contributions: A.R.: methodology, investigation, writing—review & editing, visualization; N.G.: methodology, software, investigation; A.M.: writing-original draft preparation, project administration; M.P.: data curation, funding acquisition; A.N.: writing—original draft preparation, supervision. All authors have read and agreed to the published version of the manuscript.

Funding: This research was funded by the Ministry of Science and High Education of the Russian Federation, No. AAAA-A20-120012390029-7.

References

1. CEN-EN 15280. Evaluation of A.C. corrosion likelihood of buried pipelines-Application to cathodically protected pipelines. In *Technical Specification*; BSI: London, UK, 2013.
2. Peabody, A.W. *Peabody's Control of Pipeline Corrosion*, 2nd ed.; NACE International the Corrosion Society: Houston, TX, USA, 2001; pp. 226–231.
3. Baeckmann, W.; Schwenk, W. Handbuch des kathodischen Korrosionsschutzes. In *Theorie und Praxis der Elektrochemischen Schutzverfahren*; Verlag Chemie: Weinheim, Germany, 1980; pp. 15–61.
4. Lalvani, S.B.; Lin, X.A. A theoretical approach for predicting AC-induced corrosion. *Corros. Sci.* **1994**, *6*, 1039–1046. [CrossRef]
5. Lalvani, S.B.; Lin, X.A. A revised model for predicting corrosion of materials induced by alternating voltages. *Corros. Sci.* **1996**, *10*, 1709–1719. [CrossRef]
6. Bosch, R.W.; Bogaerts, W.F. A theoretical study of AC-induced corrosion considering diffusion phenomena. *Corros. Sci.* **1998**, *2*, 323–336. [CrossRef]
7. Lalvani, S.B.; Zhang, G. The corrosion of carbon steel in a chloride environment due to periodic voltage modulation: Part I. *Corros. Sci.* **1995**, *10*, 1567–1582. [CrossRef]
8. Fu, A.Q.; Cheng, Y.F. Effects of alternating current on corrosion of a coated pipeline steel in a chloride-containing carbonate/bicarbonate solution. *Corros. Sci.* **2010**, *2*, 612–619. [CrossRef]

9. Goidanich, S.; Lazzari, L.; Ormellese, M. AC corrosion. Part 2: Parameters influencing corrosion rate. *Corros. Sci.* **2010**, *3*, 916–922. [CrossRef]

10. Xu, L.Y.; Su, X.; Yin, Z.X.; Tang, Y.H.; Cheng, Y.F. Development of a real-time AC/DC data acquisition technique for studies of AC corrosion of pipelines. *Corros. Sci.* **2012**, *61*, 215–223. [CrossRef]

11. Marshakov, A.I.; Nenasheva, T.A. The effect of alternating current on rate of dissolution of carbon steel in chloride electrolyte. Part I. Conditions of free corrosion. *Prot. Met. Phys. Chem. Surf.* **2017**, *7*, 1214–1221. [CrossRef]

12. Kim, D.-K.; Muralidharan, S.; Ha, T.-H.; Bae, J.-H.; Ha, Y.-C.; Lee, H.-G.; Scantlebury, J.D. Electrochemical studies on the alternating current corrosion of mild steel under cathodic protection condition in marine environments. *Electrochim. Acta* **2006**, *25*, 5259–5267. [CrossRef]

13. Kuang, D.; Cheng, Y.F. Understand the AC induced pitting corrosion on pipelines in both high pH and neutral pH carbonate/bicarbonate solutions. *Corros. Sci.* **2014**, *85*, 304–310. [CrossRef]

14. Brenna, A.; Ormellese, M.; Lazzari, L. A proposal of AC corrosion mechanism in cathodic protection. *NSTI Nanotech.* **2011**, *3*, 553–556.

15. Ibrahim, I.; Tribollet, B.; Takenouti, H.; Meyer, M. AC-induced corrosion of underground steel pipelines. Faradaic rectification under cathodic protection: I. Theoretical approach with negligible electrolyte resistance. *J. Braz. Chem.* **2015**, *1*, 196–208. [CrossRef]

16. Marshakov, A.I.; Nenasheva, T.A.; Kasatkin, V.E.; Kasatkina, I.V. Effect of alternating current on dissolution rate of carbon steel in chloride electrolyte. II. Cathodic potentials. *Prot. Metals Phys. Chem. Surf.* **2018**, *7*, 1236–1245. [CrossRef]

17. Buchler, M. Alternating current corrosion of cathodically protected pipelines: Discussion of the involved processes and their consequences on the critical interference values. *Mater. Corros.* **2012**, *12*, 1181–1187. [CrossRef]

18. Huo, Y.; Tan, M.Y.; Forsyth, M. Visualizing dynamic passivation and localized corrosion processes occurring on buried steel surfaces under the effect of anodic transients. *Electrochem. Commun.* **2016**, *66*, 21–24. [CrossRef]

19. Huo, Y.; Tan, M.Y. Measuring and understanding the critical duration and amplitude of anodic transients. *Corros. Eng. Sci. Technol.* **2018**, *53*, 65–72. [CrossRef]

20. Huo, Y.; Tan, M.Y. Localized corrosion of cathodically protected pipeline steel under the effects of cyclic potential transients. *Corros. Eng. Sci. Technol.* **2018**, *5*, 348–354. [CrossRef]

21. Liu, Z.Y.; Li, X.G.; Cheng, Y.F. Understand the occurrence of pitting corrosion of pipeline carbon steel under cathodic polarization. *Electrochim. Acta* **2012**, *60*, 259–263. [CrossRef]

22. Marshakov, A.I.; Nenasheva, T.A. The formation of corrosion defects upon cathodic polarization of X70 grade pipe steel. *Prot. Met. Phys. Chem. Surf.* **2015**, *7*, 1122–1132. [CrossRef]

23. Nenasheva, T.A.; Marshakov, A.I.; Kasatkina, I.V. The formation of local foci of corrosion of pipe steel under cyclic alternating polarization. *Corros. Mater. Prot.* **2015**, *5*, 9–17. (In Russian)

24. Dai, M.; Liu, J.; Huang, F.; Zhang, Y.; Cheng, Y.F. Effect of cathodic protection potential fluctuations on pitting corrosion of X100 pipeline steel in acidic soil environment. *Corros. Sci.* **2018**, *143*, 428–437. [CrossRef]

25. Parkins, R.N.; Blanchard, W.K.; Delanty, B.S. Transgranular stress corrosion cracking of high-pressure pipelines in contact with solutions of near neutral pH. *Corrosion* **1994**, *5*, 394–408. [CrossRef]

26. Gladkikh, N.A.; Maleeva, M.A.; Maksaeva, L.B.; Petrunin, M.A. A study of the initial stages of local dissolution of carbon steel in chloride solution. *Prot. Met. Phys. Chem. Surf.* **2018**, *7*, 1260–1265. [CrossRef]

27. Devanathan, M.A.V.; Stachurski, Z. The mechanism of hydrogen evolution on iron in acid solution by determination of permition rates. *J. Electrochem. Soc.* **1964**, *3*, 619–623. [CrossRef]

28. Gladkikh, N.A.; Maleeva, M.A.; Maksaeva, L.B.; Petrunin, M.A.; Rybkina, A.A.; Yurasova, T.A.; Marshakov, A.I.; Zalavutdinov, R.K. Localized dissolution of carbon steel used for pipelines under constant cathodic polarization conditions. Initial stages of defect formation. *Int. J. Corros. Scale Inhib.* **2018**, *4*, 683–696.

29. Marshakov, A.I.; Nenasheva, T.A. Kinetics of dissolution of hydrogenated carbon steel in electrolytes with pH close to neutral. *Prot. Met. Phys. Chem. Surf.* **2015**, *6*, 1018–1026.

30. Rybkina, A.A.; Kasatkina, I.V.; Gladkikh, N.A.; Petrunin, M.A.; Marshakov, A.I. Experimental determination of local corrosion damage development rate on surface of pipe steels in model soil electrolytes. *Corros. Mat. Prot.* **2019**, *3*, 1–8. (In Russian)

31. Marshakov, A.I.; Rybkina, A.A.; Maleeva, M.A.; Rybkin, A.A. The effect of atomic hydrogen on the kinetics of iron passivation in neutral solution. *Prot. Met. Phys. Chem. Surf.* **2014**, *3*, 345–351. [CrossRef]

32. Marshakov, A.I.; Rybkina, A.A.; Maksaeva, L.B.; Petrunin, M.A.; Nazarov, A.P. A study of the initial stages of iron passivation in neutral solutions using the quartz crystal resonator technique. *Prot. Met. Phys. Chem. Surf.* **2016**, *5*, 936–946. [CrossRef]

33. Cherry, B.W.; Gould, A.N. Pitting corrosion of nominally protected land–based pipelines. *Mater. Perform.* **1990**, *4*, 22–26.

34. Marshakov, A.I.; Rybkina, A.A. Dissolution of iron and ionisation of hydrogen in borate buffer under cyclic pulse polarization. *Int. J. Electrochem. Sci.* **2019**, *14*, 9468–9481. [CrossRef]

35. Morozova, N.B.; Vvedensky, A.V.; Beredina, I.P. The phase-boundary exchange and the non-steady-state diffusion of atomic hydrogen in Cu-Pd and Ag-Pd alloys. Part I. Analysis of the model. *Prot. Met. Phys. Chem. Surf.* **2014**, *6*, 699–704. [CrossRef]

36. Zakroczymski, T. Adaptation of the electrochemical permeation technique for studying entry, transport and trapping of hydrogen in metals. *Electrochim. Acta* **2006**, *11*, 2261–2266. [CrossRef]

37. Pyun, S.I.; Oriani, R.A. The permeation of hydrogen through the passivating films on iron and nickel. *Corros. Sci.* **1989**, *5*, 485–496. [CrossRef]

38. Legrand, E.; Bouhattate, J.; Feaugas, X.; Garmestani, H. Computational analysis of geometrical factors affecting experimental data extracted from hydrogen permeation tests: II-consequences of trapping and an oxide layer. *Int. J. Hydrogen Energy* **2012**, *37*, 13574–13582. [CrossRef]

39. Qin, Z.; Demko, B.; Noel, J.; Shoesmith, D.; King, F.; Worthingham, R.; Keith, K. Localized dissolution of mill scale-covered pipeline steel surfaces. *Corrosion* **2004**, *10*, 906–914. [CrossRef]

40. Marshakov, I.K. Corrosion resistance and dezincing of brasses. *Prot. Met. Phys. Chem. Surf.* **2005**, *3*, 205–210.

MDPI

St. Alban-Anlage 66

4052 Basel

Switzerland

Tel. +41 61 683 77 34

Fax +41 61 302 89 18

www.mdpi.com

Metals Editorial Office

E-mail: metals@mdpi.com

www.mdpi.com/journal/metals